Diversity, Biogeography and Community Ecology of Ants

Diversity, Biogeography and Community Ecology of Ants

Editor

Alan N. Andersen

MDPI • Basel • Beijing • Wuhan • Barcelona • Belgrade • Manchester • Tokyo • Cluj • Tianjin

Editor
Alan N. Andersen
Charles Darwin University
Australia

Editorial Office
MDPI
St. Alban-Anlage 66
4052 Basel, Switzerland

This is a reprint of articles from the Special Issue published online in the open access journal *Diversity* (ISSN 1424-2818) (available at: https://www.mdpi.com/journal/diversity/special_issues/Ants).

For citation purposes, cite each article independently as indicated on the article page online and as indicated below:

LastName, A.A.; LastName, B.B.; LastName, C.C. Article Title. *Journal Name* **Year**, *Volume Number*, Page Range.

ISBN 978-3-0365-2786-4 (Hbk)
ISBN 978-3-0365-2787-1 (PDF)

Cover image courtesy of François Brassard.

Contents

About the Editor

Alan N. Andersen (Professor). Prof. Andersen is Professorial Fellow—Research Excellence and Impact with Charles Darwin University. Previously, he was the Leader of CSIRO's Tropical Ecosystems Research Centre in Darwin for twenty years. Prof. Andersen's primary research interests are in the global ecology of ant communities, where he integrates community ecology, macroecology, historical and contemporary biogeography, and systematics to gain a predictive understanding of ant diversity, behavioural dominance, and functional composition in relation to environmental stress and disturbance. He applies this understanding to the use of ants as bioindicators of ecological change.

Preface to "Diversity, Biogeography and Community Ecology of Ants"

Globally, ants are a highly diverse and ecologically dominant faunal group. They are widely used as bio-indicators in land management and are model organisms for studies of diversity, biogeography and community ecology. They are also among the most destructive invasive species. Despite such importance, however, we have a very limited understanding of patterns of ant diversity and the factors driving ant species distributions and community dynamics.

This book is a compilation of papers in the Special Issue 'Diversity, Biogeography and Community Ecology of Ants' published in the open-access journal Diversity during 2020 and 2021 (https://www.mdpi.com/journal/diversity/special issues/Ants). The papers address many of the most pressing questions relating to ant diversity and in doing so provide valuable insights into biodiversity more generally. I am most grateful to all the authors of the papers for their contributions to the Special Issue and to Ms. Emma Li from MDPI for her editorial oversight.

Alan N. Andersen
Editor

Editorial

Diversity, Biogeography and Community Ecology of Ants: Introduction to the Special Issue

Alan N. Andersen

Research Institute for the Environment and Livelihoods, Charles Darwin University, Darwin, NT 0909, Australia; alan.andersen@cdu.edu.au; Tel.: +61-468-384-038

Citation: Andersen, A.N. Diversity, Biogeography and Community Ecology of Ants: Introduction to the Special Issue. *Diversity* 2021, 13, 625. https://doi.org/10.3390/d13120625

Received: 23 November 2021
Accepted: 26 November 2021
Published: 28 November 2021

Publisher's Note: MDPI stays neutral with regard to jurisdictional claims in published maps and institutional affiliations.

Ants are a ubiquitous, highly diverse and ecologically dominant faunal group. They represent up to half of total insect biomass globally, and their biomass exceeds that of all vertebrates combined in most terrestrial biomes. Ants are deeply connected within terrestrial ecosystems. They play key roles as soil engineers, predators and re-cyclers of nutrients. They have particularly important interactions with plants as defenders against herbivores, as seed dispersers and as seed predators; as such they have had a significant influence on plant evolution and diversification. The deep connections of ants with ecosystems means that ant-species composition can provide important insights into ecosystem health. Combined with their ubiquity and ease of sampling, this has seen ants widely used as bio-indicators in land management.

One downside to the ecological dominance of ants is that they feature on the list of the world's worst invasive species. Species such as the red imported fire ant (*Solenopsis invicta*), yellow crazy ant (*Anoplolepis gracilipes*), African big-headed ant (*Pheidole megacephala*) and Argentine ant (*Linepithema humile*) have been inadvertently transported throughout the world and have invaded both agricultural and native ecosystems, with devastating impacts.

The importance of ants can be seen in more than just their ecological impacts; ants have been a transformative taxon through their contributions to a scientific understanding of biodiversity as model organisms for studies of diversity, biogeography and community ecology. Studies of ants have been foundational to the development of ecological theory relating to diversity and its distribution, such as the taxon cycle and island biogeography. They have made seminal contributions to an understanding of niche dynamics, species co-existence, community assembly and disturbance dynamics.

Despite their ecological dominance and their role as model organisms for ecological research, ants remain remarkably understudied. A large proportion of species remain undescribed, the biogeographic histories of most taxa remain poorly known, and we have a limited understanding of the spatial patterns of diversity and composition, along with the processes driving these patterns. The ten papers in this Special Issue address all these themes.

Three papers [1–3] make a significant contribution to our understanding of undocumented ant diversity. Two of these [1,2] are from Australia, illustrating the extent to which the remarkably megadiverse ant fauna of this continent remains undocumented. One relates to a large and conspicuous 'species' of the arid-adapted formicine genus *Melophorus* that is endemic to Australia [1]. It provides an integrated genetic, morphological and distributional analysis of diversity within *M. 'rufoniger'* in a 400,000 km^2 region (the 'Top End' of the Northern Territory) of Australia's monsoonal (seasonal) tropics. Thirty species from this taxon were recognized from 120 sequenced specimens, with all but two apparently endemic to the region. The findings indicate that up to 100 or more species from this taxon occur in monsoonal Australia, and are all undescribed. The study provides further evidence that the total *Melophorus* fauna comprises well over an astonishing 1000 species, which is an order of magnitude higher than any other arid-adapted ant genus found elsewhere in the world.

The second Australian paper documents the unrecognized diversity of the ecologically dominant dolichoderine genus *Anonychomyrma* in rainforests of the Wet Tropics bioregion of North Queensland. The study drew on extensive recent collections along elevational and vertical (arboreal) gradients in five upland subregions spanning the full latitudinal range of the bioregion. Integrated morphological, genetic (CO1) and distributional analyses led to the recognition of 22 species, all but two of which appear to be undescribed.

The third paper describing undocumented diversity [3] does so in an urban context. Recently, there has been a considerable level of interest in the biodiversity values of urban areas, both in terms of their conservation significance as refuges for biodiversity, and their potential as gateways for invasive species. Ants are an ideal taxon for examining this. The study in [3] reports on a comprehensive ant survey of Macao, a highly urbanized (20,000 people/km^2) territory in subtropical China. The survey detected 55 new species records for Macao, bringing its total known ant fauna to 155 species, 18 of which are exotic. Several of the newly recorded species appear to be entirely new to science; these include species of *Strumigenys*, *Syllophopsis* and *Tetramorium* that were collected only from subterranean traps. The study also includes a compilation of ant diversity in 112 other urban areas throughout the world. The overall findings are that urban ant faunas are often highly diverse and can have high conservation value, including harboring previously uncollected species. However, urban areas are likely to facilitate the establishment and subsequent spread of exotic species.

The documentation of diversity within the *Anonychomyrma* fauna within Australia's wet tropics [2] was conducted in the context of understanding the fauna's biogeographic origins. Despite its high diversity in the humid tropics of both Australia and New Guinea, the genus is believed to have a cool-temperate Gondwanan origin. This leads to the prediction that the tropical fauna is concentrated at higher elevation. Moreover, Australia's tropical rainforest has undergone extensive contractions and expansions during climatic fluctuations over recent geological history, such that only small areas have supported rainforests throughout this period. Diversity and endemism within *Anonychomyrma* can therefore be expected to be concentrated in these refuge areas. As predicted, diversity and endemism were concentrated at a higher elevation (>900 m above sea level), and especially in refuge areas where rainforest had been maintained throughout the late quaternary.

The biogeographic theme continues in the Special Issue with an analysis of the Iberian (Spain, Portugal, Gibraltar and Andorra) ant fauna [4], one of the richest and most distinctive in the Mediterranean region with 299 recorded species, including 72 endemics. Rates of endemism (proportions of total species) vary markedly among genera, being particularly high in *Temnothorax*, *Goniomma* and *Cataglyphis*, and low in *Formica*, *Lasius* and *Tetramorium*. The high diversity and endemism can be attributed to the peninsula's complex topography and high environmental heterogeneity, which is reflected by the fauna showing pronounced biogeographic regionalization. The biogeographical origins of the fauna are primarily Palearctic, but also include central Asia (e.g., *Myrmica*, *Proformica*) and even the Nearctic (*Stenamma*). Dispersal from central Asia has likely occurred through both northern Africa and southern Europe. This analysis makes an important contribution to an understanding of the evolutionary history and biogeography of the Iberian Peninsula more generally.

The proximate mechanisms underlying species distributions are poorly understood, and this is the focus of a study of ant species occurring along an elevational gradient in Great Smoky Mountains National Park in south-eastern USA [5]. Thermal tolerance appeared to be the most important constraint on the distribution and density of ant species in colder (higher elevation) environments, whereas competition was identified as a key factor in warmer (lowland) environments. These findings indicate that species' responses to global warming are likely to be highly dependent on geographic context.

Ants can act as model organisms for understanding evolution and biogeography in a different way—as hosts of microorganisms [6]. *Wolbachia* bacteria are one of the most common and widely distributed microorganisms infecting insects, but the origins and dispersion patterns within different insect groups are poorly understood. The strains of *Wolbachia* are classified into 17 supergroups, and the analysis in [6] indicates that supergroup F is the ancestral character state for *Wolbachia* infection in ants. The infection of ants likely originated in Asia, before undergoing complex patterns of dispersal throughout the world, including through one route to South America and back to Asia before dispersing to Oceania and Madagascar.

Two papers in the Special Issue address the structure and dynamics of ant communities. The first [7] uses a wide range of sampling methods, incorporating diverse food resources along with sampling in different vertical strata, to examine niche partitioning in rainforest ant communities of French Guiana and Borneo. Many cryptobiotic species were recorded primarily or exclusively at subterranean baits, which supports findings from several recent studies (including [3]) that a diverse fauna of highly specialized hypogaeic species is characteristic of the tropics. Subterranean ants generally had more specialized food preferences than above-ground ants, and this was true for individual species foraging in both strata.

The second paper on community ecology [8] examines the extent to which a numerically dominant ant (*Formica subsericea*) affects the diversity and performance of co-occurring species. As in [5], the study was conducted in the Great Smoky Mountains National Park, where *F. subsericea* is particularly common. There was no relationship between the abundance of *F. subsericea* and the total abundance of non-dominant species across the ten sites studied, and the relationship with total species richness was weak. There was therefore little or no evidence of competitive exclusion. A comparison of colonies of two co-occurring species (*Aphaenogaster rudis* and *Nylanderia faisonensis*) located close to ($\leq$1 m) and distant (5–10 m) from *F. subsericea* nests revealed no differences in either colony size or brood production. There was, therefore, no evidence that competition from *F. subsericea* affected colony performance for either species. Markedly fewer workers of *A. rudis* occurred during the day at baits close to, compared with distant from, *F. subsericea* nests, but this was not the case for either *Nylanderia faisonensis* or *Myrmica punctiventris*. *Formica subsericea* thus had only a limited effect on resource acquisition by other species, even close to their nests. However, the number of other ants recorded at baits during the night, when *F. subsericea* was largely inactive, was found to be twice that recorded during the day, suggesting an avoidance of competition through temporal niche partitioning.

The final two papers of the Special Issue deal with ant taxa that are particularly influential in the ecosystems in which they occur. The first [9] is a comprehensive review of *Eciton* army ants. These are iconic top predators in Neotropical forests and are remarkable for the diversity of faunal taxa that are associated with them, often obligately so. Their most famous associates are the 'antbirds' that feed primarily on arthropods flushed during *Eciton* raids. Such specialization has evolved independently within multiple avian families. An astonishing variety of invertebrates have intimate associations with *Eciton* colonies, many with highly specialized morphological, behavioural and/or chemical traits that facilitate their integration into colony life. *Eciton* colonies require large forested areas, which, combined with their ecological importance, makes them ideal 'umbrella' taxa for the identification of priority areas for conservation and for assessments of disturbance impacts in Neotropical forests.

The final paper [10] describes an unwelcome impact—that of the red imported fire ant *Solenopsis invicta*, one of the world's most destructive invasive species. The gopher tortoise (*Gopherus polyphemus*) of the southeastern United States is an ecosystem engineer, whose burrows provide habitat for at least 60 vertebrate and 302 invertebrate species, many of which are of conservational interest. Gopher burrows are often colonized by fire ants, effectively concentrating their impacts on native fauna. The impact on native invertebrates was investigated at ten sites in southern Mississippi by sampling burrows before and after bait treatments to reduce fire ant populations. Baiting reduced fire ant abundance in burrows by >98%, and this had a positive effect on the abundance and diversity of other invertebrates as measured by vacuum sampling. It was concluded that targeted fire-ant management around the burrows of gopher tortoises is likely to have a disproportionate benefit for native biodiversity.

The papers in this Special Issue collectively address many of the most pressing questions relating to ant diversity. What *is* the level of ant diversity? What is the origin of this diversity, and how is it distributed at different spatial scales? What are the roles of niche partitioning and competition as regulators of local diversity? How do ants affect the ecosystems within which they occur? The papers in this Special Issue give important answers to these questions. These answers are not just relevant to ants, but provide valuable insights for biodiversity more generally.

Conflicts of Interest: The author declares no conflict of interest.

References

1. Andersen, A.N.; Hoffmann, B.D.; Oberprieler, S.K. Megadiversity in the ant genus *Melophorus*: The *M. rufoniger* Heterick, Castalanelli and Shattuck species group in the Top End of Australia's Northern Territory. *Diversity* **2020**, *12*, 386. [CrossRef]
2. Leahy, L.; Scheffers, B.R.; Williams, S.E.; Andersen, A.N. Diversity and distribution of the dominant ant genus *Anonychomyrma* (Hymenoptera: Formicidae) in the Australian Wet Tropics. *Diversity* **2020**, *12*, 474. [CrossRef]
3. Brassard, F.; Leong, C.-M.; Chan, H.-H.; Guénard, B. High diversity in urban areas: How comprehensive sampling reveals high ant species richness within one of the most urbanized regions of the world. *Diversity* **2021**, *13*, 358. [CrossRef]
4. Tinaut, A.; Ruano, F. Biogeography of Iberian ants (Hymenoptera: Formicidae). *Diversity* **2021**, *13*, 88. [CrossRef]
5. Chick, L.D.; Lessard, J.-P.; Dunn, R.R.; Sanders, N.J. The coupled influence of thermal physiology and biotic interactions on the distribution and density of ant species along an elevational gradient. *Diversity* **2020**, *12*, 456. [CrossRef]
6. Ramalho, M.O.; Moreau, C.S. The evolution and biogeography of *Wolbachia* in ants (Hymenoptera: Formicidae). *Diversity* **2020**, *12*, 426. [CrossRef]
7. Houadria, M.; Menzel, F. Digging deeper into the ecology of subterranean ants: Diversity and niche partitioning across two continents. *Diversity* **2021**, *13*, 53. [CrossRef]
8. Lessard, J.-P.; Stuble, K.L.; Sanders, N.J. Do dominant ants affect secondary productivity, behavior and diversity in a guild of woodland ants? *Diversity* **2020**, *12*, 460. [CrossRef]
9. Pérez-Espona, S. Eciton Army ants—Umbrella species for conservation in neotropical forests. *Diversity* **2021**, *13*, 136. [CrossRef]
10. Epperson, D.M.; Allen, C.R.; Hogan, K.F.E. Red imported fire ants reduce invertebrate abundance, richness, and diversity in gopher tortoise burrows. *Diversity* **2021**, *13*, 7. [CrossRef]

Article

Megadiversity in the Ant Genus *Melophorus*: The *M. rufoniger* Heterick, Castalanelli and Shattuck Species Group in the Top End of Australia's Northern Territory

Alan N. Andersen [1,*], Benjamin D. Hoffmann [2] and Stefanie K. Oberprieler [1]

[1]   Research Institute for the Environment and Livelihoods, Charles Darwin University, Darwin, NT 0909, Australia; stef_oberprieler@hotmail.com
[2]   CSIRO Tropical Ecosystems Research Centre, PMB 44 Winnellie, NT 0822, Australia; Ben.Hoffmann@csiro.au
*   Correspondence: Alan.Andersen@cdu.edu.au; Tel.: +61-468-384-038

Received: 20 September 2020; Accepted: 7 October 2020; Published: 8 October 2020

Abstract: This study contributes to an understanding of megadiversity in the arid-adapted ant genus *Melophorus* by presenting an integrated genetic, morphological and distributional analysis of diversity within the *M. rufoniger* group in the 400,000 km^2 Top End (northern region) of Australia's Northern Territory. An earlier study of the Top End's ant fauna lists eleven species from the *M. rufoniger* group, but a recent revision of *Melophorus* described the taxon as a single species occurring throughout most of the Australian mainland. CO1 sequences were obtained for 120 Top End specimens of the *M. rufoniger* group, along with a specimen from just outside the Top End. We recognize a total of 30 species among the sequenced specimens from the Top End, based on marked CO1 divergence (mean > 9%) in association with morphological differentiation and/or sympatric distribution. The sequenced specimen from just outside the Top End represents an additional species. Our unpublished CO1 data from other specimens from elsewhere in monsoonal Australia indicate that all but two of the 30 sequenced Top End species are endemic to the region, and that such diversity and endemism are similar in both the Kimberley region of far northern Western Australia and in North Queensland. The total number of species in the *M. rufoniger* group is potentially more than the 93 total species of *Melophorus* recognized in the recent revision. It has previously been estimated that *Melophorus* contains at least 1000 species, but our findings suggest that this is a conservative estimate.

Keywords: ant diversity; cryptic species; morphospecies; species delimitation; sympatric association

---

1. Introduction

Melophorus is a highly diverse, arid-adapted ant genus endemic to Australia that has been estimated to contain at least 1000 species [1,2]. However, a recent revision presents a markedly different view of diversity within the genus, recognizing only 93 species [3]. The *Melophorus rufoniger* Heterick, Castalanelli and Shattuck species group illustrates these divergent views. The group comprises polymorphic, large-bodied (4–10 mm), gracile, and highly thermophilic ants with reddish head and mesosoma that occur throughout arid, semi-arid and seasonally arid Australia (Figure 1). It belongs to the *M. aeneovirens* radiation of species where the head of minor workers has a dome-shaped occiput and an acutely angled clypeal 'apron' that projects over the base of the mandibles [1,2]. According to [1,2] (where it was referred to as the *M. aeneovirens* (Lowne) group, prior to the description of *M. rufoniger*) this group consists of many species, with diversity especially high in the monsoonal zone of northern Australia [4]. For example, a recent synopsis of the ant fauna of the 400,000 km^2 Top End (north of Katherine) of the Northern Territory (NT) listed 11 species from the group (where they are given the *Melophorus* species number codes sp. 1, sp. 2, sp. 4, sp. 35, sp. 45, sp. 46, sp. 47, sp. 52, sp. 62, sp. 63

and sp. 64) [5]. Preliminary CO1 data show that sp. 1 apparently represents several cryptic species, indicating that actual diversity within the complex is higher than is morphologically apparent [6].

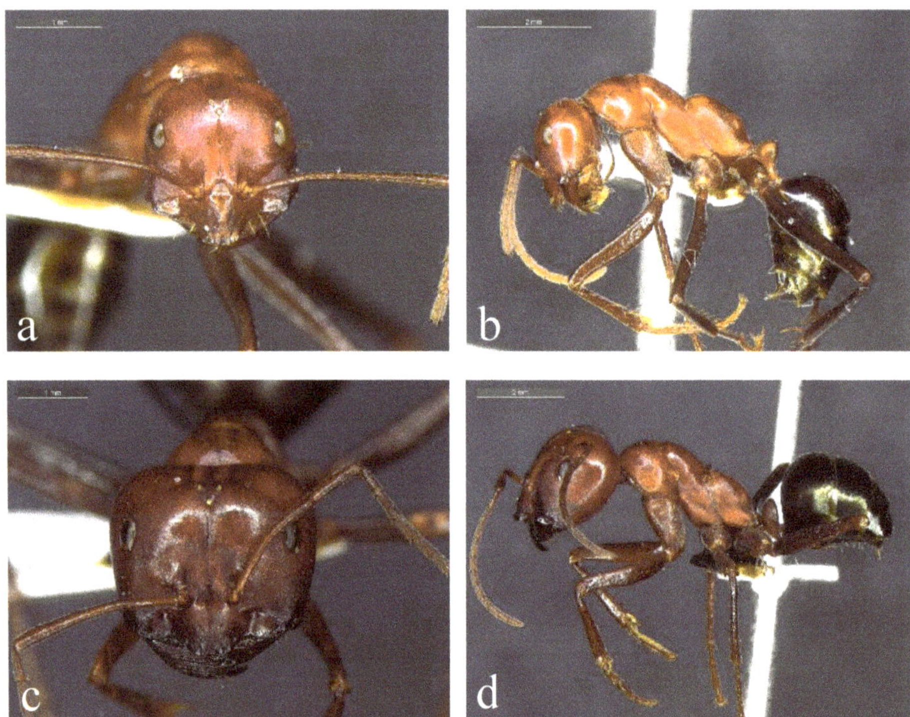

Figure 1. Representative species from the *Melophorus rufoniger* group. Minor ((**a,b**); BOLD ID: MELUS015-19) and major ((**c,d**); BOLD ID: MELPS092-18) workers of Taxon 1F, a member of the 'sp. 1' complex of the *Melophorus rufoniger* group from the Gove Peninsula in the NT.

Species within the *M. rufoniger* group show marked variation in mesosomal shape, sculpture, pilosity and configuration of the clypeal apron. In most Top End taxa, the anterior margin of the clypeal apron of both minor and major workers has a conspicuous medial notch. Such taxa can be divided into two complexes. In the first (sp. 1 and allies; Figure 1), minor workers have a deep and relatively narrow metanotal groove and the propodeum is highly asymmetrical in profile, with a distinct anterior face that curves into an oblique dorsal face (Figure 1b). The head has a matt appearance due to finely scalloped sculpture (Figure 1a), and the mesosoma never has erect hairs. The metanotal groove is particularly deep and the propodeum is especially prominently rounded anteriorly in sp. 52 (known only from sandstone country of Kakadu National Park), producing a biconvex mesosoma (Figure 2a). The head of the major workers of sp. 52 is heart-shaped, with the occipital margin V-shaped medially (Figure 2b). In sp. 46 (occurring in the southern Top End and extending further south in the NT), the profile of the pronotum rises more steeply anteriorly, and the mesosoma often has longer and denser pubescence (Figure 2c). The sp. 1 complex occurs throughout monsoonal Australia [7–11], and it is by far the most common form in the Top End of the NT [12–14].

In the second complex (sp. 2 and allies), the anterior margin of the clypeal apron is often less prominently notched, the metanotal groove is broader, the propodeum is more evenly rounded, the head (and mesosoma) is shinier, and the mesosoma often has erect hairs (Figure 3). In sp. 2 (occurring throughout the higher rainfall (northern) areas of the Top End), erect mesosomal hairs are very sparse and often restricted to the pronotum or are absent (Figure 3a,b), whereas in sp. 45

(known only from Litchfield NP), they occur throughout the mesosoma, and the first gastric segment and antennal scapes also have many erect hairs (Figure 3c,d). In sp. 62 (known only from the Tiwi Islands north of Darwin), the mesosoma is always glabrous. The sp. 2 complex appears not to extend outside the Top End into drier regions of the NT [9,11].

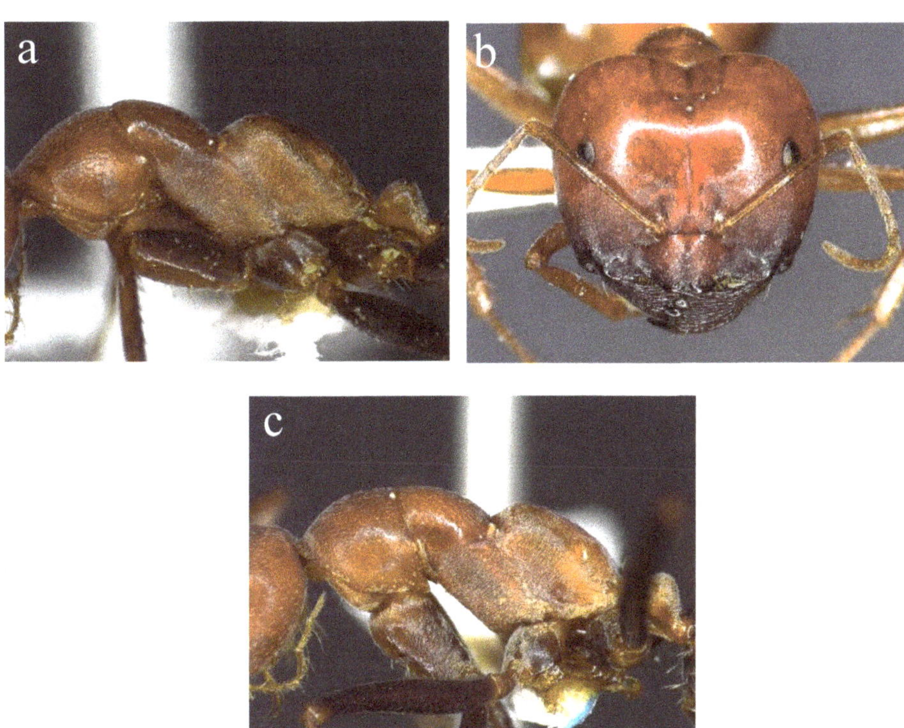

Figure 2. Morphology of sp. 52 and sp. 46 of the *Melophorus rufoniger* group. Mesosoma of minor worker ((**a**); BOLD ID: MELPS042-18) and head of major worker ((**b**); BOLD ID: MELOP008-15) sp. 52, and mesosoma of sp. 46 ((**c**); BOLD ID: MELOP079-15), both members of the sp. 1 complex.

A third complex also occurs in the Top End, represented by the slightly smaller sp. 35, where the anterior margin of the clypeal apron lacks a conspicuous medial notch (it is crenulate in minor workers and almost entire in major workers), and the metanotal groove is not so deep (Figure 4). This complex occurs most commonly outside the Top End, in lower rainfall areas of the monsoonal zone [9].

Despite the extensive morphological variation within the *M. rufoniger* group described above just within the Top End, as well as published accounts of high diversity more broadly [1], including marked genetic differentiation in sympatry [2], the recent revision of *Melophorus* [3] described the taxon as a single species. The species description provided no discussion of morphological or genetic variation within it. Here, we present an integrated morphological, genetic (CO1) and distributional analysis of diversity within the *M. rufoniger* group in the Top End of the NT, which is the most extensively collected region within its range. Our purpose is to improve an understanding of the extent of species richness within the *M. rufoniger* group, as a contribution to understanding diversity within *Melophorus* more generally.

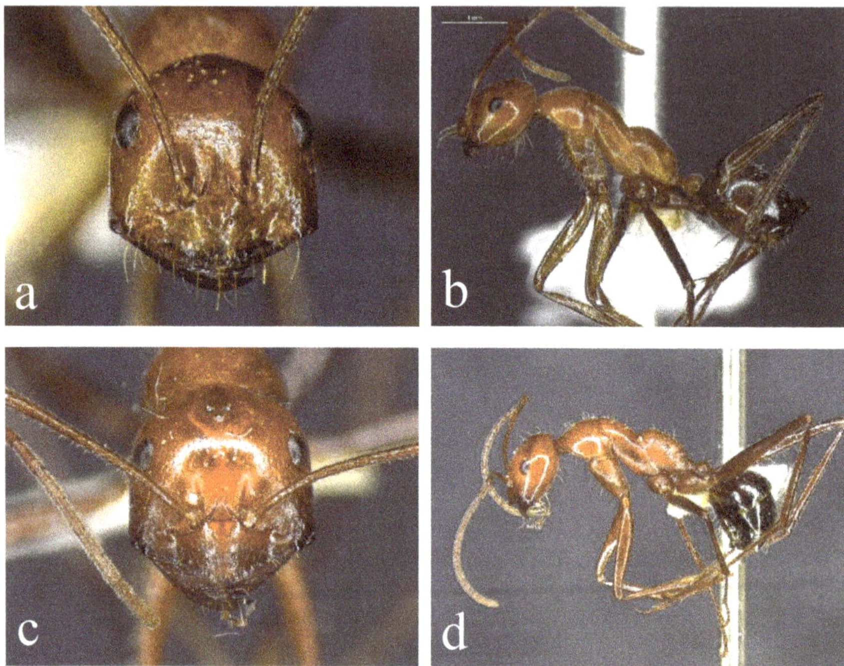

Figure 3. Morphology of the sp. 2 complex of the *Melophorus rufoniger* group. Head and lateral views of sp. 2 ((**a**,**b**); BOLD ID: MELPS048-18) and sp. 45 ((**c**,**d**); non-sequenced specimen from Litchfield NP).

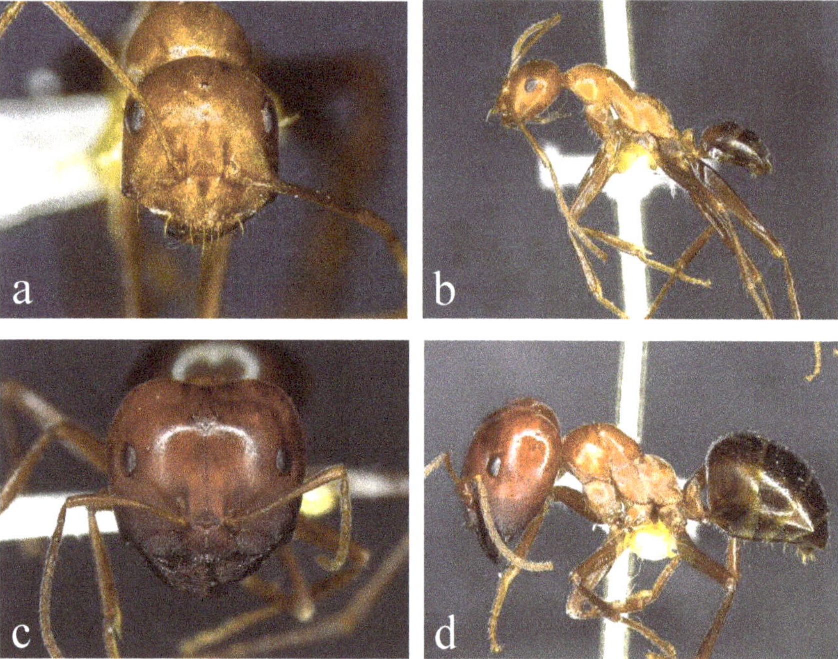

Figure 4. Morphology of sp. 35 from the Melophorus rufoniger complex. Minor (**a**,**b**) and major (**c**,**d**) workers of sp. 35 from Kalkarindji, NT (not sequenced).

2. Materials and Methods

The study was based on the >1000 pinned Top End specimens of the *M. rufoniger* group held in the ant collection at CSIRO's Tropical Ecosystems Research Centre (TERC) in Darwin, which represent the vast majority of specimens of the group collected from the Top End. The collection was inspected by B.H. Heterick when preparing the revision of the genus [3].

DNA was extracted from foreleg tissue and sequences obtained for 120 Top End specimens of the *M. rufoniger* group, along with a specimen from just outside the Top End (Lorella Springs, near Borroloola) and a specimen of *M. sulconotus* Heterick, Castalanelli and Shattuck (Supplementary Table S1). *Melophorus sulconotus* is a closely related species within the *M. aeneovirens* radiation [3], and it was used as the outgroup for building a CO1 tree. Geographic sampling within the Top End was very patchy (Figure 5). It was heavily concentrated in Kakadu and Nitmiluk National Parks in the central region, and to a lesser extent the northwest (Darwin region and Litchfield National Park), and on the Gove Peninsula in the far northeast. Most of Arnhem Land (east of Kakadu and Nitmiluk National Parks) and the southwest remain unsampled. Sequenced specimens included representatives of seven of the eleven species listed in [5]; sequences were unable to be obtained from the other species because of the old (>15 yrs) age of specimens. DNA extraction and CO1 sequencing were conducted through the Barcode of Life Data (BOLD) System (for extraction details, see http://ccdb.ca/resources). Each sequenced specimen was assigned a unique identification code that combines the batch within which it was processed and its number within the batch (e.g., MELPS087-17), and all specimens are labeled with their respective BOLD identification numbers in the TERC collection. All data sequences have been deposited in GenBank; BankIt2111481.

DNA sequences were checked and edited in MEGA 7 [15]. Sequences were aligned using the UPGMB clustering method in MUSCLE [16], and then translated into (invertebrate) proteins to check for stop codons and nuclear paralogues. The aligned sequences were trimmed accordingly, resulting in 657 base pairs.

To explore overall CO1 diversity in the samples, the mean genetic pairwise distances between sequences were calculated in MEGA 7. This was done using the Kimura-2 parameter (K2P) model [17] to ensure that results were comparable with those of most other studies of insect DNA barcoding, with 500 bootstrap replicates and the 'pairwise deletion' option of missing data (to remove all ambiguous positions for each sequences pair). Analysis involved all nucleotide sequences, excluding those of the outgroup. Codon positions included were 1st+2nd+3rd.

There is no specific level of CO1 divergence that can be used to define a species, but the level of CO1 variation within ant species is typically 1%–3% [18]. However, some ant species can show substantially higher variation (e.g., [19]), and in other cases two clear species can show no CO1 differentiation (e.g., [20]). We also note that some ant species from other genera are known to have workers that are virtually identical morphologically, and they can be separated only by detailed morphometric analysis or through reproductive castes [21]. When delimiting species, we adopted the species concept based on reproductive isolation and evolutionary independence as evidenced by morphological differentiation between sister (i.e., most closely related) clades (considering all available samples from the same collections as those of sequenced specimens) and sympatric distribution. We are not aware of any record of *M. 'rufoniger'*, or any other species of *Melophorus*, having wingless or brachypterous queens that might compromise the reliability of CO1 data for species delimitation.

Tree inference by maximum likelihood was conducted through the IQTREE web server (http://iqtree.cibiv.univie.ac.at/; [22]) using ultrafast bootstrap approximation [23]. IQTREE has been shown to be a robust algorithm for tree inference that compares favourably with other methods [24]. Model selection was inferred using a 3-codon partition file and linked branch lengths with the AutoMRE 'ModelFinder' function to find the best-fit model for tree inference [25]. Trees were viewed and edited in FigTree v1.4.3 [26] and annotated using Photoshop CS5.1.

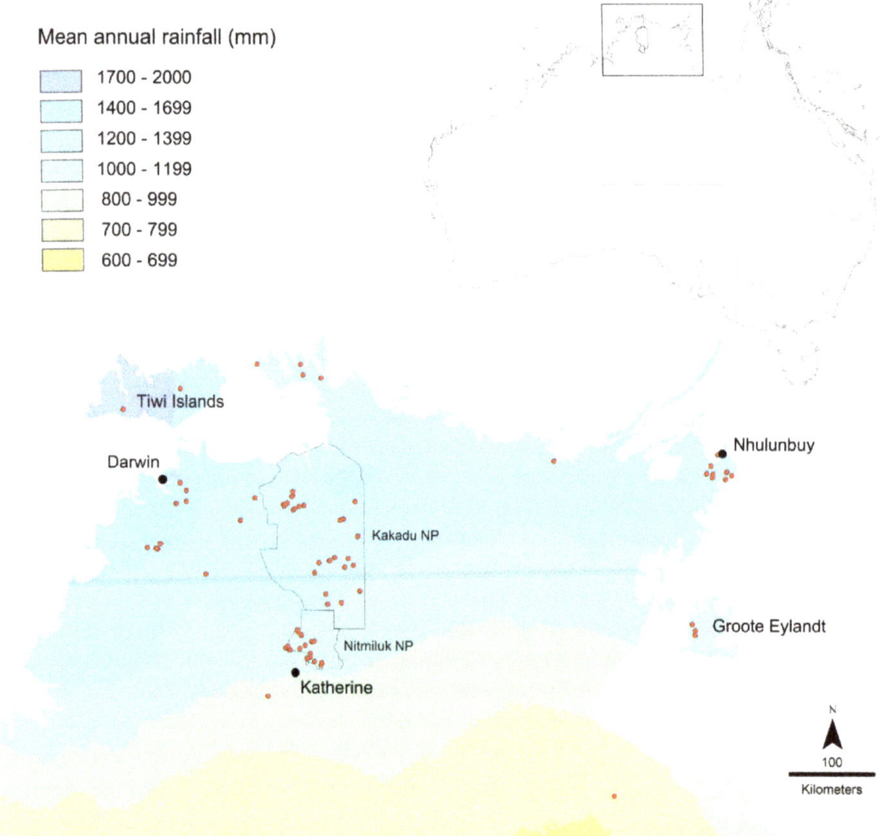

Figure 5. Map of collection localities in the Top End of the Northern Territory. Collection localities for sequenced specimens are shown in red. Collections are heavily concentrated in Kakadu and Nitmiluk National Parks, as indicated.

3. Results

The CO1 tree shows four primary clades, with 10.2–13.5% divergence among them (Figure 6). The first two clades both consist of members of the sp. 1 complex, the third clade comprises the sp. 2 complex, and the fourth clade the sp. 35 complex.

The CO1 data indicate that sp. 1 contains 22 genetically differentiated taxa across the first two primary clades (Taxa 1A—1T in Figure 6), with mean divergence among them of 9.1% and in most cases morphologically differentiated from their sister taxa. However, some sp. 1 taxa show no obvious morphological differentiation from one or more non-sister taxa, even when they occur in different primary clades.

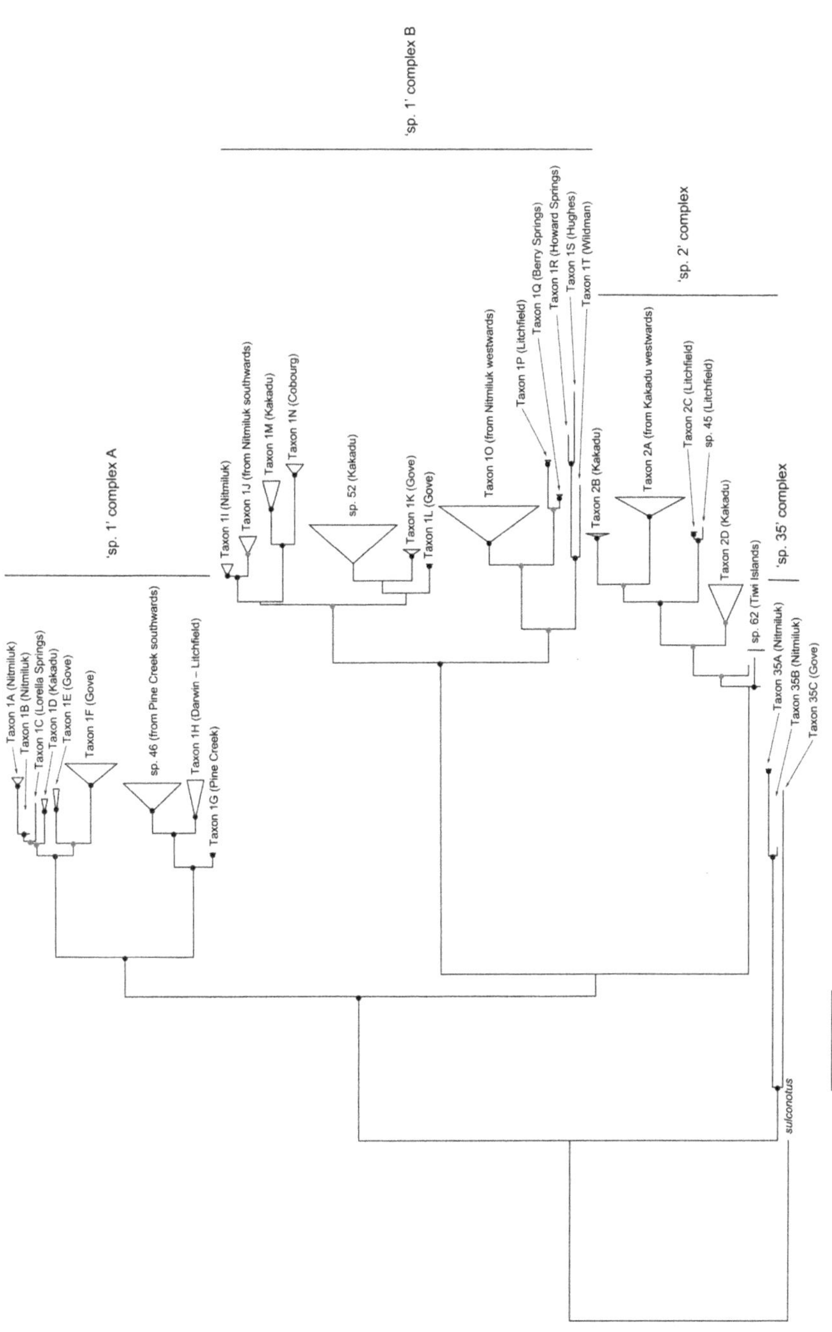

Figure 6. Summary CO1 tree of specimens from the *Melophorus rufoniger* group. Red and black circles indicate bootstrap values ≥70% and ≥90%, respectively. The full CO1 tree is shown in Supplementary Figure S2.

The first primary clade contains eight sp. 1 taxa (Taxa 1A – 1H) along with sp. 46 (Figure 2c), with 3.9% mean divergence among them. Taxon 1A and Taxon 1B are both known only from Nitmiluk NP; there are no obvious morphological differences between them, but their clear and substantial (2.2%) CO1 divergence in sympatry (in one case occurring within 100 m of each other) indicates that they represent different species. The single specimen of Taxon 1C (from Lorella Springs) can be differentiated morphologically from other sp. 1 taxa by its far denser and longer gastric pubescence (Figure 7). Taxon 1D is shown as sister to Taxon 1C but has typical gastric pubescence. In Taxon 1E (known only from the Gove Peninsula) the mesosoma of the major worker is saddle-shaped (Figure 8), lacking the prominent metanotal groove characteristic of other sp. 1 taxa (Figure 1d). Its sister taxon (Taxon 1F) is also known only from the Gove Peninsula, and the two taxa have 4.3% CO1 divergence from each other. The remaining taxa (sp. 46, Taxon 1G and Taxon 1H) occur in a separate sub-clade (Figure 6). Taxon 1G (Pine Creek region) and Taxon 1H (Darwin and Litchfield regions, approximately 150 km from Pine Creek) have 3.5% mean CO1 divergence from each other, but they show no obvious morphological differentiation, despite Taxon 1H having lower CO1 divergence from morphologically differentiated sp. 46 (Figure 2c) than from Taxon 1G.

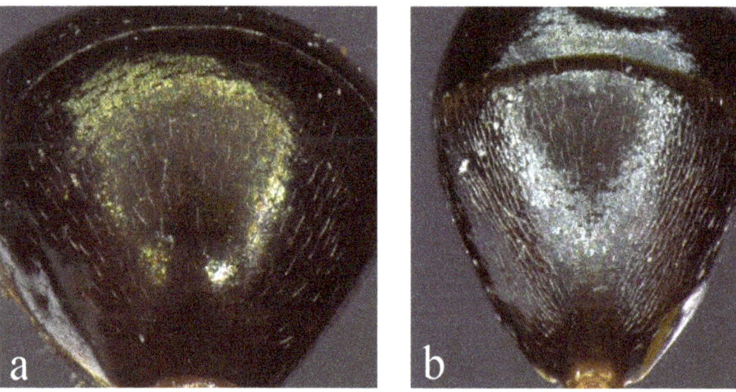

Figure 7. Variation in gastric pubescence. First gastric segments of minor workers of (**a**) Taxon 1A (BOLD ID: MELPS087-17) and (**b**) Taxon 1C (BOLD ID: DARW621-16) from the sp. 1 complex.

Figure 8. Major worker of Taxon 1E from the sp. 1 complex. Note the saddle-shaped mesosoma that lacks a conspicuous metanotal groove (BOLD ID: DARW622-15).

The second primary clade consists of 12 sp. 1 taxa along with sp. 52, with 6.2% mean CO1 divergence among them. It is also divided into two distinct sub-clades. The first sub-clade contains sp. 52 along with Taxa 1I-1N. Taxon 1I and Taxon 1J show no clear morphological differentiation, but they are differentiated genetically (2.2% CO1 divergence) despite both occurring in Nitmiluk NP (Figure 6). The former is known only from Nitmiluk, whereas the latter extends south to Newcastle Waters in the NT and across to Western Australia (A. N. Andersen, unpublished CO1 data). Taxon 1K (known only from the Gove Peninsula) is the sister taxon to sp. 52 (known only from Kakadu NP), but lacks the biconvex mesosoma. Taxon 1L is sister to Taxon 1K plus sp. 52 and, like Taxon 1K, is also known only from the Gove Peninsula. Taxon 1M (known only from Kakadu NP) and Taxon 1N (known only from Cobourg Peninsula) are sister taxa, with 3.9% mean CO1 divergence between them. The second sub-clade of the second primary clade contains Taxa 1O–1T. Taxon 1O is widespread in the western half of the Top End, and its distribution overlaps that of all the other taxa in the sub-clade. The most distinctive of these taxa morphologically is Taxon 1Q (known only from Berry Springs near Darwin), the minor workers of which have a very prominently rounded pro-mesonotum and a relatively narrow head that is longer than wide in full-face view (Figure 9). The third primary clade of the CO1 tree represents the sp. 2 complex, and strongly indicates that sp. 2 comprises multiple species (Taxa 2A -2D, with 3.2% mean CO1 divergence from each other; Figure 6). Taxon 2A is widespread in the Top End, occurring from Wildman River (approximately 100 km east of Darwin) to central Arnhem Land; its sister taxon (Taxon 2B; 3.8% CO1 divergence) is known only from Kakadu NP, where Taxon 2A is common. Taxon 2C has only 0.09% CO1 divergence from the very hairy sp. 45 (Figure 3c,d), but its mesosoma is only sparsely hairy and erect hairs are absent from its antennal scapes and gaster. Both taxa are known only from Litchfield NP. Taxon 2D is a third sp. 2 taxon occurring in Kakadu NP, and it has not been recorded elsewhere. The final species in the sp. 2 complex is sp. 62, known only from the Tiwi Islands. It has no obviously distinguishing features but is genetically distinct (shown as basal) from the rest of the complex (Figure 6).

Figure 9. Morphology of Taxon 1Q from the sp. 1 complex. Head (**a**) and lateral view (**b**) of minor worker (BOLD ID: MELPS071-18). This taxon is known only from Berry Springs, NT.

The final primary clade represents the sp. 35 complex, and it indicates that sp. 35 also contains multiple taxa, with three (Taxa 35A–C, 7.8% mean CO1 divergence) represented among the four specimens sequenced from the Top End (Figure 6). The most distinctive of these morphologically is Taxon 35C; in the minor worker the posterior half of the pronotum and the mesonotum are only feebly curved in profile and the propodeum is weakly differentiated (Figure 10).

Figure 10. Morphology of Taxon 35C from the sp. 35 complex. Minor worker (BOLD ID: MELPS118-18), known only from the Gove Peninsula.

4. Discussion

This study presents an integrated analysis of diversity within the *Melophorus rufoniger* group of ants in the Top End of the NT, a taxon that was described as a single species in a recent revision of the genus [3]. CO1 sequences were obtained for seven of the eleven species listed from the Top End in [5], and the validity of all these species has been supported. Moreover, our analyses have indicated that each of sp. 1, sp. 2 and sp. 35 represent several to many species. We recognise a total of 31 species from the sequenced specimens, based on a combination of CO1 divergence, morphological differentiation and/or sympatric distribution. One of these species is represented by the specimen from Lorella Springs, and it was not recorded from the Top End. Several of the Top End species have no obvious morphological differences between their minor workers, even when they belong to highly divergent CO1 clades. However, we note that major workers often have particularly informative morphological characters, and that major workers have not been collected for most of the apparently cryptic species.

Additional diversity is especially high in the sp. 1 complex, where our analysis revealed a remarkable 20 apparent species (Taxa 1A-T) among sequenced specimens in addition to sp. 46 and sp. 52. Multiple species from the sp. 1 complex often occurred at the same site (within 100 m); for example: Taxon 1D and sp. 52 (Kakadu NP, Biodiversity Monitoring Site 133B), Taxon 1G and sp. 46 (nr. Pine Creek), Taxon 1A and Taxon 1B (Nitmiluk NP, Biodiversity Monitoring Sites 42/42B), and Taxon 1H and Taxon 1R (Howard River) (Supplementary Table S1). Our analyses also revealed multiple species among sequenced specimens within both sp. 2 (four species) and sp. 35 (three species). Species from the sp. 2 complex were collected from only the northern (higher rainfall) half of the Top End, and there was only one case where a species occurred at the same site as one of the sp. 1 complexes (sp. 2A and sp. 52; Kakadu NP, Biodiversity Monitoring Site 69/69B; Supplementary Table S1), compared with the several cases of co-occurring species from within the sp. 1 complexes. The single sequenced specimen of Taxon 35C was collected at the same site as Taxon 1K (Gove Peninsula, Dhimurru Biodiversity Monitoring Site G1; Supplementary Table S1).

Regional diversity is particularly high in Nitmiluk National Park, where seven sequenced species have been collected, including five from the sp. 1 complex (Figure 11). Species distributions within the Park do not show strong geographic structure. None of the Nitmiluk species belong to the sp. 2 complex, which reflects the Park's southern location within the Top End. Notably, six of the seven Nitmiluk species have not been recorded outside the Park. Another centre of diversity is Kakadu National Park, which adjoins Nitmiluk to the north; six sequenced species were collected from Kakadu, including three species from the sp. 2 complex (Figure 11). The Kakadu species are strongly structured spatially, especially west versus east relating to lowland savanna and sandstone habitat, respectively (Figure 11). Only one (Taxon 2A) of the six Kakadu species has been collected outside the Park. Remarkably, none of the Kakadu or Nitmiluk species are known from both Parks; collectively, they harbour 13 sequenced species from the group, and, as far as is currently known, 11 of these are endemic to their respective Park. There are no samples from the north of Nitmiluk, and it is likely that the Kakadu species sp. 52, Taxon 1D and Taxon 2B occur there (Figure 11); nevertheless, the Kakadu and Nitmiluk faunas are remarkably disjunct. The high regional diversity in Kakadu and Nitmiluk might simply be a product of high sampling intensity (Figure 5), but we suspect that the prominence of sandstone landforms as part of the western Arnhem Plateau is an important contributing factor. Local diversity is also high on the Gove Peninsula in the far northeast of the Top End, where five sequenced species have been collected. Four of these are known only from the Gove Peninsula, and the only other known location of the fifth (Taxon 1F) is nearby Groote Eylandt. Taxon 1K and Taxon 1L are known only from small localities within the Gove Peninsula, around Gulkula and Port Bradshaw, respectively. We note that such apparent local endemism might be an artefact of low sampling intensity elsewhere in Arnhem Land (Figure 5). However, there are numerous other examples of apparent short-range endemism elsewhere in the Top End where regional sampling is more extensive: for example, despite very extensive collecting in the broader region, Taxon 1Q is known only from the Territory Wildlife Park at Berry Springs, and sp. 45 has been collected only from the Wangi Falls area of Litchfield National Park. Nine of the 30 sequenced species that we have recognized from the Top End are known from a single site, suggesting that a substantial proportion of species have very limited distributional ranges.

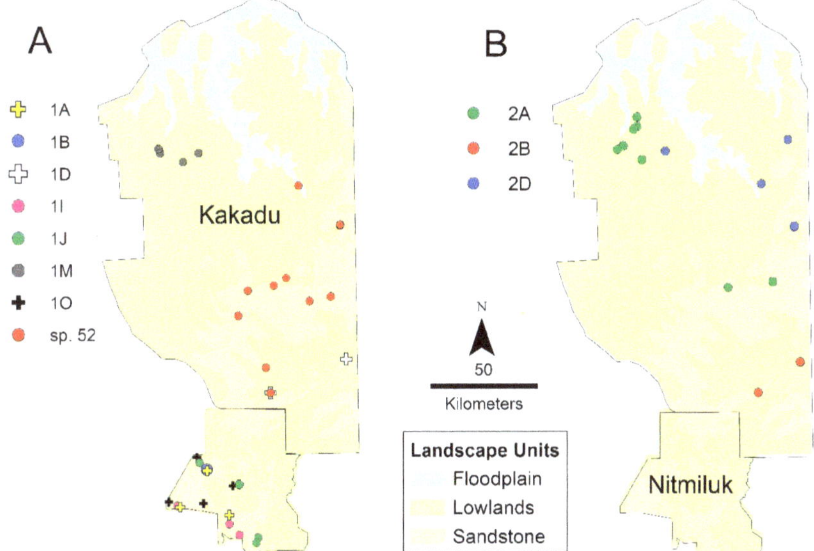

Figure 11. Distribution of species of the sp. 1 (**A**) and sp. 2 (**B**) complexes in Kakadu and Nitmiluk National Parks.

5. Conclusions

Our integrated analysis of diversity within the *Melophorus rufoniger* group indicates that the 120 specimens sequenced from the Top End of the NT represent very many and potentially 30 species. This does not include the four additional species recognised from the Top End in [5]. For example, sp. 63 is a member of the sp. 1 complex from the Darwin region with a prominently raised mesonotum and a particularly long clypeal projection, and sp. 47 is another member of the sp. 2 complex from Kakadu NP, in which the antennal scapes are clothed in short, semi-erect hairs and the head has a matt appearance due to finely punctate and scalloped sculpture. We acknowledge that the validity of all these indicated species has not been unequivocally demonstrated. However, we have clearly demonstrated that the species group is represented by very many species in the Top End. Notably, all but two of these appear to be endemic to the Top End. There are high levels of apparent short-range endemism, which suggests that many species occur in the Top End in addition to those sequenced, given the highly patchy geographic coverage of samples (Figure 4). Our finding of extensive undocumented diversity within the *M. rufoniger* gp. in the Top End is consistent with a recent phylogeographic study of ten lizard genera in the region [27], which found that many species represent multiple lineages, especially among geckos. For example, 11 lineages of *Heteronotia 'binoei'* were recorded, most of which are endemic to the Top End and have narrow distributional ranges within it. Diversity was found to be especially high on the western Arnhem Plateau (see also [28]), which is consistent with our finding of very high levels of diversity within the *M. rufoniger* gp. in Kakadu and Nitmiluk National Parks.

Diversity and endemism within the *M. rufoniger* group is also very high in the Kimberley region of northern Western Australia and in North Queensland (A. N. Andersen, unpublished data), as has also been shown for lizards [27]. This suggests that well over 50 species from the *M. rufoniger* gp., and possibly as many as 100, occur just in monsoonal Australia. None of these also occur in South Australia (A. N. Andersen, unpublished CO1 data) where the type locality for *M. rufoniger* occurs, and so all are undescribed.

Our findings have important implications for diversity more generally in *Melophorus*. There is a strong likelihood that more species are concealed under *M. 'rufoniger'* than the total of 93 species recognized by [3] for the whole genus. When [4] presented an analysis of diversity in *Melophorus* based on the >850 morphospecies sorted in the TERC collection at the time, CO1 analysis in association with closer morphological inspection of 188 of these showed that many represented multiple and often many species, as was the case here for *M. rufoniger* sp. 1, sp. 2 and sp. 35. As of April 2020, the TERC collection holds >950 morphospecies of *Melophorus*, and it is likely that these represent well over 1000 actual species. Given the high levels of short-range endemism reported here, and that only a fraction of the range of *Melophorus* across the Australian continent has been sampled for ants, total diversity in the genus is likely to be far higher.

Supplementary Materials: The following are available online at http://www.mdpi.com/1424-2818/12/10/386/s1, Figure S1: Complete fifty percent majority rule Bayesian consensus tree of CO1-barcoded specimens from the *Melophorus rufoniger* group considered in this study. A specimen of closely related *M. sulconotus* is used as a n outgroup. Red and black circles indicate bootstrap values ≥70% and ≥90%, respectively. Table S1: Collection locations of specimens from the *Melophorus rufoniger* group that were CO1-barcoded for this study. Table S2. CO1 sequence divergence (%) among taxa, calculated using the Maximum Composite Likelihood model [29] in MEGA X [15].

Author Contributions: A.N.A. conceived the study, led the development of the TERC collection, and wrote the first draft of the manuscript. B.D.H. helped develop the TERC collection and contributed to the writing of the paper. S.K.O. undertook the analysis of the CO1 data and contributed to the writing of the paper. All authors have read and agreed to the published version of the manuscript.

Funding: This research received no external funding.

Acknowledgments: We thank our many collaborators who collected the specimens analysed in this study, and especially Magen Pettit, Tony Hertog and Jodie Hayward from CSIRO, as well as staff from the Flora and Fauna Division of the NT Department of Environment and Natural Resources who conducted biodiversity monitoring in Kakadu, Nitmiluk and Litchfield National Parks. We are also most grateful to Ruben Andersen for preparing the digital images, and to Finn Andersen for assistance with image quality. Finally, we thank Brian Fisher, Tek Tay and Olivia de Sousa for their valuable comments on the draft manuscript.

Conflicts of Interest: The authors declare no conflict of interest.

References

1. Andersen, A.N. Ant diversity in arid Australia: A systematic overview. In *Advances in Ant (Hymenoptera: Formicidae) Systematics: Homage to E. O. Wilson—50 Years of Contributions*; Snelling, R.R., Fisher, B.L., Ward, P.S., Eds.; American Entomological Institute: Gainesville, FI, USA, 2007; Volume 80, pp. 19–51.
2. Andersen, A.N.; Hoffmann, B.D.; Sparks, K. The megadiverse Australian ant genus *Melophorus*: Using CO1 barcoding to inform species richness. *Diversity* **2016**, *8*, 30. [CrossRef]

3. Heterick, B.H.; Castalanelli, M.; Shattuck, S.O. Revision of the ant genus *Melophorus* (Hymenoptera, Formicidae). *ZooKeys* **2017**, *700*, 1–420. [CrossRef] [PubMed]

4. Andersen, A.N. *The Ants of Northern Australia: A Guide to the Monsoonal Fauna;* CSIRO Publishing: Collingwood, New Zealand, 2000.

5. Andersen, A.N.; Hoffmann, B.D.; Oberprieler, S. Diversity and biogeography of a species-rich ant fauna of the Australian seasonal tropics. *Ins. Sci.* **2016**, *25*, 519–526. [CrossRef] [PubMed]

6. Andersen, A.N. Ant megadiversity and its origins in arid Australia. *Austral Entomol.* **2016**, *55*, 132–147. [CrossRef]

7. Barrow, L.; Parr, C.L.; Kohen, J.L. Biogeography and diversity of ants in Purnululu (Bungle Bungle) National Park and conservation reserve. *Aust. J. Zool.* **2006**, *54*, 123–136. [CrossRef]

8. Andersen, A.N.; Lanoue, J.; Radford, I. The ant fauna of the remote Mitchell Falls area of tropical north-western Australia: Biogeography, environmental relationships and conservation significance. *J. Insect Conserv.* **2010**, *14*, 647–661. [CrossRef]

9. Andersen, A.N.; Del Toro, I.; Parr, C.L. Savanna ant species richness is maintained along a bioclimatic gradient of decreasing rainfall and increasing latitude in northern Australia. *J. Biogeogr.* **2015**, *42*, 2313–2322. [CrossRef]

10. Cross, A.T.; Myers, C.; Mitchell, C.N.A.; Cross, S.L.; Jackson, C.; Waina, R.; Mucina, L.; Dixon, K.W.; Andersen, A.N. Ant biodiversity and its environmental predictors in the North Kimberley region of Australia's seasonal tropics. *Biodivers. Conserv.* **2016**, *25*, 1727–1759. [CrossRef]

11. Del Toro, I.; Ribbons, R.R.; Hayward, J.; Andersen, A.N. Are stacked species distribution models accurate at predicting multiple levels of diversity along a rainfall gradient? *Austral Ecol.* **2019**, *44*, 105–113. [CrossRef]

12. Andersen, A.N. Responses of ground-foraging ant communities to three experimental fire regimes in a savanna forest of tropical Australia. *Biotropica* **1991**, *23*, 575–585. [CrossRef]

13. Andersen, A.N. Ants as indicators of restoration success at a uranium mine in tropical Australia. *Restor. Ecol.* **1993**, *1*, 156–167. [CrossRef]

14. Andersen, A.N.; Ribbons, R.R.; Pettit, M.; Parr, C.L. Burning for biodiversity: Highly resilient ant communities respond only to strongly contrasting fire regimes in Australia's seasonal tropics. *J. Appl. Ecol.* **2014**, *51*, 1406–1413. [CrossRef]

15. Kumar, S.; Stecher, G.; Li, M.; Knyaz, C.; Tamura, K. MEGA X: Molecular Evolutionary Genetics Analysis across computing platforms. *Mol. Biol. Evol.* **2018**, *35*, 1547–1549. [CrossRef] [PubMed]

16. Edgar, R.C. MUSCLE: Multiple sequence alignment with high accuracy and high throughput. *Nucleic Acids Res.* **2004**, *32*, 1792–1797. [CrossRef]

17. Kimura, M. A simple method for estimating evolutionary rates of base substitutions through comparative studies of nucleotide sequences. *J. Mol. Evol.* **1980**, *16*, 111–120. [CrossRef]

18. Smith, M.A.; Fisher, B.L.; Hebert, P.D.N. DNA barcoding for effective biodiversity assessment of a hyperdiverse arthropod group: The ants of Madagascar. *Philos. Trans. R. Soc. B Biol. Sci.* **2005**, *360*, 1825–1834. [CrossRef]

19. Wild, A.L. Evolution of the Neotropical ant genus *Linepithema*. *Syst. Entomol.* **2009**, *34*, 49–62. [CrossRef]

20. Schär, S.; Talavera, G.; Espadaler, X.; Rana, J.D.; Andersen, A.A.; Cover, S.P.; Vila, R. Do Holarctic ant species exist? Trans-Beringian dispersal and homoplasy in the Formicidae. *J. Biogeogr.* **2018**, *45*, 1917–1928. [CrossRef]

21. Wagner, H.C.; Gamisch, A.; Arthofer, W.; Moder, K.; Steiner, F.M.; Schlick-Steiner, B.C. Evolution of morphological crypsis in the *Tetramorium caespitum* ant species complex (Hymenoptera: Formicidae). *Sci. Rep.* **2018**, *8*, 12547. [CrossRef]

22. Trifinopoulos, J.; Nguyen, L.-T.; von Haeseler, A.; Minh, B.Q. W-IQ-TREE: A fast online phylogenetic tool for maximum likelihood analysis. *Nucleic Acids Res.* **2016**, *44*, W232–W235. [CrossRef]

23. Minh, B.Q.; Nguyen, M.A.T.; Von Haeseler, A. Ultrafast approximation for phylogenetic bootstrap. *Mol. Biol. Evol.* **2013**, *30*, 188–1195. [CrossRef] [PubMed]

24. Nguyen, L.-T.; Schmidt, H.A.; Von Haeseler, A.; Minh, B.Q. IQ-TREE: A fast and effective stochastic algorithm for estimating maximum-likelihood phylogenies. *Mol. Biol. Evol.* **2014**, *32*, 268–274. [CrossRef] [PubMed]

25. Chernomor, O.; von Haeseler, A.; Minh, B.Q. Terrace aware data structure for phylogenomic inference from supermatrices. *Syst. Biol.* **2016**, *65*, 997–1008. [CrossRef] [PubMed]

26. Ambaut, A. FigTree: Molecular Evolution, Phylogenetics and Epidemiology. 2007. Available online: http://tree.bio.ed.ac.uk/software/figtree/ (accessed on 21 February 2019).

27. Rosauer, D.F.; Blom, M.P.K.; Bourke, G.; Catalano, S.; Donnellan, S.; Gillespie, G.; Mulder, E.; Oliver, P.M.; Potter, S.; Rabosky, D.L.; et al. Phylogeography, hotspots and conservation priorities: An example from the Top End of Australia. *Biol. Conserv.* **2016**, *204*, 83–93. [CrossRef]
28. Woinarski, J.C.Z.; Russell-Smith, J.; Andersen, A.N.; Brennan, K. Fire management and biodiversity of the western Arnhem Land plateau. In *Culture, Ecology and Economy of Fire management in North Australian Savannas: Rekindling the Wurrk Tradition*; Russell-Smith, J., Whitehead, P., Cooke, P., Eds.; CSIRO Publishing: Collingwood, New Zealand, 2009; pp. 201–227.
29. Tamura, K.; Nei, M.; Kumar, S. Prospects for inferring very large phylogenies by using the neighbor-joining method. *Proc. Natl. Acad. Sci. USA* **2004**, *101*, 11030–11035. [CrossRef] [PubMed]

Article

Diversity and Distribution of the Dominant Ant Genus *Anonychomyrma* (Hymenoptera: Formicidae) in the Australian Wet Tropics

Lily Leahy [1,*], Brett R. Scheffers [2], Stephen E. Williams [1] and Alan N. Andersen [3]

[1] Centre for Tropical Environmental and Sustainability Science, College of Science and Engineering, James Cook University, Townsville, QLD 4811, Australia; stephen.williams@jcu.edu.au
[2] Department of Wildlife Ecology and Conservation, University of Florida, Gainesville, FL 32611, USA; brett.scheffers@ufl.edu
[3] Research Institute for the Environment and Livelihoods, Charles Darwin University, Darwin, NT 0909, Australia; Alan.Andersen@cdu.edu.au
* Correspondence: lily.leahy@my.jcu.edu.au

Received: 18 November 2020; Accepted: 7 December 2020; Published: 14 December 2020

Abstract: *Anonychomyrma* is a dolichoderine ant genus of cool-temperate Gondwanan origin with a current distribution that extends from the north of southern Australia into the Australasian tropics. Despite its abundance and ecological dominance, little is known of its species diversity and distribution throughout its range. Here, we describe the diversity and distribution of *Anonychomyrma* in the Australian Wet Tropics bioregion, where only two of the many putative species are described. We hypothesise that the genus in tropical Australia retains a preference for cool wet rainforests reminiscent of the Gondwanan forests that once dominated Australia, but now only exist in upland habitats of the Wet Tropics. Our study was based on extensive recent surveys across five subregions and along elevation and vertical (arboreal) gradients. We integrated genetic (CO1) data with morphology to recognise 22 species among our samples, 20 of which appeared to be undescribed. As predicted, diversity and endemism were concentrated in uplands above 900 m a.s.l. Distribution modelling of the nine commonest species identified maximum temperature of the warmest month, rainfall seasonality, and rainfall of the wettest month as correlates of distributional patterns across subregions. Our study supported the notion that *Anonychomyrma* radiated from a southern temperate origin into the tropical zone, with a preference for areas of montane rainforest that were stably cool and wet over the late quaternary.

Keywords: ant diversity; biogeography; species delimitation; Dolichoderinae; species distribution models; climatic gradients; wet tropics; climate change

---

1. Introduction

The ubiquity, high species richness, and ecological dominance of ants make them model organisms for understanding spatial patterns of diversity and community assembly [1]. Ant distribution is tightly coupled to climate [2], so that an understanding of ant species distributions across climatic gradients provides clues as to how ectothermic species coped with climatic conditions in the past and provides important insights into how they are likely to respond to a future climate.

Anonychomyrma Donisthorpe 1947 is an ecologically dominant ant genus in relatively cool and mesic habitats of Australia and New Guinea, extending northwest into Southeast Asia and east to the Solomon Islands [3]. The genus contains 27 described species and 5 subspecies, with most of these from New Guinea [4,5]. The genus is a member of an Australian dolichoderine subgroup that is thought to be of Gondwanan origin and that most likely originated in Australia approximately

30 million years ago from an ancestor shared with southern South America through a land connection via Antarctica [6]. *Anonychomyrma* is particularly common in heathlands, woodlands, and open forests of temperate southern Australia ([7,8]; in both referenced articles, the genus is referred to as the *nitidiceps* group of *Iridomyrmex*) and in rainforest of subtropical and tropical mountains of Queensland and New Guinea [5,9,10]. Most species in drier habitats nest in the ground, whereas rainforest species appear to be primarily arboreal.

The Australian Wet Tropics (AWT) bioregion of northeastern Australia represents the largest remaining remnant of the Gondwanan mesothermal rainforests that once dominated Australia [11]. Two species of *Anonychomyrma* have been described from the region. One of these is *Anonychomyrma gilberti* (Forel) (Figure 1), a widely distributed species in the region where it is a dominant ant in the rainforest canopy [10,12]. The second species is *Anonychomyrma malandana* (Forel), which was described from Malanda on the Atherton Tablelands in 1915. Since its original description, we are aware of only one other publication documenting it, where it was commonly recorded on Mt. Windsor (Figure 2) and referred to as *Anonychomyrma* sp. D [9]. Despite just two described species, the AWT *Anonychomyrma* fauna is highly diverse, with [9] 15 species reported during ant surveys of just four of the thirteen upland subregions.

Figure 1. *Anonychomyrma gilberti*, showing head (**A**) and lateral (**B**) views.

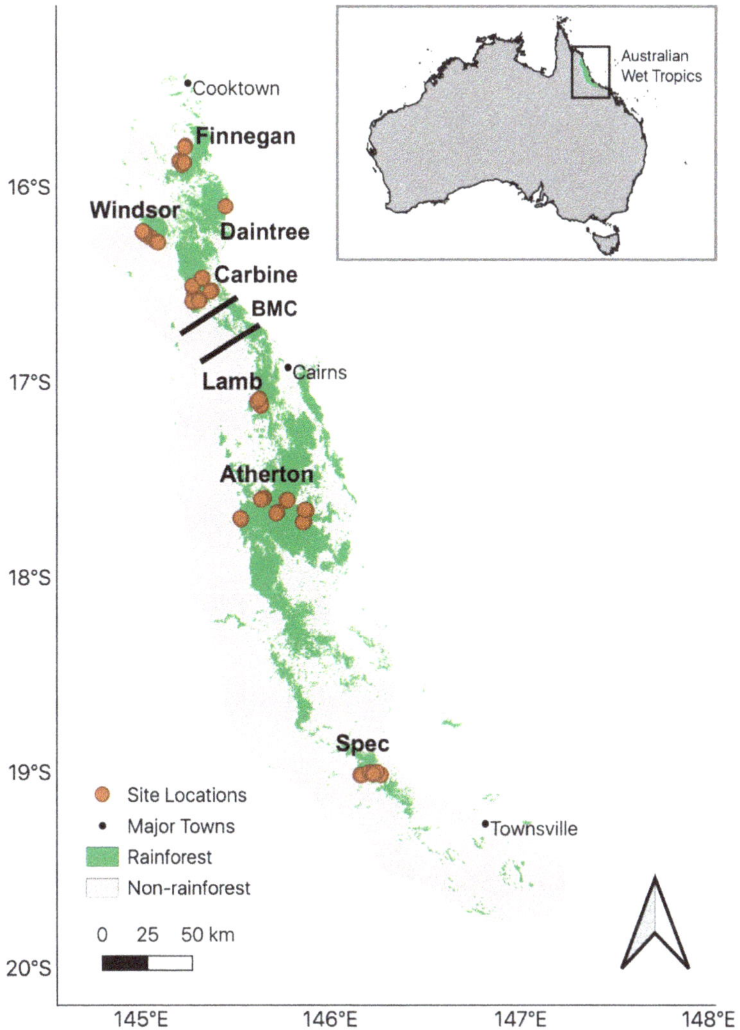

Figure 2. Map of the Australian Wet Tropics (AWT) showing survey site locations for each subregion (naming of subregions follows [13]). Rainforest distribution is shown in green. BMC is the Black Mountain Corridor and represents a well-known biogeographic barrier for numerous taxa within the AWT.

Both contemporary and historical factors drive the distribution and diversity of species and thereby, the geographic structure of biological communities [14,15]. The biogeography of the AWT is strongly driven by historic processes that occurred during climate fluctuations of the late quaternary (last ~18,000 years) when rainforests contracted and expanded several times [16–19]. As rainforests contracted to wetter uplands, many lowland species became locally extinct, leading to a substantial loss of species richness in the lowlands [20]. As a result, many vertebrate and invertebrate species that are endemic to the AWT are restricted to elevations above 300 m a.s.l. that have had historically stable climates and consistently supported rainforests throughout this period of high climatic fluctuations [21]. In addition, there is an older biogeographic barrier along the Black Mountain Corridor (BMC), just south of the Carbine uplands (Figure 2), that has possibly existed since the Pliocene more than 2.5 Ma ago [22] and did not support rainforest until ~8 ka ago [11,23]. There is evidence in multiple taxa of a phylogeographic break across the Black Mountain Corridor that structures phylogenetic diversity at interspecific as well as intraspecific levels [19,24,25], such that many species have either a northern or southern restricted distribution [24].

Biogeographic patterns in the region are also strongly influenced by contemporary climatic conditions at both regional and local scales. At a regional scale, seasonality of rainfall and temperature vary markedly among subregions [26]. Rainfall seasonality is negatively correlated with the abundance of birds in the AWT, with dry season severity creating bottlenecks in critical food resources such as insects, fruits, and nectar [26]. Similarly, there are marked seasonal fluctuations in the abundance of homopteran insects in the AWT [27], which, through their production of honeydew, provide a particularly important food resource for ants, including *A. gilberti* [28]. As such, honeydew-reliant ants, such as species of *Anonychomyrma*, might be expected to show strong biogeographic patterns in response to rainfall seasonality.

At a finer geographic scale, temperature and rainfall both vary markedly with elevation. Mean temperature in the AWT declines by approximately 1 °C per 200 m increase in elevation [29]. Elevations above 600–800 m a.s.l. receive year-round moisture via the orographic cloud layer that creates a misty and cool environment [30]. The orographic cloud layer has been linked to a significant change in ant species composition [9]. Finally, at a local scale, there are strong microclimatic gradients across vertical space from the ground to the canopy, a general feature of rainforest environments globally [31]. A recent study of ant distributions (including nine species of *Anonychomyrma*) in the AWT found a strong positive correlation between vertical niche breadth, and therefore exposure to microclimatic variation, and elevation range size [32], indicating that vertical microhabitat associations of ants influence climatic niches and distributional patterns along elevation gradients.

Here, we used extensive collections from five mountain ranges to document the AWT *Anonychomyrma* fauna and to investigate patterns of species diversity and distributions across latitude, elevation, and vertical habitat space. Like many other dolichoderine genera [3], *Anonychomyrma* is morphologically conservative and therefore taxonomically challenging, so we used an integrated taxonomic approach that included CO1 gene barcoding [33] to inform species boundaries. We identified the key climatic drivers of species distributions and employed species distribution modelling based on climatic niches to compare actual with potential distributions and thus identified factors that may have shaped contemporary subregional patterns of distribution. Specifically, we address the following questions:

1. How many species of *Anonychomyrma* can be recognised in the AWT?
2. How are the species distributed among subregions and along elevational and vertical gradients?
3. What are the patterns of species richness and endemism?
4. To what extent do species distributions correlate with contemporary climatic variables and reflect historical patterns of rainforest refugia?

We predicted that given the Gondwanan origin of *Anonychomyrma* and its prevalence in cool temperate Australia, most species will have montane distributions in the AWT, with high levels of diversity and endemism in the cool upland regions that have retained stable rainforest vegetation and climate conditions more reminiscent of the Gondwanan rainforest that once dominated the east coast of Australia [21]. This is the pattern for other faunal lineages of Gondwanan origin, such as *Terrisswalkerius* earthworms [24], rainforest possums [34], and myobatrachid frogs [35]. We also expect species distributions to be strongly influenced by the paleogeography of the AWT, with most species having restricted northern or southern distribution across the Black Mountain Corridor [23]. Finally, in addition to high temperature, we expected rainfall seasonality to be a key driver of species distributions.

2. Materials and Methods

2.1. Study Region

The Australian Wet Tropics bioregion is a World Heritage Area in far northeastern Australia covering approximately 36,000 km^2 (20–15° S and 147–145° E) consisting of approximately 12,000 km^2 of rainforest. Rainfall is highly seasonal, with 75–90% of the annual 2000–8000 mm rainfall occurring in the wet season between November and April. Elevations above 1000 m a.s.l. can also receive up to 66% of monthly water input from cloud stripping [30]. Rainforests of the AWT harbour a distinctive ant fauna that is highly disjunct from that of surrounding savanna [36,37]. The fauna has strong affinities with that of Indo-Malayan rainforests, containing many genera that are rainforest specialists and whose Australian distributions are restricted to North Queensland. The fauna notably lacks arid-adapted taxa, such as *Iridomyrmex* and *Melophorus*, that dominate ant communities of Australia's open sclerophyll habitats [38]. Tropical rain forests are generally regarded as supporting the world's richest ant faunas [39,40], but Australia's rainforest ant fauna is relatively depauperate [36], in striking contrast to its exceptionally rich savanna fauna [41].

2.2. Ant Sampling

Collections of *Anonychomyrma* were assembled from three sources. The first was the study of [9], which was based on surveys at 26 of the long-term biodiversity monitoring sites established by Stephen E. Williams at James Cook University [35]. These sites were distributed across six subregions, covering the full latitudinal and elevational range of the AWT and representing approximately 94% of the available environmental space in the region [34]. The subregions spanned approximately 500 km from north to south and sites were placed approximately every 200 m along the elevation gradient as follows: Finnegan (200, 500, 600, 800 m a.s.l.), Windsor (900, 1100, 1300 m a.s.l.), Carbine (100, 400, 600, 800, 1000, 1200 m a.s.l.), Lamb (700, 900, 1100 m a.s.l.), Atherton (200, 400, 600, 800, 1000 m a.s.l.), and Spec (350, 600, 800, 1000 m a.s.l.) (Figure 2). Sampling occurred within six plots at each site separated by a distance of 200 m; only three plots were located at each of the 350 m a.s.l. site at Spec and 100 m a.s.l. site at Atherton due to limited rainforest cover. Each site was sampled using a combination of pitfall trapping, litter extractions, and baiting (both on the ground and on trees at a height of 1.5 m). At Windsor, Carbine, Atherton, and Spec, sampling was conducted on three occasions from 2011 to 2013, covering two wet seasons (November–January) and one dry season (June–September). At Finnegan and Lamb, sampling occurred only during one wet season. The study also included ants collected in pitfall traps during previous beetle surveys at the Windsor, Carbine, Atherton, and Spec sites [42].

The second source was from [32], a study involving additional surveys at 15 of the above sites: Finnegan (200, 500, 700 m a.s.l.), Windsor (900, 1100, 1300 m a.s.l.), Carbine (100, 600, 1000, 1200 m a.s.l.), and Atherton (200, 400, 600, 800, 1000 m a.s.l.). At each of these sites, either two (1 site), three (4 sites), four (4 sites), or five (6 sites) trees were sampled, totalling 60 trees. Trees were at least 50 m apart and were chosen for surveying based on size and climbing accessibility. Trees were sampled using tuna bait traps and accessed using the single-rope climbing technique. At each tree, five bait traps were set on the ground and at every 3 m above ground to the maximum accessible height of the tree, which ranged from 15 to 27 m. Traps were set in the morning and collected 2–3 h later. Finnegan, Windsor, and Carbine were surveyed from October to December 2012 and Atherton was surveyed from December 2017 to February 2018.

The third source was data from additional surveys conducted in 2018–2019. Carbine sites were resampled in February 2018 using the same methodology as [32], and from June to October 2019, but with surveys during both day and night. At each site, five trees (at least 50 m apart) were surveyed once in the daytime and once at night. Day and night surveys of the same trees were not conducted sequentially and were at least a day apart to allow resumption of normal ant activity in the tree in case of disruption. Daytime surveys started at 10:00 h, night-time surveys started at 21:00 h, and baits were collected three hours later. Ants were also opportunistically hand collected immediately adjacent to baited vials. In addition, a collection was made over five days during the same period at the Daintree Rainforest Observatory (100 m a.s.l.) placing baited vials at 25–30 m height at ten trees (accessed by the canopy crane) for one hour and conducting one-hour hand searches on the ground.

2.3. Gene Barcoding

DNA was extracted from foreleg tissue and sequences were obtained for 97 specimens of *Anonychomyrma* collected during the study (Table S1). DNA extraction and CO1 sequencing were conducted through the Barcode of Life Data (BOLD) System (for extraction details, see http://ccdb.ca/resources). Each sequenced specimen was assigned a unique identification code that combined the batch within which it was processed and its number within the batch (e.g., ANONC006-20), and all specimens were labelled with their respective BOLD identification numbers in the ant collection held at the CSIRO Tropical Ecosystems Research Centre in Darwin.

DNA sequences were checked and edited in MEGA 7 [43]. Sequences were aligned using the UPGMB clustering method in MUSCLE [44], and translated into (invertebrate) proteins to check for stop codons and nuclear paralogues. The aligned sequences were trimmed accordingly, resulting in 657 base pairs. A sequence from the *anceps* complex of *Iridomyrmex* (sp. A, IRIDX092-18 from [45]) was used as an outgroup for rooting the gene tree. Tree inference by maximum likelihood was conducted through the IQTREE web server (http://iqtree.cibiv.univie.ac.at/; [46]) using ultrafast bootstrap approximation [47]. IQTREE has been shown to be a robust algorithm for tree inference that compares favourably with other methods [48]. Model selection was inferred using a 3-codon partition file and linked branch lengths with the AutoMRE 'ModelFinder' function to find the best-fit model for tree inference [49]. Trees were viewed and edited in FigTree v1.4.3 [50] and annotated using Photoshop CS5.1.

2.4. Species Delimitation

There is no specific level of CO1 divergence that can be used to define a species, but the level of CO1 variation within ant species is typically 1–3% [51]. However, some ant species can show substantially higher variation (for example, [51,52]), and in other cases two clear species can show no CO1 differentiation (for example, [53]). Some ant species from other genera are known to have workers that are virtually identical morphologically, and they can only be separated by detailed morphometric analysis or through reproductive castes [54]. When delimiting species, we focused on morphological differentiation between sister (i.e., most closely related) CO1 clades, considering all available samples from the same collections as sequenced specimens. A full set of voucher specimens of recognised species was held at the CSIRO Tropical Ecosystems Research Centre in Darwin and a duplicate set in the ant collection at James Cook University.

2.5. Patterns of Abundance, Diversity, and Endemism

We used the total number of records in survey plots including repeat surveys ($n = 525$ surveys) across all sites ($n = 23$) as a measure of abundance. Species were then ranked by abundance to document the species abundance pattern. We excluded the Daintree subregion from all analyses of diversity as there was only one site sampled. *Anonychomyrma* was not recorded from the Lamb subregion and these sites were also excluded from the analysis. To assess latitudinal diversity patterns, we assessed variation in species richness among subregions by plotting the mean number of species across sites per subregion along with the total number of surveys per subregion. We likewise plotted the number of subregion endemics per subregion. To assess elevation diversity patterns, we investigated differences in species richness among elevations by pooling across the five subregions and plotting the mean number of species across plots per elevation along with the total number of surveys per elevation. We tested for a correlation in the number of species with survey effort for both subregion diversity and elevation diversity using Pearson's product-moment correlation test. To assess the vertical distributions (foraging activity from ground to canopy) of species, we considered only the vertical (arboreal) surveys conducted by [32]. We selected species for which there were two or more survey records and ten or more sample records (as there were 5 samples per vertical height band). For each of the resultant eight species, we plotted the number of sample records in a 3-m band divided by the total number of sample records for that species, to provide a relative proportion of occurrence in each 3-m band from ground to canopy.

2.6. Species Distribution Modelling

We used species distribution models as an exploratory tool to three purposes: first, to investigate the potential climatic drivers of species distributions; second, to identify potential additional areas of suitable habitat that were not sampled in this study; and third, to explore how species-predicted distributions correlate with historical patterns of rainforest refugia. We included the Daintree subregion in our species records and conducted distribution modelling over only the rainforest regions of the AWT (~12,000 km^2) as we were interested in predicting the distribution of *Anonychomyrma* species within their primary habitat of rainforest vegetation. Species distribution models based on species occurrences and climatic data were derived using MaxEnt, a maximum entropy algorithm, in the program MaxEnt using default settings (version 3.4.1, [55]). We acknowledge that the use of default settings in MaxEnt may not lead to the optimal model in all cases, but given the exploratory purpose of modelling here, these settings were sufficient for our stated goals [56]. MaxEnt performs well in modelling species with small sample sizes if the ecological niche (e.g., environmental tolerance) of those species is sufficiently covered by sampling and they have small geographic ranges [57,58]. However, MaxEnt can perform poorly when species with small number of records are geographically widespread or have a wide environmental tolerance that is not sufficiently covered by sampling [59,60]. For example, [57] reported increasing model performance for species with increasing sample size, but that useful models are still produced for rare and narrow range species with 5–10 records, while [58] strongly emphasised that sample size is relevant to the total area over which modelling is being conducted. They found 14 records to be a minimum for species prevalence (fraction of raster cells occupied) of 0.1 over an area that encompassed most of tropical Africa, which is a far larger area than we model here [58]. To explore the issue of sample size, we first filtered species for ≥ 5 location points (not counting repeated surveys and counting a record in a survey plot as one occurrence/location point), which resulted in a subset of nine species. The number of location points ranged from 96 (*A. gilberti*) to 6 (sp. H). To investigate the relationship between sample size and sample coverage of ecological niche, we looked at how much of the available environmental space of the rainforest region of the AWT was covered by each species' location points. We regressed mean annual temperature against mean annual precipitation to create the environmental space of all survey locations and plotted each species' location points within that plot. Species with low (< 15) numbers of points were highly concentrated in environmental space, indicating that the number of location points sufficiently captured each species' climatic niche volume (Figure S1). As such, our species with low samples sizes were very likely to have narrow environmental niches and therefore could be modelled with the number of location records available. However, we acknowledge that caution should still be applied in interpreting the models with <15 location points [61].

We used the accuCLIM climate variables derived by [62] that were based off a distributed network of microclimate loggers in the AWT and produced using a boosted regression tree approach that statistically downscaled existing coarse weather layers to fine-scale weather layers at 250 m^2 resolution for the AWT region comparable with current best-practice climate layers (e.g., ANUCLIM, [63]). These layers are highly accurate in relation to regional topography and vegetation [62]. We clipped all climate variables to rainforest extent in the AWT. We selected nine of the seventeen climate variables that were most ecologically relevant [42,64–66]. We looked for collinearity in variables by looking at all pairwise interactions between continuous covariates using Pearson's correlation coefficient. Variables with an R^2 value of >0.7 were considered for removal [67]. Variance inflation factors were then calculated and any variables that had a value >10 were excluded from further analysis using the package *usdm* in R [68]. A final set of five variables was selected as follows: maximum temperature of the warmest month, temperature seasonality, rainfall of the wettest month, rainfall seasonality, and isothermality (an indicator of temperature variability: mean diurnal temperature range divided by annual temperature range; [62]). In MaxEnt, 10 replicates were used per species model with cross validation and 1000 iterations [69]. All models presented had values for the area under the receiver operating characteristic curve (AUC) greater than 0.9 and therefore performed adequately [70]. We then calculated summed habitat suitability for the rainforest extent by summing all nine species habitat suitability scores across all pixels and standardizing each pixel's value to a summed habitat suitability score between 0 and 1.

3. Results

3.1. The Anonychomyrma Fauna

We recognised 22 species of *Anonychomyrma* (sp. A–C and E–U, along with *A. gilberti* and *A. malandana*) among our samples, nine of which were recorded at a single subregion (Table 1). All but one of our recognised species were successfully sequenced. The CO1 tree (Figure 3) indicated that species C (Figure 4A,B) was the most phylogenetically divergent, with no close relatives (13–20% divergence from other species). It was a relatively large, gracile, and somewhat polymorphic species with an angular propodeum, and was common at high elevation at Carbine and Windsor. *Anonychomyrma malandana* was also indicated as highly distinctive phylogenetically (> 12% divergence from all other species) as well as being highly distinctive morphologically—it was an extremely shiny species with a globose head and very long antennal scapes (Figure 4C,D). It was a high-elevation species that was common at Windsor. Species H and sp. G were also phylogenetically distinct (Figure 3), both with > 11% divergence from all other species. Species H was a small, nondescript species with a biconvex mesosoma and relatively short scapes (Figure 5A,B) and was morphologically very similar to several unrelated species (see below); it was recorded at Carbine and Windsor at the highest elevations (1200 m and 1300 m, respectively). Species G had a short, prominently rounded propodeum, and its scape and first gastric tergite atypically lacked erect hairs (Figure 5C,D). These characters were shared by the smaller sp. J (Figure 5E,F), but despite their morphological similarity (they were considered conspecific in [9]), these species were widely separated in the CO1 tree (Figure 3). They had overlapping geographic distributions at Atherton: sp. G was recorded from all five subregions, whereas sp. J was recorded exclusively from Atherton (Table 1). Species J was shown as being most closely related to sp. I (Figure 6A,B) and sp. S (Figure 6C,D; Figure 3). Compared with sp. I, sp. S had a substantially larger head with a markedly concave occipital margin, more angular occipital corners, and shorter scapes. Both species were recorded only from Atherton and only from high elevation (800 m and 1000 m; Table 1).

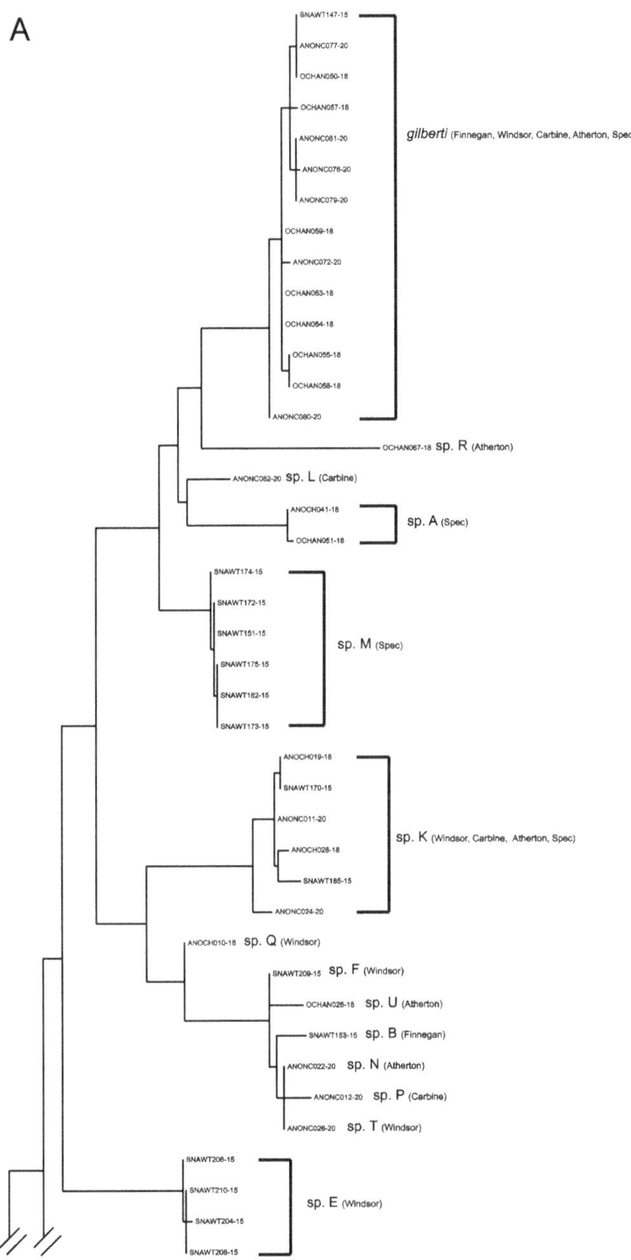

Figure 3. *Cont.*

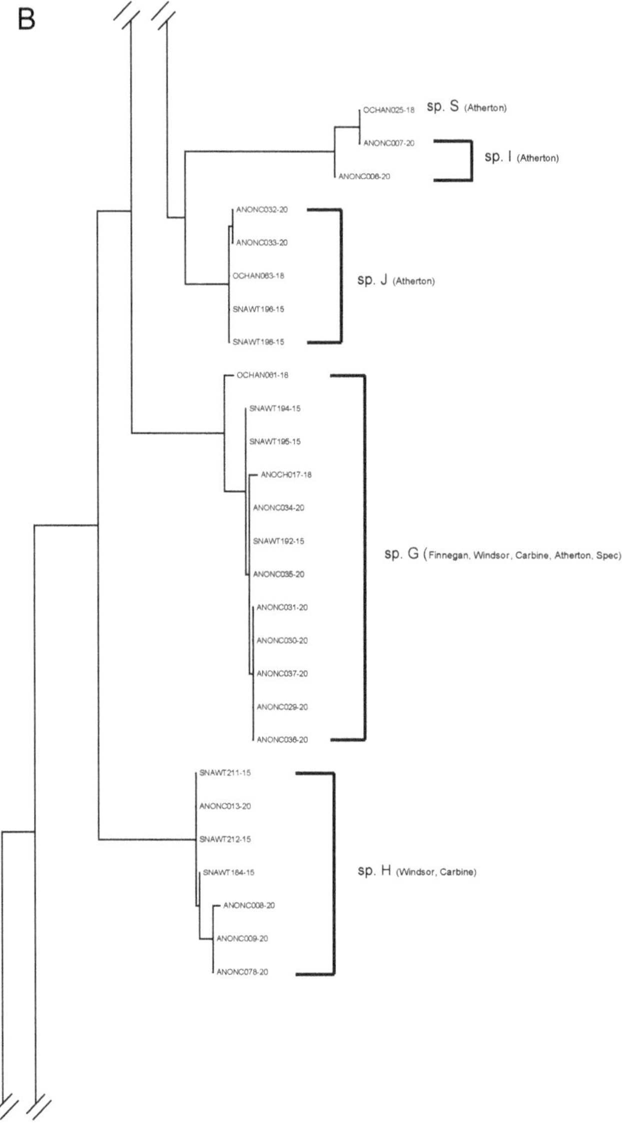

Figure 3. *Cont.*

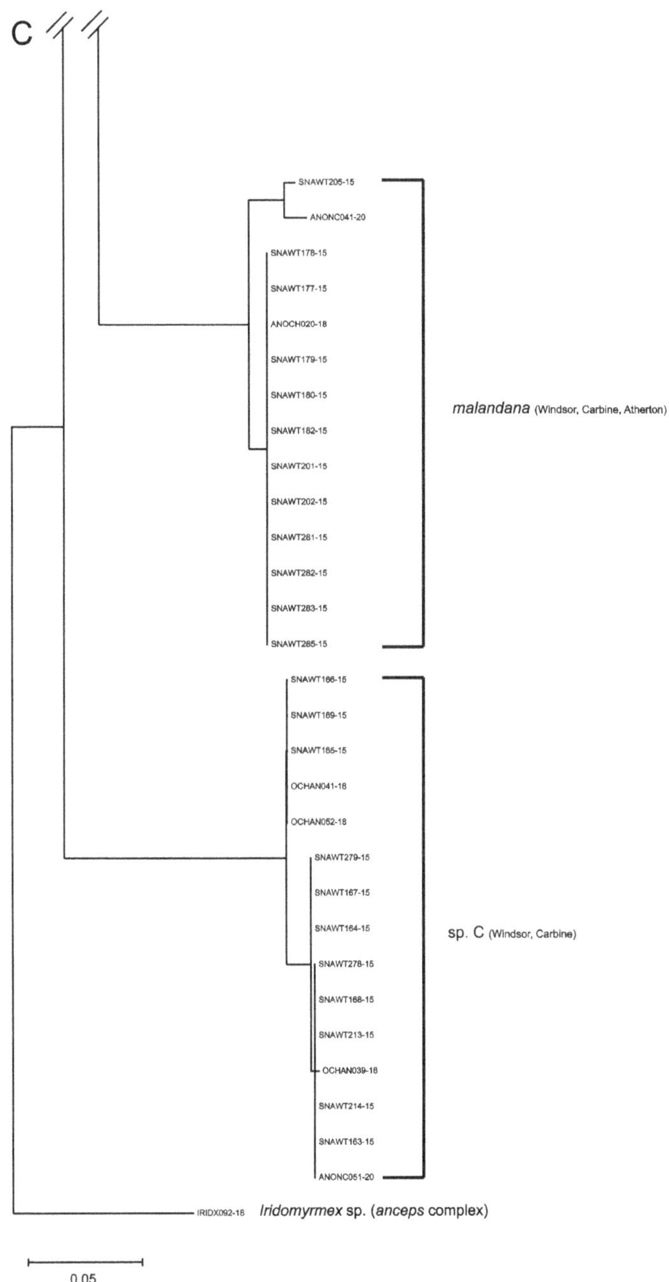

Figure 3. (**A–C**) Maximum likelihood CO1 tree of 97 specimens of *Anonychomyrma* from the Australian Wet Tropics, with a specimen of the *anceps* complex of *Iridomyrmex* as an outgroup.

Table 1. The subregion locations and elevation range (maximum–minimum elevation) and number of survey points of 22 species of *Anonychomyrma* in the Australian Wet Tropics, species are ordered from their north to south subregion distribution, and then by number of survey points, *n* indicates number of survey points. BMC position indicates the positions of each subregion north or south of the Black Mountain Corridor, a well-known biogeographic barrier. The full elevation range of rainforest habitat is shown in brackets for each subregion.

BMC Position		North	North	North	South	South
Region		Finnegan (200–800)	Windsor (900–1300)	Carbine (100–1200)	Atherton (100–1000)	Spec (350–1000)
Species	*n*					
gilberti	96	200–800	900–1300	100–1200	100–1000	350–1000
sp. G	50	200–800	900–1300	100–1200	400–1000	350–800
sp. B	1	200				
sp. C	22		1100–1300	1000–1200		
malandana	13		1100–1300	1200	1000	
sp. E	8		900–1100			
sp. K	7		900–1100	1000	800	800
sp. H	6		1300	1200		
sp. F	1		1100			
sp. Q	1		900			
sp. T	1		1100			
sp. L	1			1200		
sp. P	1			1000		
sp. J	4				200–1000	
sp. O	4				400–800	
sp. I	2				800–1000	
sp. N	1				1000	
sp. R	1				600	
sp. S	1				1000	
sp. U	1				800	
sp. M	11					800–1000
sp. A	10					350–1000

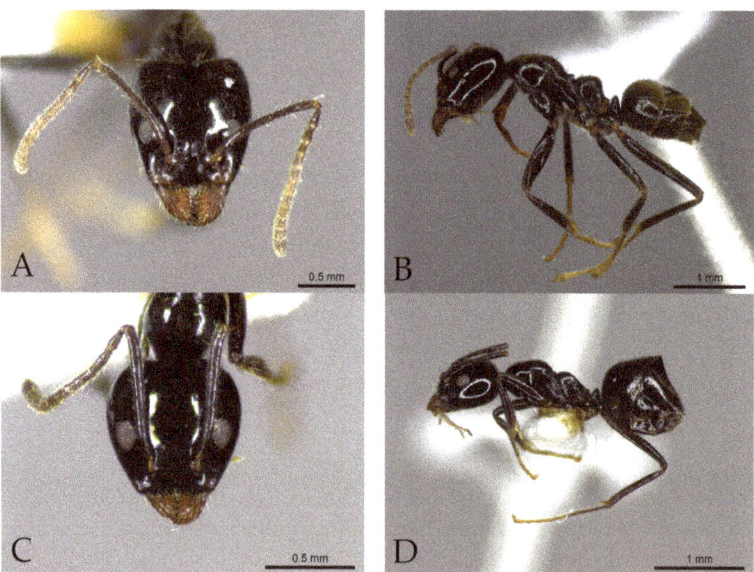

Figure 4. Head and lateral views of *Anonychomyrma* sp. C (**A,B**) and *Anonychomyrma malandana* (**C,D**).

Figure 5. Head and lateral views of *Anonychomyrma* sp. H (**A,B**), sp. G (**C,D**), and sp. J (**E,F**).

Figure 6. Head and lateral views of *Anonychomyrma* sp. I (**A,B**) and sp. S (**C,D**).

Thirteen of the remaining fourteen species formed two clades. The first contained *A. gilberti* and species A, L, M, and R, which had 6–8% mean CO1 divergence among them (Figure 3). *Anonychomyrma gilberti* occurred in all subregions, but the other four species were each recorded in a single subregion (Table 1). Species A, L, and M (Figure 7A–F) had the general appearance of *A. gilberti* (Figure 1), but without such strongly golden pubescence or deeply V-shaped occipital margin; indeed, sp. L had relatively weak pubescence and rounded occipital corners (Figure 7C,D). Species R was morphologically very different to these taxa—it was much smaller, lacked any golden pubescence, had short scapes, and the occipital margin was only feebly concave. Despite its location on the CO1 tree, it seemed to be more closely allied to members of the second clade spanning sp. P to sp. K (Figure 3). Indeed, it seemed morphologically indistinguishable from sp. N from Atherton (Figure 8A,B). All species within this second clade were similar morphologically, and CO1 divergence among them was often < 2%. Compared with sp. N, sp. T (known only from Windsor) had a more conspicuously concave occipital margin and more angular occipital corners, and in sp. P the occipital margin was even more deeply concave. Species U had a narrow head, short scapes, and a prominently rounded propodeum. The propodeum was also prominently rounded in sp. F (Figure 8C,D), but the head was markedly broader and scapes were longer. Species Q had a flattened propodeum and short scapes. Species K (Figure 8E,F) had a very similar appearance to that of sp. N (Figure 7A,B), but the two taxa were very distinct genetically (11% mean CO1 divergence).

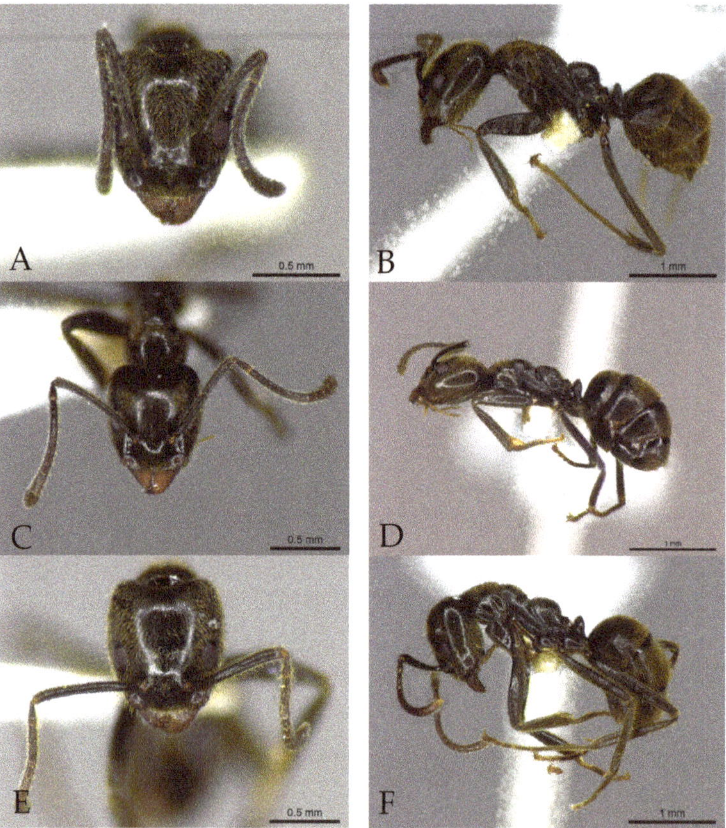

Figure 7. Head and lateral views of *Anonychomyrma* sp. A (**A,B**), sp. L (**C,D**), and sp. M (**E,F**), which are all closely related to *Anonychomyrma gilberti* (Figure 1).

Figure 8. Head and lateral views of *Anonychomyrma* sp. N (**A**,**B**), sp. F (**C**,**D**) and sp. K (**E**,**F**).

The final sequenced species was sp. E, which had a biconvex mesosoma, broad head, and long scapes (Figure 9A,B). It occurred exclusively at Windsor (Table 1). The species that was not sequenced, sp. O, was highly distinctive—it was very small, with a narrowly rectangular head and very short scapes (Figure 10). It appeared to be closely allied to *Anonychomyrma minuta* (Donisthorpe) from New Guinea. It was recorded only at Atherton, at mid elevation (400, 600, and 800 m sites).

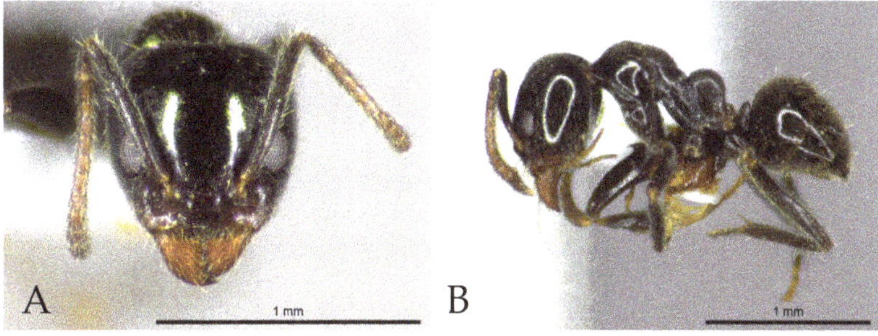

Figure 9. Head and lateral views of *Anonychomyrma* sp. E (**A**,**B**).

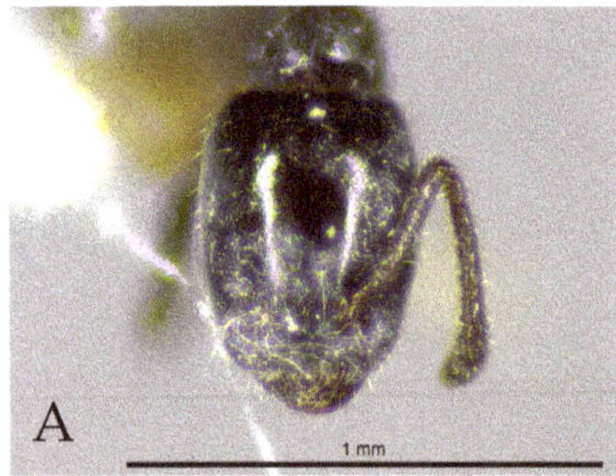

Figure 10. Head and lateral views of *Anonychomyrma* sp. O (**A**,**B**).

3.2. Patterns of Abundance, Diversity, and Endemism

Anonychomyrma gilberti (185 records) and sp. G (73 records) dominated the samples, collectively representing 81.3% of all species records (Figure 11). Nine species were recorded in only one survey (Figure 11). The number of species within a subregion ranged from 0 at Lamb Range to 11 at Atherton. There was no relationship between survey effort and mean number of species per either subregion (Pearson's correlation = −0.02, p = 0.9; Figure 12A,B) or elevation (Pearson's correlation = 0.06, p = 0.8; Figure 13). Windsor (1100–1300 m a.s.l.) had the highest mean species richness per elevational site (5.7 ± 0.7 SE; Figure 12A), and the second highest number of subregion endemics with 4 out of its 10 species (Figure 12B). Atherton (100–1000 m a.s.l.) had the highest number of subregion endemics with 7 endemics out of its total of 11 species (Figure 12B). Carbine, despite having the greatest elevation range (100–1200 m a.s.l.) and the highest survey effort, had similar mean species richness to other low diversity subregions and had only 2 endemic species out of the 8 total species recorded there (Figure 12A,B). Overall, there was high subregion endemism with 16 (72%) species recorded in only one subregion (Table 1). However, all four species (*A. gilberti*, *A. malandana*, sp. G, and sp. K) that occurred in more than two subregions had distributions spanning the BMC (Table 1). The mean number of species increased with elevation peaking at 1100 m a.s.l. and slightly declining again at 1200 and 1300 m a.s.l. (Figure 13).

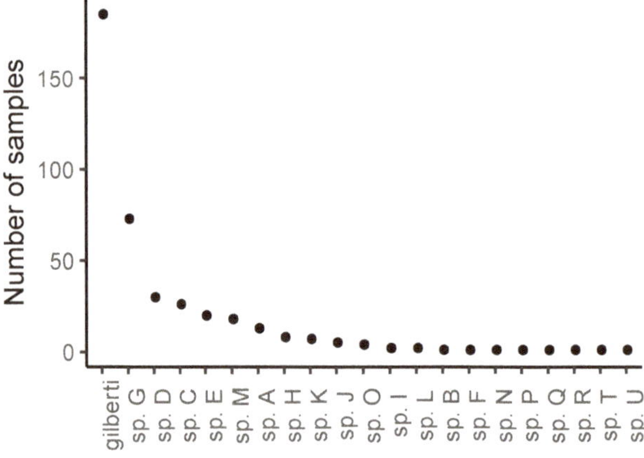

Figure 11. Ranked abundance as total number of samples for each species from pooled data for all survey sites across five subregions in the Australian Wet Tropics.

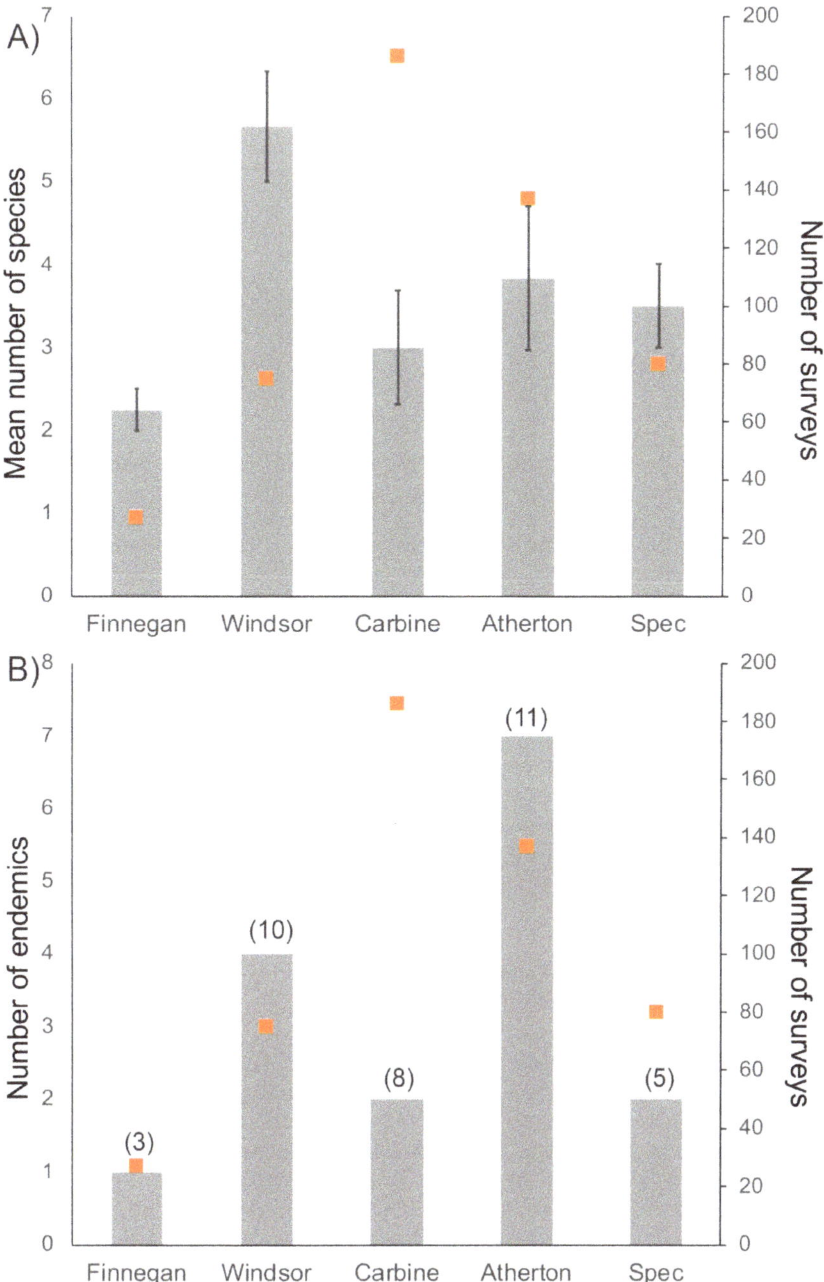

Figure 12. Variation in diversity and endemism among subregions. Subregions labelled from left to right along the north to south latitudinal gradient. (**A**) Mean (±SE) number of *Anonychomyrma* species per elevational site as grey bars and number of surveys per subregion as orange symbols. (**B**) Number of subregion endemics per subregion, and in brackets, the total species richness per subregion. Orange symbols are number of surveys per subregion.

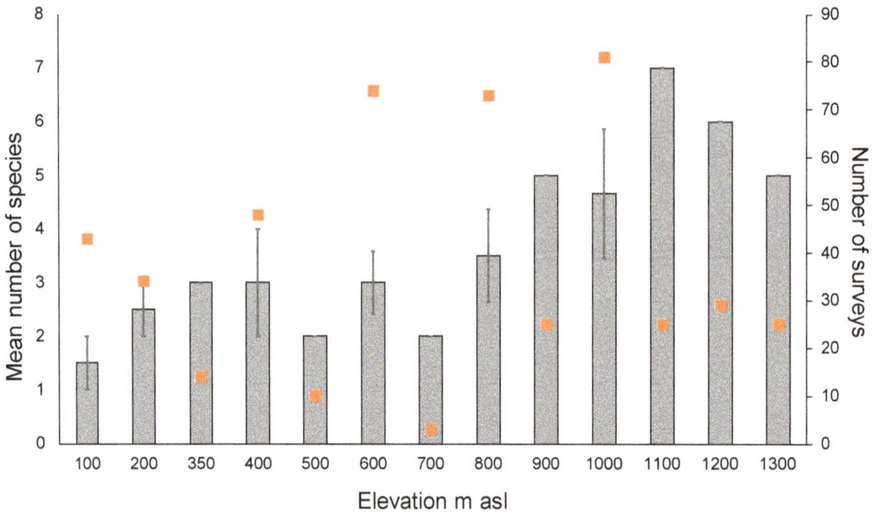

Figure 13. Diversity across elevation, pooling across five subregions. Mean number of *Anonychomyrma* species (±SE) per elevation as gray bars and number of surveys per elevation as orange symbols.

All of the eight species recorded during vertical (tree) surveys nested arboreally (L. Leahy, pers. obs.), with most species foraging high into the canopy. However, there was variation in the way different species used the vertical gradient (Figure 14). Three species, *A. gilberti*, *A. malandana*, and sp. G foraged on the ground as well as arboreally. The two most commonly sampled species, *A. gilberti* and sp. G, were most evenly distributed in their foraging along the vertical gradient, with relatively equal foraging from ground to canopy. These two species were also the most geographically and elevationally widespread species (Table 1). *Anonychomyrma malandana* and sp. C also showed relatively even foraging across the vertical gradient. The less commonly sampled species showed increasing foraging concentration in the upper parts of the tree from the subcanopy (9–12 m) up to the high canopy (18–27 m; Figure 14).

3.3. Species Distribution Modelling

Maximum temperature of the warmest month was the most important predictor for seven species and the second most important predictor for the other two species modelled (Table 2). For all species, there was a negative relationship between maximum temperature of the warmest month and predicted habitat suitability, but the temperature at which habitat suitability declined (ranging from 26 °C for the high-elevation-restricted sp. H to 34 °C for the widespread *A. gilberti* and sp. G) and the rate of decline differed substantially (Figure S2). Rainfall of the wettest month was the second most important predictor variable for *A. gilberti*, sp. E, *A. malandana*, and sp. K, being strongly negatively related to habitat suitability in all cases (Table 2, Figure S3). Rainfall seasonality was also important in models, but generally ranked as the second or third most important variable, except for two Spec endemics, sp. A and sp. M, for which rainfall seasonality was the most important predictor. These two Spec endemic species had a positive relationship between rainfall seasonality and habitat suitability, whereas all other species had higher habitat suitability at intermediate levels of rainfall seasonality or had a negative relationship with rainfall seasonality (Figure S4).

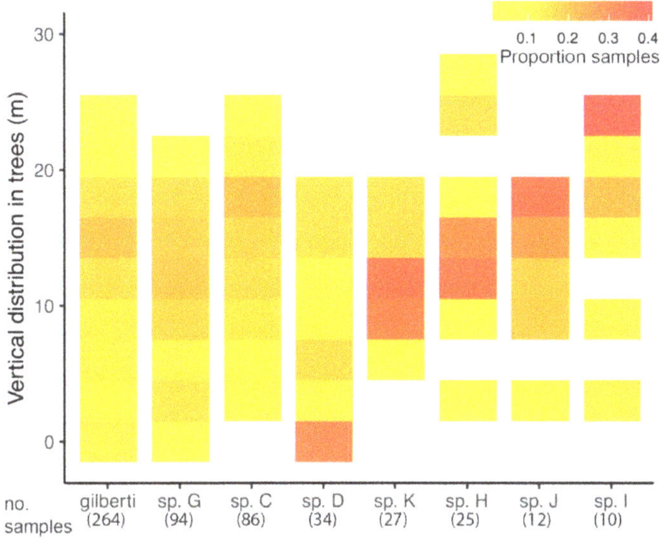

Figure 14. Vertical distribution of *Anonychomyrma* species sampled in 105 surveys (out of 525 total surveys) from ground to canopy at 15 elevation sites at four subregions in the AWT. Pooled occurrence data from four subregions across the Australian Wet Tropics, with species ordered by number of survey occurrences from left to right. Each tile represents a 3-metre vertical band, coloured by proportion of total sample occurrence for each species. Numbers beneath species labels are total number of sample occurrences for that species (total samples = 3745). The maximum potential tree survey height for each species (based on occurrence in each surveyed tree) is: *gilberti* = 27 m, sp. G = 27 m, sp. C = 24 m, *malandana* = 24 m, sp. K = 24 m, sp. H = 27 m, sp. J = 21 m, sp. I = 24 m. Red represents high proportion of samples and yellow represents low proportion of samples.

Distribution modelling predicted that six out of nine species modelled had potential distributions limited to high elevations in either the north (Windsor and Carbine) or far south (Spec) (Figure 15). Summed rainforest habitat suitability (summing each grid cell for suitability for the nine modelled species) was highest at high elevation (Figure 15A), which was consistent with the occurrence of highest species richness (Figure 13). Species-specific habitat suitability generally matched surveyed occurrences, suggesting that actual distributions were close to the potential distributions based on climatic niches (Figure 15). However, our models indicated that highly suitable (≥0.75 probability of occurrence) environments occurred in the southern region for sp. C, sp. E, and sp. H, despite these species being recorded only in the northern subregions in our study. We note that sp. E and sp. H had <10 location points and so these models should be treated with some caution. None of the species modelled had a potential distribution restricted to the centre (Atherton or Lamb) or to the northern lowlands (Figure 15B–J). However, many of the species that were not sufficiently common for modelling were recorded from Atherton, and one species (sp. B) with one record was recorded only at Finnegan at 200 m a.s.l. (Table 1). Models predicted *A. gilberti* and sp. G to have the greatest potential geographic range, followed by sp. K (Figure 15). Distribution maps showed there was a high probability that these three species occur in other rainforest subregions in the AWT not surveyed here, including Bartle Frere, Bellenden Ker, Lamb, and Kirrama (Figure 15; for map of all AWT subregions, see [18]). Although sp. K had only seven location points for modelling, these were all high elevation (Table 1) and our exploration of the environmental niche space indicated a narrow environmental tolerance; we are therefore confident that sp. K is elevationally restricted despite being geographically widespread.

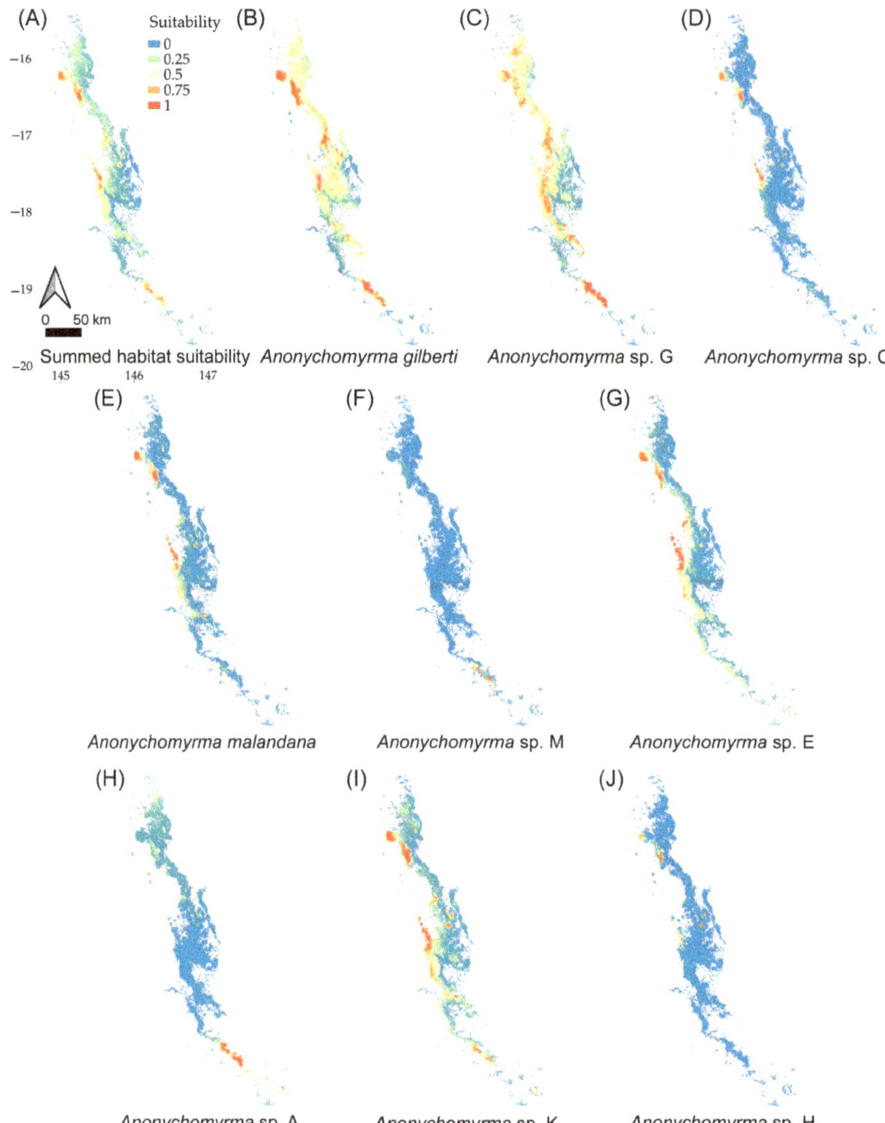

Figure 15. (**A**) Summed rainforest habitat suitability across all species in a grid cell based on the MaxEnt species distribution model for each species. (**B–J**) Modelled rainforest distributions for nine species of *Anonychomyrma* showing habitat suitability in the AWT bioregion. In all figures, blue is low habitat suitability, red is high habitat suitability, and grey background is the outline of the AWT region.

Table 2. Species distribution model outputs for nine species showing the three highest ranking environmental variables and their permutation importance in the model. AUC = area under the curve, a value between 0 and 1 indicates model performance, and higher AUC scores indicate better model fit. *n* = number of unique survey location points for each species. Max T hottest month = Maximum temperature of the hottest month, Rain seasonality = Rainfall seasonality, Rain wettest month = Rainfall of the wettest month, Temp seasonality = Temperature seasonality.

Species	*n*	AUC	Variable	Permutation Importance
gilberti	96	0.798	Max T hottest month	31.2
			Rain wettest month	21.8
			Rain seasonality	22.6
sp. G	50	0.754	Max T hottest month	64.4
			Rain seasonality	12.3
			Rain wettest month	11.4
sp. C	22	0.971	Max T hottest month	95.7
			Isothermality	3
			Rain wettest month	0.7
malandana	13	0.958	Max T hottest month	82.6
			Rain wettest month	14.7
			Rain seasonality	2.4
sp. M	11	0.992	Rain seasonality	48.9
			Max T hottest month	39.4
			Rain wettest month	10.5
sp. E	10	0.948	Max T hottest month	72
			Rain wettest month	13.5
			Isothermality	11.9
sp. A	8	0.971	Rain seasonality	73.4
			Max T hottest month	23.7
			Temp seasonality	2.8
sp. K	7	0.892	Max T hottest month	95.9
			Rain wettest month	2.1
			Isothermality	1.3
sp. H	6	0.99	Max T hottest month	98.7
			Rain seasonality	1.1
			Isothermality	0.2

4. Discussion

The Australian Wet Tropics hosts a diverse *Anonychomyrma* fauna, whose species show strong geographic patterning. Our genetic and morphological analysis delimited 22 species from our samples, far more than previously recorded in the region. Only two of the species appeared to be described. One of these, *A. gilberti*, is an ecologically dominant ant in rainforest canopies throughout the region [9,10]. The other is *A. malandana*, which was originally described from Malanda on the Atherton Uplands, but in our samples, it was recorded in Windsor and Carbine as well (Figure 2). Our samples were from a restricted range of sites in the region and many of our species are known from just one or a few records, which suggests that many additional species likely occur in the AWT.

The distribution and diversity of *Anonychomyrma* within the AWT suggest that this Gondwanan genus has retained a preference for cooler climates. This is evidenced by the concentration of species at elevations above 900 m a.s.l., with only three of the 22 species being recorded at elevation below 300 m a.s.l. The uplands of Windsor and Atherton are centres of particularly high diversity and endemism. Species distribution models also supported this pattern, and summed habitat suitability showed a general trend that species had high habitat suitability in upland areas that are also the main refugial areas of long-term rainforest stability (for AWT map of paleo-geological habitat stability, see [21]). Such a preference for high elevation is in contrast to the distributional patterns of most other ant taxa in the AWT, and more broadly across the tropics, where ant diversity peaks at low to mid elevations and drops dramatically at high elevations [9,71–73]. A similar distributional pattern for

Anonychomyrma occurs in Papua New Guinea, where six of eight species surveyed were restricted to high elevation (≥900 m a.s.l.) [5]. Similarly, in neotropical sites of Central America, several lineages of ants that have a north American temperate climate origin, such as *Temnothorax* [74], are restricted to montane rainforests [71].

All eight species surveyed for vertical distribution were strongly arboreal, often foraging from the understorey up into the high canopy. Only a few species were recorded also foraging on the ground. This suggests that arboreality is a strong trait in the tropical clades of *Anonychomyrma* in Australia. Many of the species were noted to nest arboreally in trunk cavities (L. Leahy, pers. obs.). The two most abundant species (*A. gilberti* and sp. G) had the broadest vertical ranges and also had the broadest geographic and elevation ranges, occurring in all subregions and at all elevations. Species with foraging restricted to the canopy tended to have more restricted subregional and elevational distributions. These findings are consistent with the general pattern for AWT ants that vertical niche breadth is positively related to elevation range size [32]. This pattern also follows the general macroecological rule that locally common species tend to be widely distributed [75].

Species distribution modelling of nine species strongly supported maximum temperature during the warmest month as an important predictor of distribution, indicating that high temperature is likely to be an important factor in limiting species distributions. The two most abundant species (*A. gilberti* and sp. G) were more tolerant of higher maximum temperatures, occurring in all subregions and across the full elevation range. In our modelling, the estimated maximum temperature at which predicted habitat suitability declines was based on a downscaled climatic layer specific to the AWT and was therefore more accurate than a climate layer based on the limited coverage of weather stations in the region [62,76]. However, it is important to note that climate circulation models do not account for vertical climatic gradients, whereby the canopy can experience temperatures several degrees warmer than the forest floor [31,77]. This is an important consideration because the diminutive stature of ants places them close to surface temperatures, which can be far hotter under direct solar radiation, particularly in the canopy [78]. Therefore, given the arboreal habits of *Anonychomyrma*, maximum temperature exposure is likely to be much higher than the predicted estimates from our species distribution models.

For several species, there was a negative relationship between habitat suitability and rainfall of the wettest month and rainfall seasonality. This could explain the relative depauperate species representation in the uplands of Carbine, where elevations above 1000 m a.s.l. have high rainfall seasonality [26] and very high rainfall during the wet season. Windsor, in comparison, which had the highest species richness, sits in the rain shadow of Carbine and therefore has less rainfall with less seasonality [26]. Similarly, Atherton, also with high species richness and endemism, has relatively low seasonality of both rainfall and temperature. Spec in the extreme south of the AWT had low species richness and may be too dry for most species as its dry seasons are substantially more severe than in the other subregions [26]. Rainfall seasonality is a strong driver of insect abundance throughout the tropics [27,79,80], including in the AWT [26]. High rainfall seasonality is likely to limit ant distributions through its effect on food availability, particularly honeydew from homopteran insects.

A number of our distribution models predicted species to occur in subregions that were not sampled in this study. Of particular note, although we did not record any *Anonychomyrma* species in the Lamb subregion, several species were predicted to occur there based on the availability of a suitable habitat. The absence of *Anonychomyrma* at Lamb was therefore likely to be at least partly an artifact of low sampling intensity at that subregion. Other subregions that had high predicted suitability, but were not sampled here, include Bartle Frere and Bellenden-Ker. These are the two highest mountain ranges in the Australian Wet Tropics and have not been systematically sampled for ants. They are centres of diversity and subregion endemism for several Wet Tropics species, including the Gondwanan plant species *Eucryphia wilkei*, which is endemic to elevations above 1500 m a.s.l. on Bartle Frere, and whose congeners occur only in Tasmania and Chile [35,81]. It is highly likely that the tops of these two mountains harbour additional species of *Anonychomyrma*.

There was limited evidence of a biogeographic barrier across the Black Mountain Corridor. Four out of six species that occurred in more than one subregion had distributions spanning north and south of the BMC. Invertebrate taxa, such as dung beetles [82], earthworms [24], flightless insects [83], and schizophoran flies [84], that show a biogeographic divide between northern and southern distributions, all have limited dispersal ability compared to ants with winged queens (such as that occurs in *Anonychomyrma*), which can be dispersed by wind over long distances, including across biogeographic barriers (e.g., [85]).

5. Conclusions

The biogeographic pattern of *Anonychomyrma* in the AWT supports the hypothesis that this genus has a Gondwanan origin and has radiated from a Australian southern temperate distribution into the tropics [6]. *Anonychomyrma* has persisted and diversified in rainforest areas of cool climate with only a few species reaching into the hotter lowland environments and spreading across multiple subregions [18,86,87]. This mirrors the distributional pattern of the genus in Papua New Guinea [5] and we would expect a similar pattern to occur in other parts of the genus' tropical range.

Tropical species restricted to upland habitats with a limited geographic range are considered highly vulnerable to anthropogenic climate change [88,89]. The arboreal foraging habits of *Anonychomyrma* may provide some options to behaviourally regulate climate exposure, given the strong thermal gradient along trees [90–92]. However, the increasingly warm and more seasonal climate that is forecasted for the Australian Wet Tropics [34], along with significant predicted shifts of habitat from rainforest to drier vegetation types [65], places these Gondwanan mountain-top relics under increasing threat.

Supplementary Materials: The following are available online at http://www.mdpi.com/1424-2818/12/12/474/s1, Table S1: Collection locations of specimens from the Anonychomyrma group that were CO1-barcoded for this study, Figure S1: Environmental space occupied by nine species used in species distribution modelling based on our sampling across the AWT, Figure S2: Species distribution modelling response curves showing Anonychomyrma species responses to accuCLIM 1996–2015 Variable 05: Maximum temperature of the warmest month, Figure S3: Species distribution modelling response curves showing Anonychomyrma species responses to accuCLIM 1996–2015 Variable 13: Rainfall of the wettest month, Figure S4: Species distribution modelling response curves showing Anonychomyrma species responses to accuCLIM 1996–2015 Variable 15: Rainfall seasonality.

Author Contributions: A.N.A. conceived of the study, managed the TERC collection, undertook the analysis of the CO1 and morphological data, and contributed substantially to the writing of the paper, L.L. collected specimens in the field, conducted all other analyses, took specimen photos, and wrote the first draft of the manuscript. B.R.S. collected specimens in the field and contributed to the writing of the paper, S.E.W. provided funding for field work and contributed to the writing of the paper. All authors have read and agreed to the published version of the manuscript.

Funding: This study was supported by multiple grants from L.L.: The Explorer's Club, Wet Tropics Management Authority, Skyrail Rainforest Foundation, Holsworth Wildlife Research Endowment—Equity Trustees Charitable Foundation, and Ecological Society of Australia. Funding from S.E.W.: National Environmental Research Program—Tropical Ecosystem Hub from the Australian Government. L.L. is also supported by a PhD scholarship from the Australian Government. Field research was conducted under permit WITK15811415 issued by the Queensland Government of Australia.

Acknowledgments: We thank the many different Aboriginal traditional owner groups on whose lands the study was conducted. We thank Somayeh Nowrouzi for collecting specimens analysed in this study and all volunteers involved in collections. We thank Stefanie K. Oberprieler for help with analysis of the CO1 data.

Conflicts of Interest: The authors declare no conflict of interest.

References

1. Hölldobler, B.; Wilson, E.O. *The Ants*; Harvard University Press: Cambridge, MA, USA, 1990.
2. Dunn, R.R.; Agosti, D.; Andersen, A.N.; Arnan, X.; Bruhl, C.A.; Cerdá, X.; Ellison, A.M.; Fisher, B.L.; Fitzpatrick, M.C.; Gibb, H.; et al. Climatic drivers of hemispheric asymmetry in global patterns of ant species richness. *Ecol. Lett.* **2009**, *12*, 324–333. [CrossRef] [PubMed]

3. Shattuck, S.O. Review of the dolichoderine ant genus Iridomyrmex Mayr with descriptions of three new genera (Hymenoptera: Formicidae). *Aust. J. Entomol.* **1992**, *31*, 13–18. [CrossRef]

4. AntWeb. Available online: https://www.antweb.org (accessed on 20 September 2020).

5. Plowman, N.S.; Mottl, O.; Novotny, V.; Idigel, C.; Philip, F.J.; Rimandai, M.; Klimes, P. Nest microhabitats and tree size mediate shifts in ant community structure across elevation in tropical rainforest canopies. *Ecography* **2020**, *43*, 431–442. [CrossRef]

6. Ward, P.S.; Brady, S.G.; Fisher, B.L.; Schultz, T.R. Phylogeny and biogeography of dolichoderine ants: Effects of data partitioning and relict taxa on historical inference. *Syst. Biol.* **2010**, *59*, 342–362. [CrossRef] [PubMed]

7. Fox, M.D.; Fox, B.J. Evidence for interspecific competition influencing ant species diversity in a regenerating heathland. In *Ant-Plant Interactions in Australia*; Buckley, R., Ed.; Springer: Dordrecht, The Netherlands, 1982; pp. 99–110.

8. Andersen, A.N. Patterns of ant community organization in mesic southeastern Australia. *Aust. J. Ecol.* **1986**, *11*, 87–97. [CrossRef]

9. Nowrouzi, S.; Andersen, A.N.; Macfadyen, S.; Staunton, K.M.; VanDerWal, J.; Robson, S.K.A. Ant diversity and distribution along elevation gradients in the Australian Wet Tropics: The importance of seasonal moisture stability. *PLoS ONE* **2016**, *11*, e0153420. [CrossRef]

10. Blüthgen, N.; Stork, N.E. Ant mosaics in a tropical rainforest in Australia and elsewhere: A critical review. *Austral Ecol.* **2007**, *32*, 93–104. [CrossRef]

11. Schneider, C.; Moritz, C. Rainforest refugia and Australia's Wet Tropics. *Proc. R. Soc. Lond. B Biol. Sci.* **1999**, *266*, 191–196. [CrossRef]

12. Blüthgen, N.; Stork, N.; Fiedler, K. Bottom-up control and co-occurrence in complex communities: Honeydew and nectar determine a rainforest ant mosaic. *Oikos* **2004**, *106*, 344–358. [CrossRef]

13. Williams, S.E.; Pearson, R.G.; Walsh, P.J. Distributions and biodiversity of the terrestrial vertebrates of Australia's Wet Tropics: A review of current knowledge. *Pac. Conserv. Biol.* **1995**, *2*, 327–362. [CrossRef]

14. Carnaval, A.C.; Moritz, C. Historical Climate Modelling Predicts Patterns of Current Biodiversity in the Brazilian Atlantic Forest. *J. Biogeogr.* **2008**, *35*, 1187–1201. [CrossRef]

15. Davies, T.J.; Buckley, L.B.; Grenyer, R.; Gittleman, J.L. The influence of past and present climate on the biogeography of modern mammal diversity. *Philos. Trans. R. Soc. B Biol. Sci.* **2011**, *366*, 2526–2535. [CrossRef] [PubMed]

16. Kershaw, A. Pleistocene vegetation of the humid tropics of northeastern Queensland, Australia. *Palaeogeogr. Palaeoclimatol. Palaeoecol.* **1994**, *109*, 399–412. [CrossRef]

17. Hopkins, M.S.; Ash, J.; Graham, A.W.; Head, J.; Hewett, R. Charcoal evidence of the spatial extent of the Eucalyptus woodland expansions and rainforest contractions in North Queensland during the late Pleistocene. *J. Biogeogr.* **1993**, *20*, 357–372. [CrossRef]

18. Williams, S.E.; Pearson, R.G. Historical rainforest contractions, localized extinctions and patterns of vertebrate endemism in the rainforests of Australia's wet tropics. *Proc. R. Soc. Lond. B Biol. Sci.* **1997**, *264*, 709–716. [CrossRef]

19. Hugall, A.; Moritz, C.; Moussalli, A.; Stanisic, J. Reconciling paleodistribution models and comparative phylogeography in the Wet Tropics rainforest land snail Gnarosophia bellendenkerensis (Brazier 1875). *Proc. Natl. Acad. Sci. USA* **2002**, *99*, 6112–6117. [CrossRef]

20. Colwell, R.K.; Brehm, G.; Cardelús, C.L.; Gilman, A.C.; Longino, J.T. Global warming, elevational range shifts, and lowland biotic attrition in the wet tropics. *Science* **2008**, *322*, 258–261. [CrossRef]

21. Graham, C.H.; Moritz, C.; Williams, S.E. Habitat history improves prediction of biodiversity in rainforest fauna. *Proc. Natl. Acad. Sci. USA* **2006**, *103*, 632–636. [CrossRef]

22. Ponniah, M.; Hughes, J.M. The evolution of Queensland spiny mountain crayfish of the genus Euastacus. I. Testing vicariance and dispersal with interspecific mitochondrial DNA. *Evolution* **2004**, *58*, 1073–1085. [CrossRef]

23. Moritz, C.; Hoskin, C.; MacKenzie, J.; Phillips, B.; Tonione, M.; Silva, N.; VanDerWal, J.; Williams, S.; Graham, C. Identification and dynamics of a cryptic suture zone in tropical rainforest. *Proc. R. Soc. Lond. B Biol. Sci.* **2009**, *276*, 1235–1244. [CrossRef] [PubMed]

24. Moreau, C.S.; Hugall, A.F.; McDonald, K.R.; Jamieson, B.G.; Moritz, C. An ancient divide in a contiguous rainforest: Endemic earthworms in the Australian Wet Tropics. *PLoS ONE* **2015**, *10*, e0136943. [CrossRef] [PubMed]

25. Moussalli, A.; Moritz, C.; Williams, S.E.; Carnaval, A.C. Variable responses of skinks to a common history of rainforest fluctuation: Concordance between phylogeography and paleo-distribution models. *Mol. Ecol.* **2009**, *18*, 483–499. [CrossRef] [PubMed]

26. Williams, S.E.; Middleton, J. Climatic seasonality, resource bottlenecks, and abundance of rainforest birds: Implications for global climate change. *Divers. Distrib.* **2008**, *14*, 69–77. [CrossRef]

27. Frith, C.; Frith, D. Seasonality of insect abundance in an Australian upland rainforest. *Aust. J. Ecol.* **1985**, *10*, 237–248. [CrossRef]

28. Blüthgen, N.; Fiedler, K. Competition for composition: Lessons from nectar-feeding ant communities. *Ecology* **2004**, *85*, 1479–1485. [CrossRef]

29. Shoo, L.P.; Williams, S.E.; Hero, J.-M. Climate warming and the rainforest birds of the Australian Wet Tropics: Using abundance data as a sensitive predictor of change in total population size. *Biol. Conserv.* **2005**, *125*, 335–343. [CrossRef]

30. McJannet, D.; Wallace, J.; Reddell, P. Precipitation interception in Australian tropical rainforests: II. Altitudinal gradients of cloud interception, stemflow, throughfall and interception. *Hydrol. Process.* **2007**, *21*, 1703–1718. [CrossRef]

31. De Frenne, P.; Zellweger, F.; Rodriguez-Sanchez, F.; Scheffers, B.R.; Hylander, K.; Luoto, M.; Vellend, M.; Verheyen, K.; Lenoir, J. Global buffering of temperatures under forest canopies. *Nat. Ecol. Evol.* **2019**, *3*, 744–749. [CrossRef]

32. Leahy, L.; Scheffers, B.R.; Andersen, A.N.; Hirsch, B.T.; Williams, S.E. Vertical niche and elevation range size in tropical ants: Implications for climate resilience. *Divers. Distrib.* In press.

33. Schlick-Steiner, B.C.; Steiner, F.M.; Moder, K.; Seifert, B.; Sanetra, M.; Dyreson, E.; Stauffer, C.; Christian, E. A multidisciplinary approach reveals cryptic diversity in Western Palearctic Tetramorium ants (Hymenoptera: Formicidae). *Mol. Phylogenet. Evol.* **2006**, *40*, 259–273. [CrossRef]

34. Williams, S.E.; Falconi, L.; Moritz, C.; Fenker Antunes, J. *State of Wet Tropics Report 2015-2016. Ancient, Endemic, Rare and Threatened Vertebrates of the Wet Tropics*; Wet Tropics Management Authority: Cairns, Australia, 2016.

35. Williams, S.E.; VanDerWal, J.; Isaac, J.; Shoo, L.P.; Storlie, C.; Fox, S.; Bolitho, E.E.; Moritz, C.; Hoskin, C.J.; Williams, Y.M. Distributions, life-history specialization, and phylogeny of the rain forest vertebrates in the Australian Wet Tropics. *Ecology* **2010**, *91*, 2493-2493. [CrossRef]

36. Taylor, R. Biogeography of insects of New Guinea and Cape York Peninsula. In *Bridge and Barrier: The Natural and Cultural History of Torres Strait*; Walker, D., Ed.; Australian National University Press: Canberra, Australia, 1972; pp. 213–230.

37. van Ingen, L.T.; Campos, R.I.; Andersen, A.N. Ant Community Structure along an Extended Rain Forest: Savanna Gradient in Tropical Australia. *J. Trop. Ecol.* **2008**, 445–455. [CrossRef]

38. Reichel, H.; Andersen, A.N. The rainforest ant fauna of Australia's Northern Territory. *Aust. J. Zool.* **1996**, *44*, 81–95. [CrossRef]

39. Brühl, C.A.; Gunsalam, G.; Linsenmair, K.E. Stratification of ants (Hymenoptera, Formicidae) in a primary rain forest in Sabah, Borneo. *J. Trop. Ecol.* **1998**, *14*, 285–297. [CrossRef]

40. Longino, J.T.; Coddington, J.; Colwell, R.K. The ant fauna of a tropical rain forest: Estimating species richness three different ways. *Ecology* **2002**, *83*, 689–702. [CrossRef]

41. Andersen, A.N. *The Ants of Northern Australia: A Guide to the Monsoonal Fauna*; CSIRO Publishing: Clayton, Australia, 2000.

42. Staunton, K.M.; Robson, S.K.; Burwell, C.J.; Reside, A.E.; Williams, S.E. Projected distributions and diversity of flightless ground beetles within the Australian Wet Tropics and their environmental correlates. *PLoS ONE* **2014**, *9*, e88635. [CrossRef]

43. Kumar, S.; Stecher, G.; Tamura, K. MEGA7: Molecular evolutionary genetics analysis version 7.0 for bigger datasets. *Mol. Biol. Evol.* **2016**, *33*, 1870–1874. [CrossRef]

44. Edgar, R.C. MUSCLE: Multiple sequence alignment with high accuracy and high throughput. *Nucleic Acids Res.* **2004**, *32*, 1792–1797. [CrossRef]

45. Andersen, A.N.; Hoffmann, B.D.; Oberprieler, S.K. Integrated morphological, CO1 and distributional analysis confirms many species in the Iridomyrmex anceps (Roger) complex of ants. *Zool. Syst.* **2020**, *45*, 219–230.

46. Trifinopoulos, J.; Nguyen, L.-T.; von Haeseler, A.; Minh, B.Q. W-IQ-TREE: A fast online phylogenetic tool for maximum likelihood analysis. *Nucleic Acids Res.* **2016**, *44*, W232–W235. [CrossRef]

47. Minh, B.Q.; Nguyen, M.A.T.; von Haeseler, A. Ultrafast approximation for phylogenetic bootstrap. *Mol. Biol. Evol.* **2013**, *30*, 1188–1195. [CrossRef]

48. Nguyen, L.-T.; Schmidt, H.A.; Von Haeseler, A.; Minh, B.Q. IQ-TREE: A fast and effective stochastic algorithm for estimating maximum-likelihood phylogenies. *Mol. Biol. Evol.* **2015**, *32*, 268–274. [CrossRef]

49. Chernomor, O.; Von Haeseler, A.; Minh, B.Q. Terrace aware data structure for phylogenomic inference from supermatrices. *Syst. Biol.* **2016**, *65*, 997–1008. [CrossRef]

50. Rambaut, A. FigTree: Molecular Evolution, Phylogenetics and Epidemiology. Available online: http://tree.bio.ed.ac.uk/software/figtree/ (accessed on 20 May 2020).

51. Smith, M.A.; Fisher, B.L.; Hebert, P.D. DNA barcoding for effective biodiversity assessment of a hyperdiverse arthropod group: The ants of Madagascar. *Philos. Trans. R. Soc. B Biol. Sci.* **2005**, *360*, 1825–1834. [CrossRef]

52. Wild, A.L. Evolution of the Neotropical ant genus Linepithema. *Syst. Entomol.* **2009**, *34*, 49–62. [CrossRef]

53. Schär, S.; Talavera, G.; Espadaler, X.; Rana, J.D.; Andersen Andersen, A.; Cover, S.P.; Vila, R. Do Holarctic ant species exist? Trans-Beringian dispersal and homoplasy in the Formicidae. *J. Biogeogr.* **2018**, *45*, 1917–1928. [CrossRef]

54. Wagner, H.C.; Gamisch, A.; Arthofer, W.; Moder, K.; Steiner, F.M.; Schlick-Steiner, B.C. Evolution of morphological crypsis in the Tetramorium caespitum ant species complex (Hymenoptera: Formicidae). *Sci. Rep.* **2018**, *8*, 1–10. [CrossRef]

55. Phillips, S.J.; Anderson, R.P.; Dudík, M.; Schapire, R.E.; Blair, M.E. Opening the black box: An open-source release of Maxent. *Ecography* **2017**, *40*, 887–893. [CrossRef]

56. Merow, C.; Smith, M.J.; Silander Jr, J.A. A practical guide to MaxEnt for modeling species' distributions: What it does, and why inputs and settings matter. *Ecography* **2013**, *36*, 1058–1069. [CrossRef]

57. Hernandez, P.A.; Graham, C.H.; Master, L.L.; Albert, D.L. The effect of sample size and species characteristics on performance of different species distribution modeling methods. *Ecography* **2006**, *29*, 773–785. [CrossRef]

58. van Proosdij, A.S.; Sosef, M.S.; Wieringa, J.J.; Raes, N. Minimum required number of specimen records to develop accurate species distribution models. *Ecography* **2016**, *39*, 542–552. [CrossRef]

59. Wisz, M.S.; Hijmans, R.; Li, J.; Peterson, A.T.; Graham, C.; Guisan, A.; Group, N.P.S.D.W. Effects of sample size on the performance of species distribution models. *Divers. Distrib.* **2008**, *14*, 763–773. [CrossRef]

60. Elith, J.; Graham, C.H.; Anderson, R.P.; Dudík, M.; Ferrier, S.; Guisan, A.; Hijmans, R.J.; Huettmann, F.; Leathwick, J.R.; Lehmann, A. Novel methods improve prediction of species' distributions from occurrence data. *Ecography* **2006**, *29*, 129–151. [CrossRef]

61. Papeş, M.; Gaubert, P. Modelling ecological niches from low numbers of occurrences: Assessment of the conservation status of poorly known viverrids (Mammalia, Carnivora) across two continents. *Divers. Distrib.* **2007**, *13*, 890–902. [CrossRef]

62. Storlie, C.; Phillips, B.; VanDerWal, J.; Williams, S. Improved spatial estimates of climate predict patchier species distributions. *Divers. Distrib.* **2013**, *19*, 1106–1113. [CrossRef]

63. McMahon, J.; Hutchinson, M.; Nix, H.; Ord, K. *ANUCLIM User's Guide. Version 1*; Centre for Resource and Environmental Studies, Australian National University: Canberra, Australia, 1995.

64. VanDerWal, J.; Shoo, L.P.; Johnson, C.N.; Williams, S.E. Abundance and the Environmental Niche: Environmental Suitability Estimated from Niche Models Predicts the Upper Limit of Local Abundance. *Am. Nat.* **2009**, *174*, 282–291. [CrossRef]

65. Nowrouzi, S.; Bush, A.; Harwood, T.; Staunton, K.M.; Robson, S.K.; Andersen, A.N. Incorporating habitat suitability into community projections: Ant responses to climate change in the Australian Wet Tropics. *Divers. Distrib.* **2019**, *25*, 1273–1288. [CrossRef]

66. Brandt, L.A.; Benscoter, A.M.; Harvey, R.; Speroterra, C.; Bucklin, D.; Romañach, S.S.; Watling, J.I.; Mazzotti, F.J. Comparison of climate envelope models developed using expert-selected variables versus statistical selection. *Ecol. Model.* **2017**, *345*, 10–20. [CrossRef]

67. Dormann, C.F.; Elith, J.; Bacher, S.; Buchmann, C.; Carl, G.; Carré, G.; Marquéz, J.R.G.; Gruber, B.; Lafourcade, B.; Leitao, P.J. Collinearity: A review of methods to deal with it and a simulation study evaluating their performance. *Ecography* **2013**, *36*, 27–46. [CrossRef]

68. Naimi, B.; Hamm, N.A.; Groen, T.A.; Skidmore, A.K.; Toxopeus, A.G. Where is positional uncertainty a problem for species distribution modelling? *Ecography* **2014**, *37*, 191–203. [CrossRef]

69. Elith, J.; Leathwick, J.R. Species distribution models: Ecological explanation and prediction across space and time. *Annu. Rev. Ecol. Evol. Syst.* **2009**, *40*, 677–697. [CrossRef]

70. Pearce, J.; Ferrier, S. Evaluating the predictive performance of habitat models developed using logistic regression. *Ecol. Model.* **2000**, *133*, 225–245. [CrossRef]

71. Longino, J.T.; Branstetter, M.G. The truncated bell: An enigmatic but pervasive elevational diversity pattern in Middle American ants. *Ecography* **2019**, *42*, 272–283. [CrossRef]

72. Longino, J.T.; Branstetter, M.G.; Colwell, R.K. How ants drop out: Ant abundance on tropical mountains. *PLoS ONE* **2014**, *9*, e104030. [CrossRef]

73. Burwell, C.J.; Nakamura, A. Can changes in ant diversity along elevational gradients in tropical and subtropical Australian rainforests be used to detect a signal of past lowland biotic attrition? *Austral Ecol.* **2015**, *41*, 209–218. [CrossRef]

74. Prebus, M. Insights into the evolution, biogeography and natural history of the acorn ants, genus Temnothorax Mayr (hymenoptera: Formicidae). *BMC Evol. Biol.* **2017**, *17*, 250. [CrossRef]

75. Gaston, K.J.; Blackburn, T.M.; Lawton, J.H. Interspecific abundance-range size relationships: An appraisal of mechanisms. *J. Anim. Ecol.* **1997**, *66*, 579–601. [CrossRef]

76. Storlie, C.; Merino-Viteri, A.; Phillips, B.; VanDerWal, J.; Welbergen, J.; Williams, S. Stepping inside the niche: Microclimate data are critical for accurate assessment of species' vulnerability to climate change. *Biol. Lett.* **2014**, *10*, 20140576. [CrossRef]

77. Scheffers, B.R.; Phillips, B.L.; Laurance, W.F.; Sodhi, N.S.; Diesmos, A.; Williams, S.E. Increasing arboreality with altitude: A novel biogeographic dimension. *Proc. R. Soc. B* **2013**, *280*, 20131581. [CrossRef]

78. Kaspari, M.; Clay, N.A.; Lucas, J.; Yanoviak, S.P.; Kay, A. Thermal adaptation generates a diversity of thermal limits in a rainforest ant community. *Glob. Chang. Biol.* **2015**, *21*, 1092–1102. [CrossRef]

79. Janzen, D.H.; Schoener, T.W. Differences in insect abundance and diversity between wetter and drier sites during a tropical dry season. *Ecology* **1968**, *49*, 96–110. [CrossRef]

80. Lowman, M.D. Seasonal variation in insect abundance among three Australian rain forests, with particular reference to phytophagous types. *Aust. J. Ecol.* **1982**, *7*, 353–361. [CrossRef]

81. Forster, P.I.; Hyland, B.P.M. Two new species of Eucryphia Cav.(Cunoniaceae) from Queensland. *Austrobaileya* **1997**, *4*, 589–596.

82. Aristophanous, M. *Understanding Patterns of Endemic Dung Beetle (Coleoptera: Scarabaeidae: Scarabaeinae) Biodiversity in the Australian Wet Tropics Rainforest: Implications of Climate Change*; James Cook University: Townsville, Australia, 2014.

83. Yeates, D.; Bouchard, P.; Monteith, G. Patterns and levels of endemism in the Australian Wet Tropics rainforest: Evidence from flightless insects. *Invertebr. Syst.* **2002**, *16*, 605–619. [CrossRef]

84. Wilson, R.D.; Trueman, J.W.; Williams, S.E.; Yeates, D.K. Altitudinally restricted communities of Schizophoran flies in Queensland's Wet Tropics: Vulnerability to climate change. *Biodivers. Conserv.* **2007**, *16*, 3163–3177. [CrossRef]

85. Levins, R.; Pressick, M.L.; Heatwole, H. Coexistence Patterns in Insular Ants: In which it is shown that ants travel a lot from island to island, and what they do when they get there. *Am. Sci.* **1973**, *61*, 463–472.

86. Pianka, E. Latitudinal Gradients in Species Diversity: A Review of Concepts. *Am. Nat.* **1966**, *100*, 33–46. [CrossRef]

87. VanDerWal, J.; Shoo, L.P.; Williams, S.E. New approaches to understanding late Quaternary climate fluctuations and refugial dynamics in Australian wet tropical rain forests. *J. Biogeogr.* **2009**, *36*, 291–301. [CrossRef]

88. Williams, S.E.; Bolitho, E.E.; Fox, S. Climate change in Australian tropical rainforests: An impending environmental catastrophe. *Proc. R. Soc. Lond. B Biol. Sci.* **2003**, *270*, 1887–1892. [CrossRef]

89. Williams, S.E.; Shoo, L.P.; Isaac, J.L.; Hoffmann, A.A.; Langham, G. Towards an integrated framework for assessing the vulnerability of species to climate change. *PLoS Biol.* **2008**, *6*, e235. [CrossRef]

90. Scheffers, B.R.; Edwards, D.P.; Macdonald, S.L.; Senior, R.A.; Andriamahohatra, L.R.; Roslan, N.; Rogers, A.M.; Haugaasen, T.; Wright, P.; Williams, S.E. Extreme thermal heterogeneity in structurally complex tropical rain forests. *Biotropica* **2017**, *49*, 35–44. [CrossRef]

91. Scheffers, B.R.; Shoo, L.; Phillips, B.; Macdonald, S.L.; Anderson, A.; VanDerWal, J.; Storlie, C.; Gourret, A.; Williams, S.E. Vertical (arboreality) and horizontal (dispersal) movement increase the resilience of vertebrates to climatic instability. *Glob. Ecol. Biogeogr.* **2017**, *26*, 787–798. [CrossRef]
92. Huey, R.B.; Kearney, M.R.; Krockenberger, A.; Holtum, J.A.M.; Jess, M.; Williams, S.E. Predicting organismal vulnerability to climate warming: Roles of behaviour, physiology and adaptation. *Philos. Trans. Biol. Sci.* **2012**, *367*, 1665–1679. [CrossRef] [PubMed]

Publisher's Note: MDPI stays neutral with regard to jurisdictional claims in published maps and institutional affiliations.

diversity

Article

High Diversity in Urban Areas: How Comprehensive Sampling Reveals High Ant Species Richness within One of the Most Urbanized Regions of the World

François Brassard [1,2,*], Chi-Man Leong [1,3], Hoi-Hou Chan [4] and Benoit Guénard [1]

[1] The Insect Biodiversity and Biogeography Laboratory School of Biological Sciences, The University of Hong Kong, Pok Fu Lam Rd, Lung Fu Shan, Hong Kong SAR, China; chimanleo@gmail.com (C.-M.L.); zeroben@gmail.com (B.G.)

[2] Research Institute for the Environment and Livelihoods, Charles Darwin University, Darwin, NT 0810, Australia

[3] Macao Science Center, Avenida Dr. Sun Yat-Sen, Macao SAR, China

[4] Division of Nature Conservation Studies, Instituto Para Os Assuntos Municipaís, Macao SAR, China; hhchan@iam.gov.mo

* Correspondence: francois.brassard.bio@gmail.com; Tel.: +61-423-200-842

Abstract: The continuous increase in urbanization has been perceived as a major threat for biodiversity, particularly within tropical regions. Urban areas, however, may still provide opportunities for conservation. In this study focused on Macao (China), one of the most densely populated regions on Earth, we used a comprehensive approach, targeting all the vertical strata inhabited by ants, to document the diversity of both native and exotic species, and to produce an updated checklist. We then compared these results with 112 studies on urban ants to illustrate the dual roles of cities in sustaining ant diversity and supporting the spread of exotic species. Our study provides the first assessment on the vertical distribution of urban ant communities, allowing the detection of 55 new records in Macao, for a total of 155 ant species (11.5% being exotic); one of the highest species counts reported for a city globally. Overall, our results contrast with the dominant paradigm that urban landscapes have limited conservation value but supports the hypothesis that cities act as gateways for exotic species. Ultimately, we argue for a more comprehensive understanding of ants within cities around the world to understand native and exotic patterns of diversity.

Keywords: biological invasions; biodiversity; species checklist; urban ecology; conservation

Citation: Brassard, F.; Leong, C.-M.; Chan, H.-H.; Guénard, B. High Diversity in Urban Areas: How Comprehensive Sampling Reveals High Ant Species Richness within One of the Most Urbanized Regions of the World. *Diversity* **2021**, *13*, 358. https://doi.org/10.3390/d13080358

Academic Editor: Michael Wink

Received: 2 July 2021
Accepted: 29 July 2021
Published: 4 August 2021

Publisher's Note: MDPI stays neutral with regard to jurisdictional claims in published maps and institutional affiliations.

1. Introduction

Over the past century, urbanization has increased drastically in most regions around the world [1–3]. This increase threatens biodiversity [4–6], with pollution [7–9], habitat loss [10], and the spread of invasive species [11] being major causes of local species extinction or population decline. As such, urban habitats have historically been considered as species-poor concrete jungles [12]. However, urban environments are not necessarily depauperate ecosystems, and may, in fact, have some degree of conservation value by harboring native species [13]. Recent studies suggest that urban habitats can harbor a high diversity of both native (including endemic species) and non-native species, and even, sometimes, surpass surrounding rural areas in terms of species richness [12–16]. The survey and monitoring of biodiversity within cities may thus allow the identification of novel habitats and species worthy of protection within urban matrices. In particular, urban habitats including large and high quality patches of green spaces and forest fragments may still support high species diversity [17–20]. How much biodiversity these areas can contain is still open for debate, but it is paramount to understand the potential conservation value of urban centers.

Beyond assessing the number of species in cities, another crucial component to consider is the composition and identity of the species present. Indeed, cities can facilitate invasions by non-native species in part due to their high level of disturbance, which provides ecological niches suitable for many exotic species [21–23]. Moreover, the constant flux of merchandise in and out of urban centers through airports, harbors, and train stations make them the ideal gateways for exotic species introductions with the common arrival of new propagules [24–26]. Consequently, surveying urban centers and their surroundings to detect new arrivals is essential to limit their spread and mitigate their potential impact on native biodiversity. This is especially true for coastal regions, which host the highest richness of exotic species [27], and, in the case of China, may represent a source of spread towards more inland regions [28].

To evaluate the biodiversity value of urban environments, surveying all flora and fauna would be ideal, but unrealistic, especially for tropical and subtropical regions where there is limited data and taxonomic knowledge. As such, ecological surveys must select a subset of taxa representing useful biodiversity proxies. For conservation monitoring purposes, ants represent an ideal taxon [29]. Indeed, their taxonomy is relatively well-resolved in comparison with most other diverse insect groups, and they can be sampled through the use of standardized and replicable protocols [30]. They are also ecologically and taxonomically diverse, abundant, and ubiquitous [31]. Moreover, they are adequate bioindicators [32,33], play key ecological roles as predators, scavengers, and herbivores [31,32,34,35], with some species acting as ecosystem engineers by modifying soil properties [34].

Ants also include some of the most damaging invasive species, impacting native ants [35–37], non-ant invertebrates [38], vertebrates [39], and plant communities [40], ultimately causing ecosystem disruptions [41]. Invasive ant species can also negatively impact human socioeconomic activities such as farming or education [42], whereas others are considered household pests [43,44], with some exotic species even acting as vectors of pathogens in hospitals [45,46]. Thus, surveying the ant fauna of urban regions and producing ant species checklists should be an important tool not only to evaluate the conservation value of cities, but also to record the worldwide spread of exotic species.

To date, the majority of studies have been limited to ecological studies (Table 1), limited in time and space, and using only a subset of sampling methods to characterize urban ant communities. However, establishing an exhaustive list of a region's ant fauna presents multiple concerns and challenges. For instance, within a specific habitat, and especially within tropical regions, distinct ant species are stratified along the vertical strata (i.e., from underground to the canopy) [47–51]. Generally, ants can be classified into three broad categories: arboreal, epigeic (i.e., ground surface-dwelling, including ants living in leaf litter), and hypogeic (i.e., subterranean). To perform a complete inventory of ants within a region, surveys should, thus, include the different dimensions of this vertical stratification by using methods targeting species from each microhabitat. Unfortunately, most studies in urban habitats use sampling methods focusing mainly or solely on epigeic ants [52], thereby potentially under-estimating the species richness and composition of local communities, as well as the magnitude of invasions. Additionally, this may misrepresent the diversity of hypogeic and arboreal ants within urban habitats and overlook the potential discovery of undescribed species [53–55].

Considering the current expansion of urban habitats [1], describing patterns in urban ant diversity is of great urgency. This is essential to foster biodiversity in cities, but also to gain a better understanding of which factors may facilitate the spread of exotic species. To the best of our knowledge, however, few cities have comprehensive ant species checklists, especially in tropical Asia (Figure 1, Table 1), which makes meaningful comparisons in urban biodiversity challenging at best. More biodiversity surveys and checklists are, thus, required to address this issue.

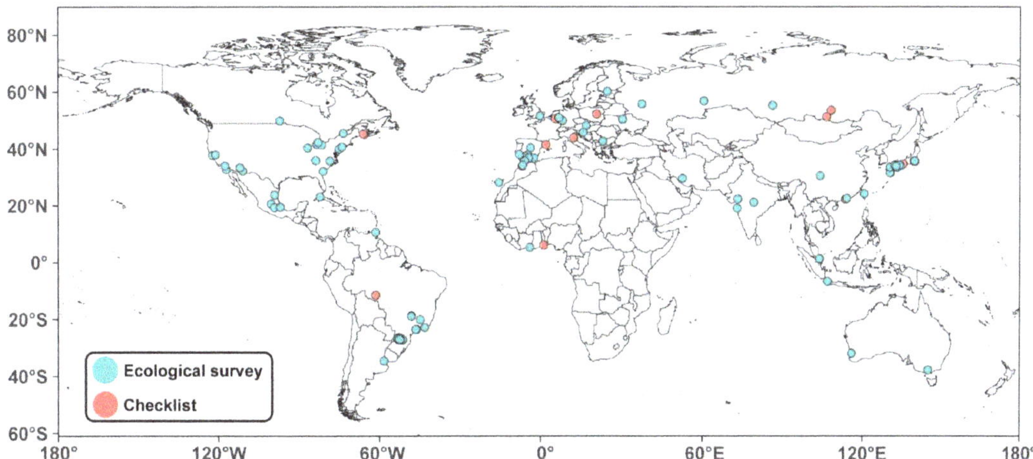

Figure 1. World map showcasing the locations of studies compiled in Table 1. Points show urban locations for which ant species richness estimates were available. Color indicates the type of study (i.e., ecological survey or checklist, see legend).

One such region is the Guangdong–Hong Kong–Macao Greater Bay Area, a major Asian megalopolis [56]. Located in subtropical China, it covers an area of 56,000 km^2 and has a combined population of 68 million people [56]. Within the Greater Bay Area, Macao can be distinguished by its human density, which exceeds 20,000 hab./km^2, making it the most densely populated region on the planet [57,58]. Macao also represents an historical global hotspot in its role within the global trade exchange, first within the Portuguese network and then within China [59,60], making it particularly vulnerable to biological invasions. Finally, the development of Macao has led to a complex matrix of habitats with various levels of disturbance. As such, Macao represents a unique opportunity to better understand how much biodiversity a city characterized by extreme urban development may contain and assess the role of cities as gateways for exotic species.

Following the publication of a preliminary checklist of the ant species of Macao [53] and the discovery of a new subterranean ant species [61], a new survey was conducted across Coloane Island, in the southern half of Macao. To our knowledge, our survey is the first to use an exhaustive sampling approach covering all vertical strata inhabited by ants within an urban area (i.e., arboreal, ground-dwelling, and subterranean). We hypothesized that this sampling coverage would uncover a substantial amount of new species records, give a fair representation of the ant species richness of Macao, and detect new introductions of exotic species. Moreoever, we expected that this methodology, which could be replicated across cities around the world, would be particularly useful for finding cryptic and potentially undescribed species. Finally, we compare our results with previous published studies on urban ants to illustrate the potential that cities may represent for ant diversity, but also more broadly for other insect groups.

Table 1. Ecological and taxonomic studies that produced ant species richness values for a city. Studies are classified by a function of their study region and ranked by the function of the overall species richness retrieved. Studies not providing a complete species list or without enough information on habitats sampled were not included. (*) Complete list of species not provided in article, (**) a combination of 3 articles by the same author and in the same city, and (***), richness value is a combination of more than one urban region.

References	Study Type	Locality	Latitude	Area (km^2)	Natives	Exotics	Total
Africa (7)							
Kouakou et al. 2018 [62]	ecological survey	Abidjan, Ivory Coast	5.3	422	170	6	176
Yeo et al. 2016 [63]	ecological survey	Abidjan, Ivory Coast	5.3	422	83	8	91
Kasseney et al. 2019 [64]	checklist	Lomé (Togo)	6.1	99.1	41	2	43
Taheri et al. 2017 [65]	ecological survey	Tangier, Morocco	35.76	116	30	8	38
Bernard 1958 [66]	checklist	Rabat, Kenitra and Tangier, Morocco	34.3 to 35.8	905 (multiple cities included)	5 ***	11 ***	16 ***
Bernard 1974 [67]	ecological survey	Kenitra, Morocco	34.3	112	13	0	13
Reyes-Lopez and Carpintero 2014 [68]	ecological survey	Las Palmas, Canary Islands	28.2	100.6	6	6	12
Asia (42)							
Ito et al. 2001 [69]	checklist	Bogor, Java, Indonesia	−6.6	118.5	202	11	213
This study	checklist	Macao, SAR, China	22.2	32.9	137	18	155
Leong et al. 2017 [61]	checklist	Macao, SAR, China	22.2	32.9	88	10	98
Rizali et al. 2008 [70]	ecological survey	Bogor, West Java, Indonesia	−6.6	118.5	82	12	94
Terayama 2005a,b and 2014 [71–73]	ecological survey **	Tokyo, Japan	35.7	2194.1	54	0	54
Matsumura and Yamane 2012 [74]	ecological survey	Kagoshima City, Japan	31.6	547.6	48	3	51
Liu et al. 2019 [75]	ecological survey	Taichung City, Taiwan	24.2	2215	?	?	50 *
Natuhara 1998 [76]	checklist	Osaka City, Japan	34.7	225.2	44	5	49
Tan and Corlett 2012 [77]	ecological survey	Singapore	1.3	728.3	38	4	42
Harada et al. 2012 [78]	checklist	Isa City, Japan	32.1	392.4	40	0	40
Iwata, Eguchi and Yamane 2005 [79]	ecological survey	Kagoshima City, Japan	31.6	547.6	37	2	39
Yamaguchi 2004 [80]	ecological survey	Chiba, Japan	35.6	271.8	37	1	38
Park et al. 2014a [81]	ecological survey	Fukuoka City, Japan	33.6	343.4	35	3	38
Matsumura and Yamane 2012 [74]	ecological survey	Kagoshima City, Japan	31.6	547.6	36	2	38
Wang et al. 2012 [82]	ecological survey	Shenzhen, Guangdong, China	22.5	2050	29	6	35

Table 1. *Cont.*

References	Study Type	Locality	Latitude	Area (km²)	Natives	Exotics	Total
Khot et al. 2014 [83]	ecological survey	Mumbai, Maharashtra, India	19.1	603	25	3	28
Yamaguchi 2004 [80]	ecological survey	Tokyo, Japan	35.7	2194.1	27	1	28
Hiroyuki 2012 [84]	checklist	Matsuyama, Japan	33.8	429.4	28	0	28
Harada et al. 2021 [85]	ecological survey	Kagoshima City, Japan	31.6	547.6	21	6	27
Miyake et al. 2002 [86]	ecological survey	Hatsukaichi City, Japan	34.4	489.4	22	4	26
Tan et al. 2009 [87]	ecological survey	Chengdu, Sichuan, China	30.7	885.6	24	2	26
Park et al. 2014b [88]	ecological survey	Hiroshima city, Japan	34.4	906.7	21	4	25
Touyama, Ogata and Sugiyama 2003 [89]	ecological survey	Hatsukaichi and Hiroshima city, Japan	34.4	1395 (multiple cities included)	23 ***	1 ***	24 ***
Harada and Yamashita, 2019 [90]	ecological survey	Tokushima, Japan	34.1	191	21	1	22
Yasuda and Koike 2009	ecological survey	Matsudo, Japan	35.8	61.4	?	?	22 *
Malozemova and Malozemov 1999 [91]	ecological survey	Yekaterinburg, Russia	56.8	495	20	1	21
Harada 2020 [92]	ecological survey	Hioki City, Japan	31.6	253.1	16	3	19
Roshanak et al. 2017 [93]	ecological survey	Shiraz, Iran	29.6	240	19	0	19
Hosoishi et al. 2019 [94]	ecological survey	Fukuoka City, Japan	33.6	343.4	17	1	18
Terayama et al. 2006 [95]	ecological survey	Iwakuni City, Japan	34.2	873.7	14	2	16
Putyatina et al. 2017 [96]	ecological survey	Moscow, Russia	55.8	2511	16	0	16
Kumar and Archana 2008 [97]	ecological survey	Vadodara, Gujarat, India	22.3	220	15	0	15
Harada and Yamashita, 2019 [90]	ecological survey	Kochi, Japan	33.6	309.2	15	0	15
Terayama et al. 2006 [95]	ecological survey	Iwakuni City, Japan	34.2	873.7	12	2	14
Antonov 2008 [98]	checklist	Irkutsk, Baikal, Russia	52.3	277	14	0	14
Harada and Yamashita, 2019 [90]	ecological survey	Takamatsu, Japan	34.4	375.4	14	0	14
Harada and Yamashita, 2019 [90]	ecological survey	Matsuyama City, Japan	33.8	429.4	12	1	13
Hosaka et al. 2019 [99]	ecological survey	Tokyo, Japan	35.7	2194.1	11	0	11
Antonov 2008 [98]	checklist	Gusinoozersk, Baikal, Russia	51.3	?	7	0	7
Blinova 2008 [100]	ecological survey	Kemerovo, Russia	55.4	282.3	7	0	7

Table 1. *Cont.*

References	Study Type	Locality	Latitude	Area (km^2)	Natives	Exotics	Total
Meshram et al. 2015 [101]	ecological survey	Nagpur, Maharashtra, India	21.2	393.5	25 (genera)	3	NA *
Yong et al. 2017 [102]	checklist	Pulau Aubin, Singapore	1.3	10.2	35 (genera)	2 (no total mentioned)	NA *
Australia (6)							
Ossola et al. 2015 [103]	ecological survey	Melbourne, Australia	−37.8	9993	59 *	1 *	60
Heterick et al. 2013 [104]	ecological survey	Perth, Australia	−32.0	6418	54	6	60
Majer and Brown 1986 [105]	ecological survey	Perth, Australia	−32.0	6418	45	2	47
Callan and Majer 2009 [106]	ecological survey	Perth, Australia	−32.0	6418	32	4	36
Heterick et al. 2000 [107]	ecological survey	Perth, Australia	−32.0	6418	19	8	27
May and Heterick 2000 [108]	ecological survey	Perth, Australia	−32.0	6418	18	8	26
Europe (28)							
Radchenko et al. 2019 [109]	checklist	Kyiv, Ukraine	50.5	839	55	4	59
Ordóñez-Urbano, Reyes-López and Carpintero-Ortega 2008 [110]	ecological survey	Córdoba, Sevilla, Málaga and Cádiz, Spain	36.5–37.9	>1800 (multiple cities included)	?	?	59 *,***
Antonova and Penev 2006 [111–113]	ecological survey	Sofia, Bulgaria	42.7	492	54	0	46
Dauber 1997 [114]	ecological survey	Mainz, Germany	50.0	97.7	49	0	49
Dauber and Eisenbeis [114]	ecological survey	Mainz, Germany	50.0	97.7	46	0	46
Reyes-Lopez and Carpintero 2014 [68]	ecological survey	Cordoba and Seville, Spain	37.4 and 37.9	1393 (multiple cities included)	39 ***	5 ***	44 ***
Pisarski and Czechowski 1978 [115]	checklist	Warsaw, Poland	52.2	517.2	36	1	37
Pisarski 1982 [116]	checklist	Warsaw, Poland	52.2	517.2	35	2	37
Ruiz Heras et al. 2011 [117]	ecological survey	Madrid, Spain	40.4	604.3	36	1	37
Klesniakova et al. 2016 [118]	ecological survey	Bratislava, Slovakia	48.1	367.6	?	1 *	36 *
Trigos-Peral et al. 2020 [119]	ecological survey	Warsaw, Poland	52.2	517.2	32	2	34
Reyes-Lopez and Carpintero 2014 [68]	ecological survey	Almeria, Cadiz, Huelva and Malaga, Spain	36.5 to 37.3	859.1 (multiple cities included)	24 ***	8 ***	32 ***
Ješovnik and Bujan 2021 [120]	ecological survey	Zagreb, Croatia	45.8	641	30	0	30
Behr, Lippke and Cölln 1996 [121]	ecological survey	Köln (Cologne), Germany	51.0	405.1	25	3	28

Table 1. *Cont.*

References	Study Type	Locality	Latitude	Area (km^2)	Natives	Exotics	Total
Ślipiński et al. 2012 [122]	ecological survey	Warsaw, Poland	52.2	517.2	27	0	27
Behr and Cölln 1993 [123]	checklist	Gönnersdorf, Germany	50.5	5.2	27	0	27
Rigato and Wetterer 2018 [124]	checklist	San Marino	43.9	61.2	23	0	23
Espadaler and López-Soria 1991 [125]	checklist	Sant Cugat, Barcelona, Spain	41.5	48.2	22	1	23
Vepsäläinen, Ikonen and Koivula 2008 [126]	ecological survey	Helsinki, Finland	60.2	213.8	17	2	19
Trigos Peral and Reyes Lopez 2018 [127]	ecological survey	Beja, Portugal	38.0	1146	17	0	17
Vepsäläinen, Ikonen and Koivula 2008 [126]	ecological survey	Helsinki, Finland	60.2	185	16	0	16
Stukalyuk 2017 [128]	ecological survey	Kyiv, Ukraine	50.5	839	16	0	16
Reyes López and Taheri 2018 [129]	checklist	Cádiz, Andalusia, Spain	36.5	12.1	6	9	15
Smith et al. 2006 [130]	ecological survey	London, UK	51.5	1572	6	0	6
Gaspar and Thirion 1978 [131]	checklist	Liege, Belgium	50.6	69.4	6	0	6
N. America (22)							
Guénard et al. 2015 [16]	ecological survey	Raleigh, NC, USA	35.8	380	77	12	89
Nuhn and Wright 1979 [132]	checklist	Raleigh, NC, USA	35.8	380	50	6	56
Baena et al. 2019 [133]	ecological survey	Coatepec, Mexico	19.5	255.8	51	4	55
Menke et al. 2011 [134]	ecological survey	Raleigh, NC, USA	35.8	380	49	5	54
Miguelena and Baker 2019 [135]	ecological survey	Tucson, AZ, USA	32.2	623.6	45	3	48
Rocha-Ortega and Castano-Meneses 2015 [136]	ecological survey	Santiago de Querétaro, Mexico	20.6	363	45	3	48
Toennisson et al. 2011 [137]	ecological survey	Knoxville, TN, USA	36.0	270	44	2	46
Suarez et al. 1998 [138]	ecological survey	San Diego, CA, USA	32.7	964.6	42	4	46
Gochnour et al. 2019 [139]	ecological survey	Garden City, Georgia, USA	32.1	37.6	32	13	45
Savage et al. 2015 [140]	ecological survey	New York, NY, USA	40.7	778	36	6	42
Baena et al. 2019 [133]	ecological survey	Xalapa, Mexico	19.5	124.4	36	4	40
Uno S. pers. Comm. In Friedrich and Philpott 2009 [141]	ecological survey	Toledo, USA	39.9	217.1	?	?	35 *
García-Martínez et al. 2019 [142]	ecological survey	Ciudad Victoria, Mexico	23.7	188	28	4	32
Fairweather et al. 2020 [143]	checklist	St-John, NB, Canada	45.3	316	30	0	30

Table 1. *Cont.*

References	Study Type	Locality	Latitude	Area (km^2)	Natives	Exotics	Total
Uno, Cotton and Philpott 2010 [144]	ecological survey	Toledo, USA	41.7	217.1	28	2	30
Ivanov and Keiper 2010 [145]	ecological survey	Cleveland, Oh, USA	41.5	201	28	1	29
Uno, Cotton and Philpott 2010 [144]	ecological survey	Detroit, USA	42.3	370.1	26	1	27
Lessard and Buddle 2005 [146]	ecological survey	Montreal, CAN	45.5	431.5	23	1	24
Clarke, Fisher and LeBuhn 2008 [147]	ecological survey	San Francisco, CAL, USA	37.8	600.6	18	3	21
Buczkowski and Richmond 2012 [10]	ecological survey	West Lafayette, Indiana, USA	40.4	35.8	19	1	20
King and Green 1995 [148]	ecological survey	Philadelphia, PA, USA	40.0	369.6	18	1	19
Staubus et al. 2015 [149]	ecological survey	Claremont, CAL, USA	34.1	34.9	12	6	18
Pećarević et al. 2010 [150]	ecological survey	New York, NY, USA	40.7	778	10	3	13
Thompson and McLachlan 2007 [151]	ecological survey	Winnipeg, Manitoba, Canada	49.8	464.1	10	0	10
Villar and Ríos-Casanova [152]	ecological survey	La Cantera Oriente, Mexico city, Mexico	19.3	0.1	8	2	10
Stahlschmidt and Johnson 2018 [153]	ecological survey	Stockton, CAL, USA	38.0	169	4	5	9
Marussich and Faeth 2009 [154]	ecological survey	Phoenix, AZ, USA	33.4	1341	7	1	8
S. and C. America (31)							
Pacheco and Vasconcelos 2007 [155]	ecological survey	Uberlândia, Brazil	−18.9	4116	137	6	143
Santos et al. 2019 [156]	ecological survey	Rio de Janeiro, Brazil	−22.9	1221	116	4	120
Santos-Silva et al. 2016 [157]	checklist	Cacoal, Rondônia, Brazil	−11.4	3793	98	4	102
De Souza et al. 2012 [158]	ecological survey	Mogi das Cruzes, Brazil	−23.5	713	91	1	92
Lutinski et al. 2013 [159]	ecological survey	Chapecó, Santa Catarina, Brazil	−27.1	624.3	89	2	91
Munhae et al. 2014 [160]	ecological survey	Alto Tietê region, São Paulo, Brazil	−23.5	1455 (multiple cities included)	82 ***	5 ***	87 ***
Lutinski et al. 2013 [159]	ecological survey	Palmitos, Santa Catarina, Brazil	−27.1	350.7	81	4	85
Lutinski et al. 2013 [159]	ecological survey	Campo Erê, Santa Catarina, Brazil	−26.4	478.7	82	3	85
Lutinski et al. 2013 [159]	ecological survey	Xanxerê, Santa Catarina, Brazil	−26.9	377.8	80	3	83
Lutinski et al. 2013 [159]	ecological survey	São Miguel do Oeste, Santa Catarina, Brazil	−26.7	234.4	81	2	83

Table 1. *Cont.*

References	Study Type	Locality	Latitude	Area (km^2)	Natives	Exotics	Total
Lutinski et al. 2013 [159]	ecological survey	Abelardo Luz, Santa Catarina, Brazil	−26.6	953.6	81	2	83
Lutinski et al. 2013 [159]	ecological survey	Concórdia, Santa Catarina, Brazil	−27.2	800	80	2	82
Lutinski et al. 2013 [159]	ecological survey	Pinhalzinho, Santa Catarina, Brazil	−26.8	128.3	76	4	80
Lutinski et al. 2013 [159]	ecological survey	Joaçaba, Santa Catarina, Brazil	−27.1	232.4	76	3	79
Morini et al. 2007 [161]	ecological survey	São Paulo, Brazil	−23.6	1521.1	76	3	79
Lutinski et al. 2013 [159]	ecological survey	Seara, Santa Catarina, Brazil	−27.1	312.5	75	3	78
Iop et al. 2009 [162]	checklist	Xanxerê, Santa Catarina, Brazil	−26.9	377.8	45	2	67
Caldart et al. 2012 [163]	ecological survey	Chapecó, Santa Catarina, Brazil	−27.1	624.3	63	3	66
Ilha et al. 2017 [164]	ecological survey	Chapecó, Santa Catarina, Brazil	−27.1	624.3	60	3	63
Josens et al. 2017 [165]	ecological survey	Buenos Aires, Argentina	−34.6	203	57	3	60
Kamura et al. 2007 [166]	ecological survey	Mogi das Cruzes, São Paulo, Brazil	−23.5	713	49	9	58
Santiago et al. 2018 [167]	ecological survey	Divinópolis, Minas Gerais, Brazil	−20.1	192	55	0	55
De Souza-Campana et al. 2016 [168]	checklist	São Paulo, Brazil	−23.6	1521	46	1	47
Piva and de Carvalho Campos 2012 [169]	ecological survey	São Paulo, Brazil	−23.6	1521	38	6	44
Simonetti, Brito and Luis 2010 [170]	ecological survey	Havana City, Cuba	23.1	728.3	?	?	37*
Ribeiro et al. 2012 [171]	ecological survey	São Paulo, Brazil	−23.6	1521	33	3	36
Lutinski and Mello Garcia 2005 [172]	ecological survey	Chapecó, Santa Catarina, Brazil	−27.1	624.3	32	0	32
Lange et al. 2015 [173]	ecological survey	Araguari, Minas Gerais, Brazil	−18.6	2730.6	21	2	23
Starr and Ballah 2017 [174]	ecological survey	Port of Spain, Trinidad	10.7	12	23	0	23
Soares et al. 2006 [175]	ecological survey	Uberlândia, Minas Gerais, Brazil	−18.9	4116	10	4	14

2. Materials and Methods

2.1. Geographic and Climatic Characteristics of Macao

Macao is a special administrative region on the southern coast of China. It is located 60 km south-west of the Hong Kong special administrative region, separated from it by the pearl river delta. Macao's climate is characterized by dry winters and hot summers [176], with an average daily temperature of 22.8 °C and an annual rainfall of 1967 mm [177].

In the 19th century, Macao's land surface was only 10.28 km^2 but, following numerous reclamation projects, it now covers around 32.9 km^2 [58,178]. Despite its high urbanization, Macao still retains several nature parks consisting of young secondary forests, most of which are on Coloane Island. Macao's government started managing these forest patches in 1980, protecting them from wild-fires and establishing restauration plantations of *Pinus massoniana* and *Acacia confusa* [179].

2.2. Sampling Effort and Collection Methods

Most ant specimens examined were collected during a survey conducted in 2019, from March to October, across 21 plots in Coloane Island, Macao (Figures S1 and S2, Table S1). The survey focused on collecting ants within Coloane's nature parks, which consist of secondary forests, but also covered two golf courses and a mangrove site. To extensively sample the hypogeic, epigeic, and arboreal ants of Macao, we used a range of sampling methods during the 2019 survey. Across the 21 sites, we used 225 ground baits, hand collection, and 42 leaf litter extractions with Winkler bags. Half the Winkler extractions consisted of combining the leaf litter of $4 \times 1\ m^2$ quadrats taken at each corner of a plot of 20×20 m (i.e., standard area method), and half consisted of combining a few handfuls of leaf litter taken at 12 random locations within the same plot (i.e., species pool method). For a subset of 16 sites, we used 256 subterranean and 320 arboreal baits, and 1024 artificial nests. For more details on traps, baits, and nest design, see Brassard et al. (2020) [54]. Note that the nests were built following Booher et al. (2017) [180], and were mainly used to obtain sociometric data for the species collected (i.e., colony size and composition). The remainder of the specimens included are from collections made by hand or leaf litter extractions between 2015 and 2020 from different locations across Macao, with detailed collection information presented in the species accounts section.

2.3. Sample Processing

We processed samples by first sorting specimens to morphospecies, which we then stored in ethanol 70%. For each morphospecies, we point-mounted at least one individual and labeled it with a locality and collection label. All specimens are currently located in the Insect Biodiversity and Biogeography Laboratory (IBBL) at The University of Hong Kong.

2.4. Imaging

We used a Leica DFC450 camera mounted on a Leica M205 C dissecting microscope to image mounted specimens of each species and morphospecies. We used the Leica Application suite v. 4.5 to take, stack, and enhance image montages. When necessary, we used Adobe Photoshop Lightroom to make final color corrections and diminish ghosting effects.

2.5. Mapping Species Distributions and Urban Studies

We used *R* to produce all maps [181]. The maps shown at the south-east Asia scale use records at the country level, or the administration level for larger countries (e.g., China, India, and Japan). Following previous work [182,183], we used island boundaries instead of political boundaries for large islands. For maps centered on Macao, we used the GPS coordinates associated with each specimen to add their collection localities.

2.6. Analyses

We produced maps, bar graphs, heatmaps, species accumulation curves, and Venn diagrams using *ggplot2* [184], whereas we used Adobe Illustrator to assemble the species account figures (Figures A1–A158). We produced species accumulation curves and diversity estimates using the package *iNEXT* [185].

2.7. Literature Search

To compile studies that produced a species checklist for cities, we performed a Scopus search using the following formula on the 9th of February 2021: "formicidae" AND "checklist" AND "city" OR "urban". We then pruned the resulting dataset manually by reading the abstracts and only keeping the studies that produced a total number of species for a city. We further added appropriate studies known by authors that were not present within the Scopus search. In particular, we use the literature information combined in GABI [186] to identify suitable articles on urban ants. We classified studies as either ecological or checklists. If a study was primarily hypotheses driven, with limited sampling efforts in time, habitats, or in the methods used, it was classified as ecological, whereas studies solely producing a species checklist, including records from previous published studies, were classified as checklists. Studies including other non-urban habitats outside the main city area, and for which detailed information did not allow to separate species composition and richness, were not considered.

2.8. Notes on Invasion Status

An understanding of the native and introduced ranges of species represents a fundamental step in the detection and management of biological invasions. However, for many species of ants, clear geographic boundaries between those ranges remain undetermined, either at global (e.g., uncertainty in the realm of origin) or regional scales (e.g., native vs. introduced range within a particular realm). Here, we thus distinguish three categories between native, exotic, and tramp species. For species that we could establish with some confidence whether or not they were introduced, we used the exotic and native status, respectively. We used the tramp status for species whose biogeographic origins were more uncertain in Macau or south China. Note that all species here labelled as tramps have been previously transported in other regions of the world and have established populations in non-native habitats. This demonstrates their potential to colonize new regions. Furthermore, these tramp species often occur within anthropogenic habitats. As such, tramp species, regardless of their potential non-native status, are important to consider from a management perspective, as they have the potential to invade non-native localities. The establishment of the native and exotic ranges for each species was based on the maps available on antmaps.org [186,187].

2.9. Notes on Records

Since the last publication of a species checklist for Macao [53], two studies with a focus on specific genera (i.e., *Polyrhachis* and *Strumigenys*) published new species records for the region [54,55]. Since all but one species record—*Polyrhachis tyrannica* [55]—are from specimens collected during our 2019 sampling, we here report these specimens, with the exception of *P. tyrannica*, as new records for Macao.

2.10. Notes on Taxonomy

To identify our specimens at species-level, we used the Insect Biodiversity and Biogeography Lab (IBBL) ant collection as a reference. For the especially challenging species, we relied on the taxonomic knowledge of Dr. Benoit Guénard. To verify our identification of the genera *Nylanderia* and *Carebara*, we shared stacked images with specialists familiar with their taxonomy, Dr. Jason L. Williams and Dr. Georg Fischer, respectively. When a species could not be identified at species level with certainty, we labelled it as "nr." the morphologically closest known species (e.g., *Colobopsis* nr. *nipponica*). If the unknown species was morphologically distinct but not identifiable, we used the unique morphospecies code their collector used to label it (e.g., *Camponotus* sp.1 FB).

3. Results

Our 2019 survey collected a total of 112 species and morphospecies from 46 genera and nine subfamilies. Among these, 51 species (46%), 10 genera (22%), and one subfamily (11%) represented new records for Macao. We also found four additional species and one new genus record among the specimens opportunistically collected between 2017 and 2020. In total, the new records reported here include one subfamily, 11 genera, and 55 species and morphospecies (Figure 2, Table 2). The new genera reported for Macao are: *Brachymyrmex* (exotic), *Buniapone*, *Dilobocondyla*, *Gesomyrmex*, *Iridomyrmex*, *Mayriella*, *Probolomyrmex*, *Proceratium*, *Pseudolasius*, *Rotastruma* and *Vollenhovia*, while the Proceratiinae subfamily is here recorded for the first time. The overall number of ant species known from Macao thus increases by 55%, from 100 to 155 species and morphospecies (Figure 2, Table 2), which represents the third highest urban ant diversity out of 123 entries (see Figure 1, Table 1). For images of these species and maps of their distribution in Macao and SE Asia, see Appendix A (Figures A1–A158).

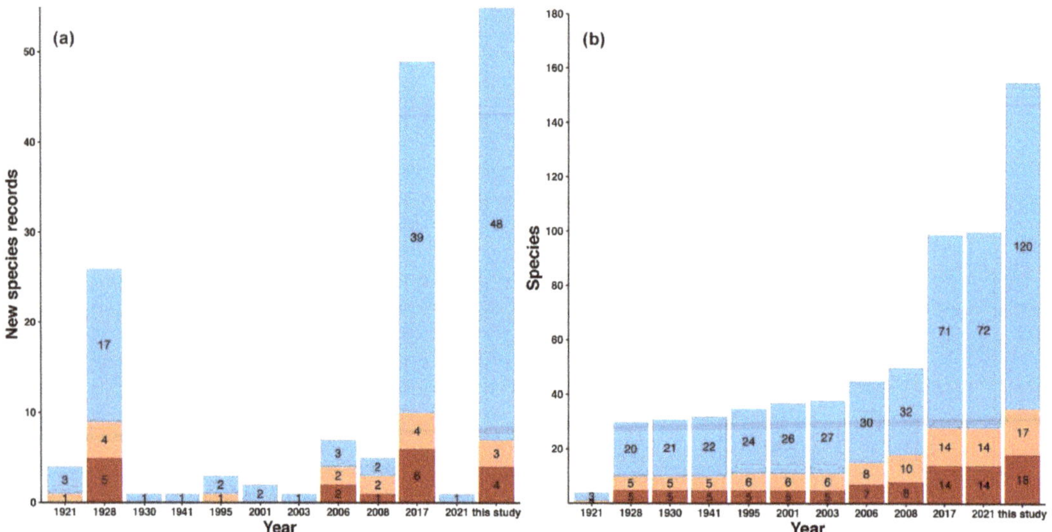

Figure 2. Number of new records per year in Macao. Bar plots showing (**a**) the number of new ant species record for Macao based on literature records and this study and (**b**) bar plots showing the accumulation of ant species found in Macao based on literature records and this study. The proportion of native species (blue), tramp species (light orange), and exotic species (dark red) are denoted within bar plots.

Table 2. Summary table of all new ant species records made in Macao over time with details on the number of reported ant species, subspecies, and morphospecies.

Studies	Species	Subspecies	Morphospecies	Total per Year	Cumulative Total
Wheeler 1921 [188]	4	0	0	4	4
Wheeler 1928 [189]	24	2	0	26	30
Wheeler 1930 [190]	1	0	0	1	31
Wu 1941 [191]	0	1	0	1	32
Tang et al. 1995 [192]	3	0	0	3	35
Zhou 2001 [193]	2	0	0	2	37
Xu 2003 [194]	1	0	0	1	38
Hua 2006 [195]	6	1	0	7	45
Eguchi 2008 [196]	5	0	0	5	50
Leong, Shaijo and Guénard 2017 [53]	41	1	7	49	99
Wong and Guénard 2021 [55]	1	0	0	1	100
This study	40	0	15	55	155

The sampling methods varied in their overlap for the species they collected (Figure 3). Of the 112 species collected during the 2019 survey, 10 species were only found in leaf litter extractions, five in ground baits, 10 in subterranean baits, 13 in hand collections, and eight in arboreal baits, whereas the other 66 were collected with more than one method. Artificial nests did not collect new species records nor unique species, but they did provide sociometric data for 15 species from a total of 913 nests recovered, for a colonization rate of 3% (Table S2).

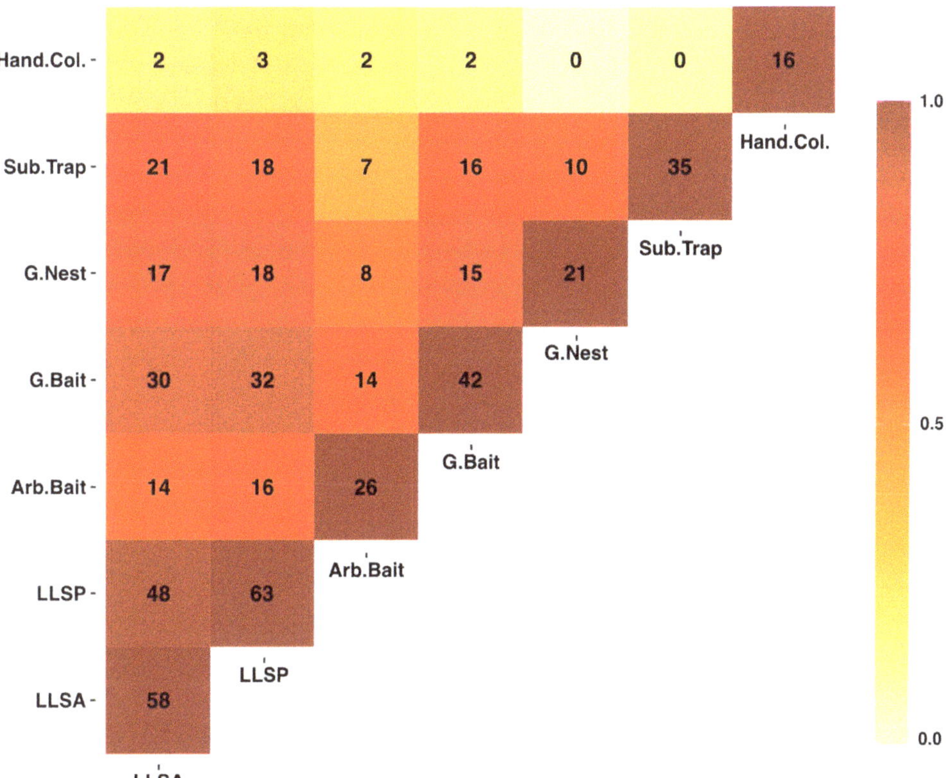

Figure 3. Heatmap showing the number of shared species between collection methods used in the 2019 survey. Color illustrates the strength of the correlation between sampling methods in the shared species they collect. The diagonal represents the total number of species collected for each method. Abbreviations are for hand collection (Hand. Col.), subterranean trap (Sub. Trap.), ground nest (G. Nest), ground bait (G. Bait), arboreal bait (Arb. Bait), leaf litter extraction with the species pool technique (LLSP), and leaf litter extraction with the standard area technique (LLSA).

We found few habitat generalists, with only seven species collected in all strata (Table 3, Figure 4)—*Monomorium intrudens*, *M. floricola*, *Nylanderia sharpii*, *Pheidole megacephala*, *P. tumida*, *Tapinoma indicum*, and *T. melanocephalum*—three of which (43%) are exotics. Ten species were only collected within the subterranean stratum and 14 species only in the arboreal stratum (Table 3, Figure 4). In contrast, most species were ground-dwelling ($n = 52$) or collected within two strata ($n = 36$). The highest proportion of new records found solely within one stratum were in the subterranean (9/10 species: 90%), arboreal (9/14 species: 64%), and then ground (25/52 species: 48%) strata (Figure 4).

Table 3. Summary checklist of the species recorded during previous studies and the current study in Macao. Status of species are mentioned as native, tramp, or exotic. The asterisk symbol (*) denotes new records. The dagger symbol (†) denotes morphospecies collected in 2019 that probably belong to previous records but could not be assigned a species name due to incomplete taxonomic descriptions (as such, they were not counted in the total number of species). The diesis symbol (‡) denotes a species collected previously, but with a mislabeled status. An "X" under the column Arboreal, Ground, or Subterranean indicates that this species was collected within this stratum during the 2019 survey. We left blanks for species not collected during the 2019 survey. For images of species and maps of their distribution in Macao and SE Asia, see Figures A1–A158. For detailed accounts of the material examined, see the species account section of the supplementary material.

Subfamily	Genus	Species	Status	Year of First Published Record	Arboreal	Ground	Subterranean
AMBLYOPONINAE	*Stigmatomma*	nr. *rothneyi*	native	1928	-	X	-
DOLICHODERINAE	*Chronoxenus*	*dalyi*	native	1928			
		walshi	native	2006			
		wroughtonii	native	1995			
		wroughtonii formosensis	native	2006			
		† morpho1	native	2021	-	X	X
		† morpho2	native	2021	-	X	X
	Dolichoderus	*taprobanae*	native	1928			
		nr. *sibiricus*	native	2017	X	-	-
	Iridomyrmex	sp. *anceps* cplx *	tramp	2021	-	X	-
	Ochetellus	*glaber*	tramp	1928	-	X	-
	Tapinoma	*indicum*	native	2017	X	X	X
		melanocephalum	tramp	1921	X	X	X
		sp. 1 FB *	native	2021	X	-	-
	Technomyrmex	*brunneus*	tramp	1995	-	X	-
		horni *	native	2021	-	X	-
DORYLINAE	*Ooceraea*	*biroi*	exotic	2017	-	X	X
FORMICINAE	*Acropyga*	*acutiventris*	native	2017			
		sauteri	native	1928			
		sp. mo02	native	2017			
	Anoplolepis	*gracilipes*	exotic	1928	X	X	-
	Brachymyrmex	*patagonicus* *	exotic	2021	X	-	-
	Camponotus	*albosparus*	native	1928			
		carin *	native	2021			
		irritans *	native	2021	-	X	-
		lighti	native	2017			
		mitis	native	1928	X	X	-
		nicobarensis	native	2017	X	X	-
		parius	native	1921			
		vitiosus	native	2017	X	-	-
		variegatus	tramp	2006			
		variegatus dulcis	native	1928			
		variegatus proles	native	1941			
		sp. 1 FB *	native	2021	-	X	-
	Colobopsis	nr. *nipponica*	native	2017			
		nr. *vitrea*	native	2017	X	-	-
	Gesomyrmex	*howardi* *	native	2021			
	Lepisiota	*rothneyi*	native	1921	-	X	-
	Nylanderia	*amia*	tramp	1928			
		bourbonica	tramp	2006	-	X	-
		indica	native	1928	-	X	-

Table 3. *Cont.*

Subfamily	Genus	Species	Status	Year of First Published Record	Arboreal	Ground	Subterranean
		sharpii	native	2021	X	X	X
		taylori *	native	2021	-	X	-
		vividula	exotic	2006			
		yerburyi	native	1928			
		sp. 3 BG *			-	X	-
		sp. 6 BG *	native	2021	-	X	-
	Paraparatrechina	*sauteri*	native	2006			
		sp.1 BG *	native	2021	-	X	X
	Paratrechina	*longicornis*	exotic	1928	-	X	X
	Plagiolepis	*alluaudi* *	exotic	2021	X	-	-
	Polyrhachis	*confusa* *	native	2021	-	X	-
		demangei	native	2017			
		dives	native	1995	X	-	-
		illaudata	native	2017	X	X	-
		latona *	native	2021	-	X	-
		tyrannica	native	2021			
	Pseudolasius	*risii* *	native	2021			
LEPTANILLINAE	*Leptanilla*	*macaoensis*	native	2017			
MYRMICINAE	*Cardiocondyla*	*minutior*	exotic	2017	-	X	-
		wroughtonii *	tramp	2021	X	-	-
	Carebara	*affinis* *	native	2021	-	X	X
		capreola	native	2003			
		diversa	native	1921	-	-	X
		diversa laotina	native	2017			
		melasolena *	native	2021	-	X	X
		sangi *	native	2021	-	-	X
		zengchengensis	native	2017	-	X	X
	Crematogaster	*binghamii* *	native	2021	-	X	-
		biroi	native	1928			
		dohrni	native	1928			
		ferrarii	native	2017	X	X	-
		macaoensis	native	1928			
		quadriruga	native	2017	-	X	-
		rogenhoferi	native	2017	X	-	-
	Dilobocondyla	*propotriangulata* *	native	2021	X	-	-
	Mayriella	*granulata* *	native	2021	-	X	-
	Meranoplus	sp. mo01 nr. *bicolor*	native	2017			
	Monomorium	*chinense* *	native	2021	X	X	X
		intrudens *	tramp	2021	X	X	-
		floricola	tramp	2017	X	X	X
		pharaonis	exotic	2017	-	X	X
		sp. psw-cn01	native	2021	-	X	-
	Myrmecina	*nomurai* *	native	2021	-	X	-
		sinensis	native	2017			
	Pheidole	*elongicephala* *	native	2021	-	-	X
		fervens	tramp	2008	-	X	X
		hongkongensis	native	2008	-	X	-
		indica	tramp	1928			

Table 3. *Cont.*

Subfamily	Genus	Species	Status	Year of First Published Record	Arboreal	Ground	Subterranean
		megacephala	exotic	2008	X	X	X
		ochracea	native	2017	-	X	X
		parva	tramp	2008	-	X	X
		*pieli**	native	2021	-	X	X
		taipoana	native	2008	-	X	X
		tumida	native	2017	X	X	X
		nodus	tramp	2017	-	X	-
		vulgaris *	native	2021	-	X	-
		zoceana *	native	2021	-	X	-
		nr. *ryukyuensis* *	native	2021	-	-	X
	Recurvidris	*recurvispinosa*	native	2017	-	X	-
	Rotastruma	*stenoceps* *	native	2021	-	X	-
	Solenopsis	*geminata*	exotic	1928			
		invicta	exotic	2006	-	X	-
		jacoti	native	2017	-	X	X
	Strumigenys	*emmae*	exotic	2017	-	X	-
		elegantula	native	2020	-	X	-
		exilirhina	native	2017	-	X	-
		feae	native	2020	-	X	-
		membranifera	exotic	1928	-	X	-
		minutula	native	2017	-	X	-
		‡*nepalensis*	exotic	2017	-	X	-
		sauteri	native	2020	-	X	-
		subterranea	native	2020	-	-	X
	Syllophopsis	nr. *cryptobia* *	native	2021	-	-	X
		sp. mo01 nr *sechellensis*	native?	2017	-	X	X
		sp. 1 BMW *	native	2021	-	X	X
		sp. 2 BMW *	native	2021	-	X	-
	Tetramorium	*bicarinatum*	tramp	2017	X	X	-
		indicum *	native	2021	X	-	-
		insolens *	exotic	2021	-	X	-
		kraepelini	tramp	2017	-	X	X
		lanuginosum	exotic	1928	-	X	-
		nipponense	native	2017	X	X	-
		parvispinum	native	2017			
		simillimum	exotic	2017			
		tonganum *	exotic	2021	X	-	-
		wroughtonii *	native	2021	-	X	-
		nr. *elisabethae* *	native	2021	-	-	X
		sp.1 *obseum gr.*	native	2017	-	X	-
		sp. 2 JF *	native	2021	X	-	-
		sp. 9 JF *	native	2021	-	-	X
	Vollenhovia	sp.1 BG *	native	2021	-	X	-
		sp. 2 BG *	native	2021	-	X	-
PONERINAE	*Anochetus*	*risii*	native	2017	-	X	-
	Bothroponera	*rubiginosa*	native	1928			
	Brachyponera	*luteipes*	tramp	1928			
		obscurans	native	1928	-	X	X

Table 3. *Cont.*

Subfamily	Genus	Species	Status	Year of First Published Record	Arboreal	Ground	Subterranean
	Buniapone	*amblyops* *	native	2021	-	-	X
	Diacamma	sp. 1	native	2017	X	X	-
	Ectomomyrmex	*annamitus* *	native	2021	-	-	X
		astutus	native	2001			
		leeuwenhoecki	native	2017	-	X	-
	Euponera	*pilosior*	native	2017	-	-	X
		sharpi	native	1928			
	Harpegnathos	*venator*	native	2001	-	X	-
		venator rugosus	native	1928			
	Hypoponera	*exoecata*	native	2017	-	X	-
		sp. mo01	native	2017	X	X	-
	Leptogenys	*chinensis*	native	2017			
		peuqueti	native	1928	-	X	-
	Odontoponera	*denticulata*	native	2017	-	X	-
	Pseudoneoponera	*rufipes*	native	1930	-	X	-
PROCERATIINAE	*Probolomyrmex*	*dabermanii**	native	2021			
	Proceratium	sp. cf. *bruelheidei* *	native	2021	-	X	-
PSEUDOMYR-MICINAE	*Tetraponera*	*allaborans*	native	2017			
		binghami *	native	2021	X	X	-
		nitida *	native	2021	X	-	-

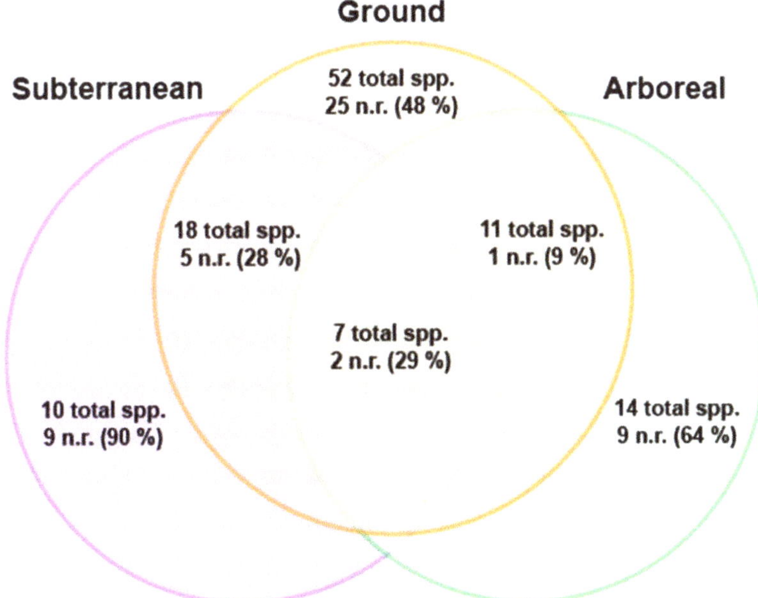

Figure 4. Venn diagram showing the total number of species (112 spp.) and new species records (51 n.r.) collected in each stratum during the 2019 survey. Numbers in parentheses represent the proportion of species collected within a stratum that are new records from the last 2017 checklist (Leong et al. 2017). Note that this figure does not include the four records added from specimens collected besides the main 2019 survey.

At least five of the species collected during our 2019 survey, which belong to the genera *Strumigenys*, *Syllophopsis*, *Tetramorium*, and *Vollenhovia*, were considered potentially novel to science at the time of collection. We found three of the undescribed species in subterranean traps (i.e., *Strumigenys subterranea*, *Syllophopsis* nr. *Cryptobia*, and *Tetramorium* sp. 9 JF), one in leaf litter samples (i.e., *Vollenhovia* sp. 2 BG), and one in arboreal traps (i.e., *Tetramorium* sp. 2 JF).

Several of the new records have rarely been reported in the literature and represent extensions of their known range. First, we found workers of *Dilobocondyla propotriangulata*, an arboreal species described from Vietnam [197], at two different sites in Macao, which represents the third and fourth records of the species worldwide. Second, we found workers and a queen of *Mayriella granulata*, also described from Vietnam [198], which represents the first record of this species in China. Lastly, we found a worker of *Probolomyrmex* (*P. dammermani*), a pantropical but rarely collected genus [199–201], which represents the first record of this species in China.

Before our survey, 14 tramp and 14 exotic species were known to occur in Macao. Here, we report four additional exotic species records: *Brachymyrmex patagonicus*, *Plagiolepis alluaudi*, *Tetramorium insolens*, and *T. tonganum*. We also report three additional tramp species records: *Cardiocondyla wroughtonii*, *Iridomyrmex* sp. *anceps* complex, and *Monomorium intrudens*. Moreover, we report new localities in Macao for several exotic species: *Anoplolepis gracilipes*, *Cardiocondyla minutior*, *Monomorium pharaonis*, *Ooceraea biroi*, *Paratrechina longicornis*, *Pheidole megacephala*, *Solenopsis invicta*, *Strumigenys emmae*, *S. membranifera*, *S. nepalensis*, and *Tetramorium lanuginosum*.

Nevertheless, despite achieving a high sampling coverage (i.e., between 80 to 98% depending on the method), species accumulation curves indicate that further sampling should uncover several more species on Coloane Island (Figure 5, Table 4). Indeed, estimates predict that each sampling method could collect from 2 to 25 additional species each.

Table 4. Summary of the species richness collected, the sampling completeness, and the richness estimates for each sampling method used during the sampling done in Coloane in 2019.

Sampling Method	Observed Richness	Sampling Completeness	Estimated Richness
Arboreal Bait	26	0.98	28.49
Ground Bait	42	0.95	49.17
Ground Nest	21	0.80	41.14
Leaf Litter Sampling (Standardized area)	59	0.93	64.70
Leaf Litter Sampling (Species pool)	64	0.86	88.80
Subterranean trap	35	0.88	56.25

We identified 112 studies, representing 109 cities, that focused on ants within urban environments (Figure 1, Table 1). Among those, 23 studies provided species checklists, while 88 represented ecological surveys. The studies were unevenly distributed across biogeographic regions. The highest numbers were from North America (n = 41), Asia (n = 34), Europe (n = 24), and Central and South America (n = 21), whereas Australia (n = 6) and Africa (n = 5) had the lowest number of studies. Although Asia had the second highest number of studies, most of these originated from temperate regions, with only eight studies (23.5%) conducted within tropical or subtropical regions.

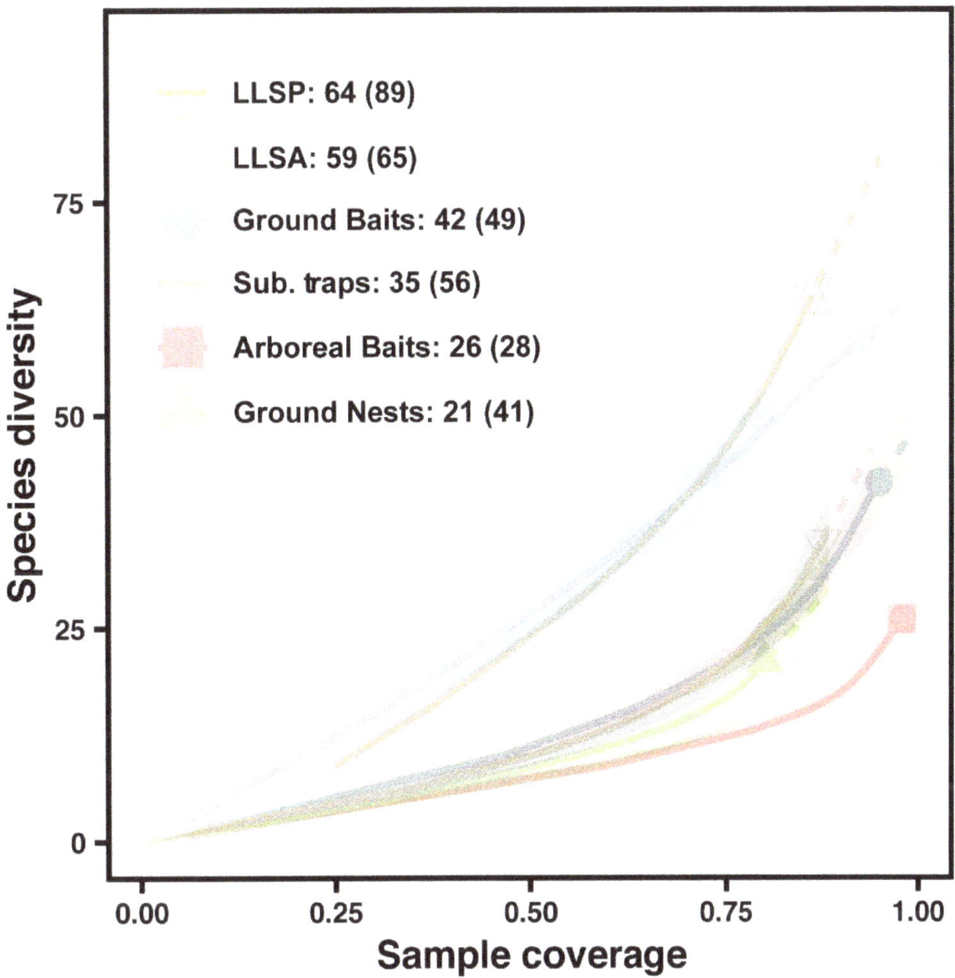

Figure 5. Species diversity in relation to sampling coverage for each standardized technique used in Macao in 2019. Values in the legend represent the number of species collected, and, in brackets, the estimated number of species that would be collected by this method if sampling coverage would reach 100% (asymptotic estimates of order q = 0 obtained using the function ChaoRichness in the package iNEXT). The dotted line shows the extrapolation for the predicted number of new records if sampling completeness would reach 100% (i.e., a value of 1.00 on the graph). Shaded areas represent the 95% confidence intervals of each curve. Calculation method used species incidence frequency. Leaf litter samples were considered as four units of 1 m^2 per Winkler sack (which pooled 4 m^2 of leaf litter). Abbreviations are (LLSP) leaf litter extraction with the species pool technique and (LLSA) leaf litter extraction with the standard area technique.

4. Discussion

A common perception of urban biodiversity is that it is characterized by low species richness and dominated by exotic species [12]. This perception may be induced by the excess of local scale studies (α diversity) compared to the limited number of studies at larger scale (γ diversity) encompassing the full diversity encountered within of a city (Table 1). Here, our results contrast with the former assumption but agree with the latter. Indeed, we found that Macao hosts a diverse ant fauna, but that a high number of that fauna consists of exotic and tramp species.

Macao's ant fauna presents one the highest known ant richness reported for an urban region (Table 1). Our results indicate that, while there are few comprehensive studies for tropical regions—most studies on urban ants have been conducted within temperate regions where species diversity is usually much lower than in tropical and subtropical regions [202]— several cities, including Macao, offer potential conservation values for ants. For instance, a study limited to the botanical garden of Bogor (Java), a small green oasis within an urban area, captured 216 ant species [69]. Similarly, ecological studies in Abidjan (Ivory Coast, 176 species) and Uberlândia (Brazil, 143 species), among others, also presented high ant species richness [62,155]. Altogether, these results highlight the potential conservation value of urban habitats, but also their potential to increase the biogeographic and taxonomic knowledge on ants. Indeed, contrary to most natural habitats, urban habitats are characterized by their easy access, which facilitates continuous and thorough sampling. As for tropical forests, the vertical stratification of ants within cities does exist, and, as a result, researchers should consider diversifying their sampling approach to include subterranean and arboreal communities as well as epigeic ants.

Ant assemblages are known to be highly structured along a vertical gradient ranging from the top soil layer (first 50 cm) to the tree canopy [203,204], but such stratification had not been shown for urban environments prior to this study. Our results show that ignoring these strata may lead to an underestimation of species richness estimates. Indeed, although most of the new species records were collected within the ground stratum, we found several previously unrecorded arboreal and subterranean specialists. In particular, of the 35 species collected with our subterranean trapping, 10 were found only within that stratum, nine of which were new records, and three represented undescribed species. Remarkably, this parallels the results of previous surveys focusing on multiple strata but conducted within natural ecosystems [205]. For instance, in Ecuador, Wilkie and collaborators collected 47 species in subterranean probes, nine of which were exclusively subterranean, and two were undescribed [205]. It is also worth noting that 14 species collected during the 2019 survey were unique to the arboreal stratum, nine of which were new records, and one was an undescribed species. This shows the importance of using sampling methods targeting the ant communities of all vertical strata, instead of focusing solely on ants found within a single stratum as is commonly done in urban studies [52].

Other methods captured fewer novel records, but, nonetheless, provided ecological and biogeographic information for a wide range of species. For instance, ground baits often collected large series of workers, including multiple worker castes for polymorphic species, which is often essential for their identification. As for ground nests, they had the lowest rate of capture, but provided important and rarely collected sociometric data, including new colony size information for seven species (Guénard, unpublished). Since we still lack information on the sociometry of most ant species [206,207], ground nests proved especially useful in collecting this valuable data. However, it is worth noting that the colonization rate of the ground nests, with 3%, was neatly inferior to the rate observed in previous studies using similar devices (e.g., 8% in [180]), or from other urban areas [141]. While the type of nests may be suboptimal for the ant community present in Macao, it is surprising that they were not more heavily exploited by ants, especially because urban habitats are usually characterized by limited nesting resources [141]. Perhaps the subtropical conditions characterized by heavy rains prevented the establishment of ants. Indeed, several nests had their openings clogged with mud and some had their inner cavities filled with fungal growths. Just as with temperate regions [141], the testing and deployment of different artificial nest apparatuses may represent an interesting opportunity to census urban ants and their sociometric characteristics.

Even though we used an exhaustive sampling approach, and our results substantially increased our knowledge on Macanese ants, species accumulation curves indicate a substantial number of unrecorded species to be collected using a similar methodology (Figure 4). This is supported by the several species previously recorded within Macao, but not collected in this study (Table 2). Previous studies conducted within more natural or relatively undisturbed habitats showed that achieving an exhaustive sampling of local diversity represents a challenging task [125,208,209]. After two separate survey programs ([53], this study), our results confirm this is similar for urban areas. Furthermore, other sampling methods not used here could have been added, such as pitfall traps to collect larger ground-dwelling ants [210], canopy fogging to collect arboreal species [211], Malaise traps to collect alates [212], and multiple soil sampling methods to collect subterranean ants [50]. Thus, while our study contributes to a better understanding of ant species richness and composition in Macao, it represents a steppingstone and not a final outcome, with future sampling likely to provide additional new records, and potentially more undescribed species.

Despite the geographic limitations of our sampling, being restricted to Coloane island, this allowed us to considerably increase the list of Macanese ant species. These new records highlight the potential for urban forest fragments and other urban habitats to maintain a significant portion of ant diversity. Similar examples can be found in Asia, such as in in Bogor and Singapore [69,102]. Likewise, Hong Kong harbors a high ant richness (Lee et al. in press, Guénard unpublished), with numerous new records and species recently reported (e.g., [55,192]). Additionally, comparative studies focusing on old growth and secondary forests in both cities also reported no significant difference in ant richness between the types of forests ([213,214], Nooten et al. submitted). As such, past studies and the current one should motivate the conservation of forest patches in and around urban matrices in Asia, whether they are primary or secondary.

Nevertheless, a key difference between disturbed and undisturbed habitats lies within their species composition instead of in the number of species they harbor, and a significant part of Macao's fauna consists of non-native species. The previous survey reported several new records of exotic species in Macao [53], which are here completed with the additions of four exotic and three tramp species (Figure 1), totaling in 18 exotic species and representing 11.5% of the Macanese ant fauna. The most notable newly reported exotic is *Brachymyrmex patagonicus*, a major pest in south-east USA [215]. This represents the second record of *B. patagonicus* in continental Asia, with the first report from Hong Kong [216]. Its presence in Macao is worrisome, and clear plans to determine the extent of its distribution, with programs to destroy established populations, should be developed quickly. We also report the spread of several alien and potentially harmful species. We found the three exotic species *Anoplolepis gracilipes*, *Monomorium pharaonis*, and *Paratrechina longicornis* in five, seven, and six forested sites, respectively. They can, thus, establish populations and persist in forested habitats, where their effects on native ants are unknown. Through our standardized sampling, we also found the notorious invasive species *Solenopsis invicta* and *Pheidole megacephala* but, interestingly, rarely found it in Coloane's forested sites. Despite finding several workers and queens of *S. invicta* through hand collection in open and urban habitats, we found this species in only one forested site near a hiking trail: one individual was found in a leaf litter sample and several workers colonized two ground baits at that site. This reflects previous findings that the distribution of *S. invicta* is mainly restricted to disturbed habitats [217–219]. Similarly, while we found a high abundance of workers of *P. megacephala* on one golf course, we collected only two individuals in forested sites. For both of these species, notorious for their destructive impacts on native communities [220–222], this would indicate that, for now, they are mainly restricted to more open and disturbed habitats. This suggests that forested habitats may help preserve local biodiversity while simultaneously limiting the distribution of some invasive ant species, reflecting results from Hong Kong and Hainan, where less disturbed habitats harbored relatively fewer exotic species than did more disturbed ones [223].

Nonetheless, the occurrence of exotic species in Macao is high. In comparison, the checklist of the ants of Yunnan reports that only 2% of its species are introduced [224], a proportion 5.75 times lower than in Macao. Based on current knowledge, Macao appears to have one of the highest numbers and proportions of exotic species encountered within cities (Table 1). Alas, we also report the presence of 17 tramp species, some of which may very well be introduced, as our biogeographic understanding and the region of origin for several species remains limited; thus, the overall number may be even higher. The high proportion of exotic and tramp species in Macao supports the hypothesis that coastal cities act as gateways for the introduction of exotic species through high propagule pressure [225]. In addition, since the occurrence of non-native species is an indicator of reduced ecological integrity [226], the high proportion of non-native species we found in Macao suggests these habitats have suffered substantial damage. It is, thus, crucial to make periodical biodiversity surveys in Macao and other cities to monitor their habitats' health through time, as well as to identify new arrivals of exotics and prevent their further spread.

An historic baseline is lacking for Macao's ant assemblages, but we suspect there may be substantial differences compared to the ant assemblages before urbanization. Indeed, several genera known to occur in southern China and within neighboring cities of the Greater Bay Area, such as Hong Kong, are missing from Macao. These include *Aenictus* and *Dorylus* army ants, *Discothyrea*, *Odontomachus*, *Ponera*, and weaver ants (*Oecophylla smaragdina*). The absence of army ants in Coloane (except *O. biroi*, an exotic species), such as species of *Aenictus* and *Dorylus*, may be explained by the morphology of the queen caste in these genera: they lack wings and, thus, cannot disperse by flight [227]. As such, if these ants disappeared from Coloane during a disturbance event, we may expect that it would be arduous for these species to recolonize the island. Of course, native army ant species may have been absent from Coloane throughout Macao's urbanization history due to its insularity or of its small size ([228], but see [229]). However, the presence of native army ants on several of Hong Kong' islands—for example, the *Aenictus* species found on Lamma Island (13.55 km^2, François Brassard pers. obs.)—suggests they could have been extirpated from Coloane Island and then failed to recolonize. More puzzling is the absence of genera such as *Discothyrea*, *Odontomachus*, *Oecophylla*, and *Ponera*. Small and relatively uncommon species of *Discothyrea* and *Ponera* may have escaped our sampling or may be especially sensitive to disturbance. However, for unknown reasons, large and conspicuous species such as *Odontomachus* and especially *Oecophylla*, frequently encountered on forest edges or in disturbed habitats within Hong Kong, are probably truly absent from the island.

5. Conclusions

In summary, this study highlights the importance of conducting holistic biodiversity surveys in cities to discover new records as well as potential new species for science, and to monitor the introduction of new exotic species. Our results suggest that forest patches in cities can harbor a diverse ant fauna and may have a significant conservation value. However, exhaustive ant diversity surveys in cities are rare, and are often based on incomplete sampling approaches. Thus, until the completion of several more surveys in cities around the world, particularly within tropical regions, a clear understanding of how urban environments may act as biodiversity refuges and gateways for exotic species will be lacking. As such, we advocate that conservation management practices should implement regular biodiversity surveys using an exhaustive sampling approach in urban regions worldwide. Whenever possible, we also recommend that urban biodiversity assessments be combined with surveys done in less urbanized habitats nearby to compare the diversity and composition of each habitat.

Supplementary Materials: The following are available online at https://www.mdpi.com/article/10.3390/d13080358/s1. Figure S1: Map of Coloane showing the 21 sites sampled in the 2019 survey. White dots mark sites where the full protocol was done (i.e., standardized and species pool leaf litter extractions, ground baiting, ground nests, subterranean traps, and arboreal traps), whereas grey dots mark preliminary sites where only ground baiting and leaf litter extractions were done. Hand collection was also opportunistically used at each site. Figure S2: Design of a 20 × 20 m sampling plot. Each of the 1 × 1 m quadrats where subterranean traps and leaf litter extraction (standardized area) was performed were placed at a corner of the plot. Black dots show the emplacement of nest bundles. For the species pool leaf litter extraction, 12 microhabitats were sampled within the light gray area. For the arboreal baiting protocol, four trees measuring a minimum of 5 m height were sampled within the grey area. Table S1: List of the localities of each sampling sites, their associated number, and their geolocation. The date refers to the first sampling event made at a site, which corresponded to the leaf litter extraction and placement of subterranean traps. Sampling protocols are defined as follows: the letter (P) signifies a partial sampling protocol (i.e., leaf litter extraction, ground baiting, and hand collection), whereas the letter (F) signifies a full protocol (i.e., leaf litter extraction, ground baiting, ground nests, subterranean traps, arboreal traps, and hand collection). Table S2: Sociometry data collected using ground nests. Macao species: material examined.

Author Contributions: Conceptualization, F.B. and B.G.; methodology, F.B. and B.G.; validation, F.B. and B.G.; formal analysis, F.B.; investigation, F.B.; resources, B.G., H.-H.C., and C.-M.L.; data curation, F.B. and B.G.; writing—original draft preparation, F.B.; writing—review and editing, F.B., B.G., C.-M.L., and H.-H.C.; visualization, F.B.; supervision, B.G.; project administration, B.G.; funding acquisition, B.G. All authors have read and agreed to the published version of the manuscript.

Funding: F.B. was supported by the Instituto para os Assuntos Municipais, Macao SAR, China, CML was supported by the Macao Foundation and the Direcção dos Serviços do Ensino Superior, Macao SAR, China, and CML and FB were supported by The University of Hong Kong, Hong Kong SAR, China.

Institutional Review Board Statement: Ethical review and approval were waived for this study, due to our work being conducted on invertebrates.

Data Availability Statement: The data presented in this study are available in tables within the main text and within the supplementary material section.

Acknowledgments: We thank Jason L. Williams and Georg Fischer for their help with the identification of species within the genera *Nylanderia* and *Carebara*, respectively. We acknowledge that the images we shared with these specialists are not optimal for identification and, as such, if some misidentifications are later revealed, these errors should be considered our own, not theirs. We thank Siu Yiu for her help conducting fieldwork, as well as with sorting, mounting, and imaging specimens. We thank Carly McGregor for her help conducting fieldwork. We thank Sabine Nooten for her help producing species accumulation curves. We also thank the Environmental Protection Bureau, Caesars Golf Macau, and the Macau Golf and Country Club for allowing us to sample on their premises.

Conflicts of Interest: The authors declare no conflict of interest. The funders had no role in the design of the study; in the collection, analyses, or interpretation of data; in the writing of the manuscript, or in the decision to publish the results.

Appendix A

The appendix shows all species for which we had specimens within the IBBL collection. These are shown in lateral, dorsal, and face view. Within the figure, we include a map of south-east Asia showing the distribution of the species. We colored polygons according to whether a species was recorded as a native or exotic species, or unrecorded for a specific region. Note that we could not use the label tramp because most studies do not distinguish beyond native or exotic. We also include an inset map showing where the species was found in Macao. Black triangles indicate collection locations made during the 2019 survey, whereas white dots indicate collection locations not done during the survey.

Appendix A.1 AMBLYOPONINAE

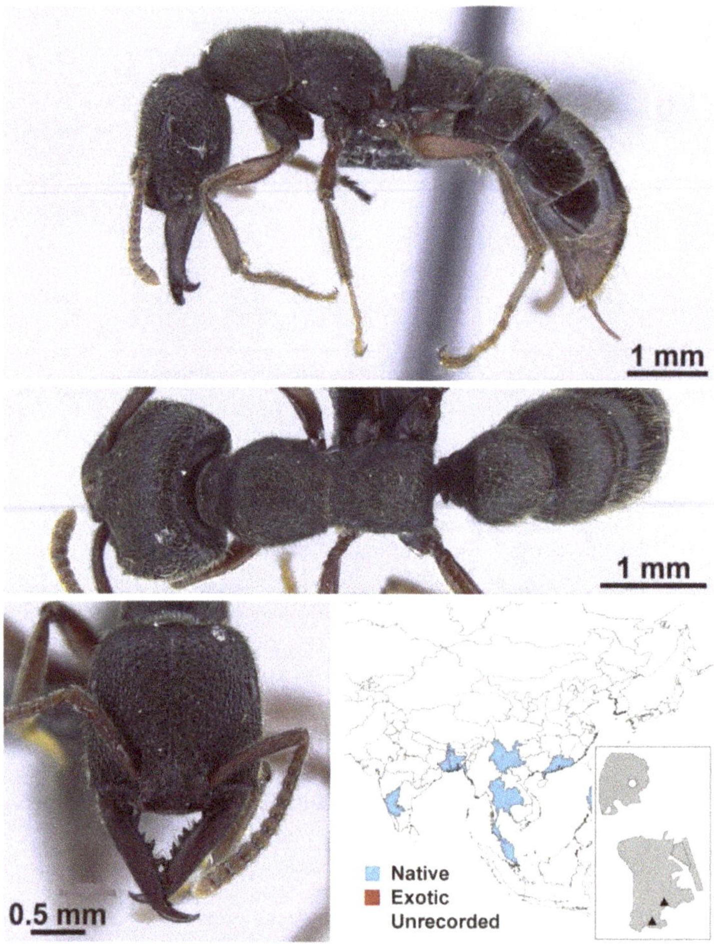

Figure A1. *Stigmatomma rothneyi* Forel, 1900 worker (MAC_S11_LLSA_Sp.2, IBBL).

Appendix A.2 DOLICHODERINAE

Figure A2. *Chronoxenus morpho 1* worker (MAC_S14_LLSP, IBBL).

Figure A3. *Chronoxenus morpho 2* worker (MAC_S21_q2_50_sp.2, IBBL).

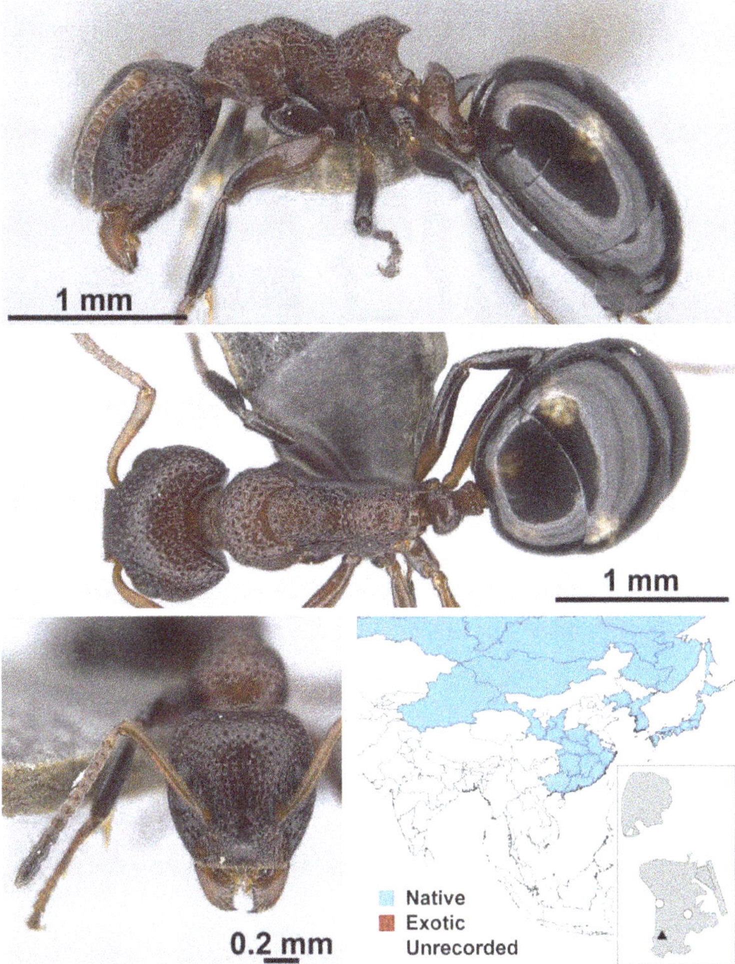

Figure A4. *Dolichoderus* nr. *sibiricus* Emery, 1889 worker (FB19279, IBBL).

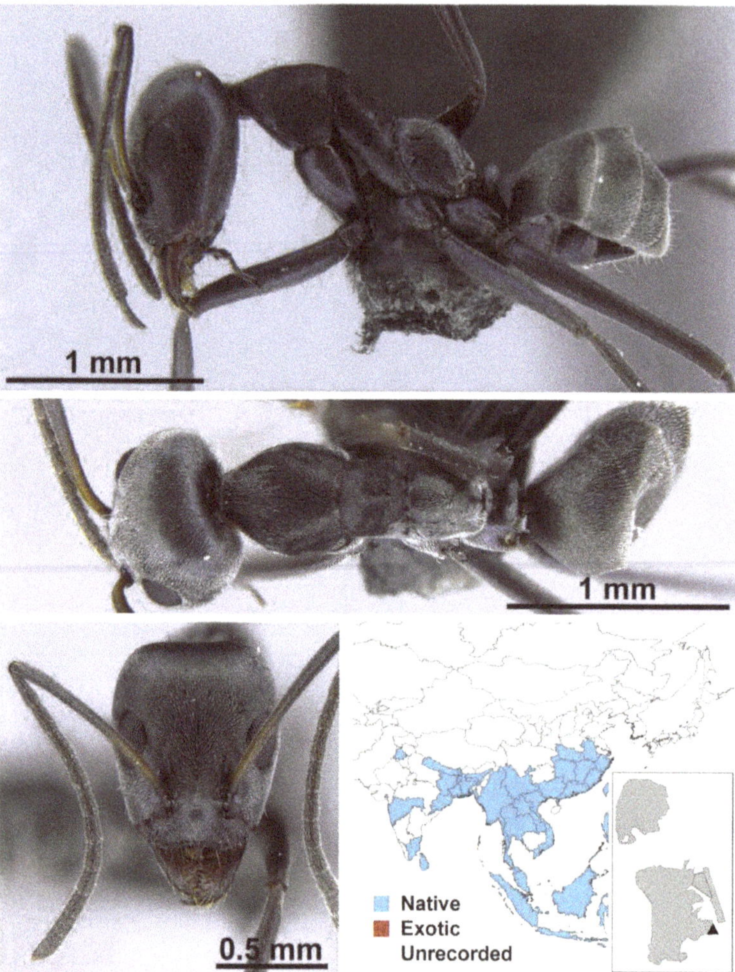

Figure A5. *Iridomyrmex anceps* grp. Roger, 1863 worker (FB19166, IBBL).

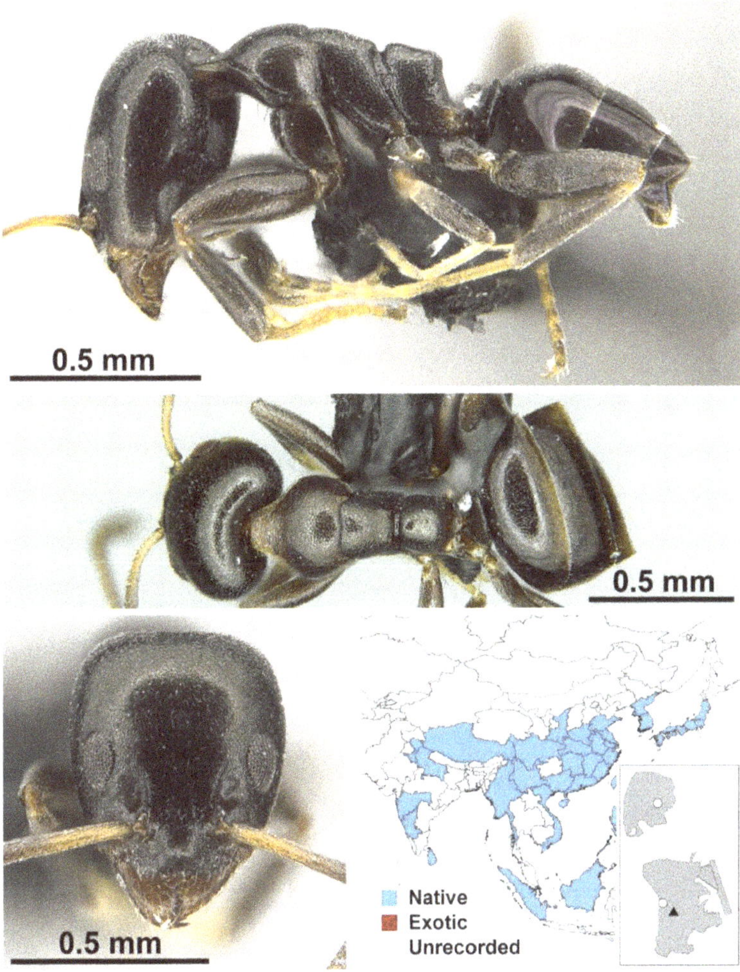

Figure A6. *Ochetellus glaber* Mayr, 1862 worker (MAC_S19_LLSP_sp.5, IBBL).

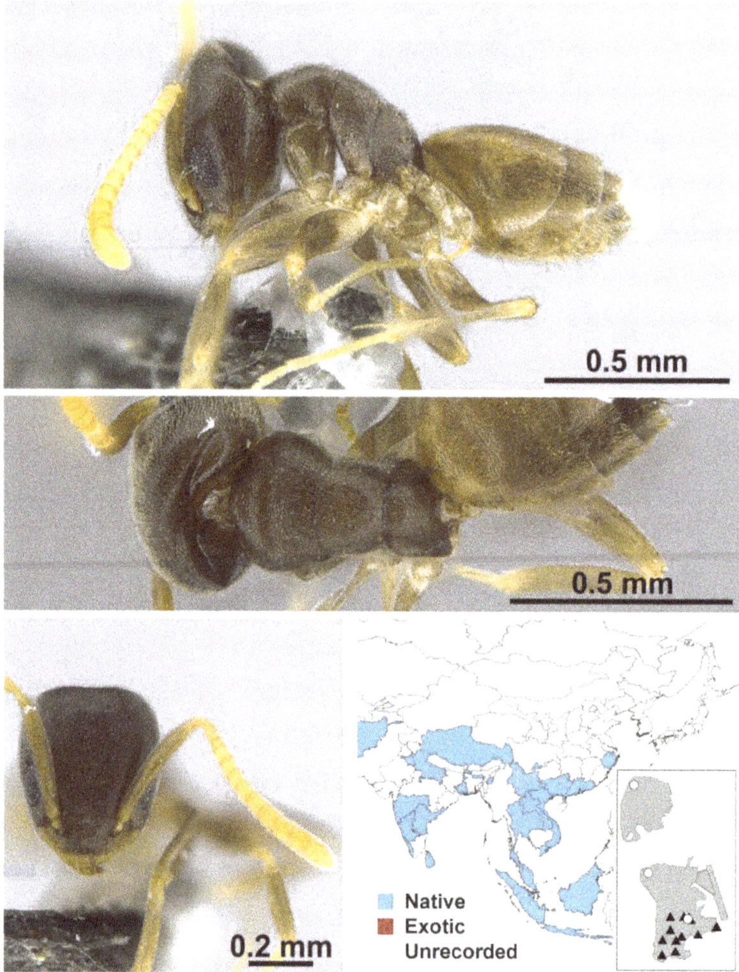

Figure A7. *Tapinoma indicum* Forel, 1895 worker (MAC_S01_LLSP_Sp.9, IBBL).

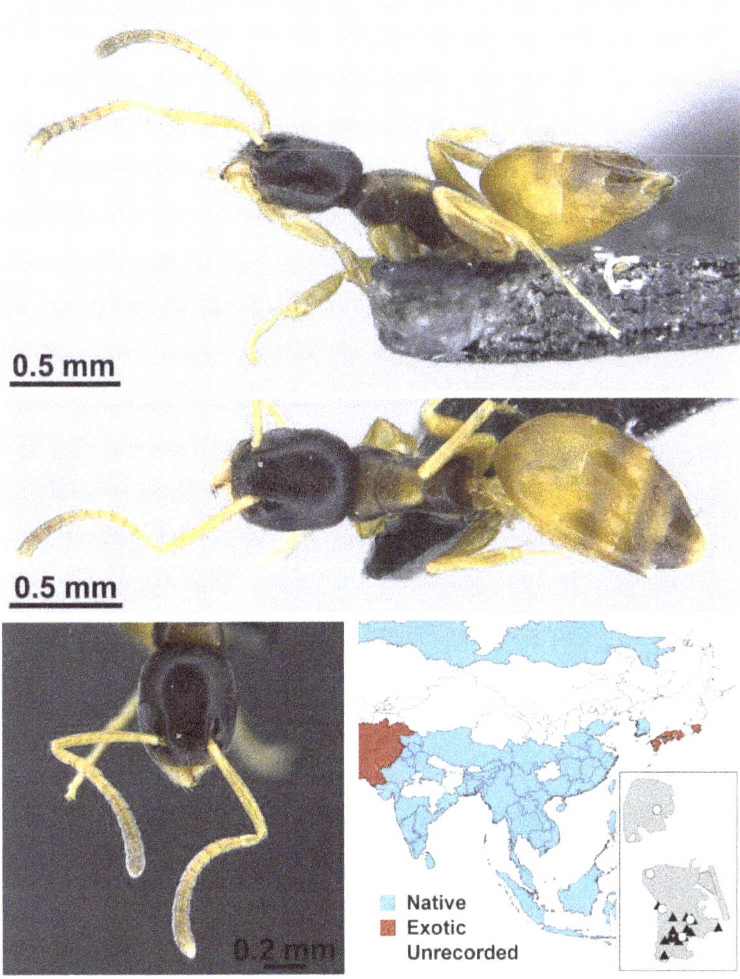

Figure A8. *Tapinoma melanocephalum* Fabricius, 1793 worker (MAC_S01_LLSP_Sp.6, IBBL).

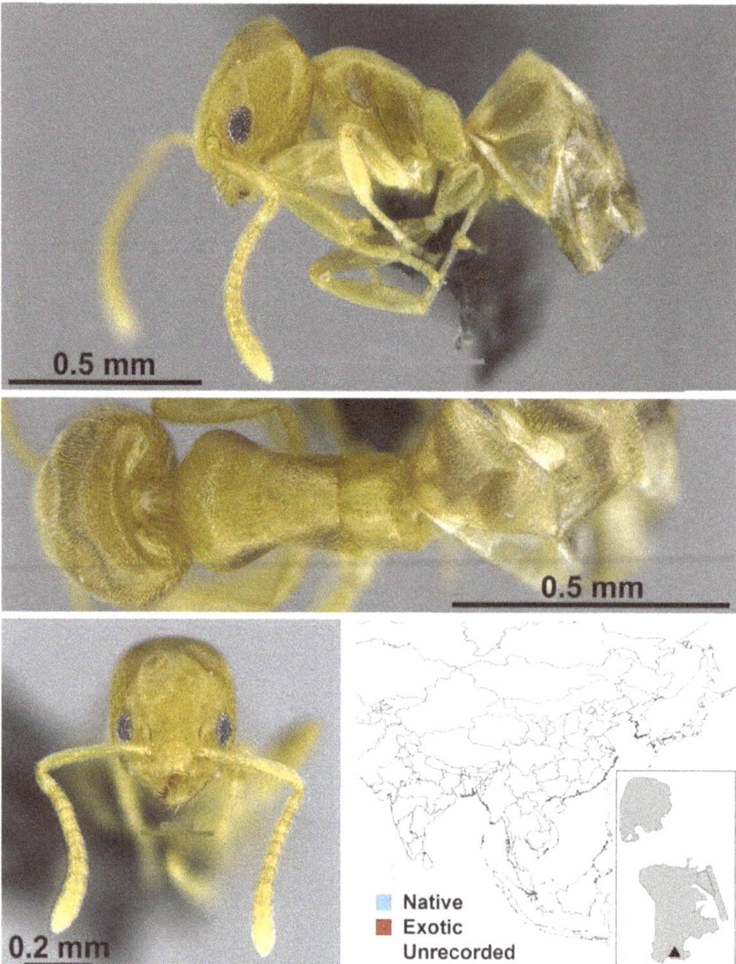

Figure A9. *Tapinoma* sp. 1 FB worker (MAC_S11_T4_1m_sp.2, IBBL).

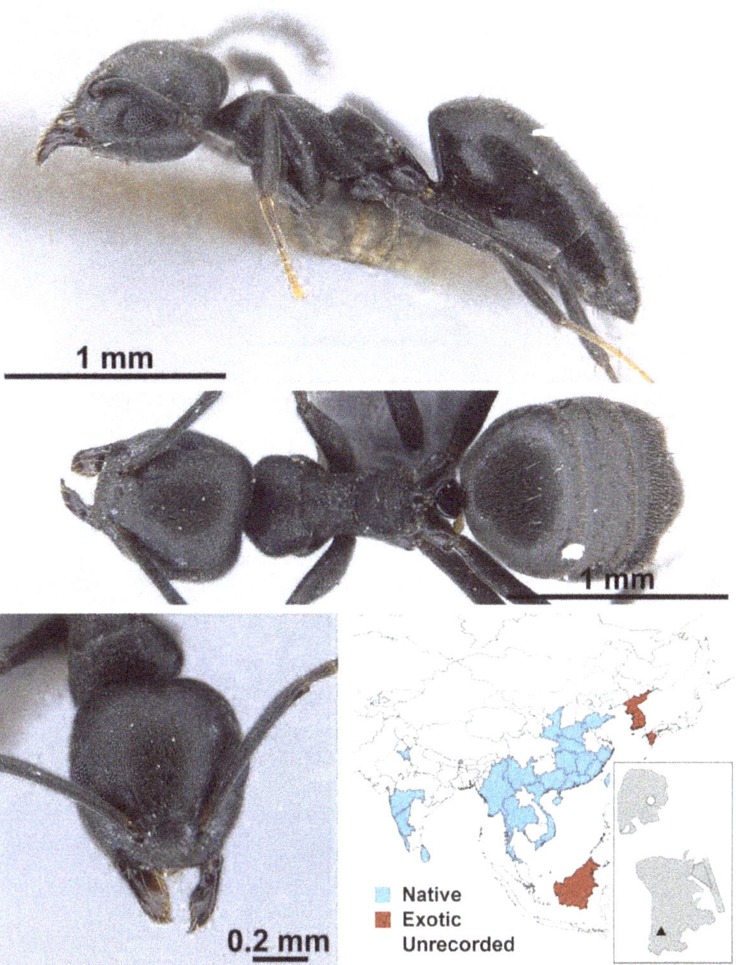

Figure A10. *Technomyrmex brunneus* Forel, 1895 worker (FB19281, IBBL).

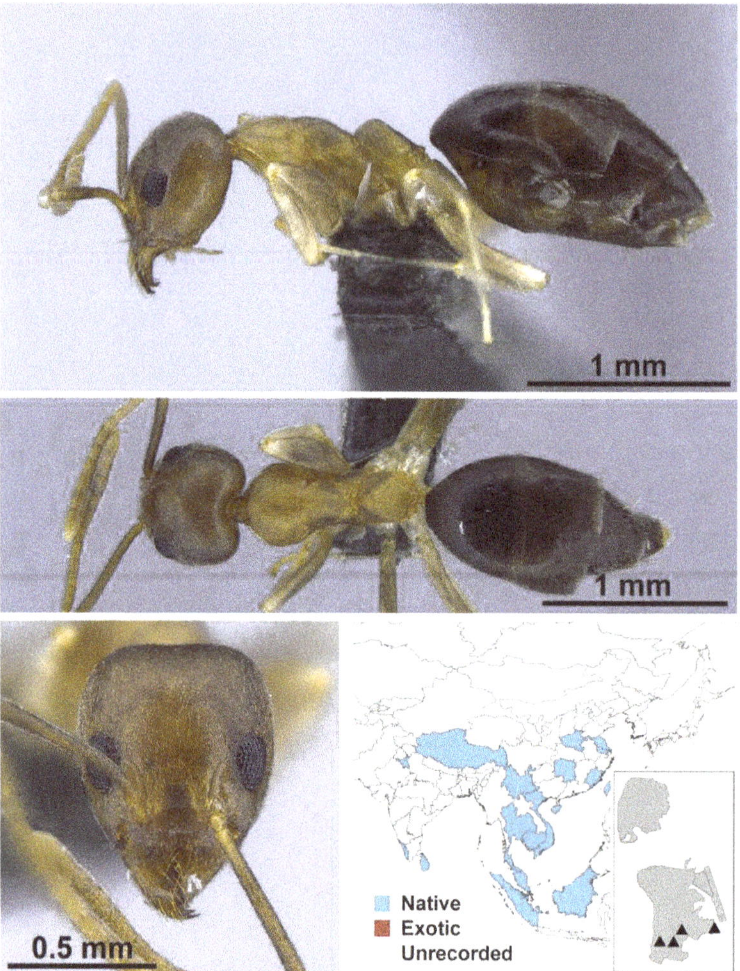

Figure A11. *Technomyrmex horni* Forel, 1912 worker (MAC_S01_LLSP_Sp.2, IBBL).

Appendix A.3 DORYLINAE

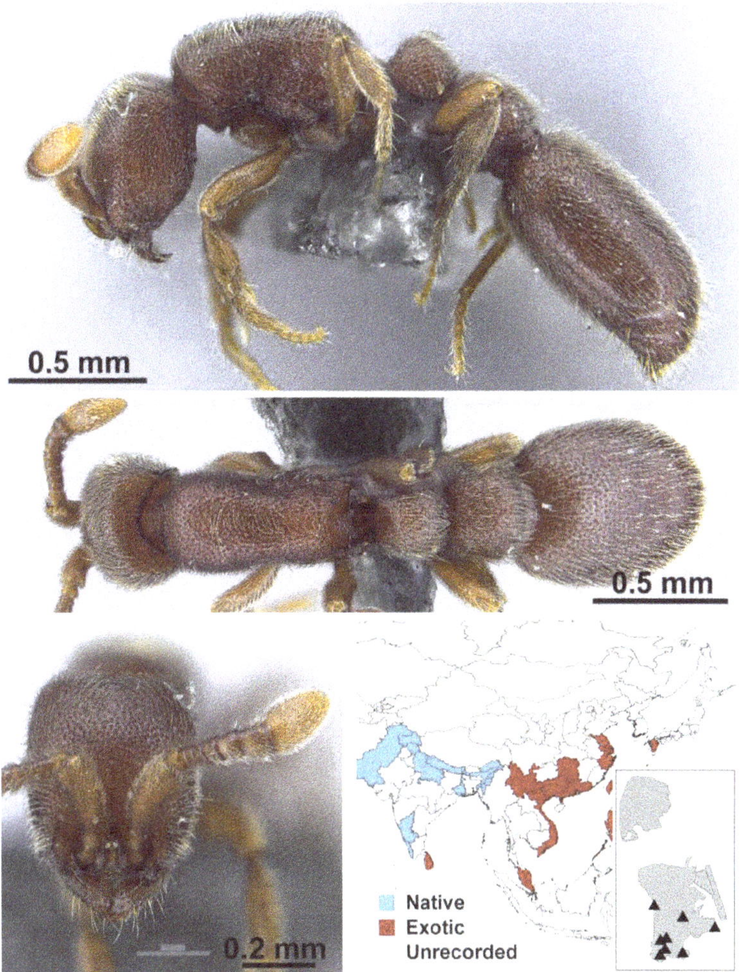

Figure A12. *Ooceraea biroi* Forel, 1907 worker (MAC_S02_LLSP_Sp.4, IBBL).

Appendix A.4 FORMICINAE

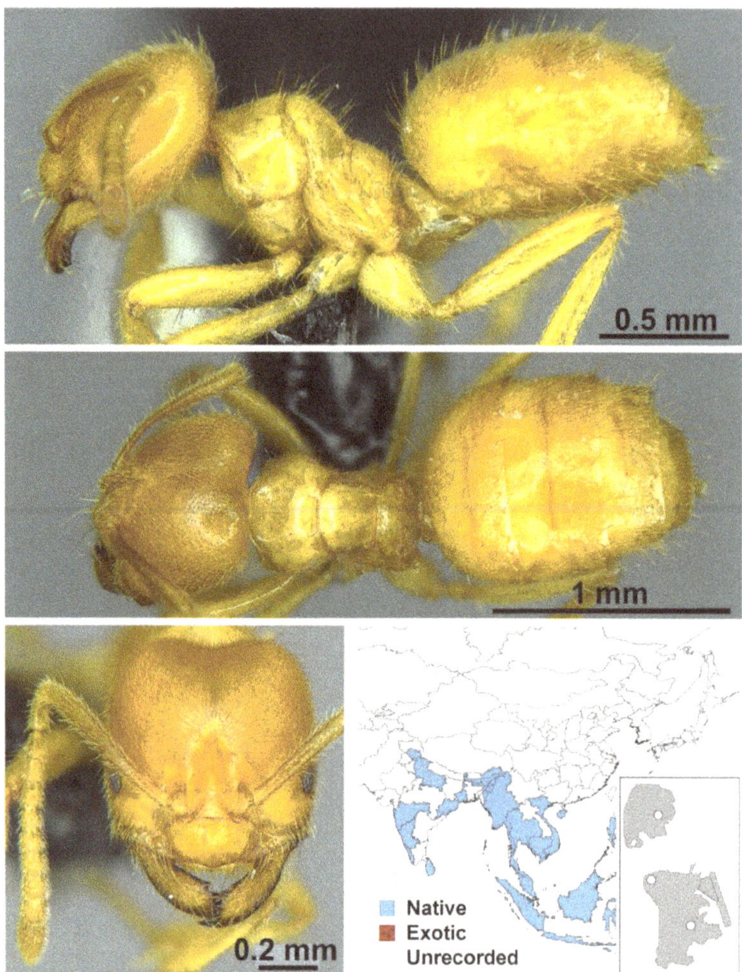

Figure A13. *Acropyga acutiventris* Roger, 1962 worker (*Acropyga acutiventris*, CML collection).

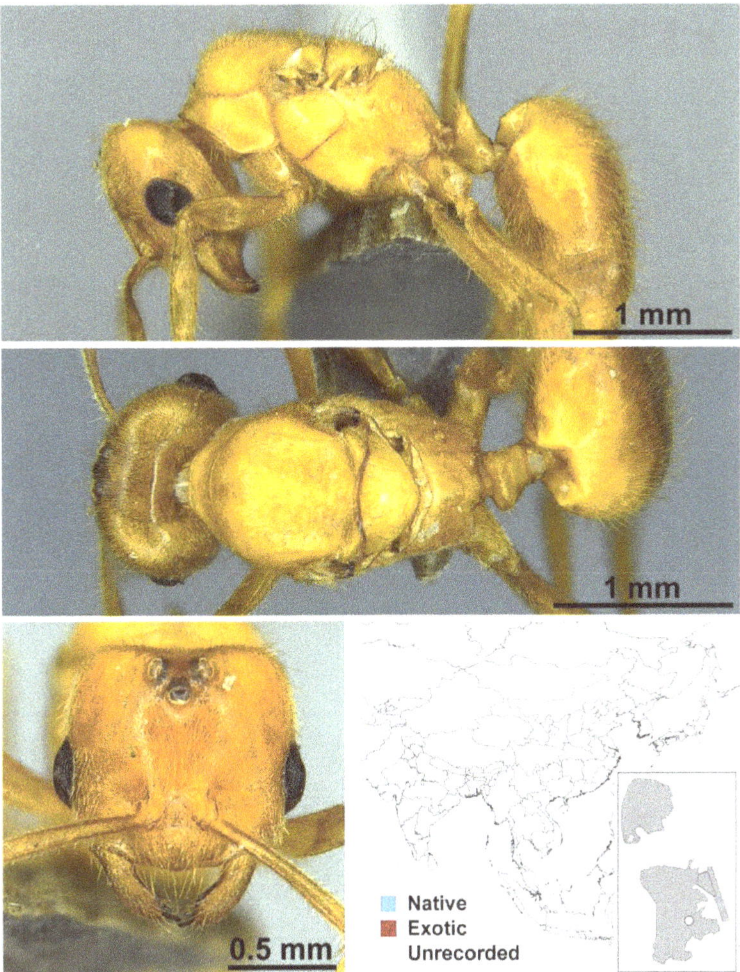

Figure A14. *Acropyga* sp. mo02 gyne (*Acropyga* sp. mo02, CML collection).

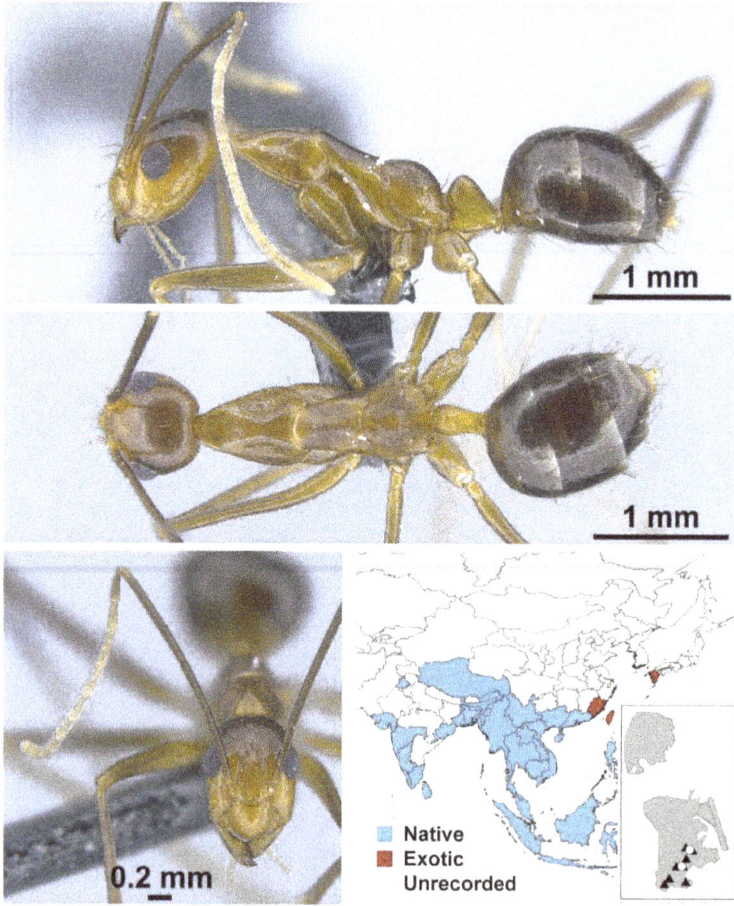

Figure A15. *Anoplolepis gracilipes* Smith, 1857 worker (MAC_S14_LLSA_Sp.14, IBBL).

Figure A16. *Brachymyrmex patagonicus* Mayr, 1868 worker (FB19202, IBBL).

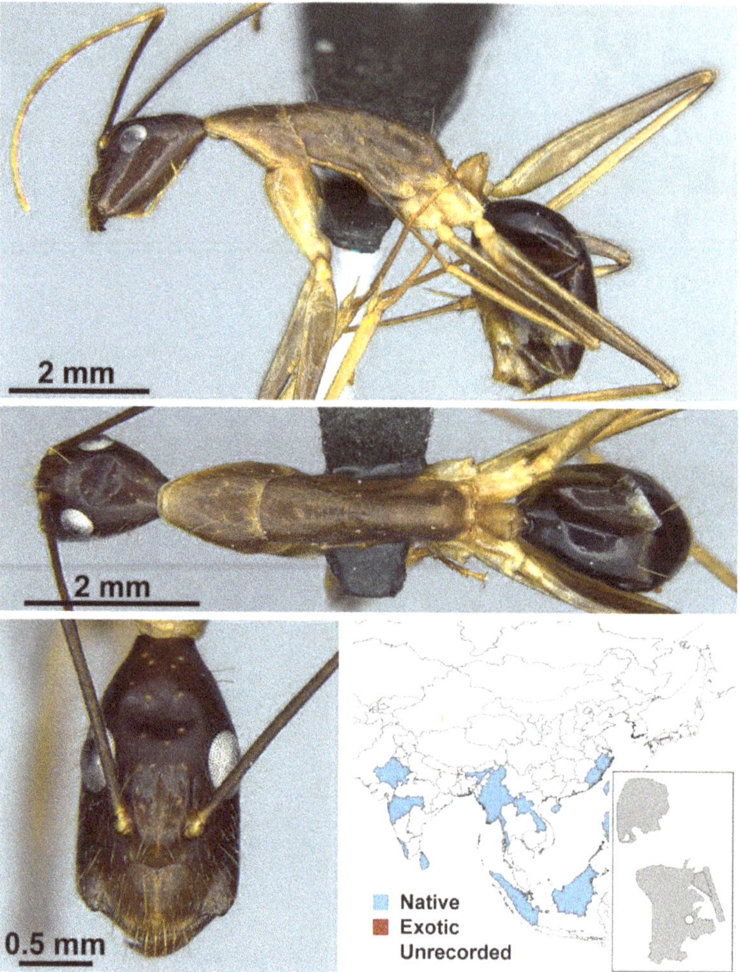

Figure A17. *Camponotus carin* Emery, 1889 worker (*Camponotus carin*, CML collection).

Figure A18. *Camponotus* nr. *irritans* Smith, F., 1857 worker (MAC_S20_B08, IBBL).

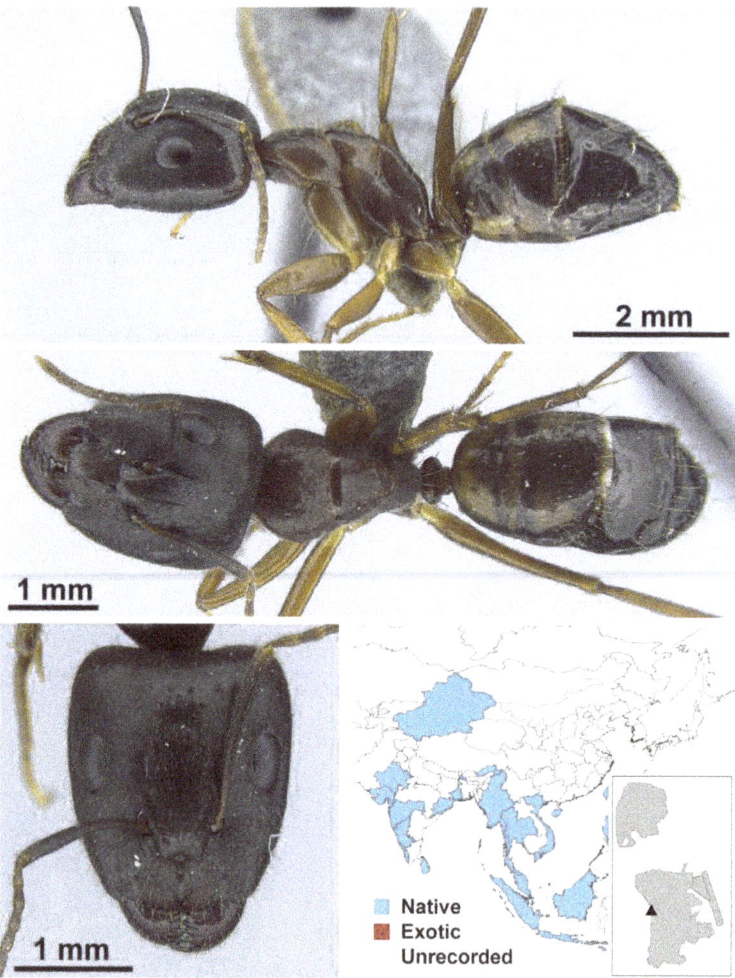

Figure A19. *Camponotus* nr. *irritans* Smith, F., 1857 major (MAC_S20_B08, IBBL).

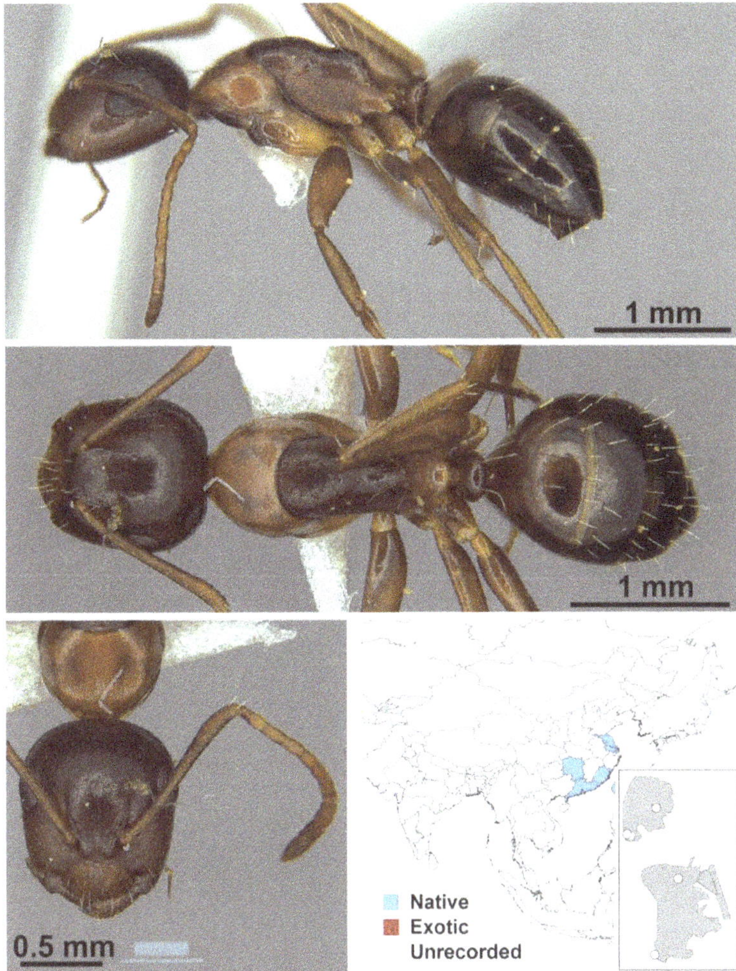

Figure A20. *Camponotus lighti* Wheeler, 1927 worker (*Camponotus lighti*, CML collection).

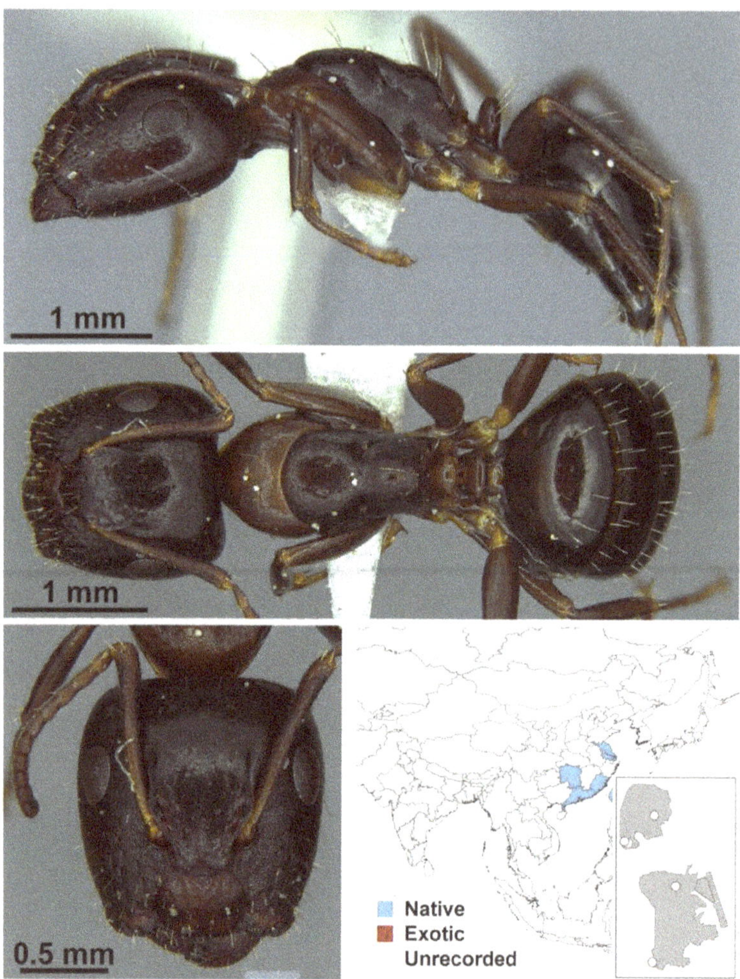

Figure A21. *Camponotus lighti* Wheeler, 1927 major (*Camponotus lighti*, CML collection).

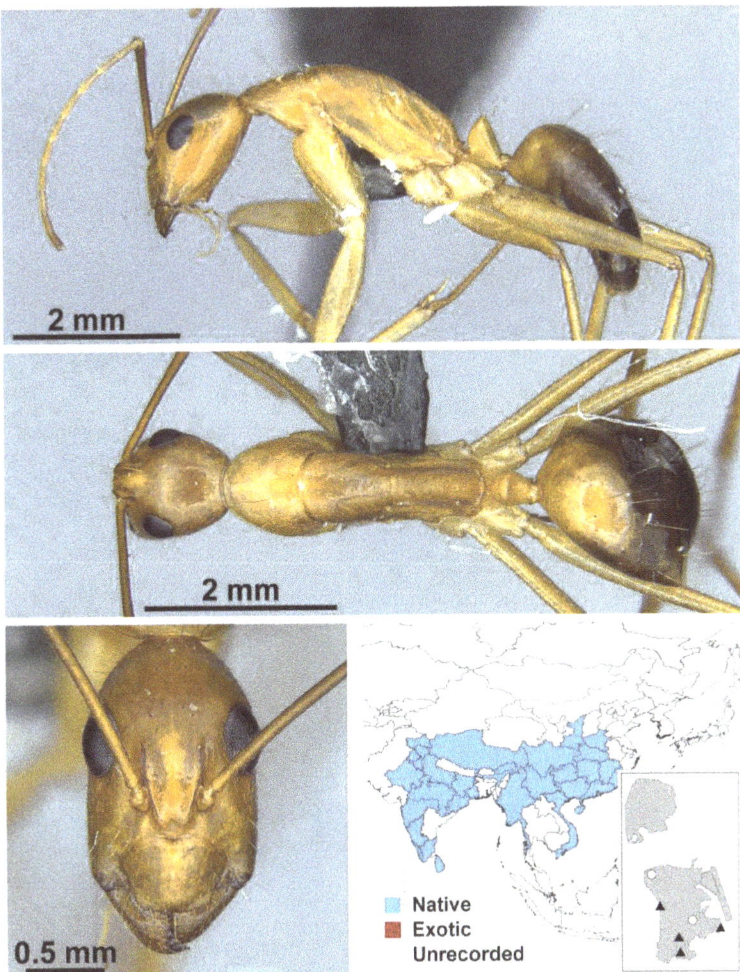

Figure A22. *Camponotus mitis* Smith, 1858 worker (MAC_S07_LLSA_sp.1, IBBL).

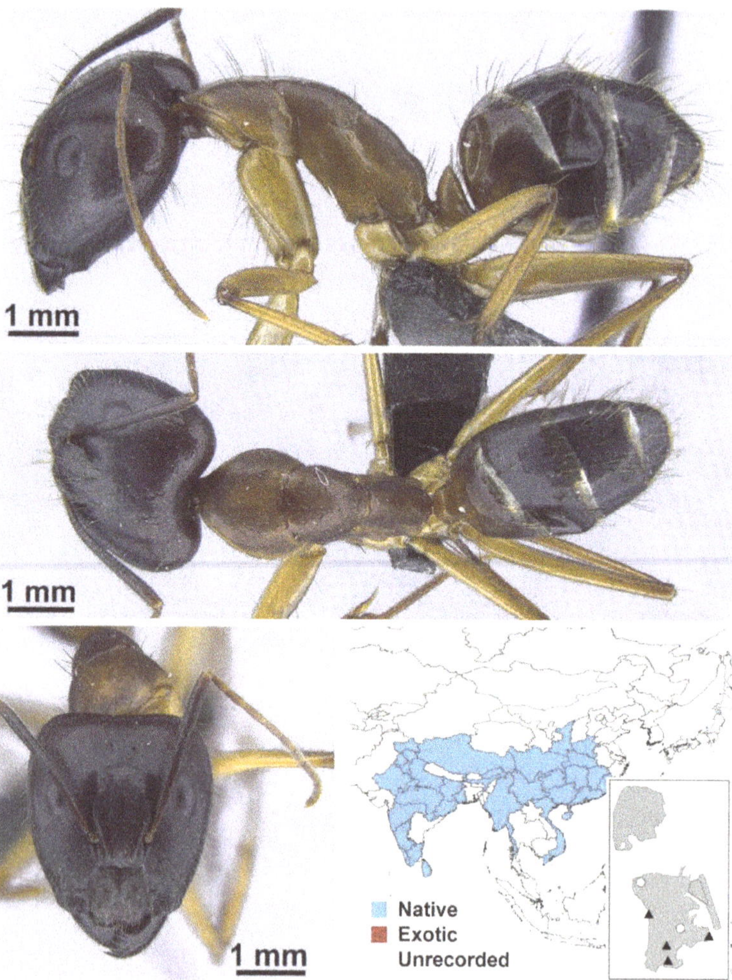

Figure A23. *Camponotus mitis* Smith, 1858 major (MAC_S11_LLSP_Sp.12, IBBL).

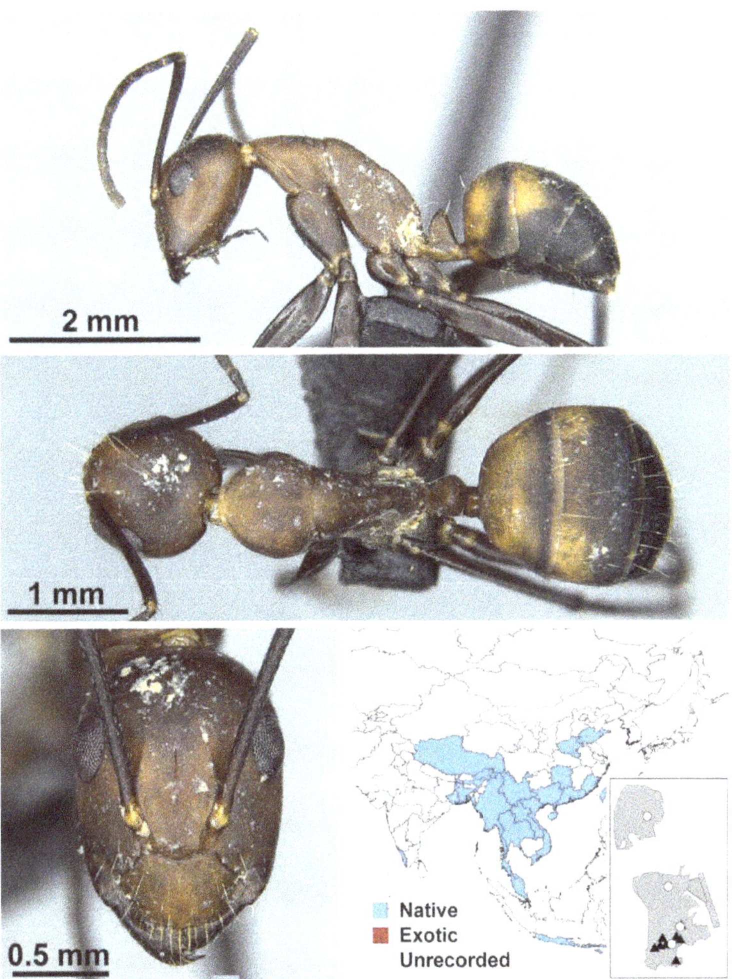

Figure A24. *Camponotus nicobarensis* Mayr, 1865 worker (MAC_S21_LLSP, IBBL).

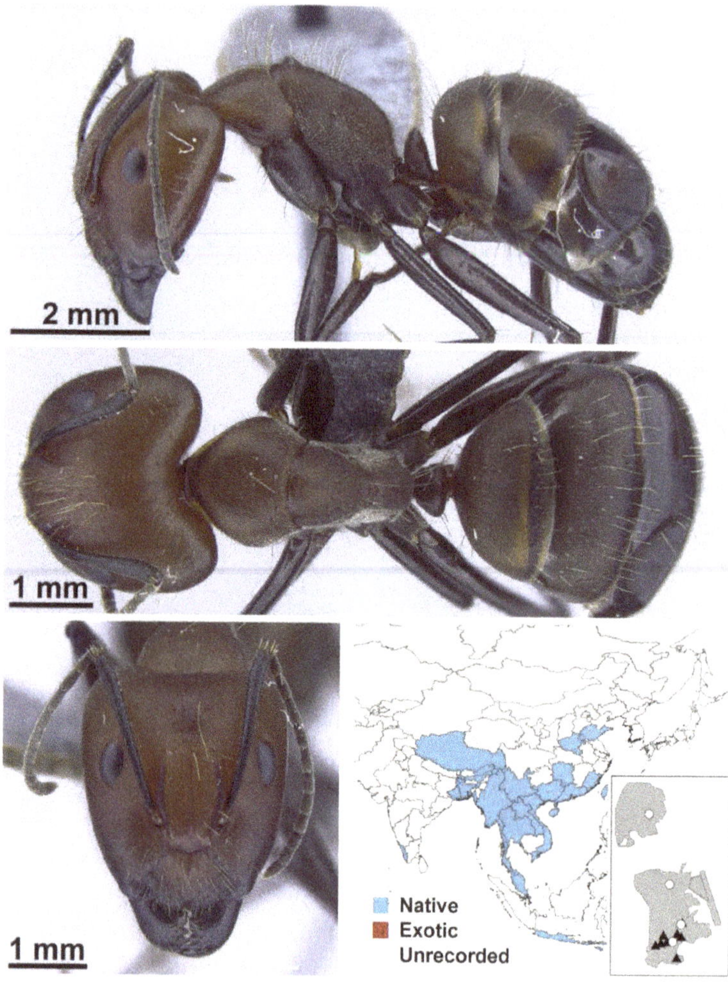

Figure A25. *Camponotus nicobarensis* Mayr, 1865 major (MAC_S16_T1_5m_sp.3, IBBL).

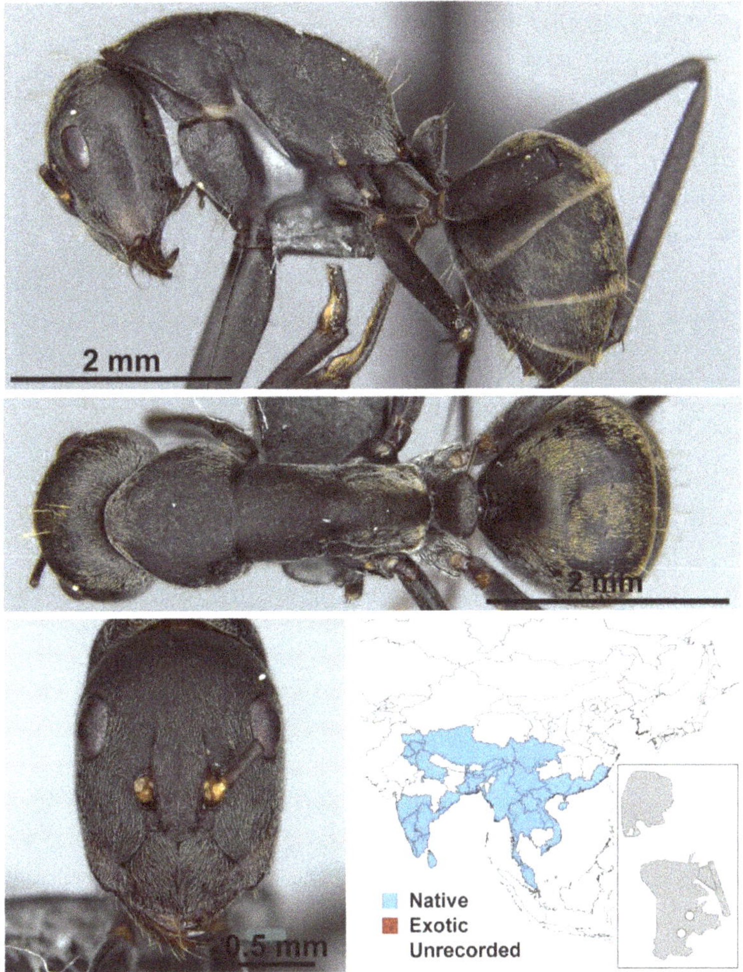

Figure A26. *Camponotus parius* Emery, 1889 worker (*Camponotus parius*, CML collection).

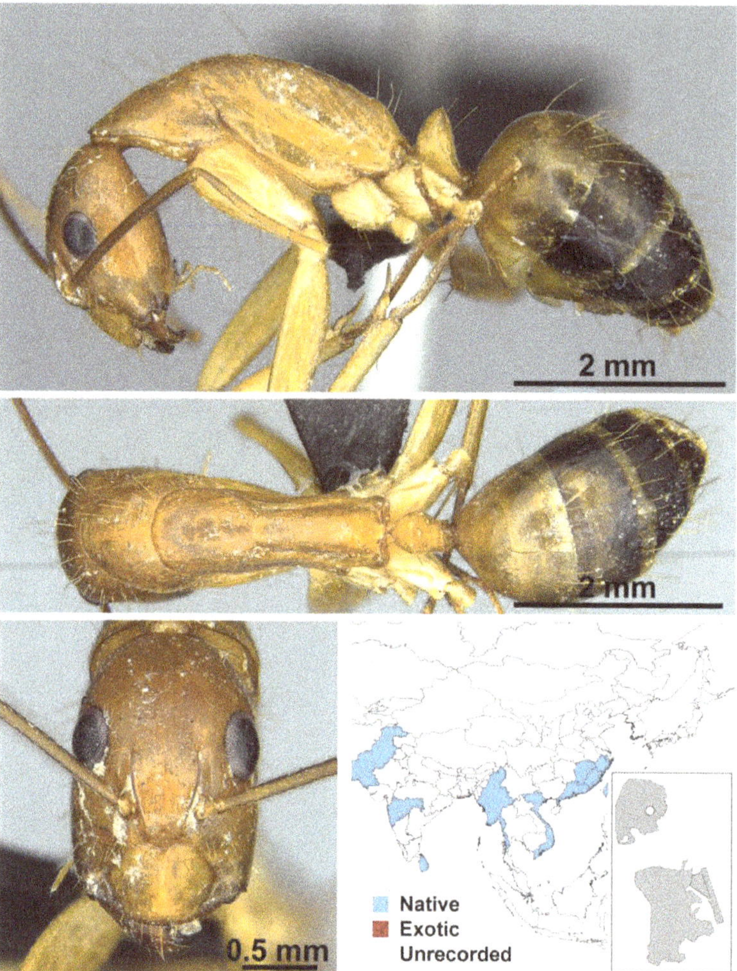

Figure A27. *Camponotus variegatus* dulcis Dalla Torre, 1893 worker (*Camponotus variegatus* dulcis, CML collection).

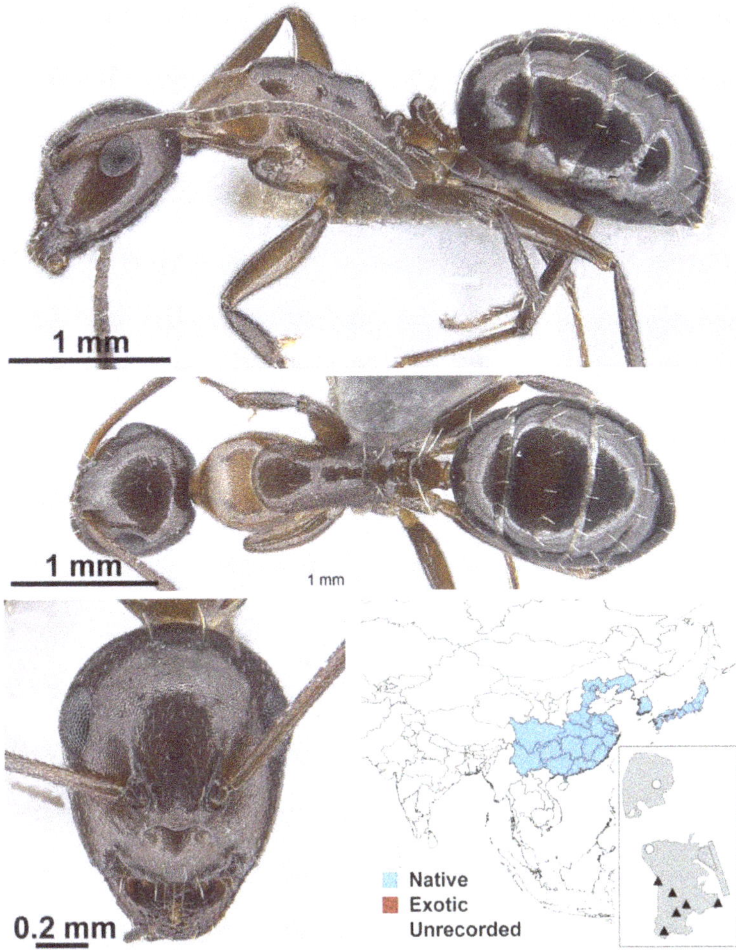

Figure A28. *Camponotus vitiosus* Smith, 1874 worker (MAC_S20_T4_2m_Sp.1, IBBL).

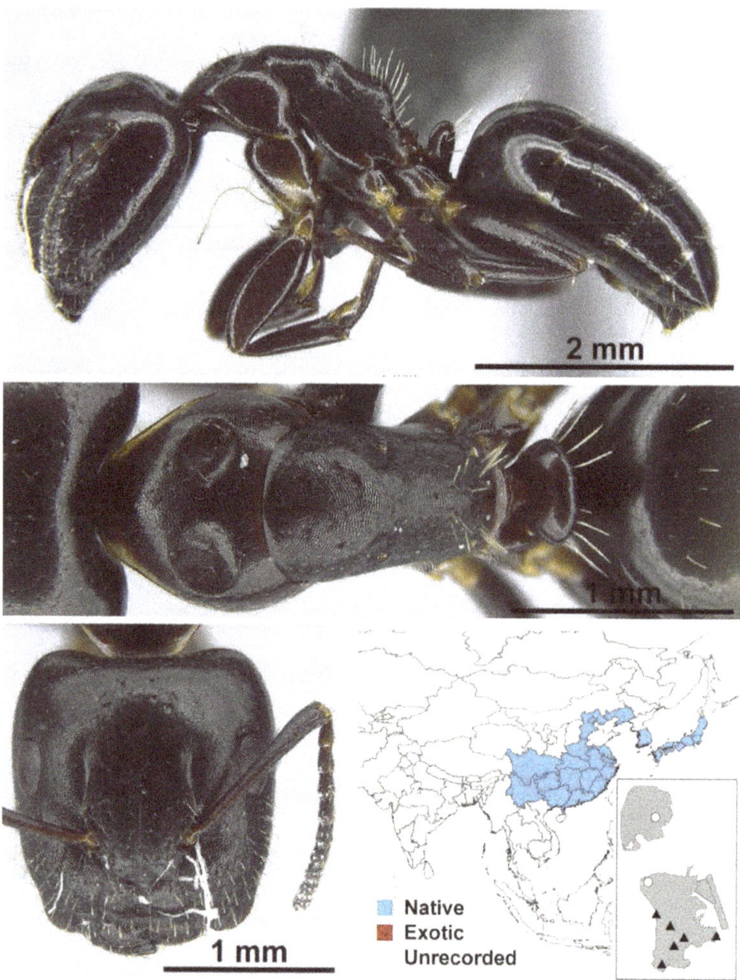

Figure A29. *Camponotus vitiosus* Smith, 1874 major (MAC_S14_T1_3m_Sp.1, IBBL).

Figure A30. *Camponotus* sp.1 FB worker (MAC_FB19182, IBBL).

Figure A31. *Colobopsis* nr. *nipponica* Wheeler, 1928 worker (*Colobopsis* nr. *nipponica*, CML collection).

Figure A32. *Colobopsis* nr. *vitrea* Smith, 1860 worker (*Colobopsis* nr. *vitrea*, CML collection).

Figure A33. *Colobopsis* nr. *vitrea* Smith, 1860 major (FB19268, IBBL).

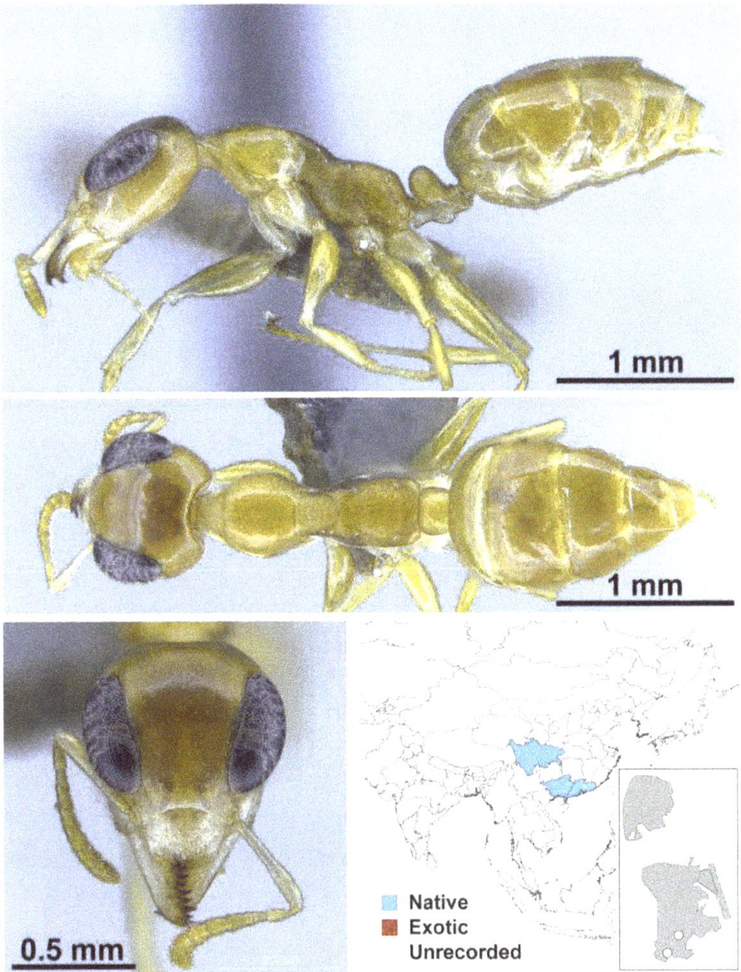

Figure A34. *Gesomyrmex howardi* Wheeler, W. M., 1921 worker (MWong_MaiPo_7viii2018, IBBL).

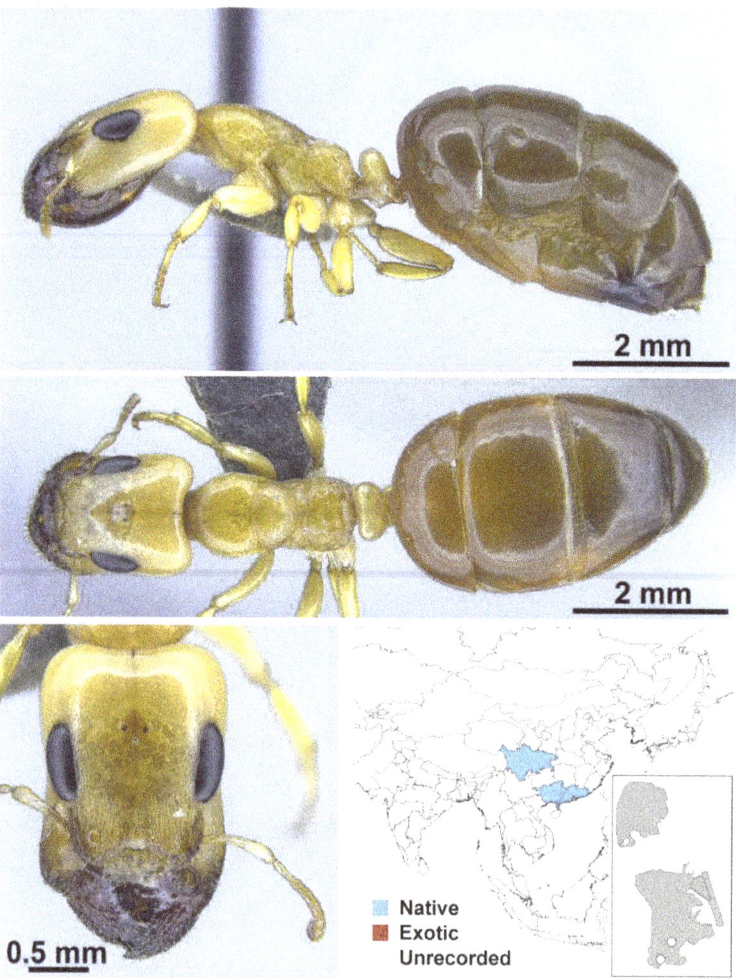

Figure A35. *Gesomyrmex howardi* Wheeler, W. M., 1921 supermajor (MWong_MaiPo_7viii2018_Colony4_6_2.6x12.5, IBBL).

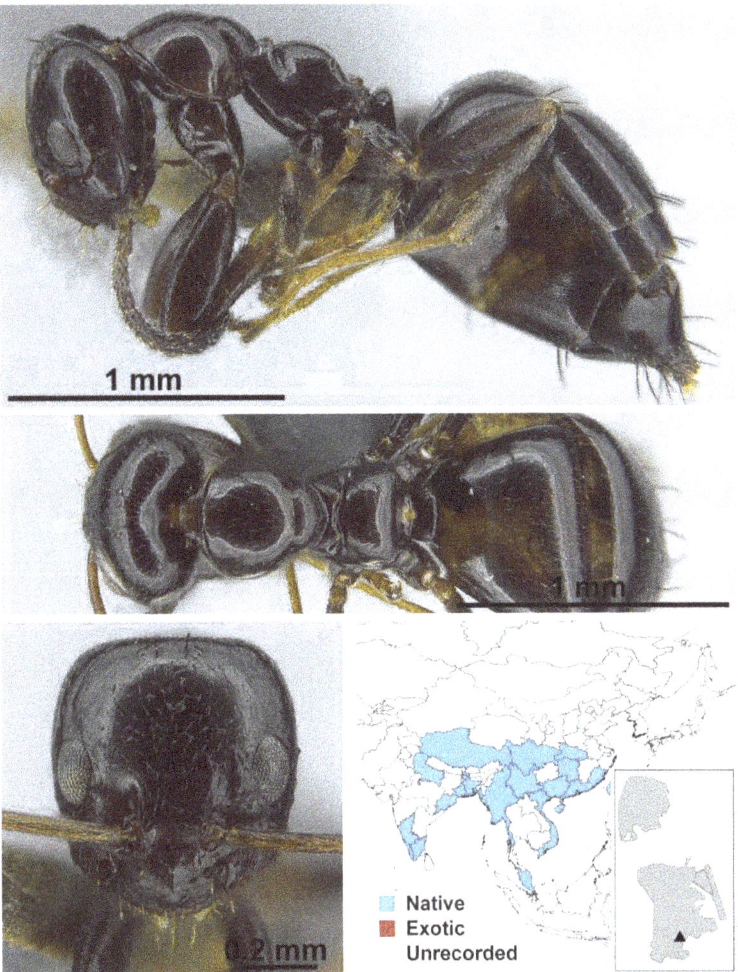

Figure A36. *Lepisiota rothneyi* Forel, 1894 worker (MAC_S06_B08_Sp.1_top, IBBL).

Figure A37. *Nylanderia amia* Forel, 1913 worker (ANTWEB1016677, IBBL).

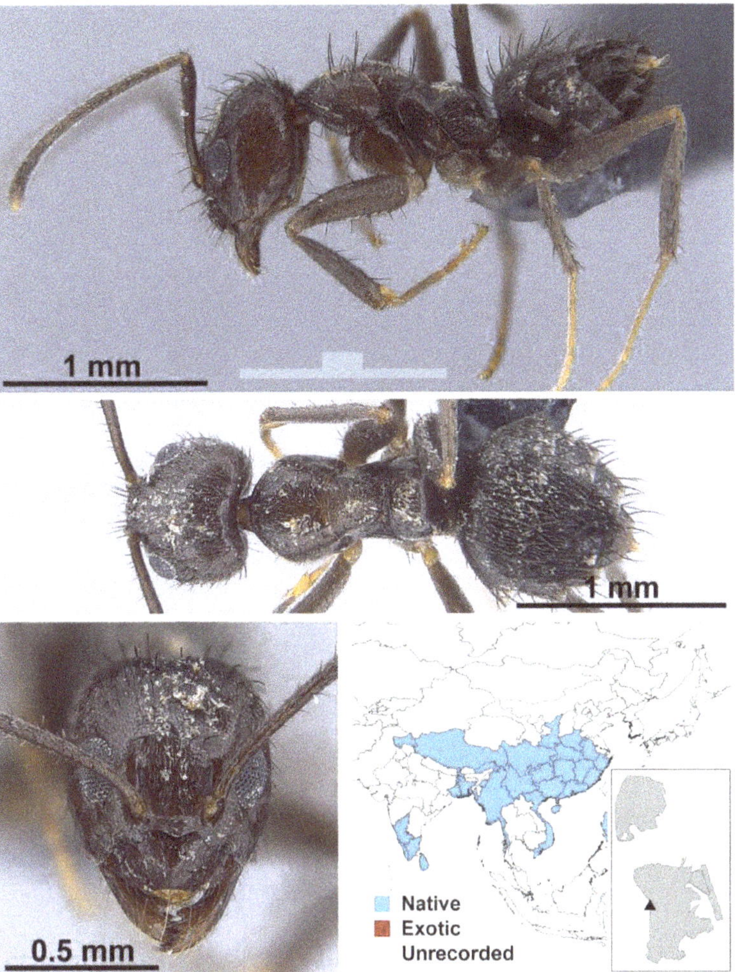

Figure A38. *Nylanderia bourbonica* Forel, 1886 worker (MAC_S20_LLSP_Sp.4, IBBL).

Figure A39. *Nylanderia indica* Forel, 1894 worker (MAC_S12_LLSP_sp.4, IBBL).

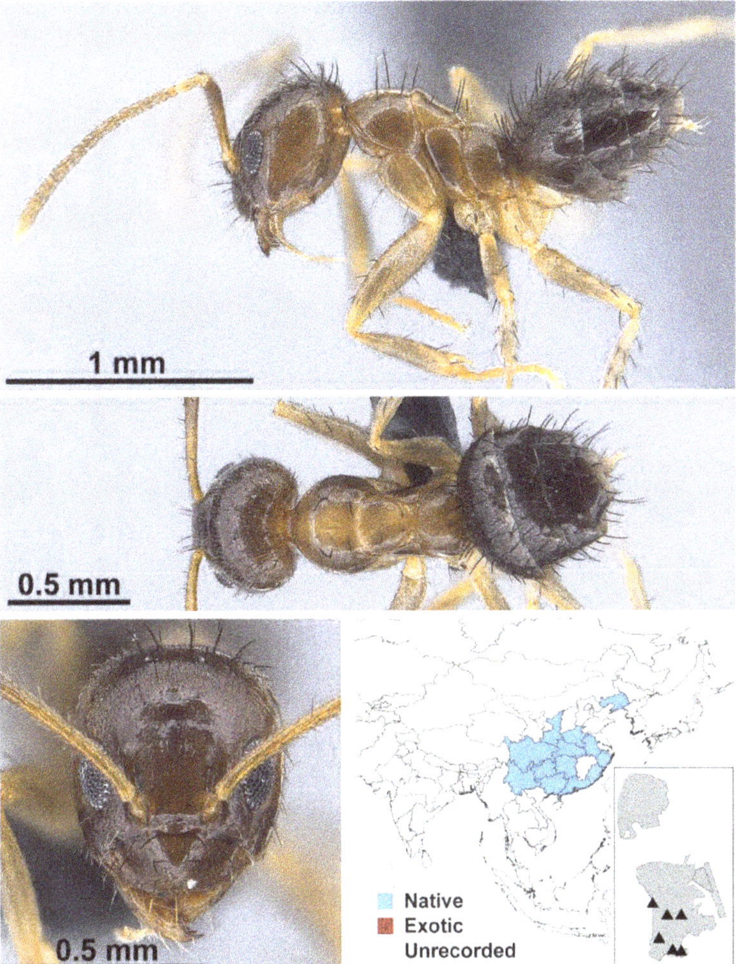

Figure A40. *Nylanderia sharpii* Forel, 1899 worker (MAC_S19_LLSA_Sp.5, IBBL).

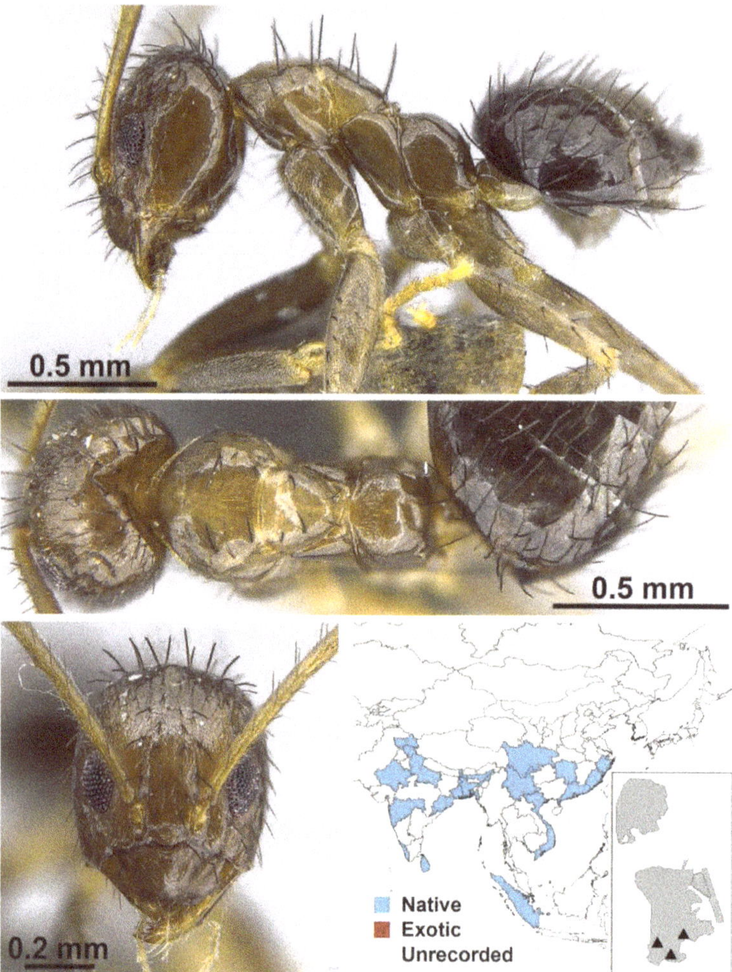

Figure A41. *Nylanderia taylori* Forel, 1894 worker (MAC_S21_GN1_H4_n1_bottom, IBBL).

Figure A42. *Nylanderia* sp. 3 BG worker (MAC_S15_B03_sp.3, IBBL).

Figure A43. *Nylanderia* sp. 6 BG worker (MAC_S15_LLSP_sp.10, IBBL).

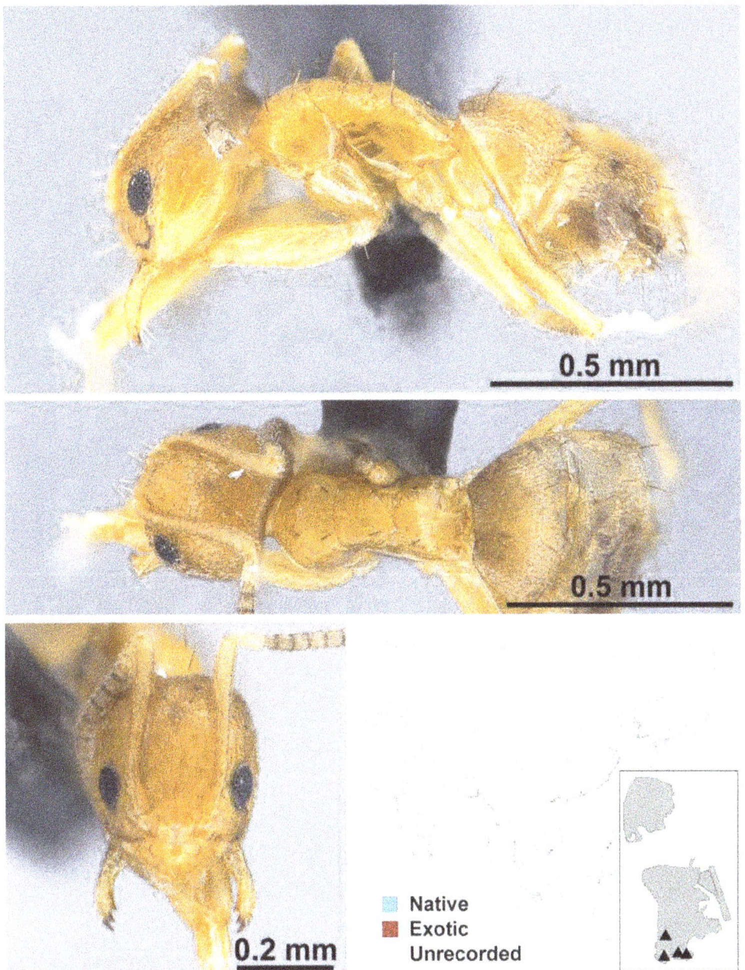

Figure A44. *Paraparatrechina* sp.1 BG Forel, 1913 worker (MAC_S18_q2_37.5_Sp.2, IBBL).

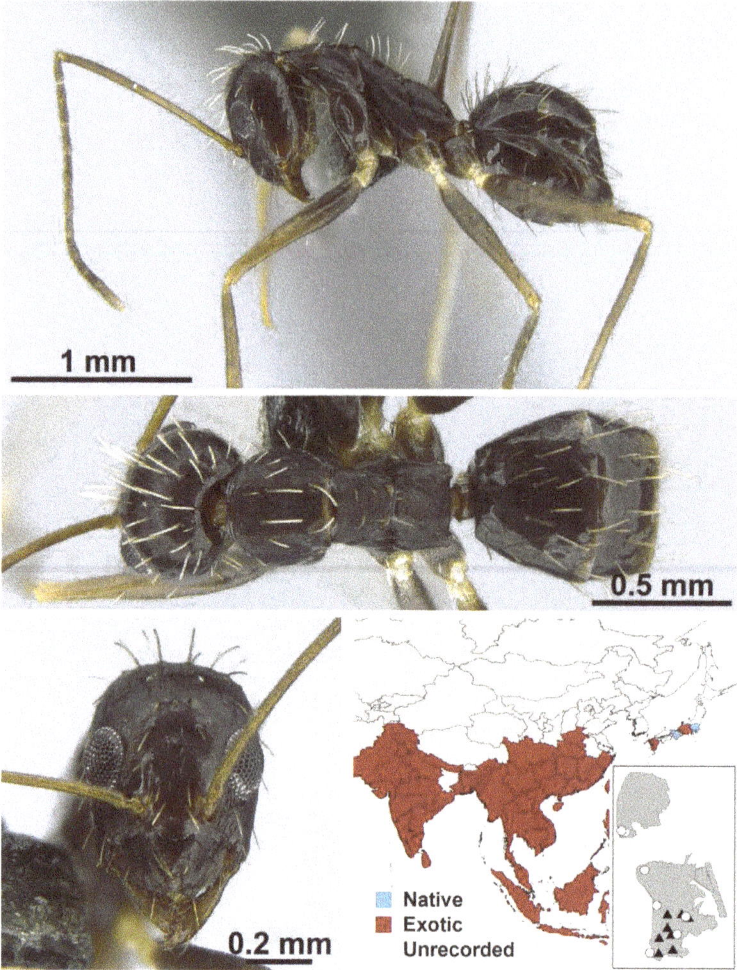

Figure A45. *Paratrechina longicornis* Latreille, 1802 worker (MAC_S11_LLSP_Sp.6, IBBL).

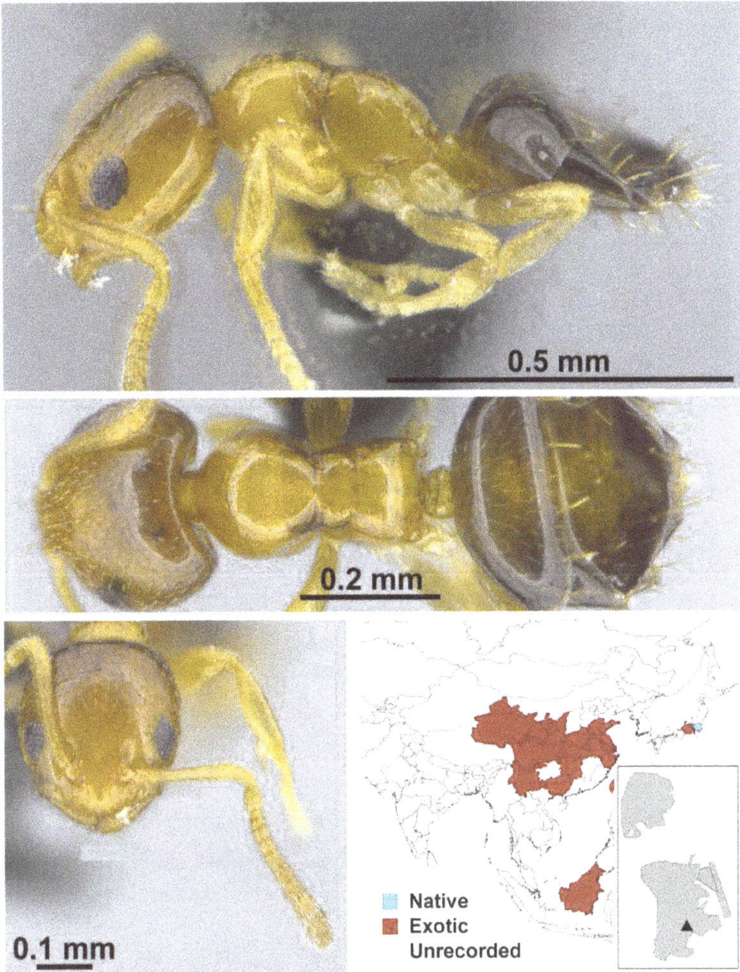

Figure A46. *Plagiolepis alluaudi* Emery, 1894 worker (MAC_S14_T2_2m_Sp.1, IBBL).

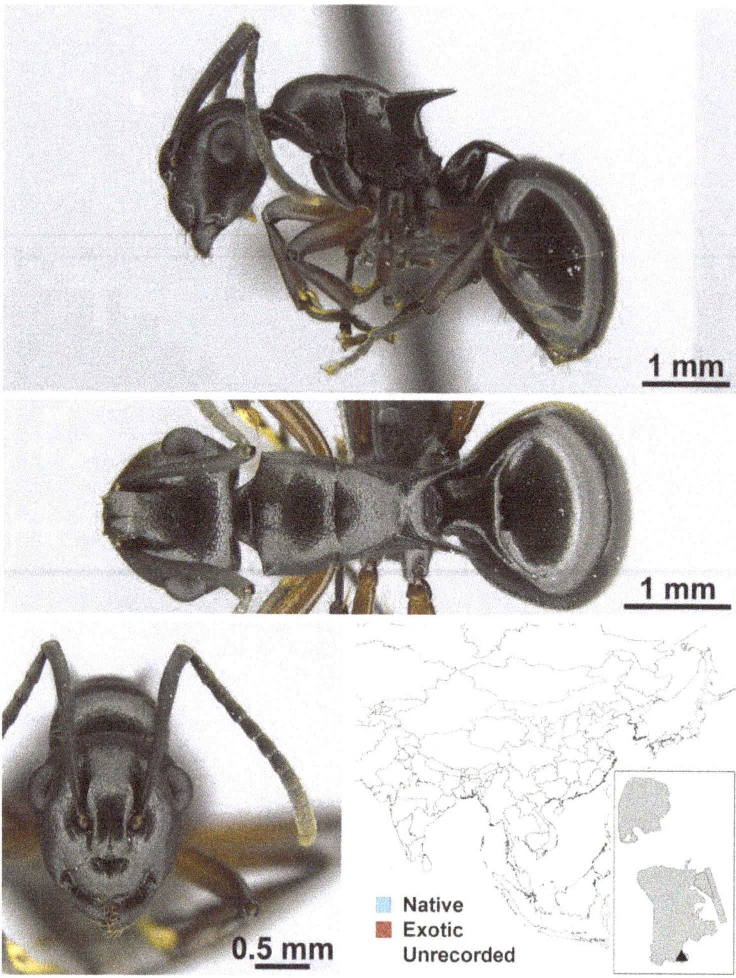

Figure A47. *Polyrhachis confusa* Emery, 1893 worker (FB19152, IBBL).

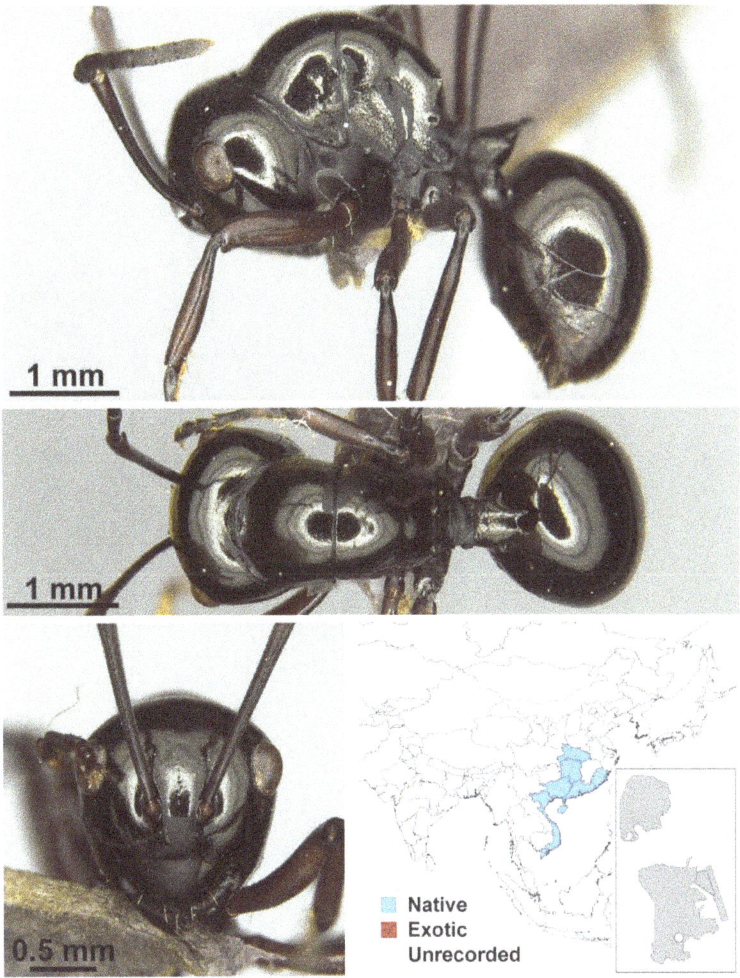

Figure A48. *Polyrhachis demangei* Santschi, 1910 worker (*Polyrhachis demangei*, CML collection).

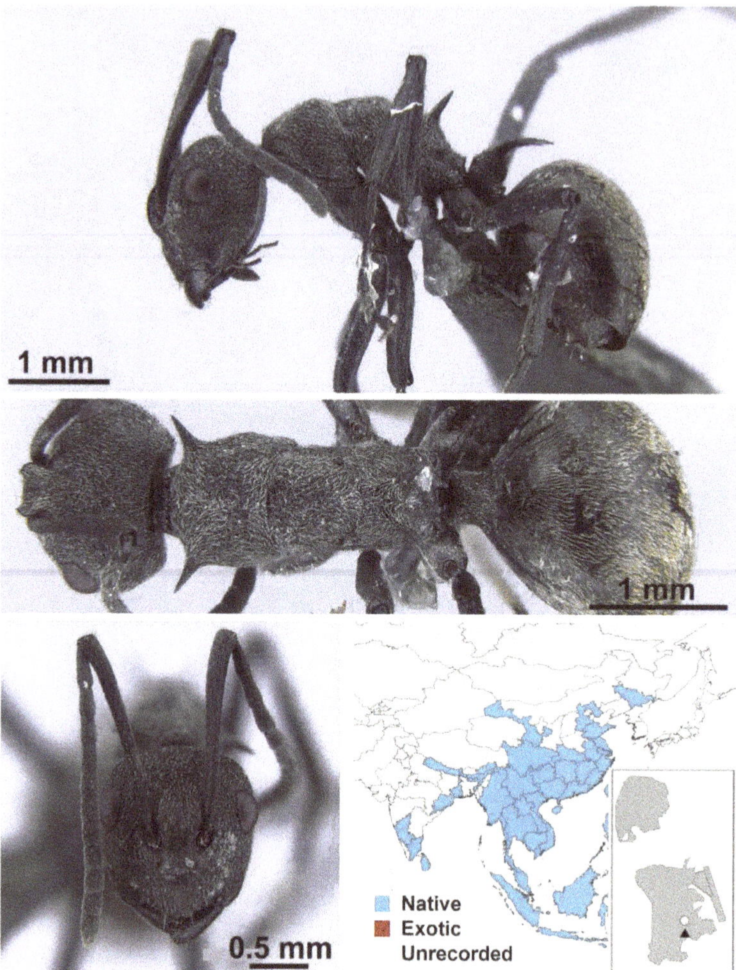

Figure A49. *Polyrhachis dives* Smith, 1857 worker (MAC_S03_HC_01_Sp.2, IBBL).

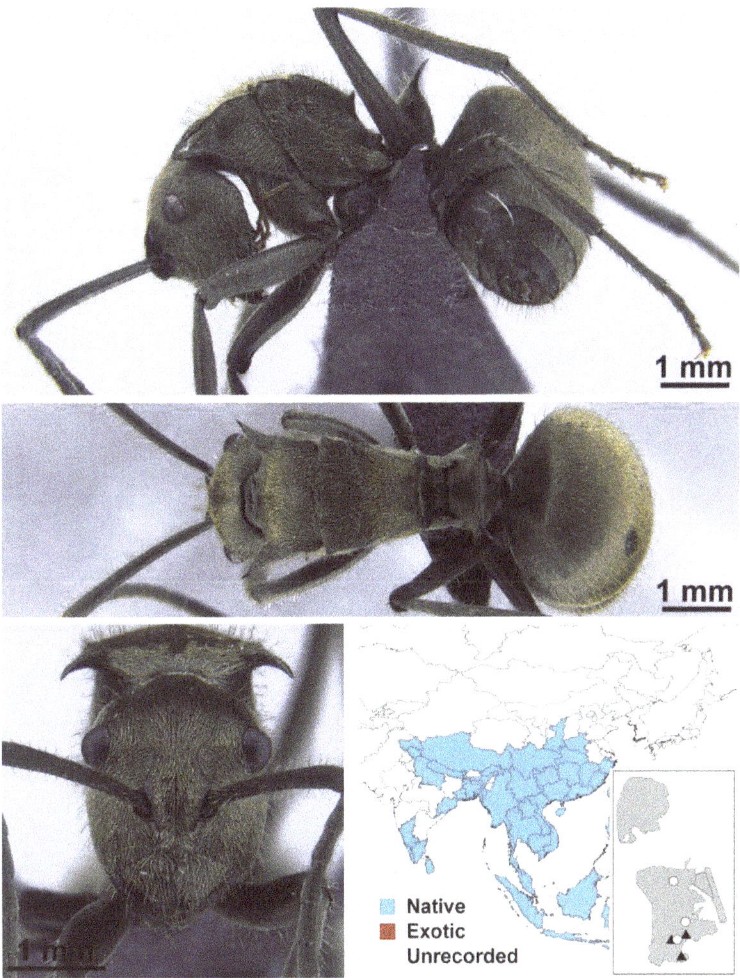

Figure A50. *Polyrhachis illaudata* Walker, 1859 worker (MAC_S03_HC_01_Sp.1, IBBL).

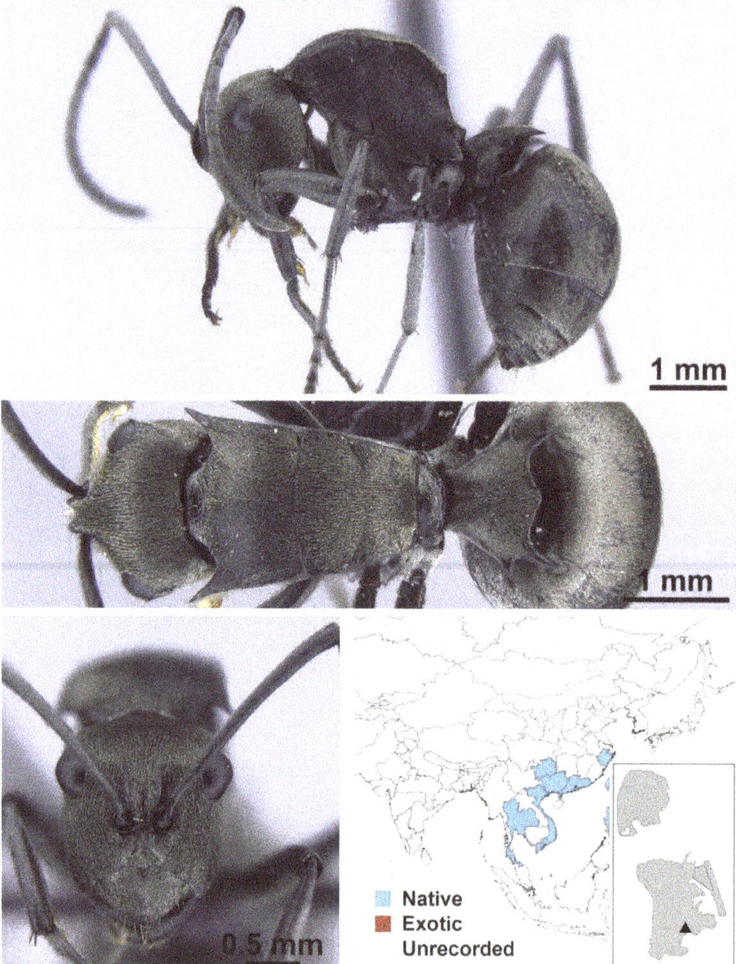

Figure A51. *Polyrhachis latona* Wheeler, 1909 worker (MAC_S03_LLSA_Sp.5, IBBL).

Figure A52. *Polyrhachis tyrannica* Smith, 1858 worker (K6558(2)). Species images taken from Wong and Guénard 2020 [55] with permission.

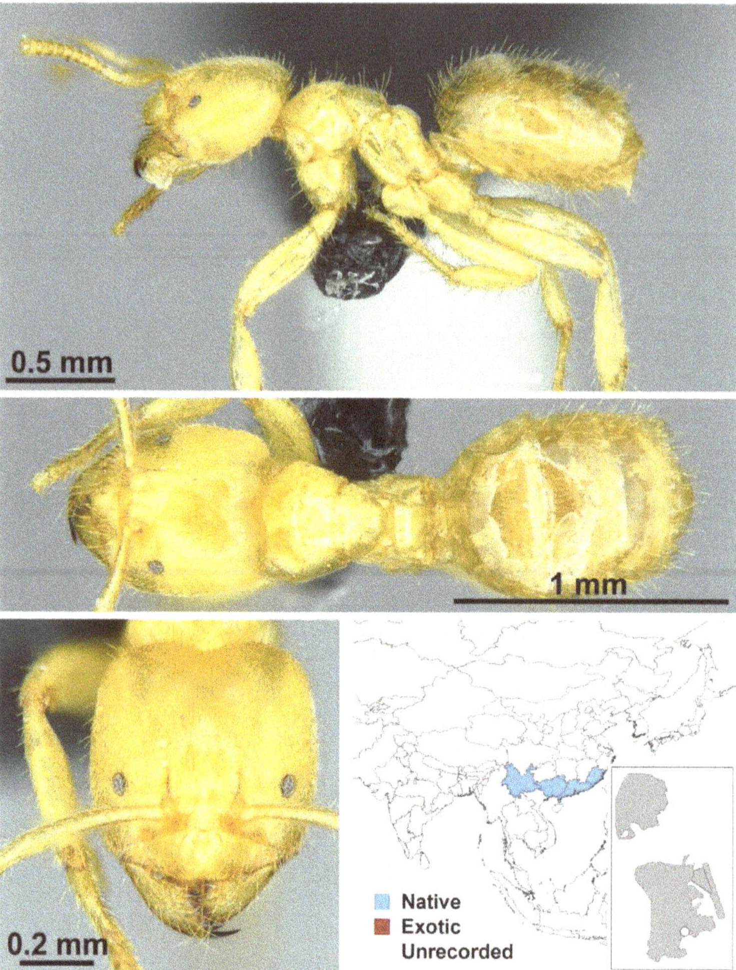

Figure A53. *Pseudolasius risii* Forel, 1894 worker (*Pseudolasius risii*, CML collection).

Appendix A.5 LEPTANILLINAE

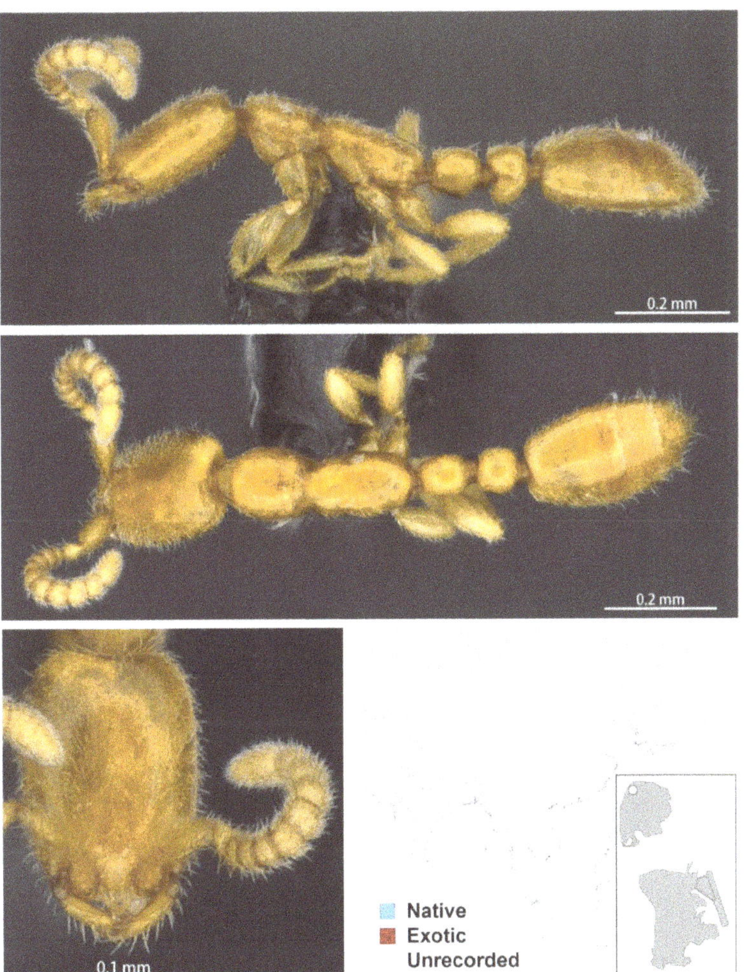

Figure A54. *Leptanilla macaoensis* Leong, Yamane & Guénard, 2018 worker (LCM00039, IBBL). Species images taken from Leong, Yamane & Guénard, 2018 [61] with permission.

Appendix A.6 MYRMICINAE

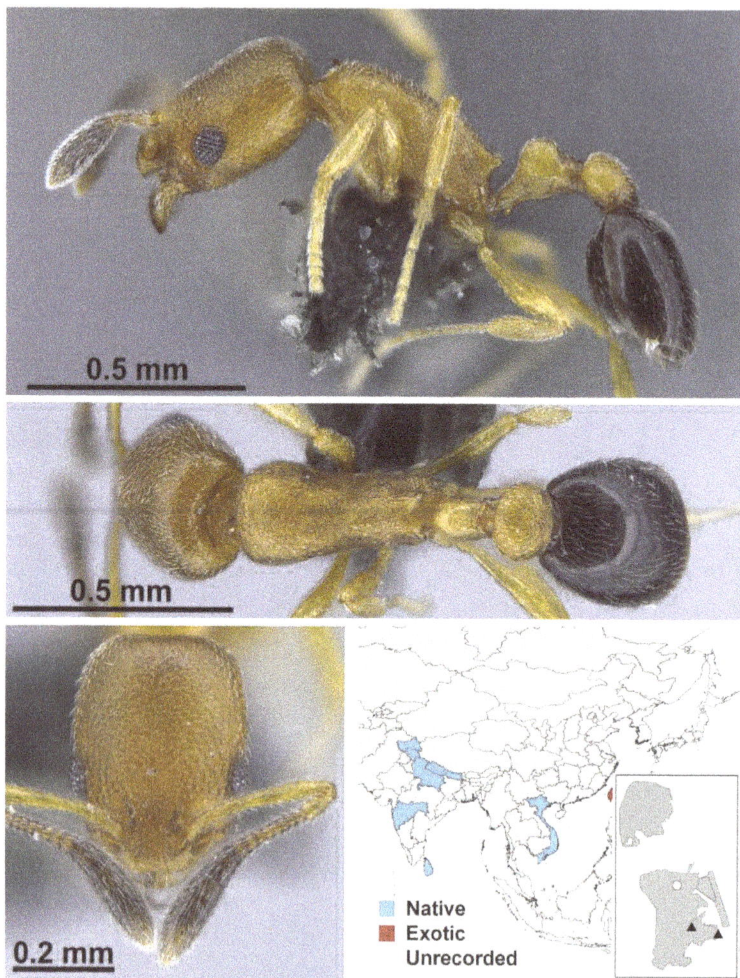

Figure A55. *Cardicondyla minutior* Forel, 1899 worker (MAC_S04_LLSP_sp.6, IBBL).

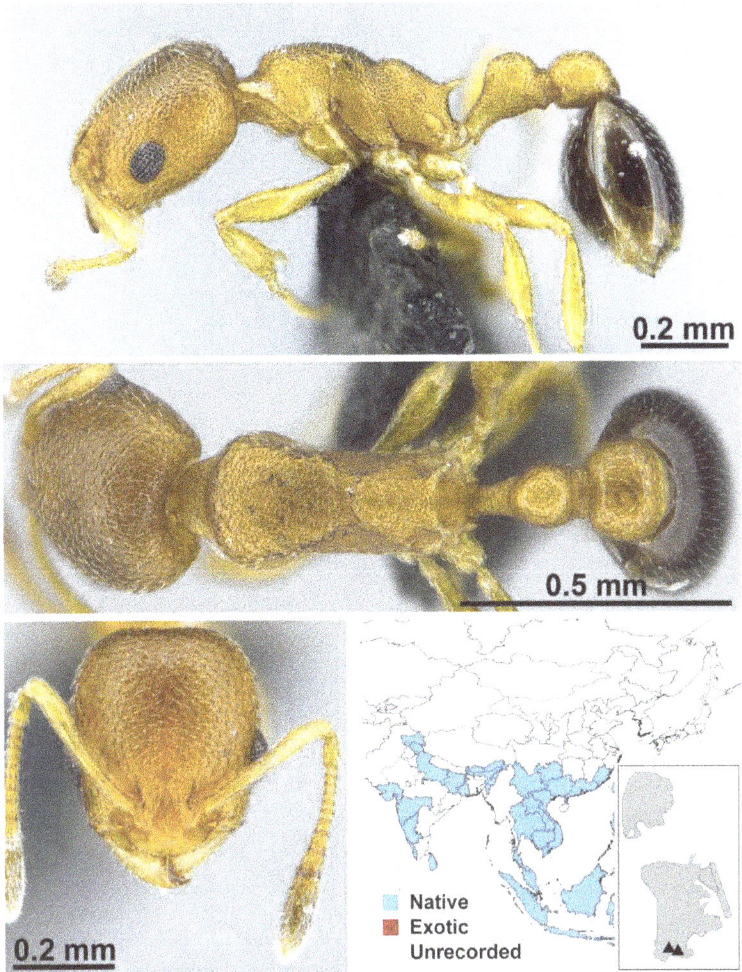

Figure A56. *Cardiocondyla wroughtonii* Forel, 1890 worker (MAC_S11_T3_3m_sp.4, IBBL).

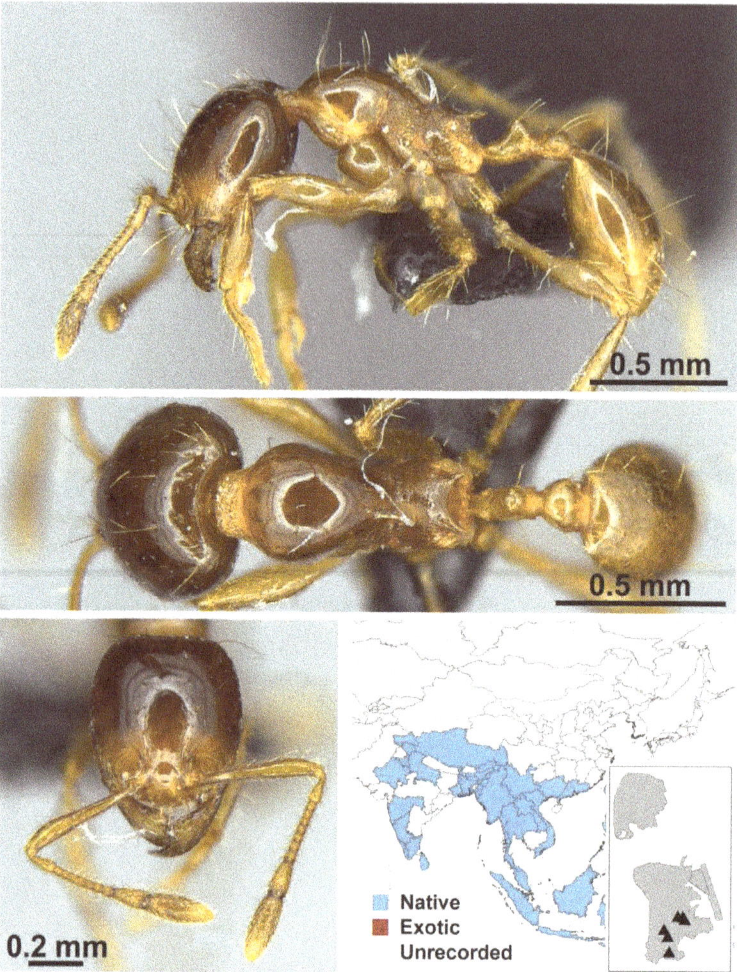

Figure A57. *Carebara affinis* Jerdon, 1851 worker (MAC_S9_37.5_q2_sp.1, IBBL).

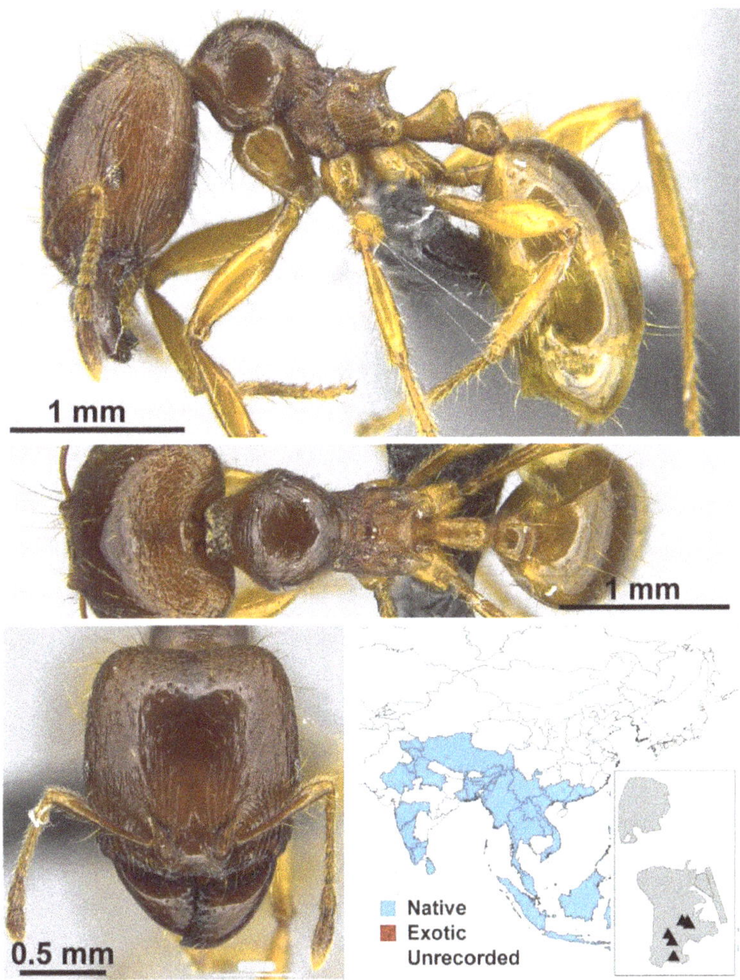

Figure A58. *Carebara affinis* Jerdon, 1851 major (MAC_S8_25_q1_sp.1, IBBL).

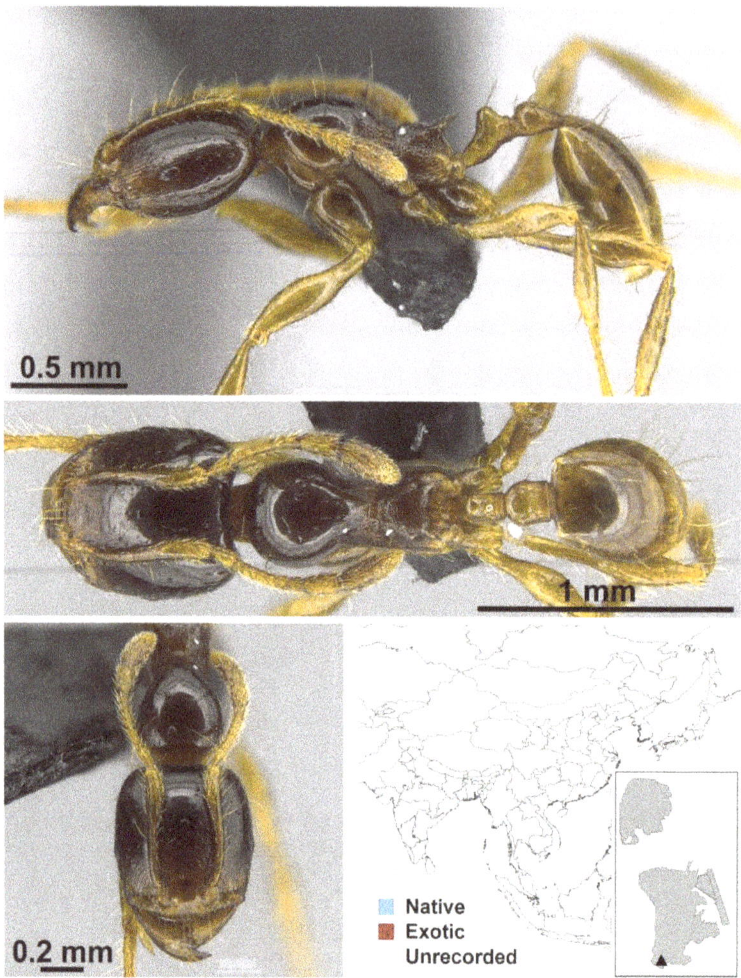

Figure A59. *Carebara* nr. *diversa* Jerdon, 1851 worker (MAC_S18_q2_37.5_sp.3, IBBL).

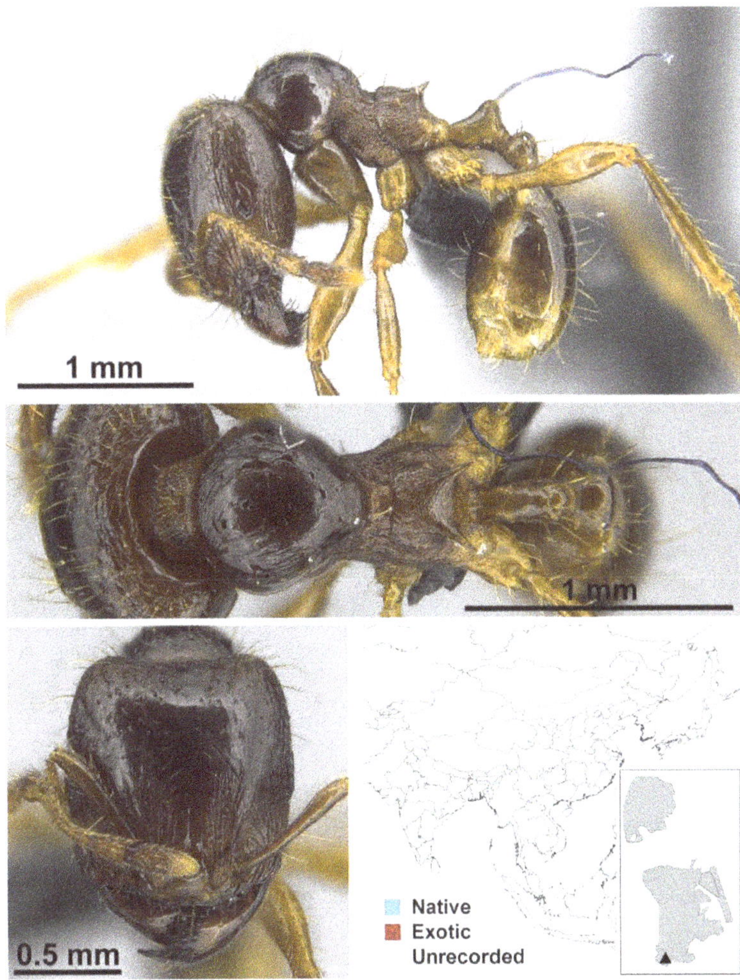

Figure A60. *Carebara* nr. *diversa* Jerdon, 1851 major (MAC_S18_q2_37.5_sp.3, IBBL).

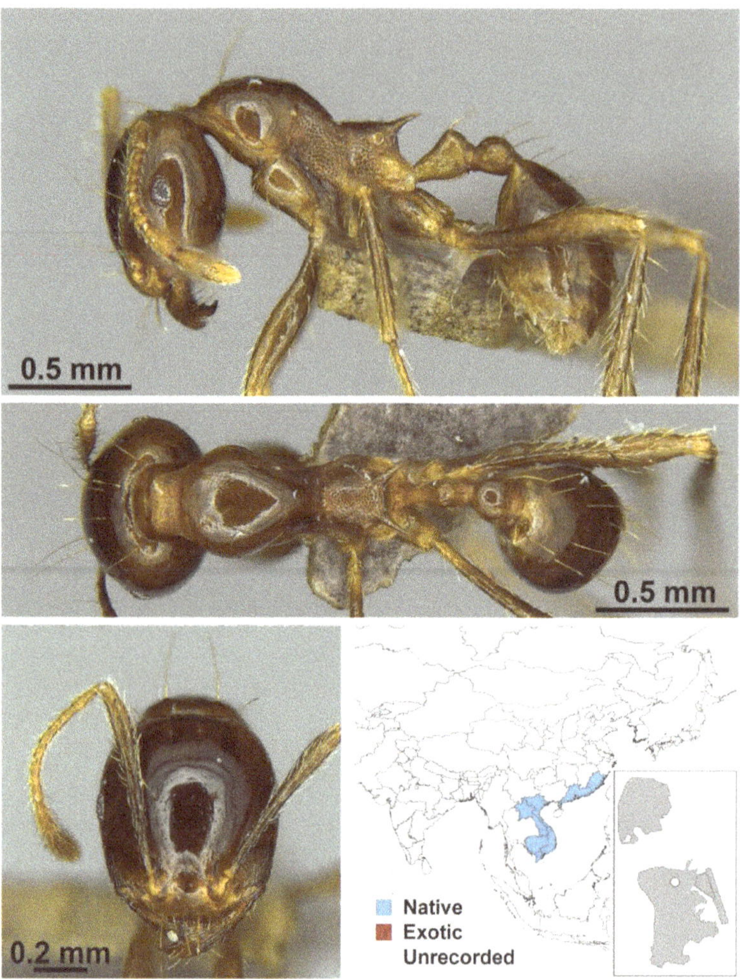

Figure A61. *Carebara diversa laotina*, Santschi, 1921 worker (*Carebara diversa laotina*, CML collection).

Figure A62. *Carebara diversa laotina*, Santschi, 1921 major (*Carebara diversa laotina*, CML collection).

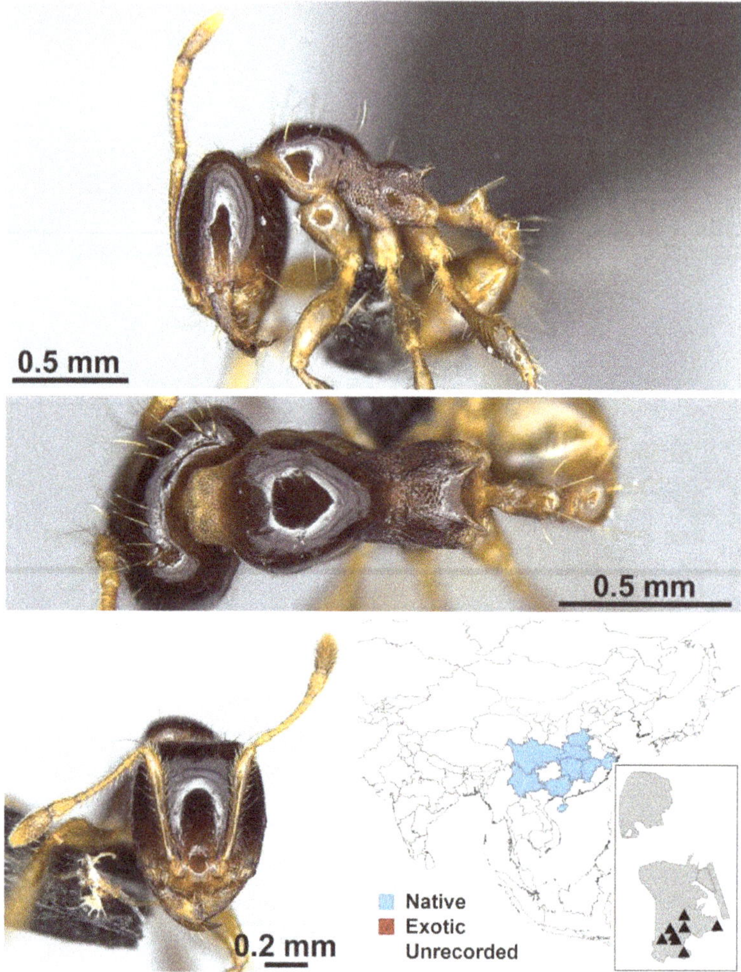

Figure A63. *Carebara melasolena* Zhou & Zheng, 1997 worker (MAC_S12_LLSA_sp.1, IBBL).

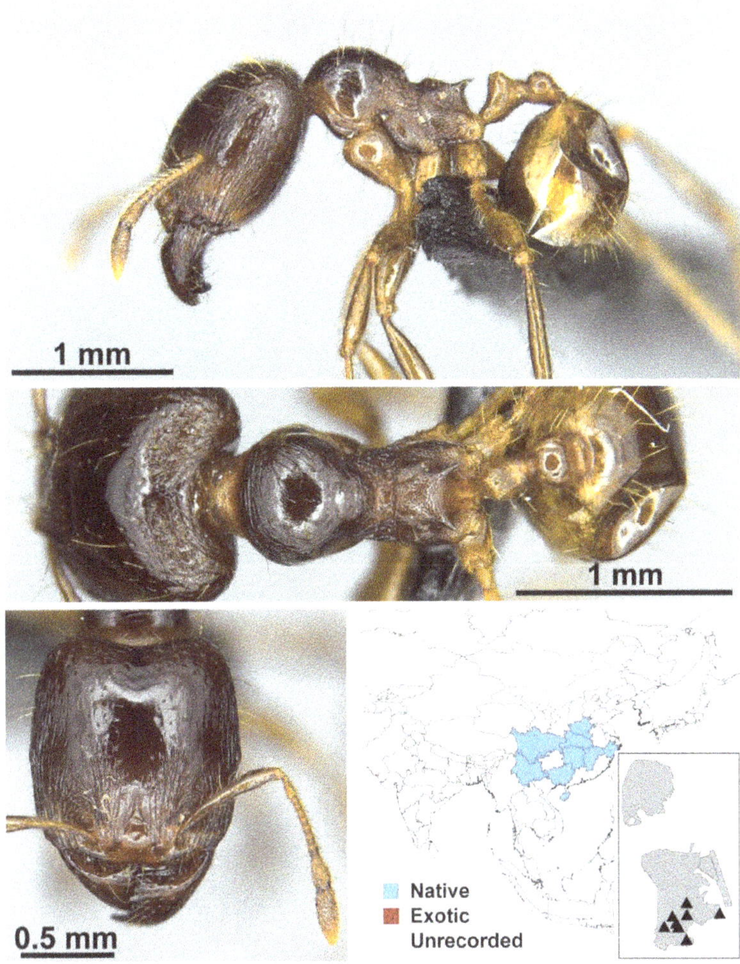

Figure A64. *Carebara melasolena* Zhou & Zheng, 1997 major (MAC_S12_LLSA_sp.1, IBBL).

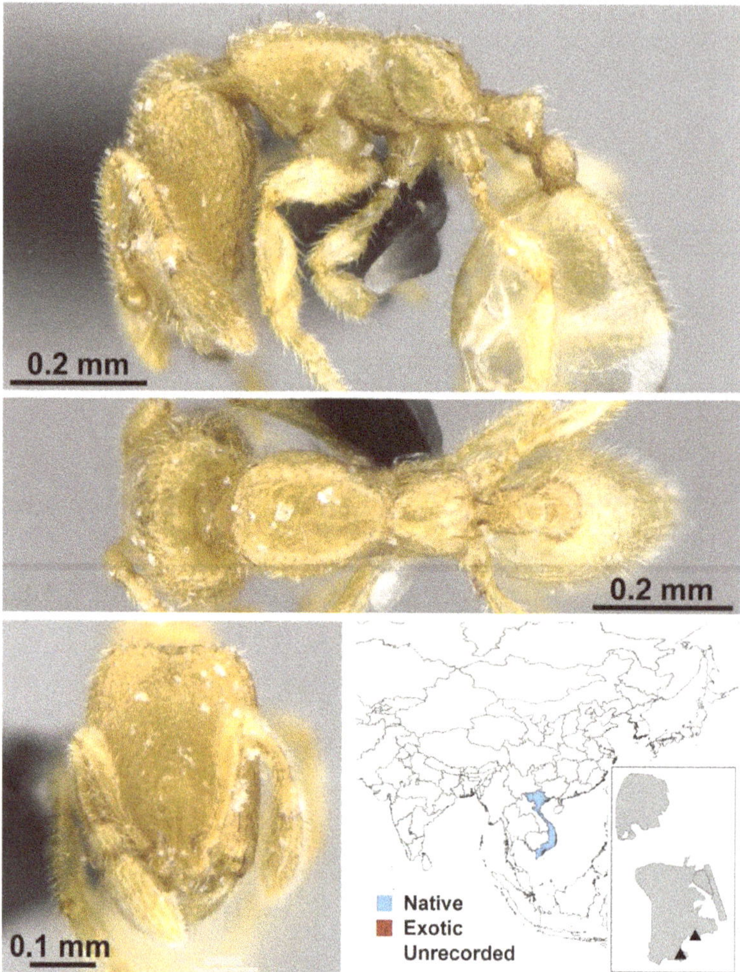

Figure A65. *Carebara sangi* Eguchi & Bui, 2007 worker (MAC_S13_q1_25_Sp.2, IBBL).

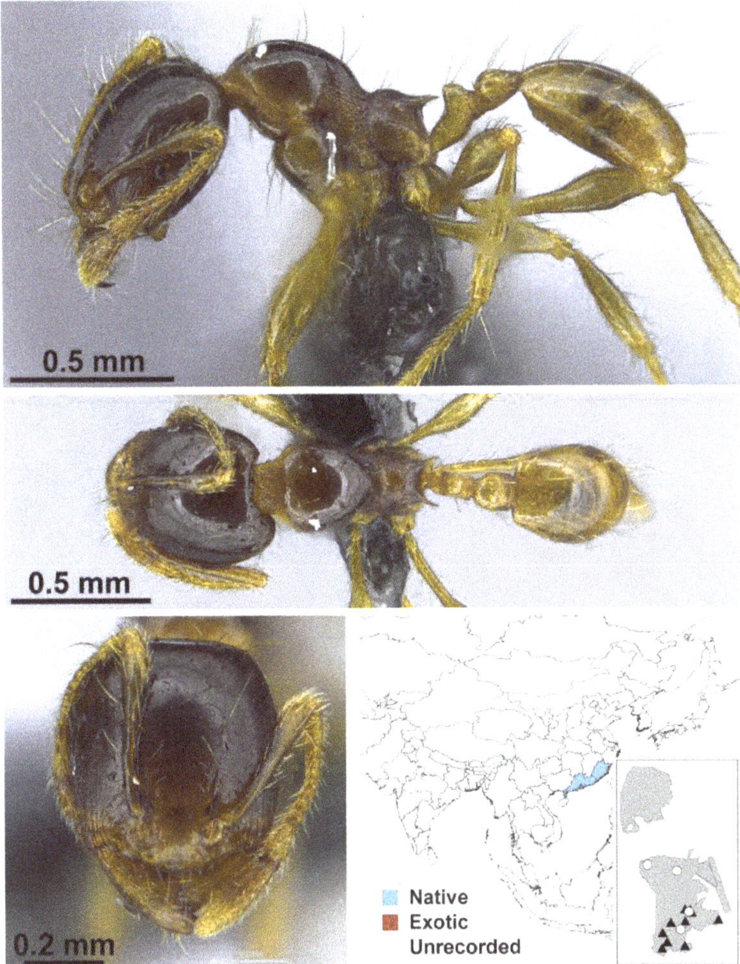

Figure A66. *Carebara zengchengensis* Zhou, Zhao & Jia, 2006 worker (MAC_S12_q3_37.5_Sp.4, IBBL).

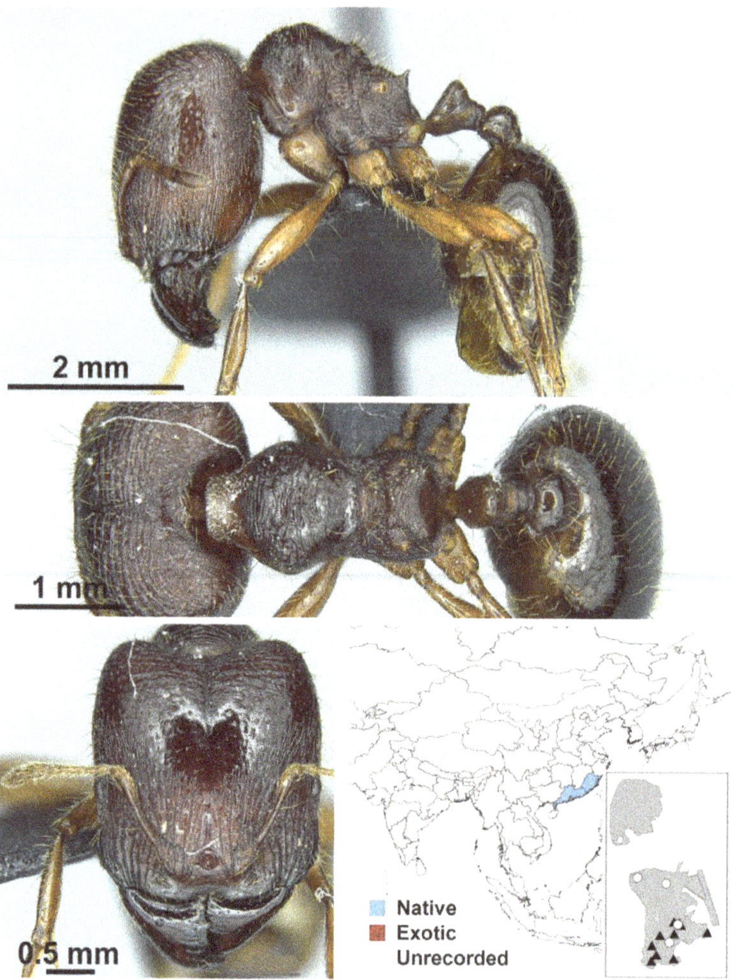

Figure A67. *Carebara zengchengensis* Zhou, Zhao & Jia, 2006 worker (MAC_S12_q3_37.5_Sp.4, IBBL).

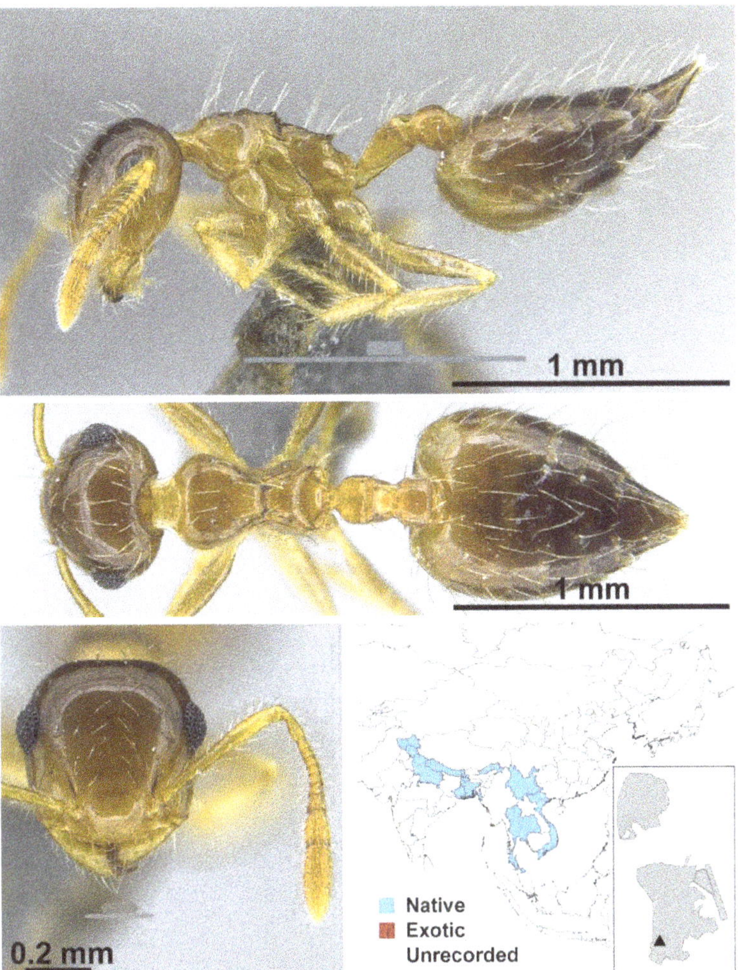

Figure A68. *Crematogaster binghamii* Forel, 1904 worker (MAC_S21_B08_sp.1, IBBL).

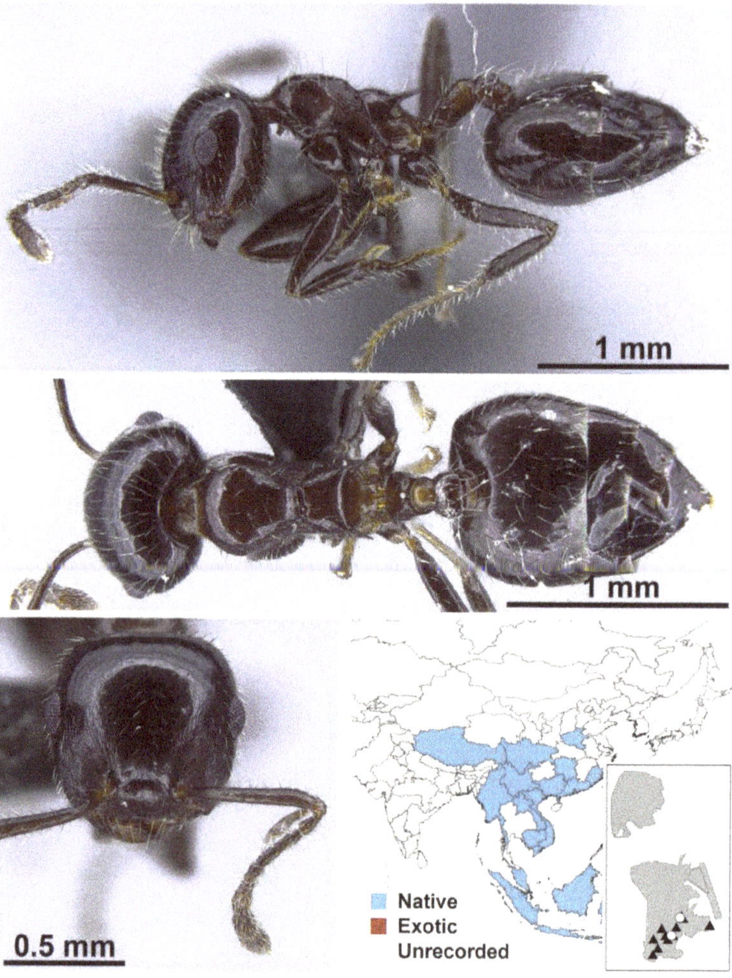

Figure A69. *Crematogaster ferrarii* Emery, 1888 worker (MAC_S01_LLSP_Sp.7, IBBL).

Figure A70. *Crematogaster quadriruga* Forel, 1911 worker (MAC_S06_LLSA_sp.1, IBBL).

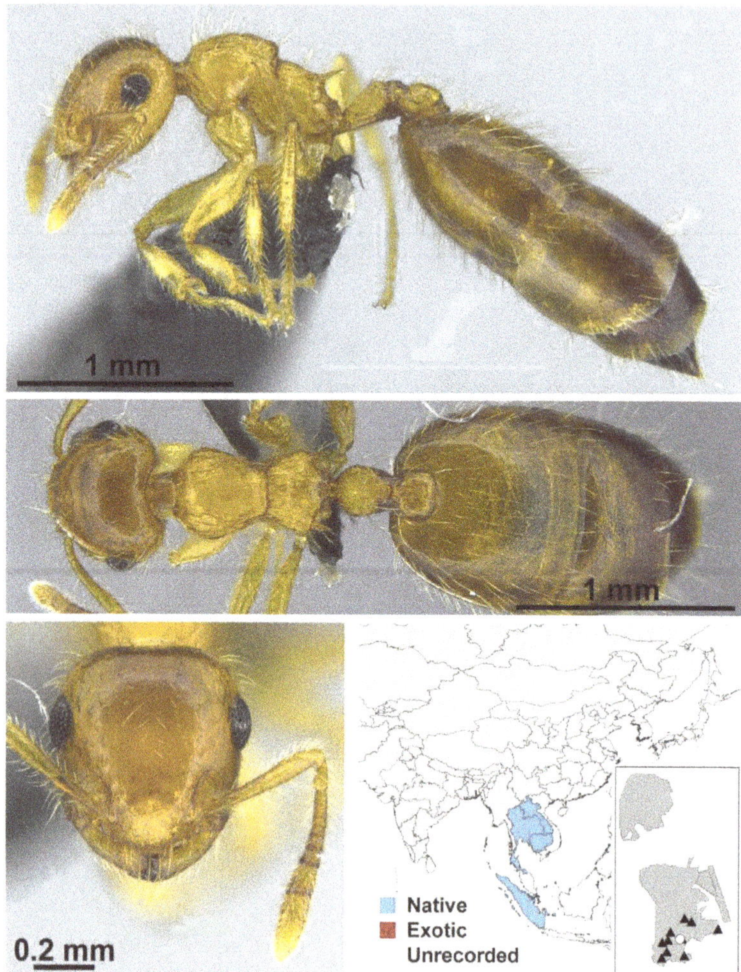

Figure A71. *Crematogaster quadriruga* Forel, 1911 intercaste (MAC_S17_LLSA_sp.3, IBBL).

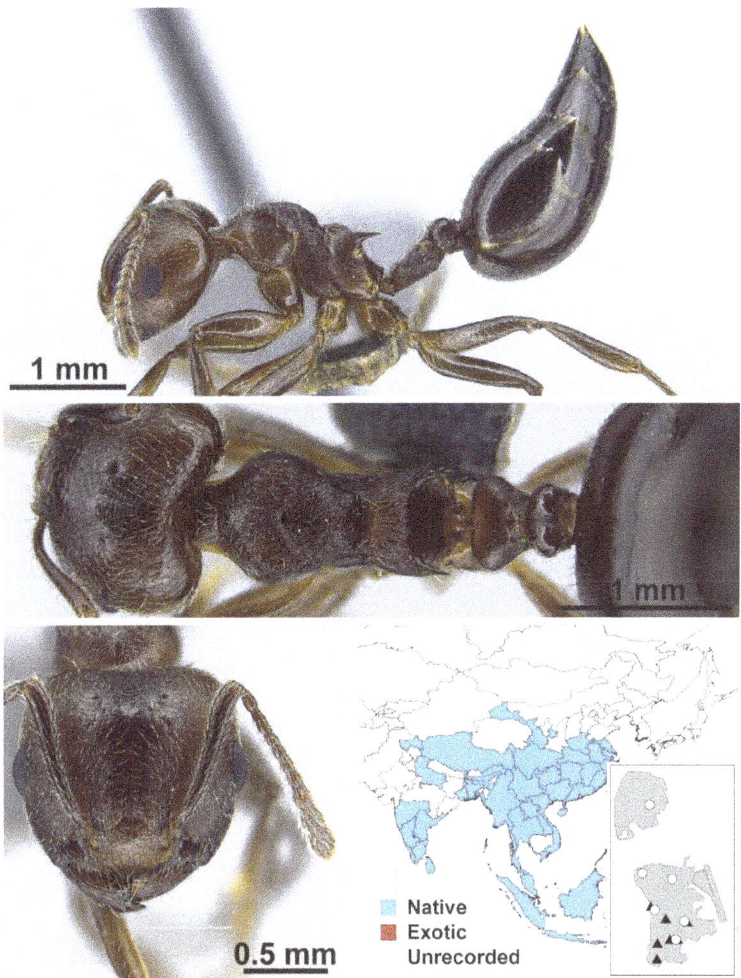

Figure A72. *Crematogaster rogenhoferi* Mayr, 1879 worker (MAC_S19_T4_1m_sp.1, IBBL).

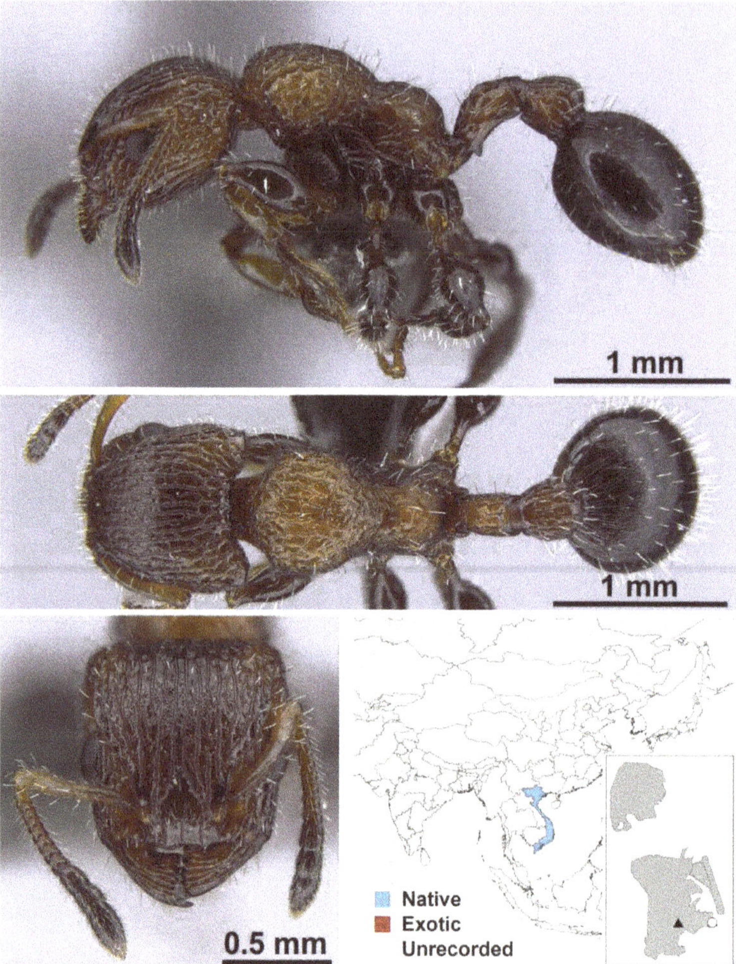

Figure A73. *Dilobocondyla propotriangulata*, Bharti & Kumar, 2013 worker (FB19145, IBBL).

Figure A74. *Mayriella granulata*, Dlussky & Radchenko, 1990 worker (MAC_S18_LLSA_Sp.4, IBBL).

Figure A75. *Meranoplus* sp. mo01 nr. *bicolor* Guérin-Méneville, 1844 worker (*Meranoplus* sp. mo01 nr. *Bicolor*, CML collection).

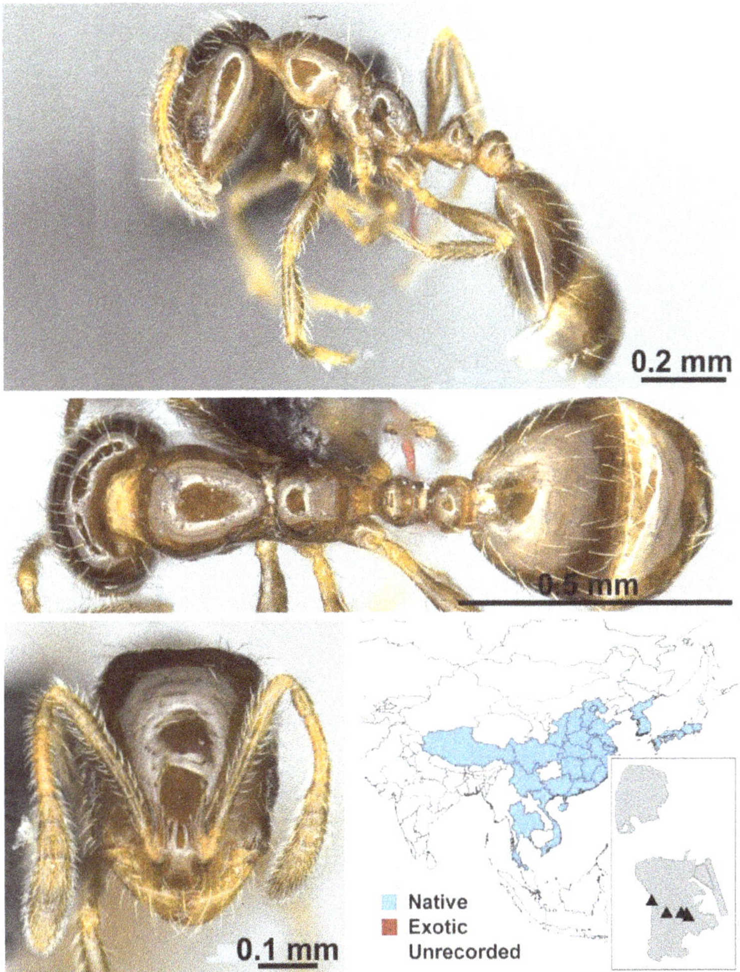

Figure A76. *Monomorium chinense* Santschi, 1925 worker (MAC_S20_LLSP_sp.12, IBBL).

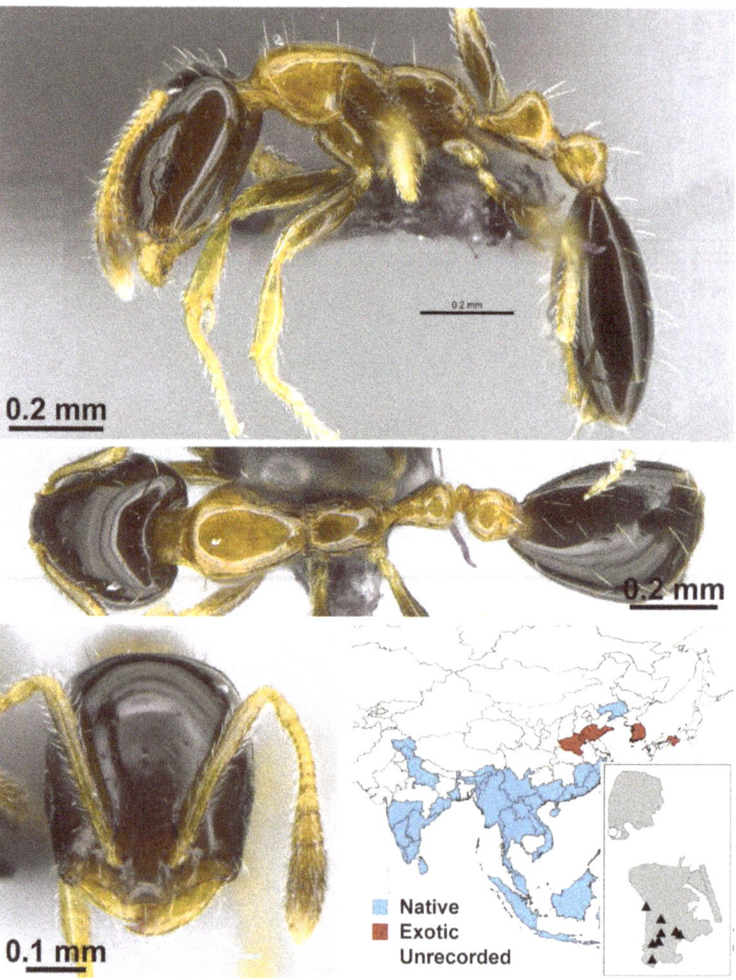

Figure A77. *Monomorium floricola* Jerdon, 1851 worker (MAC_S03_LLSA_Sp.8, IBBL).

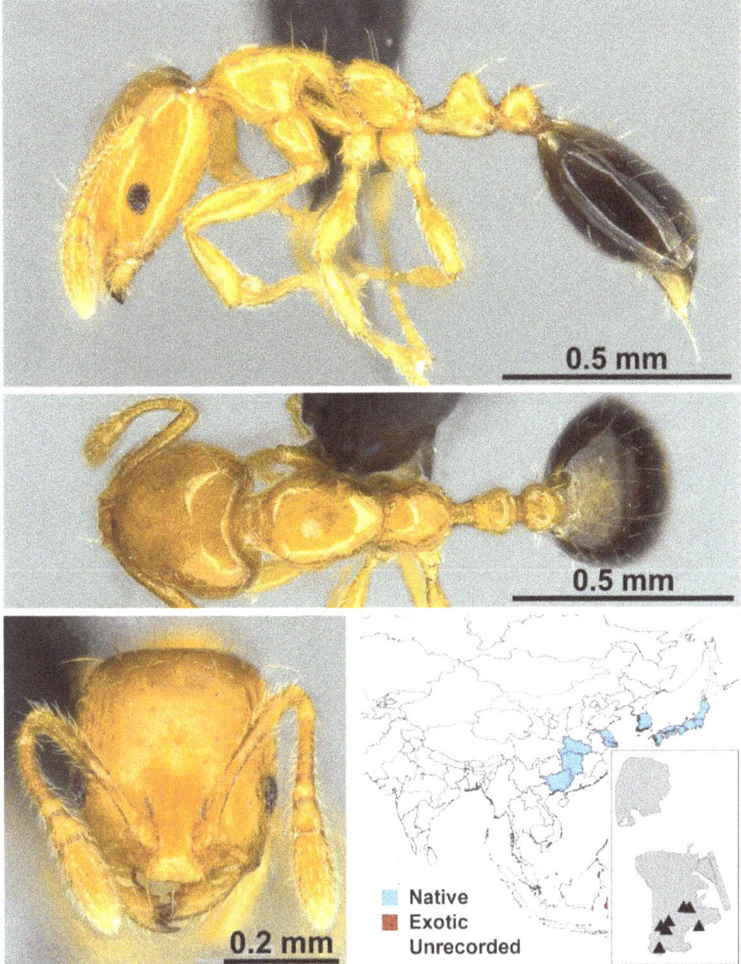

Figure A78. *Monomorium intrudens* Smith, 1874 worker (MAC_S18_LLSP_sp.3, IBBL).

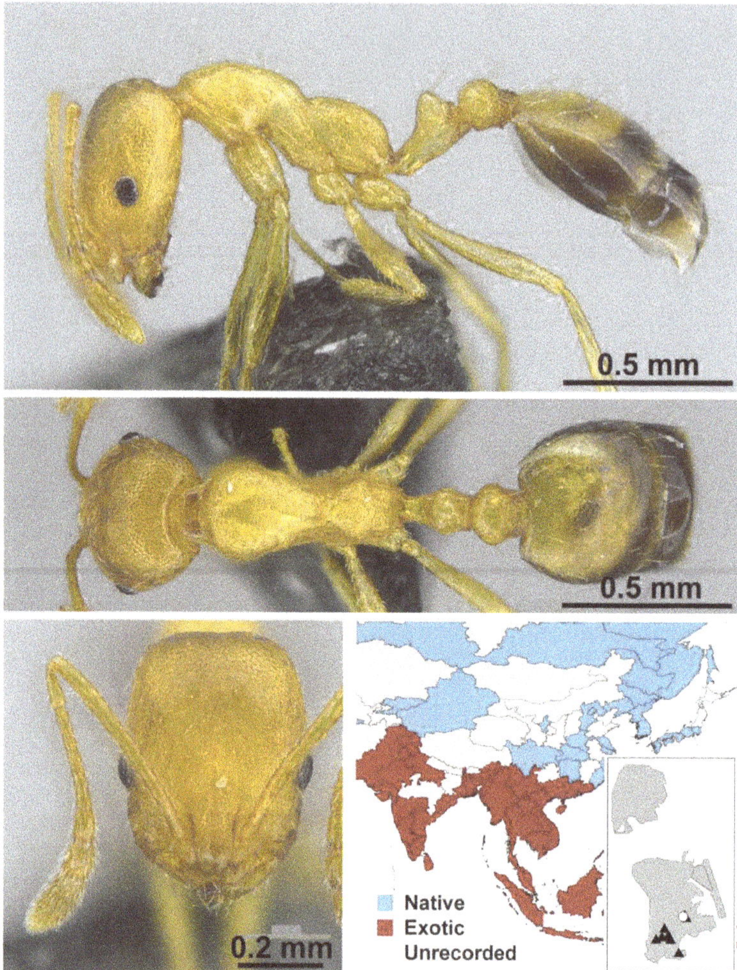

Figure A79. *Monomorium pharaonis* Linnaeus, 1758 worker (MAC_S09_LLSA_sp.9, IBBL).

Figure A80. *Monomorium* sp. psw-cn01 worker (MAC_S21_LLSA_bottom_sp.2, IBBL).

Figure A81. *Myrmecina nomurai* Okido, Ogata & Hosoishi, 2020 worker (MAC_S05_LLSA_Sp.1, IBBL).

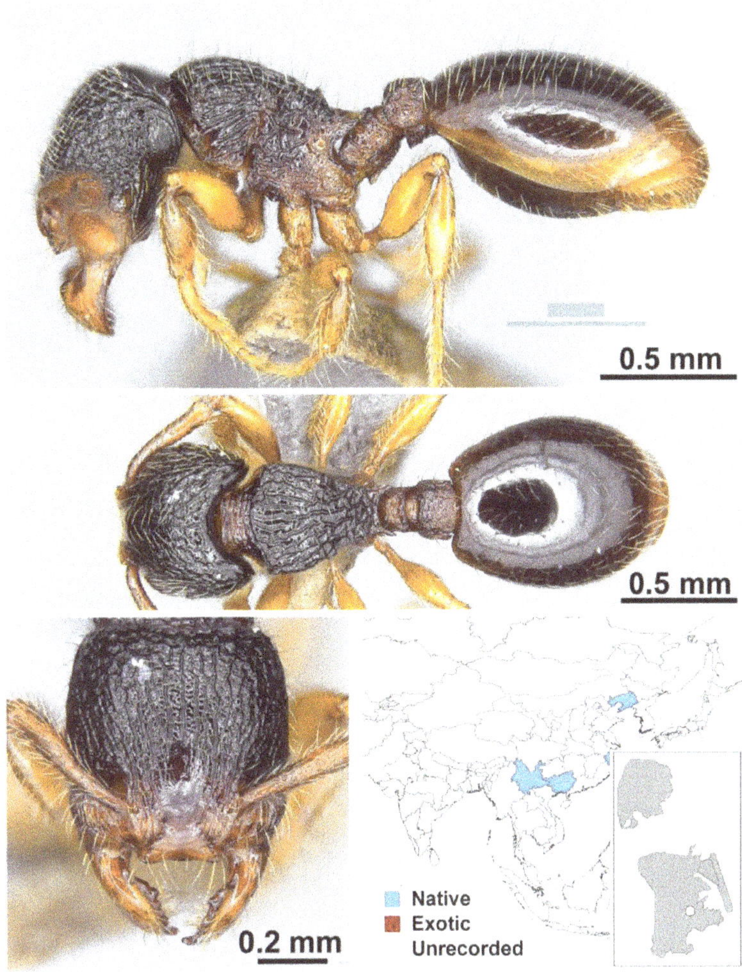

Figure A82. *Myrmecina sinensis* Wheeler, W. M., 1921 worker (*Myrmecina sinensis*, CML collection).

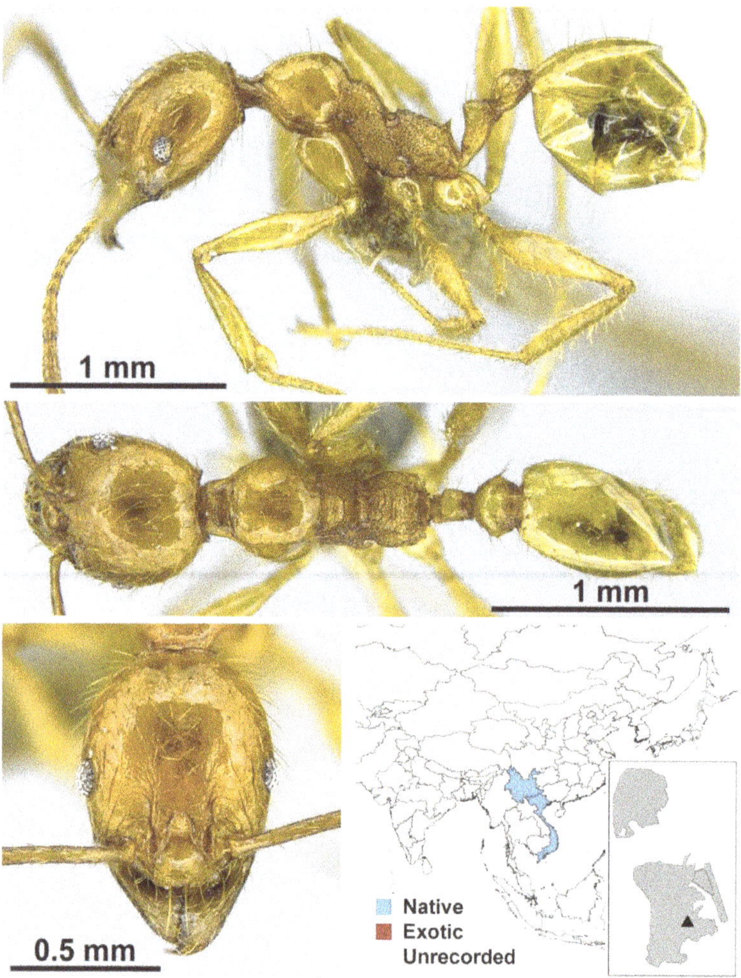

Figure A83. *Pheidole elongicephala* Eguchi, 2008 worker (MAC_S09_q2_25_sp.2, IBBL).

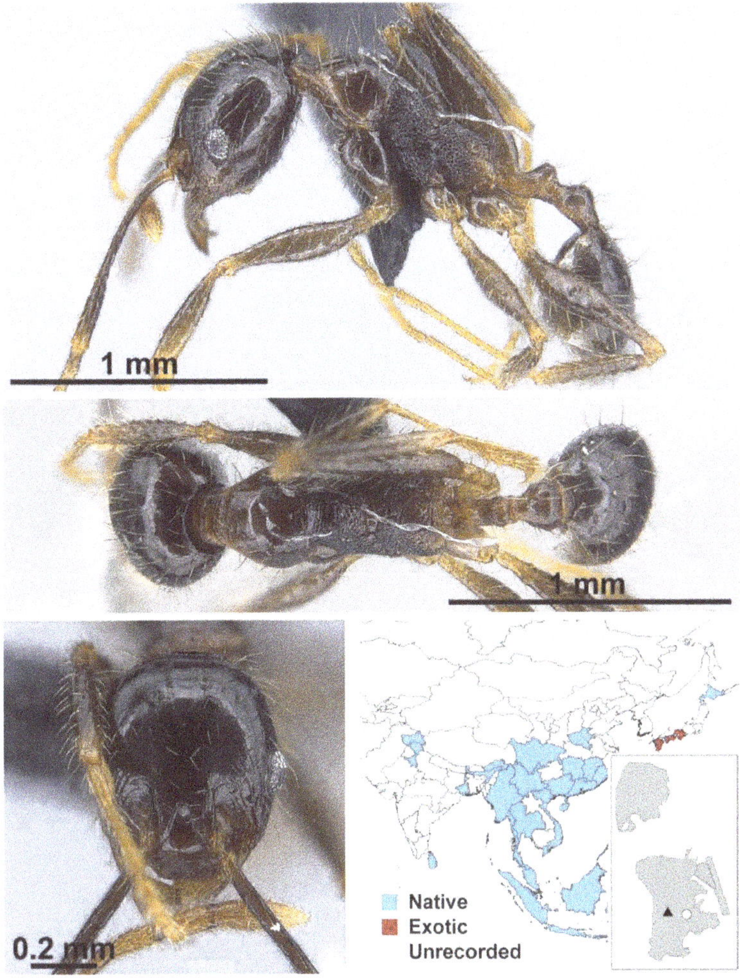

Figure A84. *Pheidole fervens* Smith, 1858 worker (MAC_S19_q4_GL_03_Sp.2, IBBL).

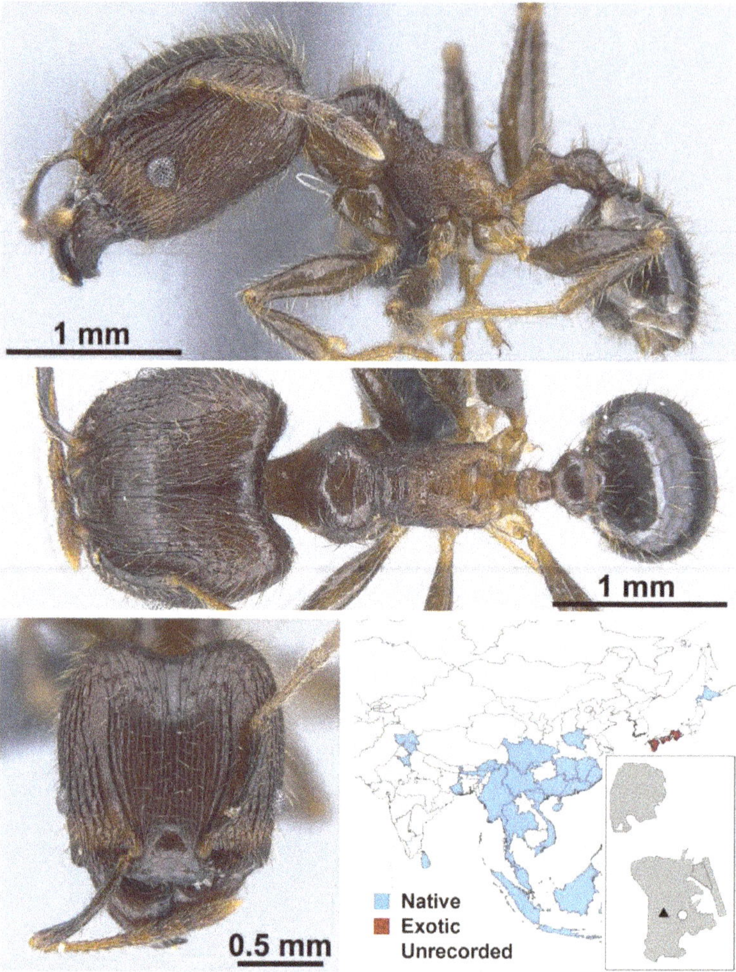

Figure A85. *Pheidole fervens* Smith, 1858 major (MAC_S19_q4_GL_03_Sp.2, IBBL).

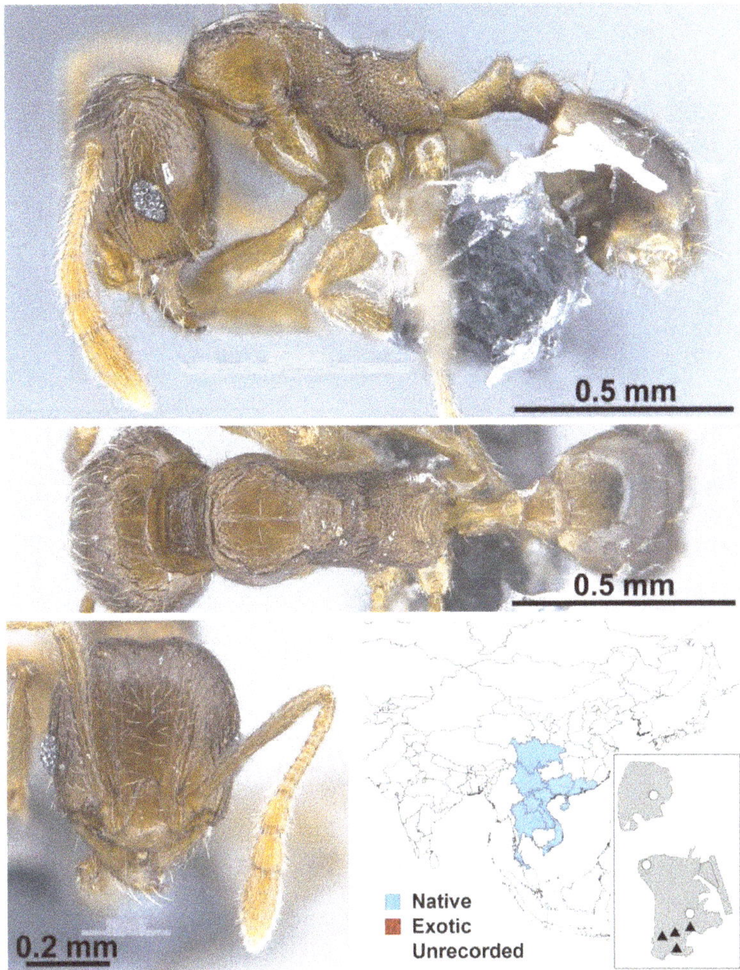

Figure A86. *Pheidole hongkongensis* Wheeler, 1928 worker (MAC_S21_LLSP_Sp.9, IBBL).

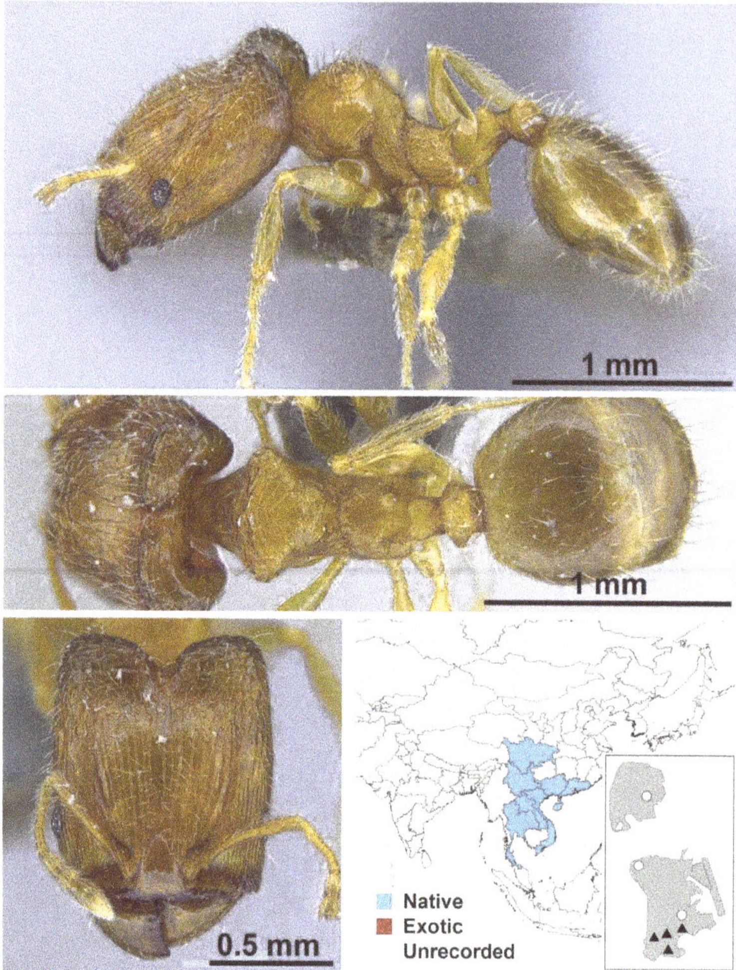

Figure A87. *Pheidole hongkongensis* Wheeler, 1928 major (MAC_S07_B08_sp.1, IBBL).

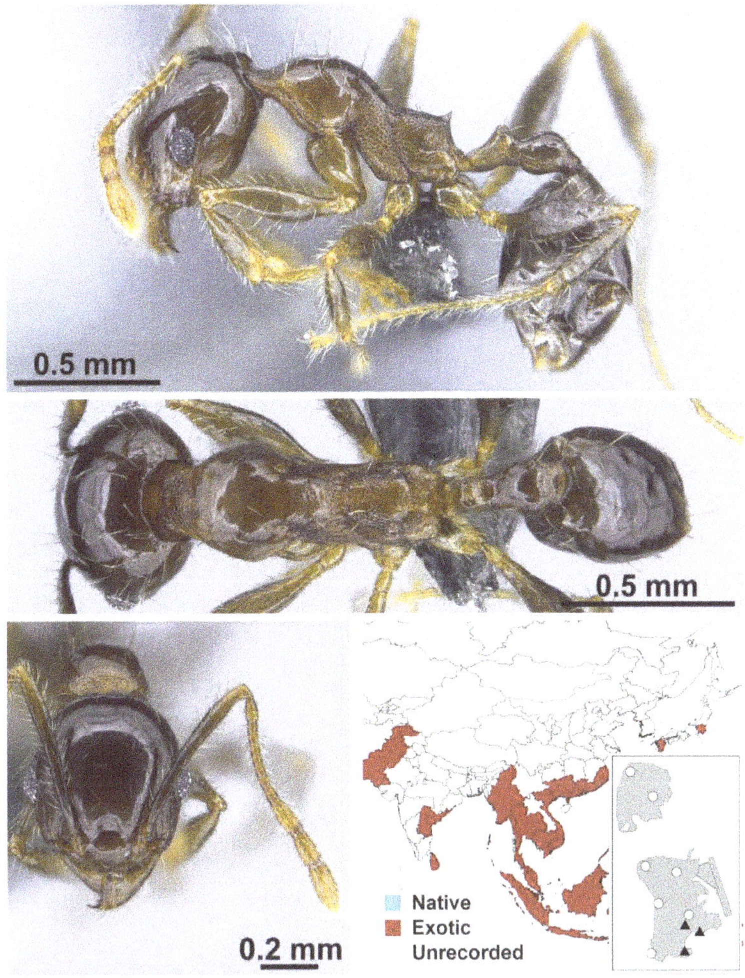

Figure A88. *Pheidole megacephala* Fabricius, 1793 worker (MAC_S13_LLSP_Sp.1, IBBL).

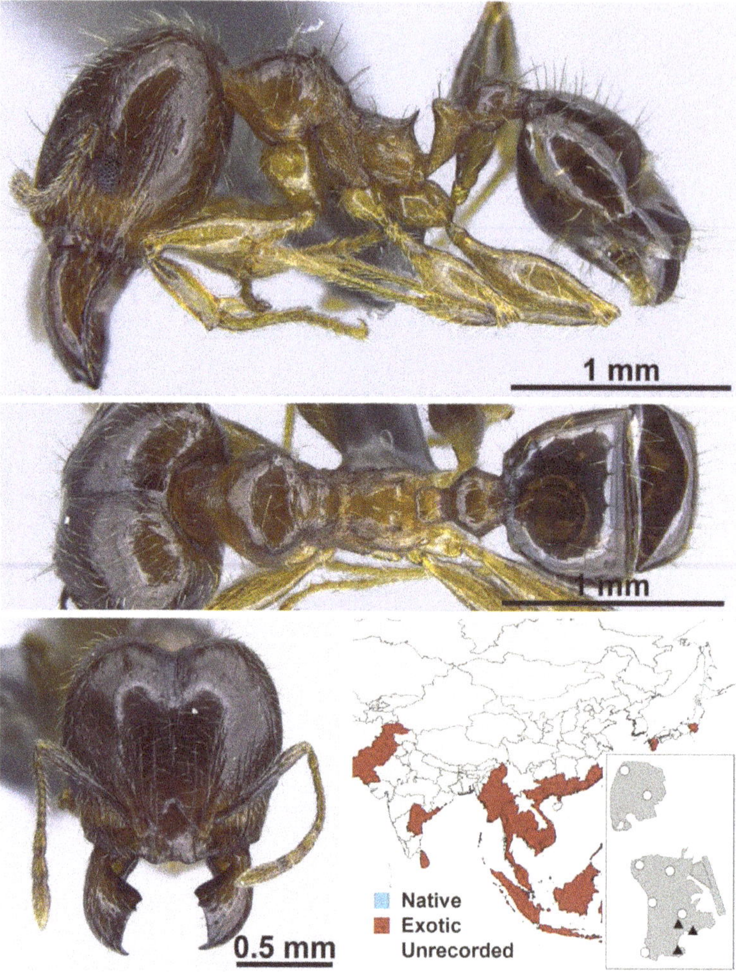

Figure A89. *Pheidole megacephala* Fabricius, 1793 major (MAC_S13_LLSP_Sp.1, IBBL).

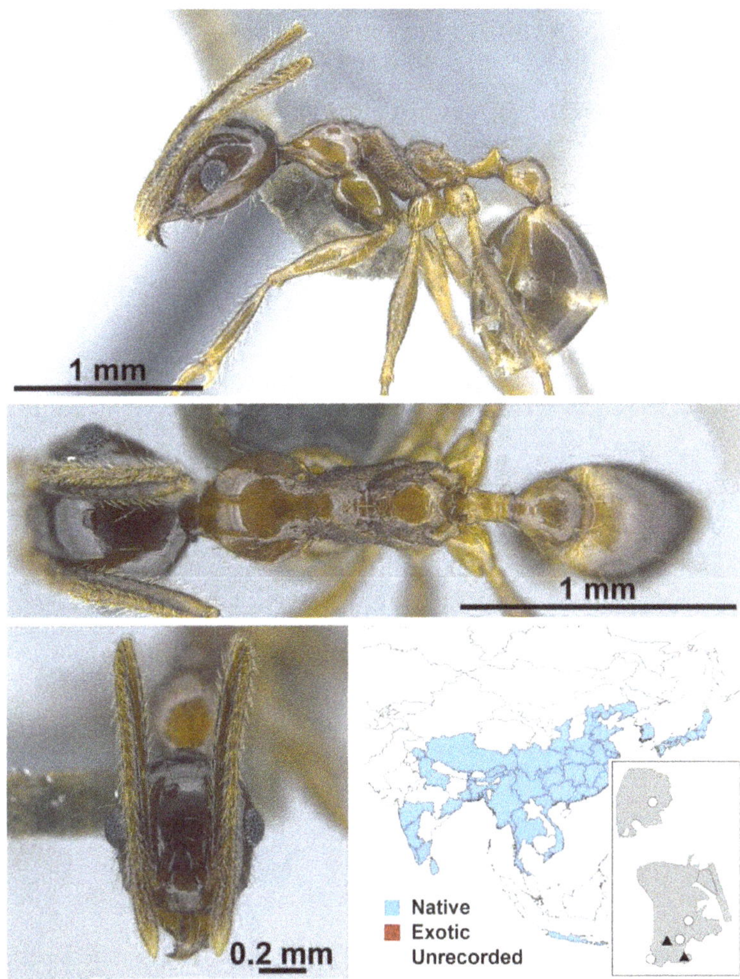

Figure A90. *Pheidole nodus* Smith, 1874 worker (MAC_S02_B09_sp.1_top, IBBL).

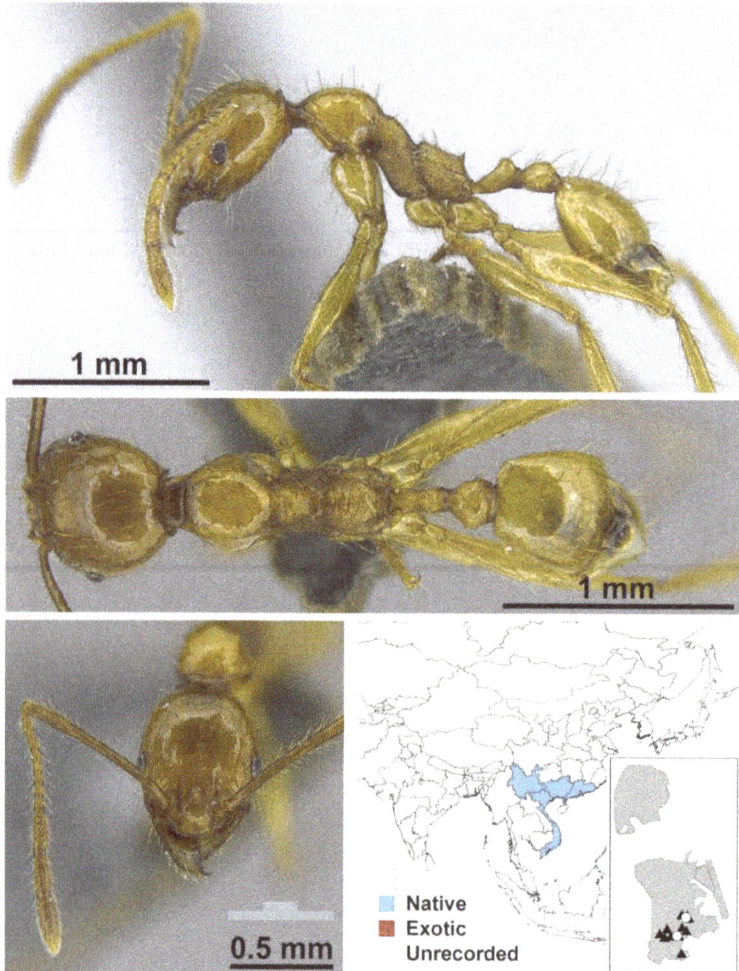

Figure A91. *Pheidole ochracea* Eguchi, 2008 worker (MAC_S03_B03_sp.1, IBBL).

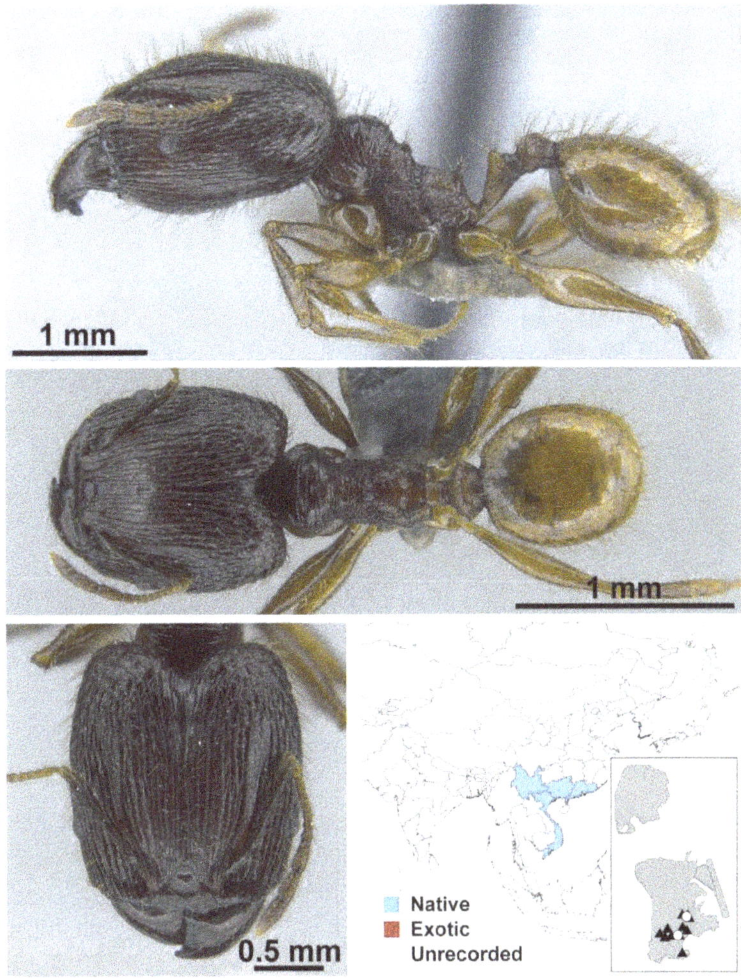

Figure A92. *Pheidole ochracea* Eguchi, 2008 major (MAC_S03_B03_sp.1, IBBL).

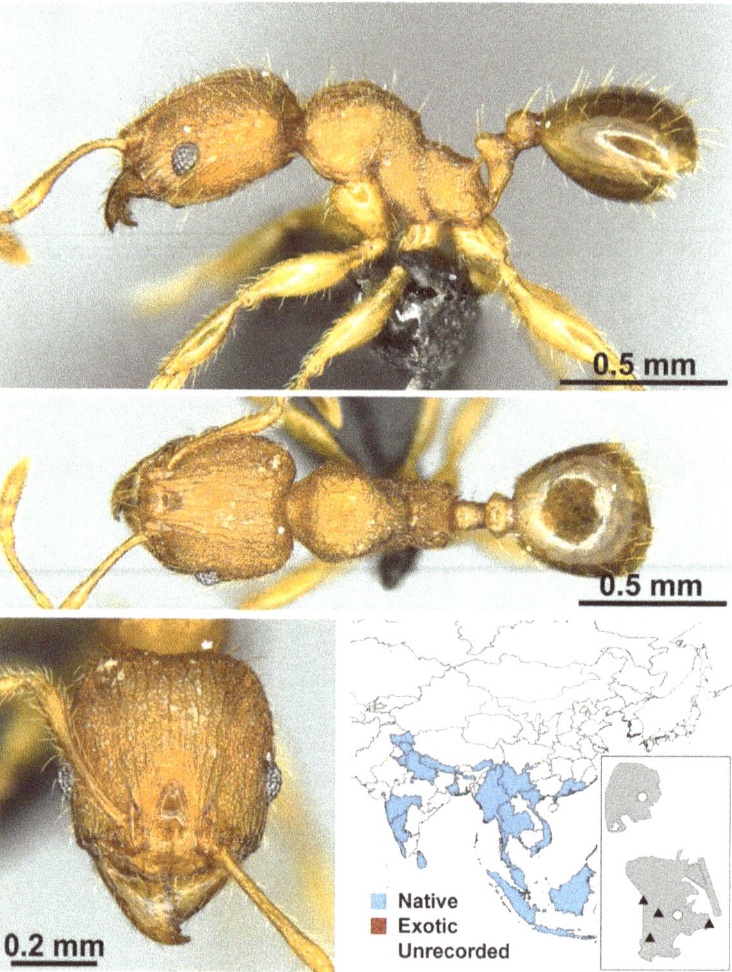

Figure A93. *Pheidole parva* Mayr, 1865 worker (MAC_S20_LLSA_sp.4, IBBL).

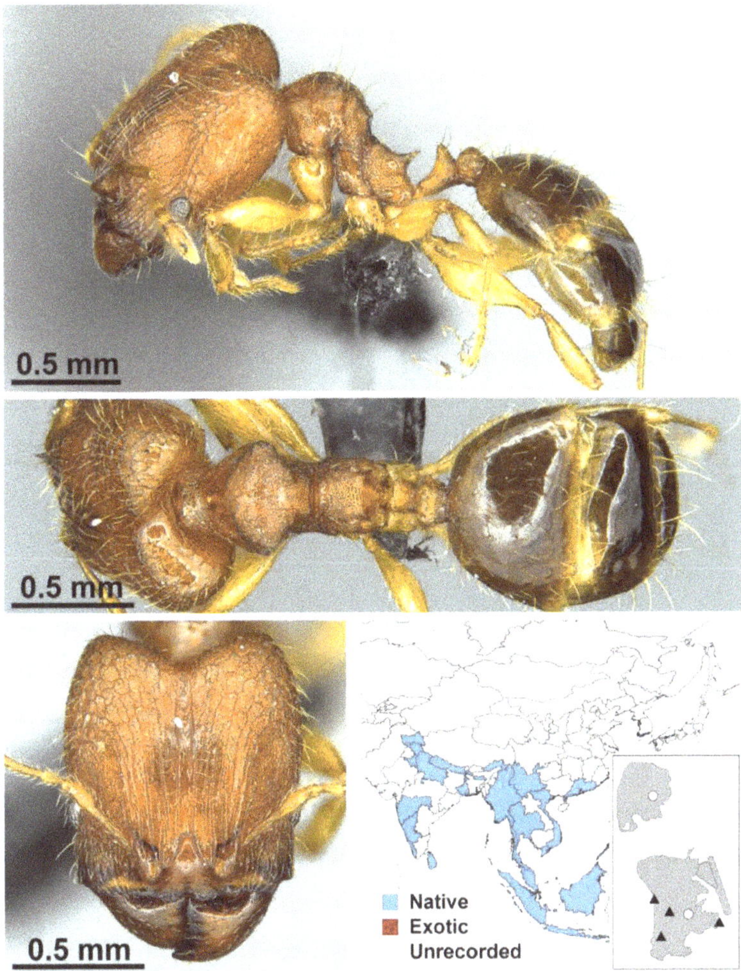

Figure A94. *Pheidole parva* Mayr, 1865 major (MAC_S20_LLSA_sp.4, IBBL).

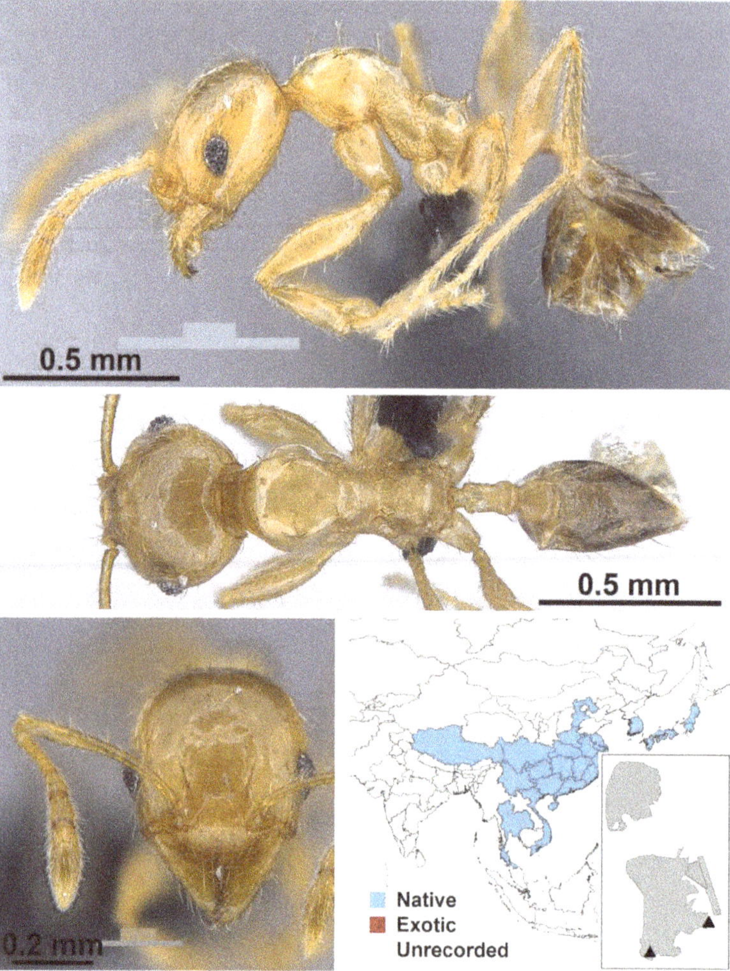

Figure A95. *Pheidole pieli* Santschi, 1925 worker (MAC_S17_LLSA_Sp.4, IBBL).

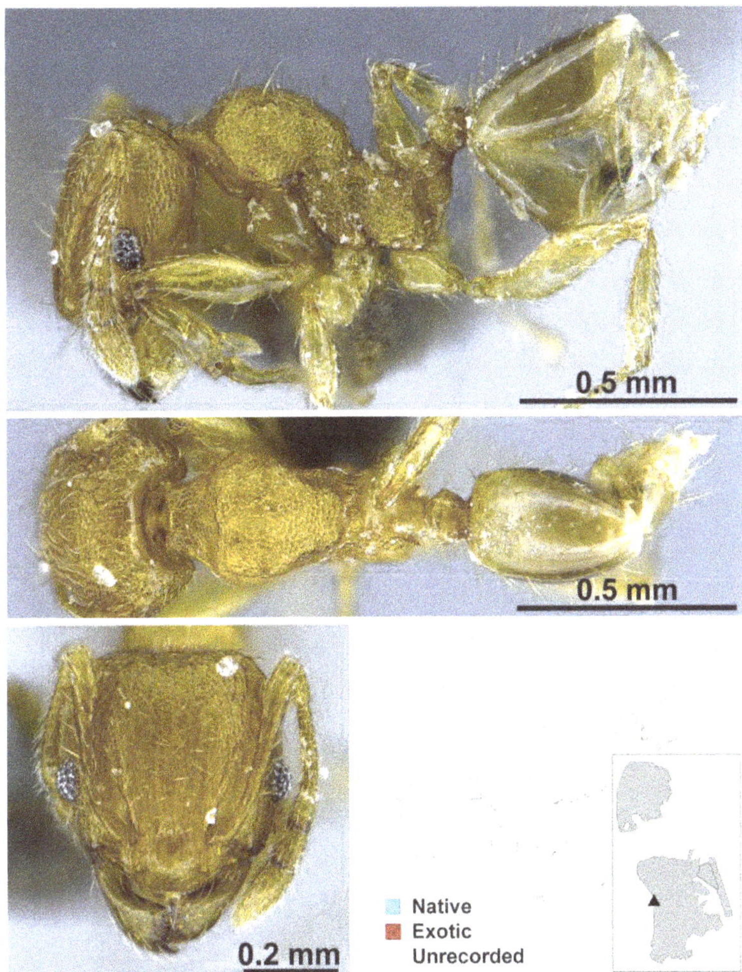

Figure A96. *Pheidole* nr. *ryukyuensis* Ogata, 1982 worker (MAC_S20_12.5_q4_sp.2, IBBL).

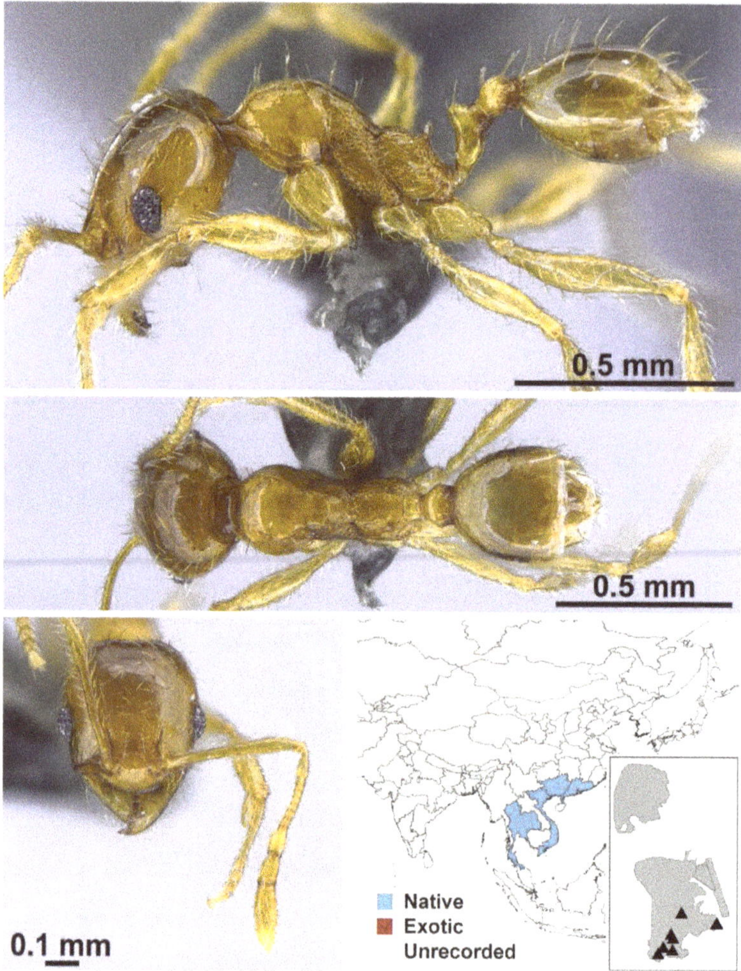

Figure A97. *Pheidole taipoana* Wheeler, 1928 worker (MAC_S04_LLSA_Sp.4, IBBL).

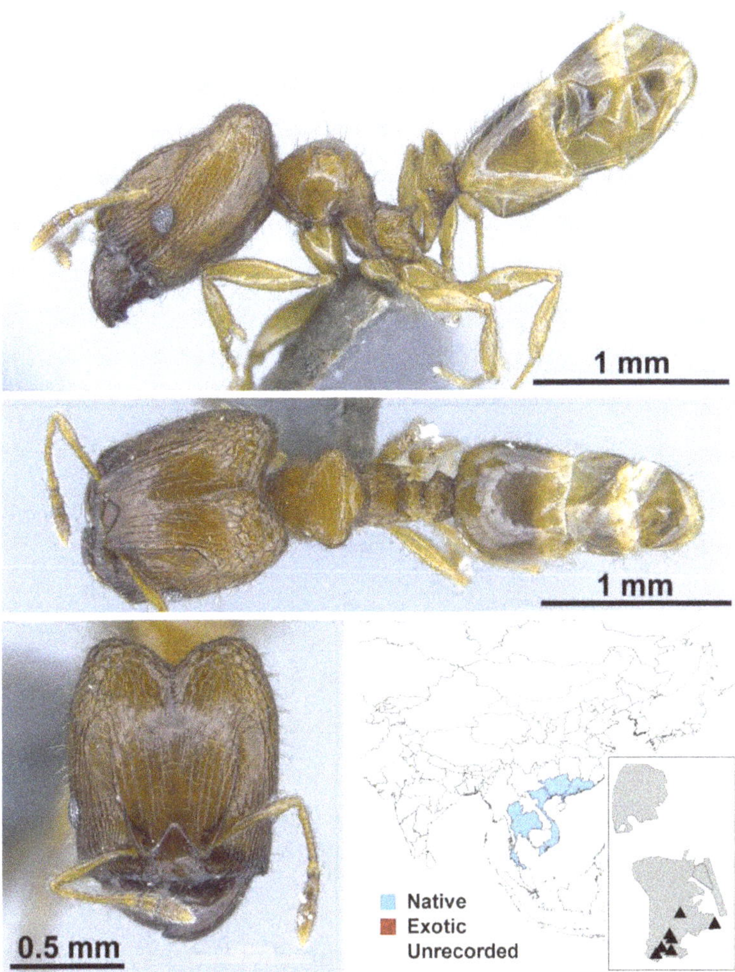

Figure A98. *Pheidole taipoana* Wheeler, 1928 major (MAC_S04_B07_sp.1_top, IBBL).

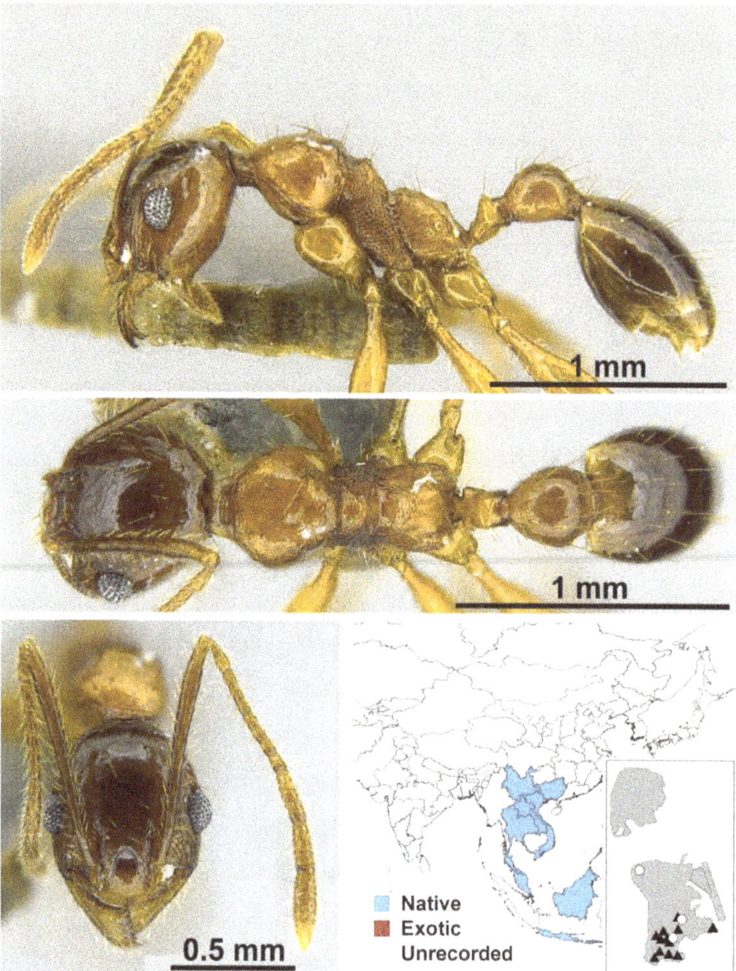

Figure A99. *Pheidole tumida* Eguchi, 2008 worker (MAC_S04_B06_sp.2_top, IBBL).

Figure A100. *Pheidole tumida* Eguchi, 2008 major (MAC_S7_GN1_H4_n1, IBBL).

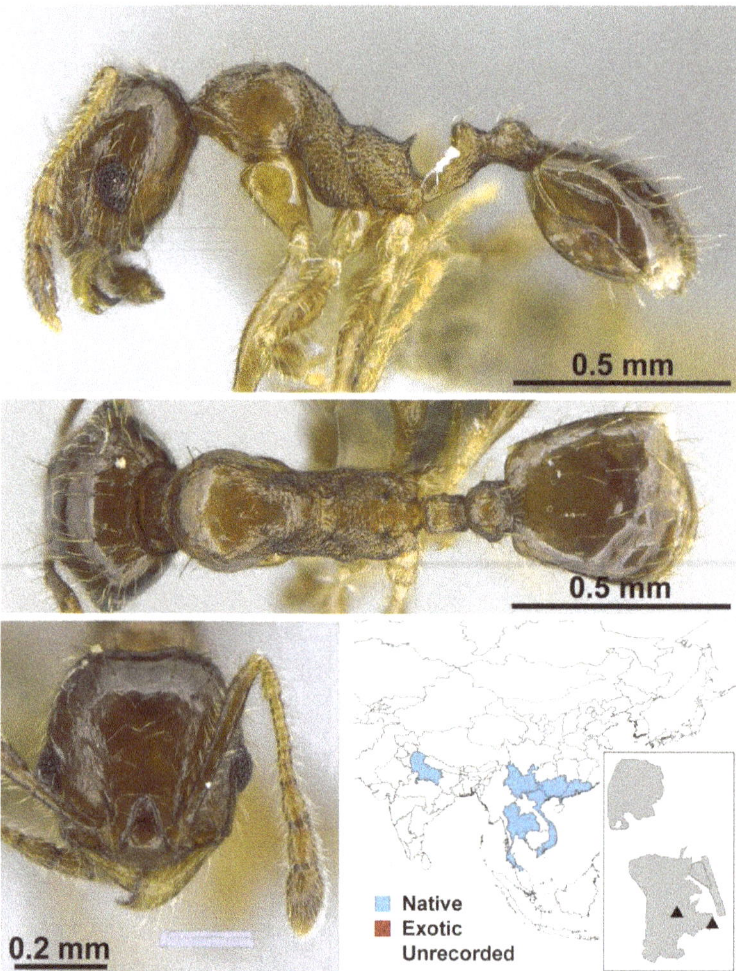

Figure A101. *Pheidole vulgaris* Eguchi, 2006 worker (MAC_S12_B04_sp.2_bottom, IBBL).

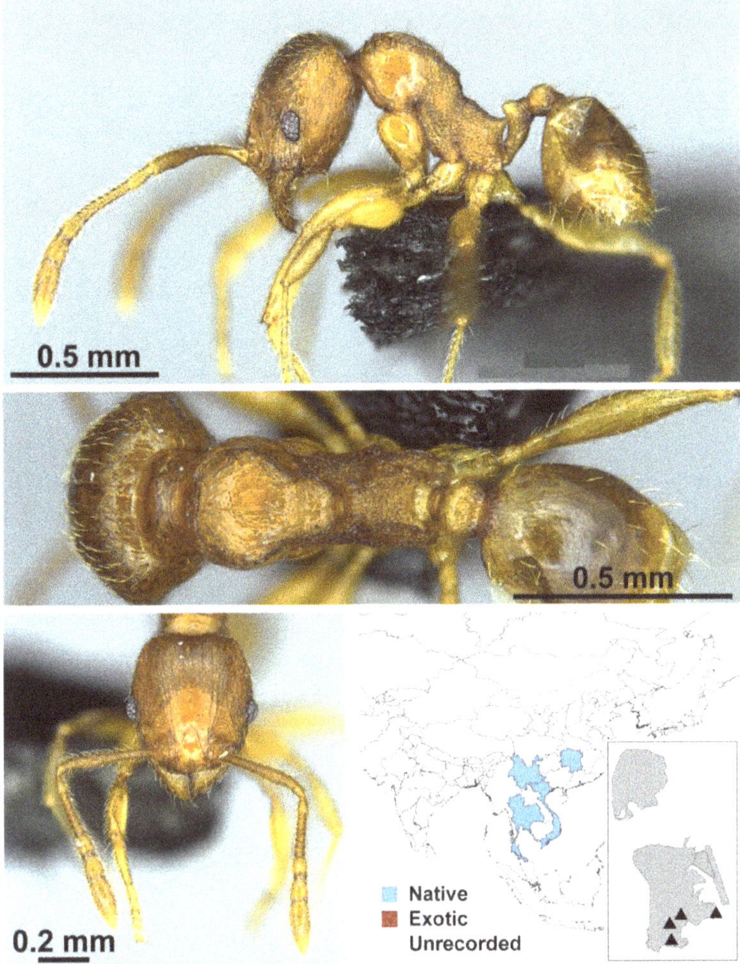

Figure A102. *Pheidole zoceana* Santschi, 1925 worker (MAC_S11_LLSA_sp.6, IBBL).

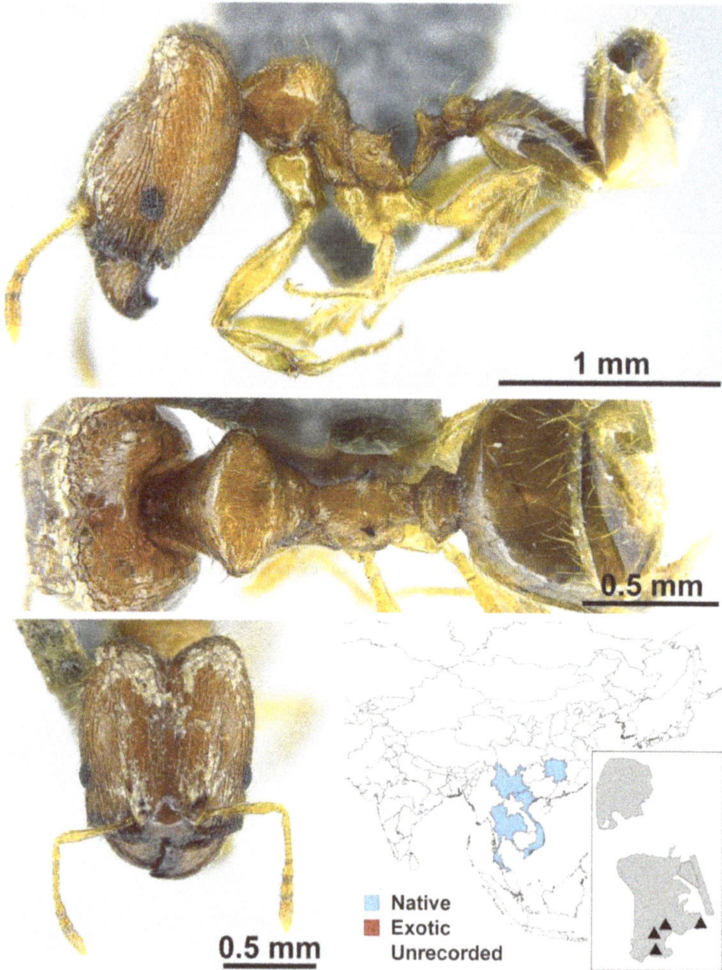

Figure A103. *Pheidole zoceana* Santschi, 1925 major (MAC_S17_LLSA_sp.1, IBBL).

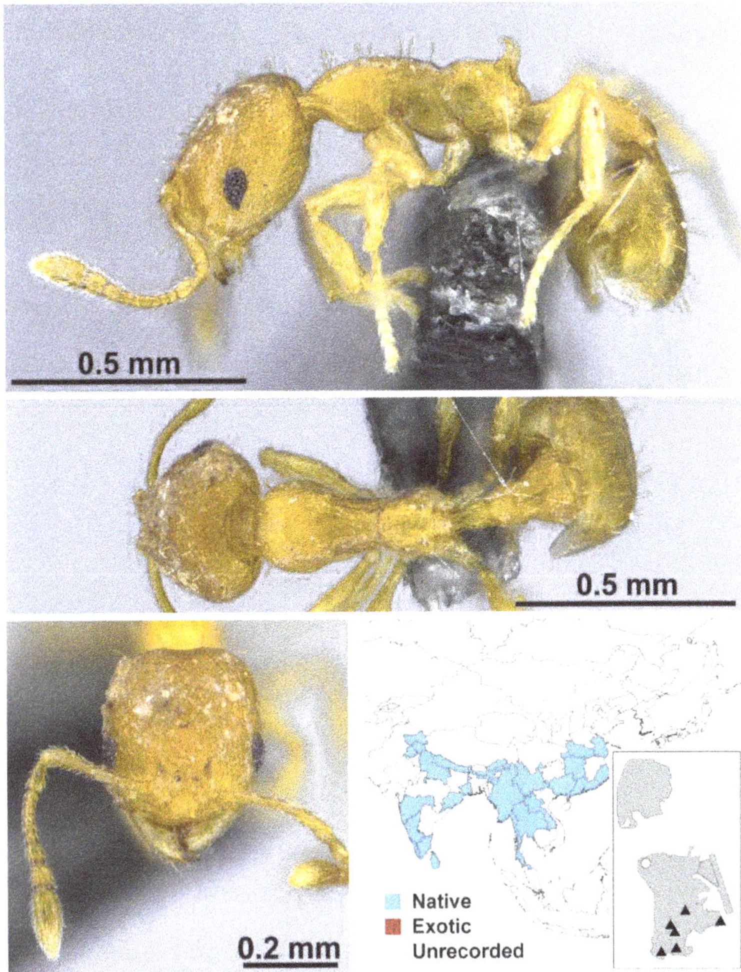

Figure A104. *Recurvidris recurvispinosa* Forel, 1890 (MAC_S12_LLSA_Sp.9, IBBL).

Figure A105. *Rotastruma stenoceps* Bolton, 1991 worker (MAC_S15_LLSA__sp.6, IBBL).

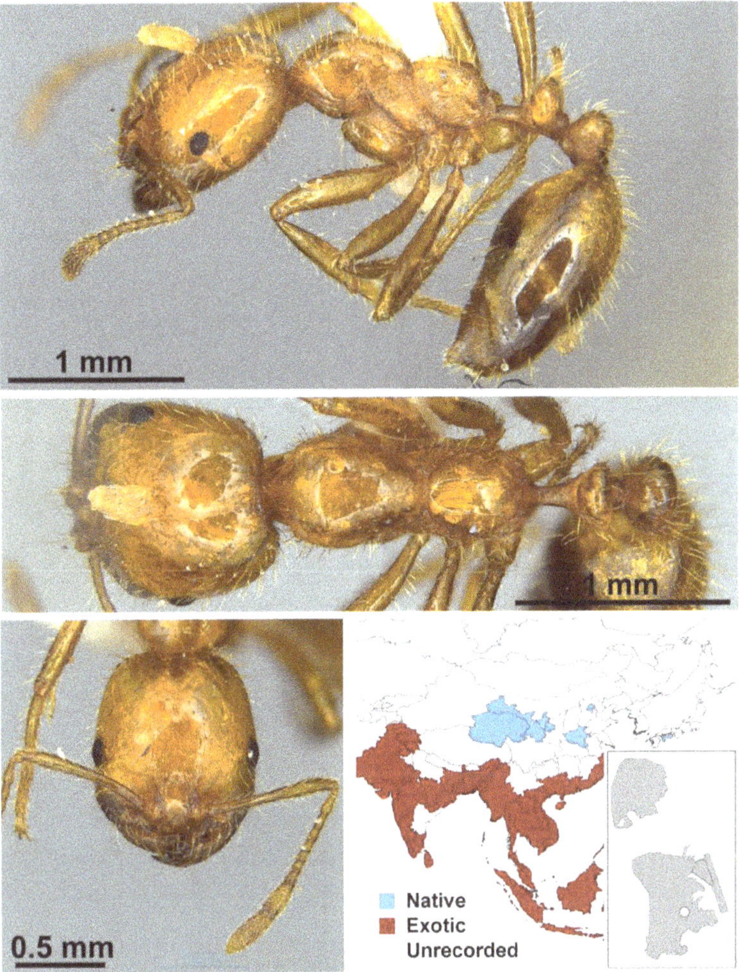

Figure A106. *Solenopsis geminata* Fabricius, 1804 worker (*Solenopsis geminata*, IBBL).

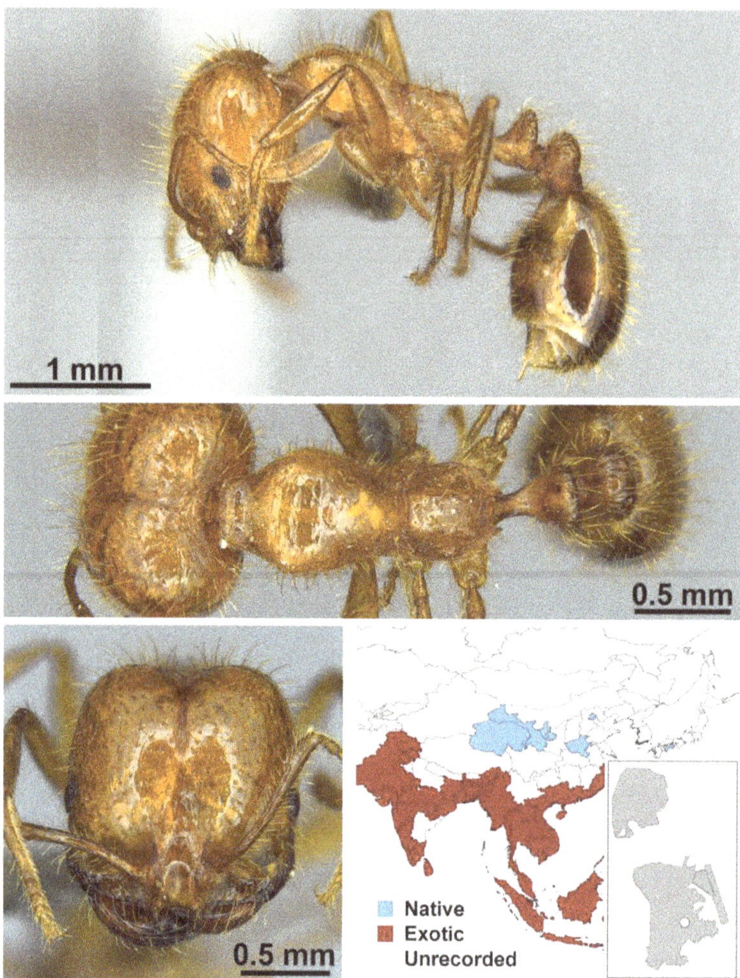

Figure A107. *Solenopsis geminata* Fabricius, 1804 major (*Solenopsis geminata*, IBBL).

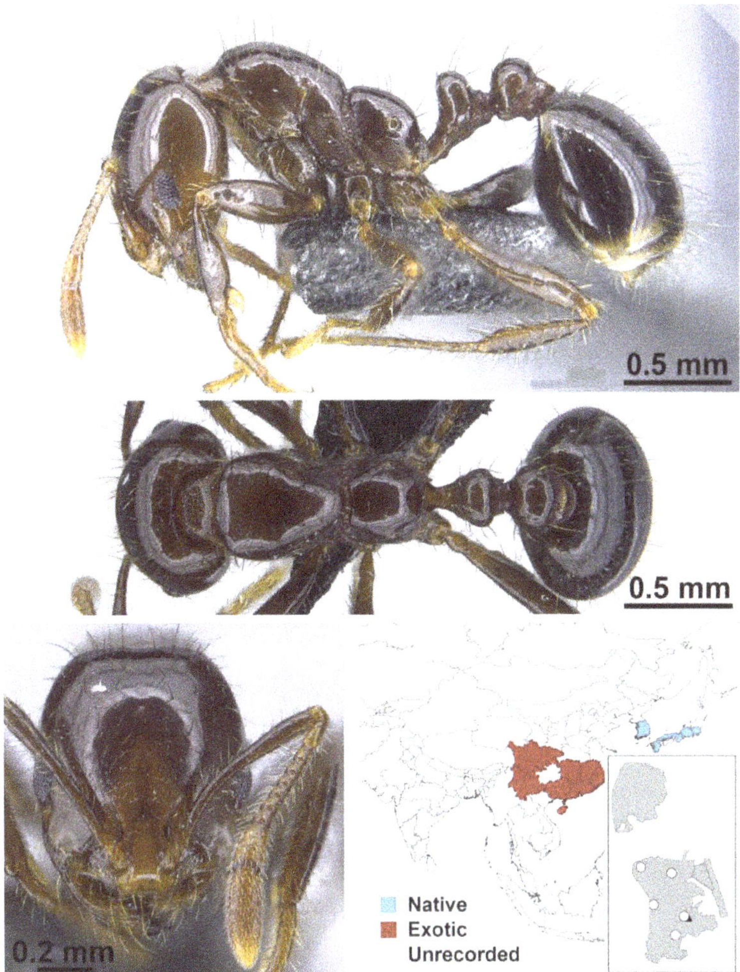

Figure A108. *Solenopsis invicta* Buren, 1972 worker (MAC_S09_LLSP_Sp.5, IBBL).

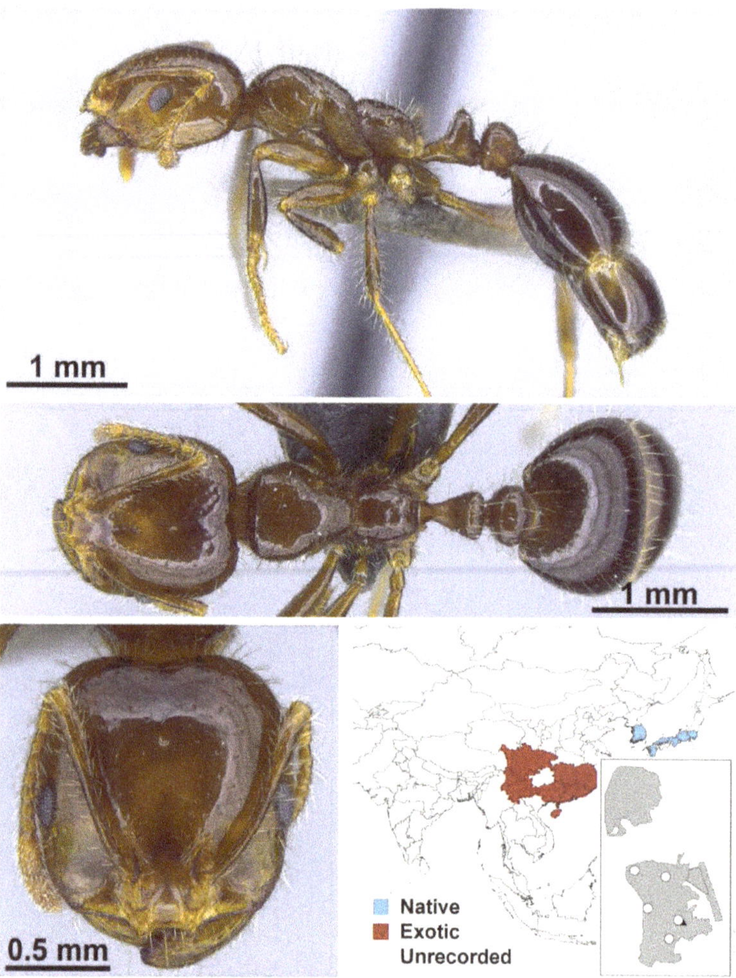

Figure A109. *Solenopsis invicta* Buren, 1972 major (MAC_S09_B05_sp.1_bottom, IBBL).

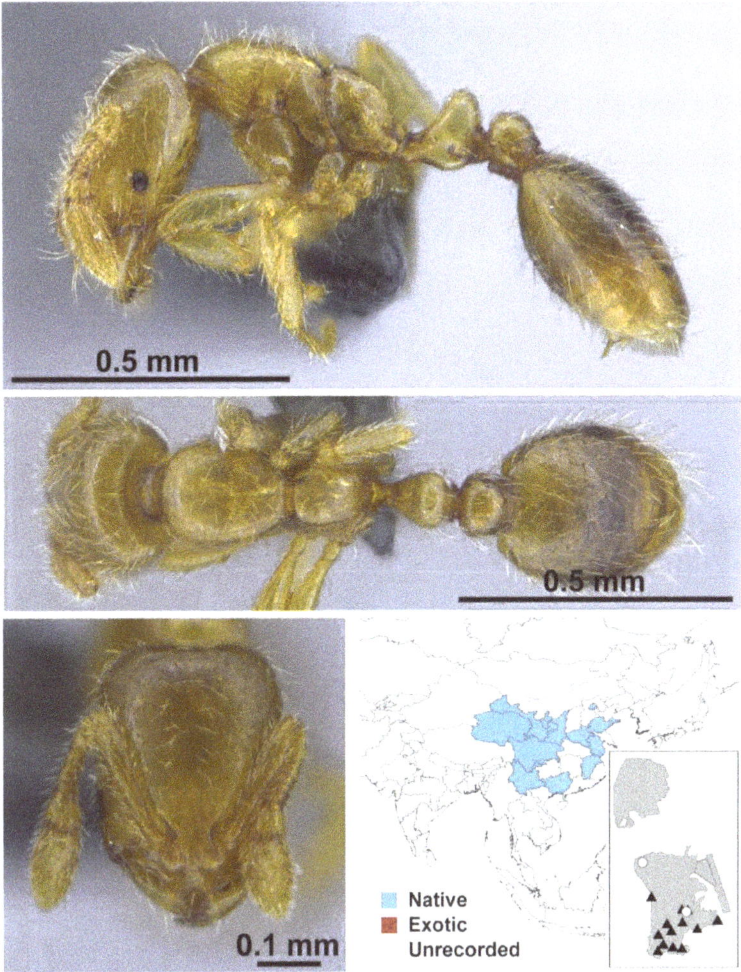

Figure A110. *Solenopsis jacoti* Wheeler, 1923 worker (MAC_S12_q3_37.5_sp.2_top, IBBL).

Figure A111. *Strumigenys elegantula* Terayama & Kubota, 1989 worker (MAC_S04_LLSP_sp.9, IBBL).

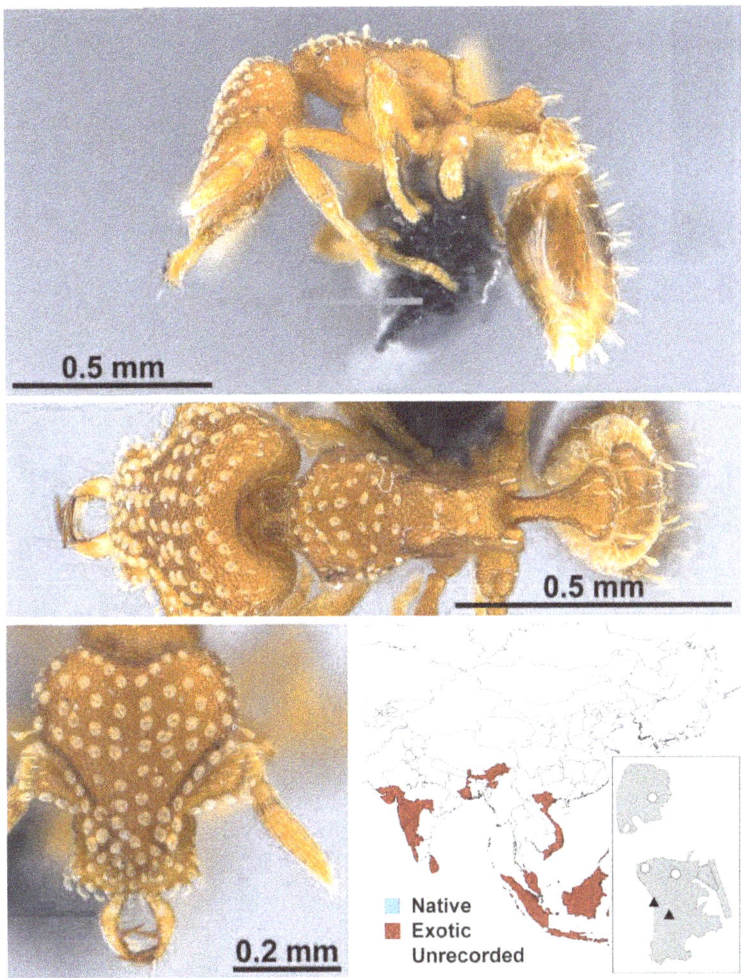

Figure A112. *Strumigenys emmae* Emery, 1890 worker (MAC_S20_LLSP_Sp.7, IBBL).

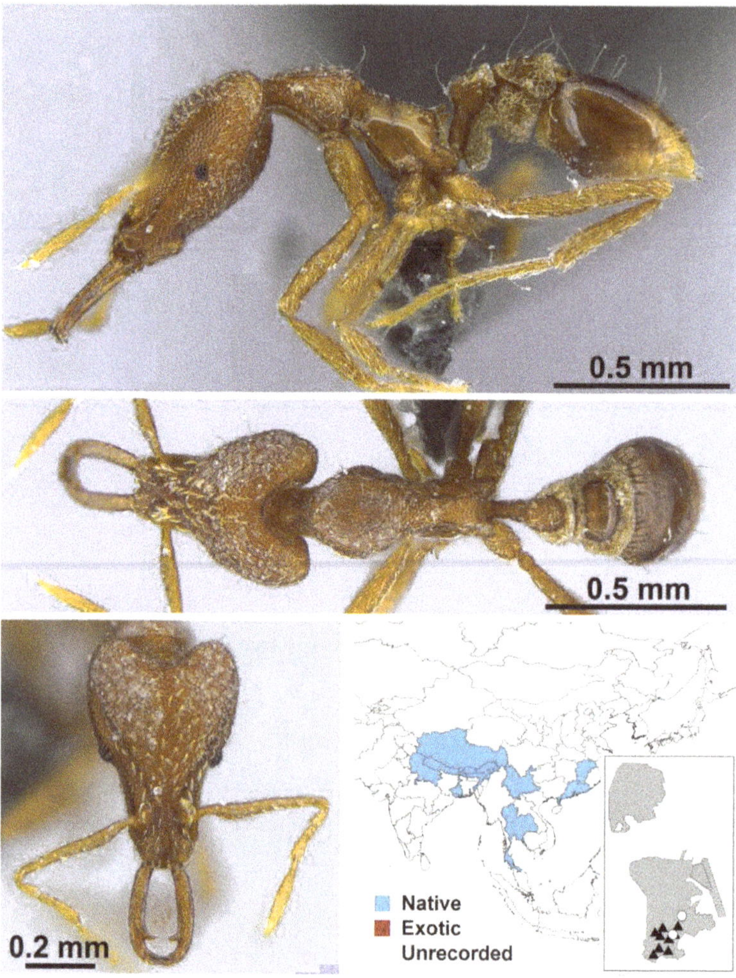

Figure A113. *Strumigenys exilirhina* Bolton, 2000 worker (MAC_S01_LLSA_Sp.3, IBBL).

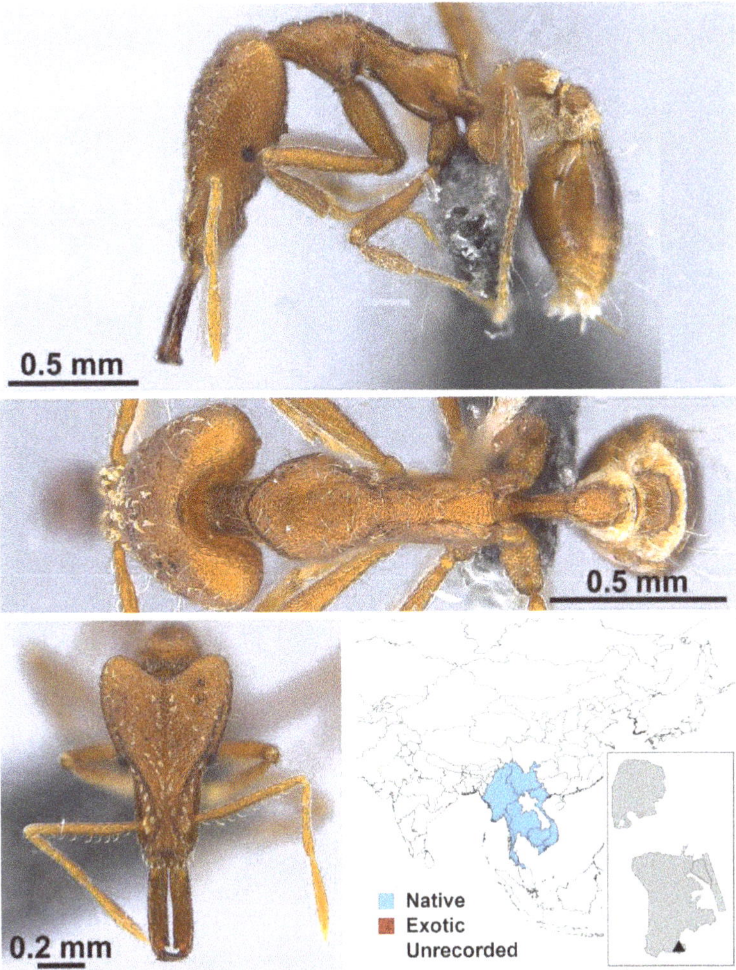

Figure A114. *Strumigenys feae* Emery, 1895 worker (MAC_S15_LLSP_Sp.8, IBBL).

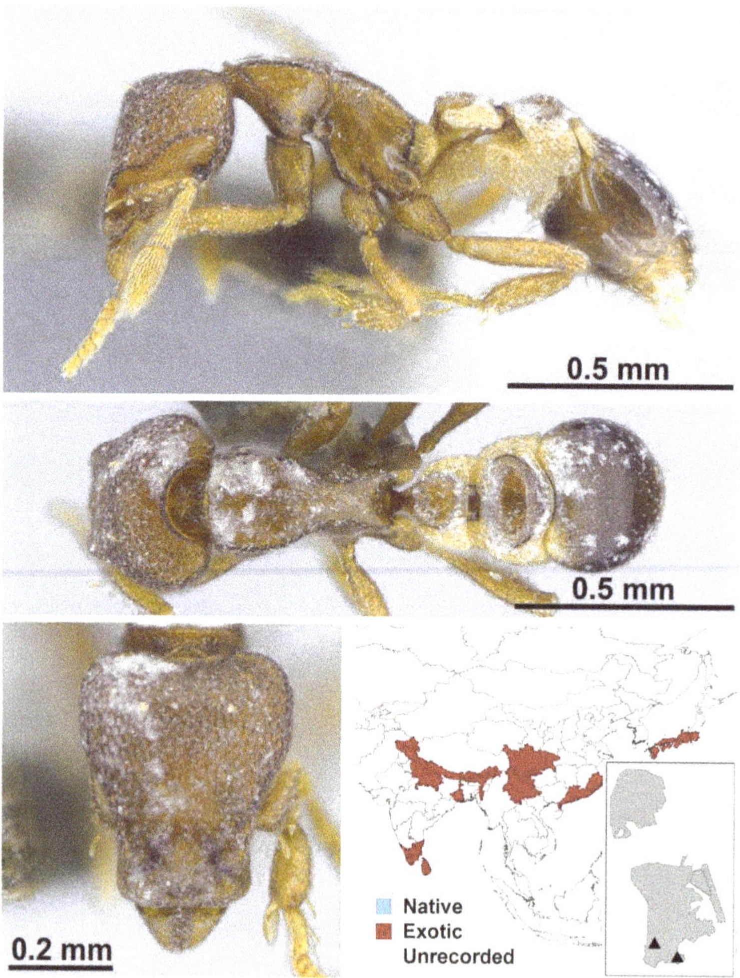

Figure A115. *Strumigenys membranifera* Emery, 1869 worker (MAC_S15_GN3_H3_n1_top, IBBL).

Figure A116. *Strumigenys minutula* Terayama & Kubota, 1989 worker (MAC_S14_LLSP_Sp.4, IBBL).

Figure A117. *Strumigenys nepalensis* Baroni Urbani & De Andrade, 1994 worker (MAC_S19_LLSP_Sp.3, IBBL).

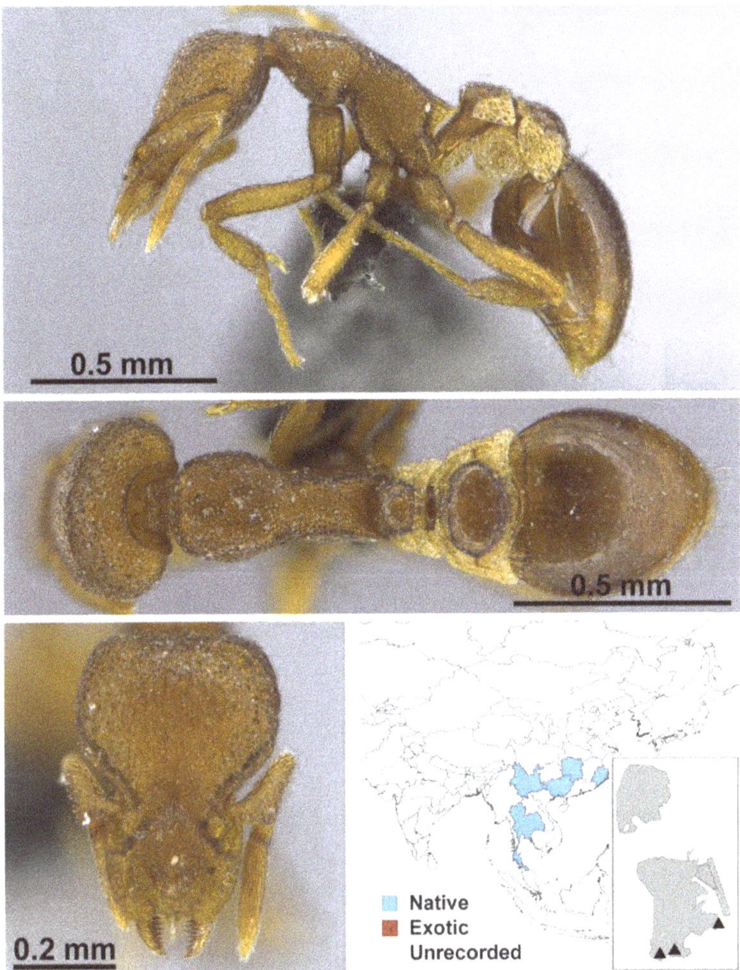

Figure A118. *Strumigenys sauteri* Forel, 1912 worker (MAC_S04_LLSP_sp.2, IBBL).

Figure A119. *Strumigenys subterranea* Brassard, Leong & Guénard, 2020 worker (MAC_S12_q4_12.5_sp.2, IBBL).

Figure A120. *Syllophopsis* nr. *cryptobia* worker (MAC_S12_q3_37.5_sp.3, IBBL).

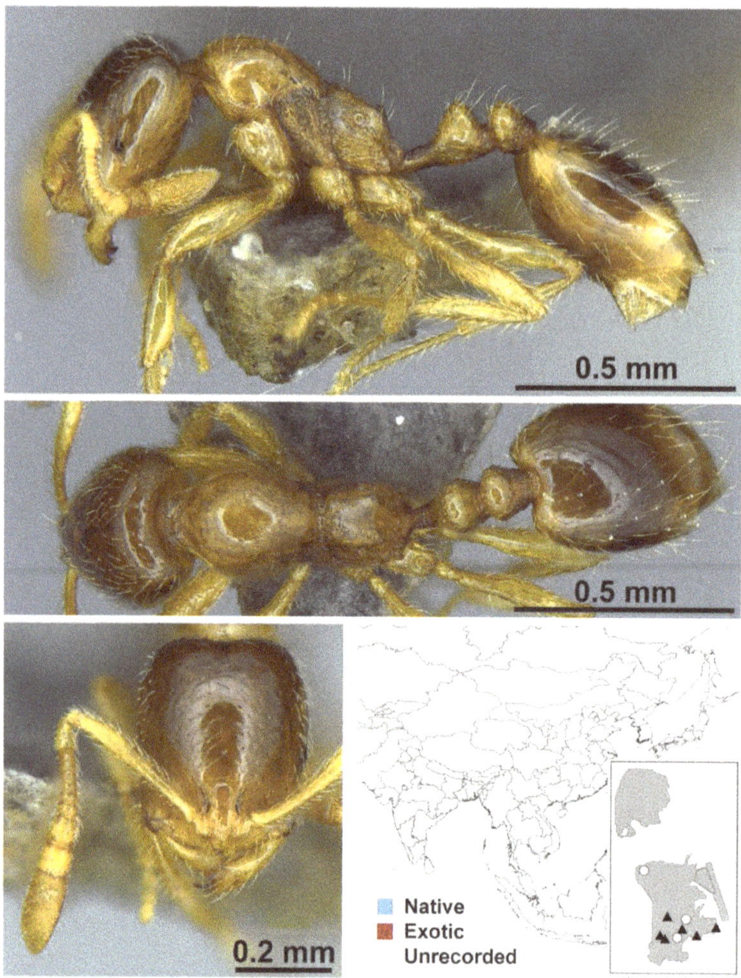

Figure A121. *Syllophopsis* sp. mo01 nr. *sechellensis* Emery, 1894 worker (*Syllophopsis* sp. mo01 nr. *Sechellensis*, CML collection).

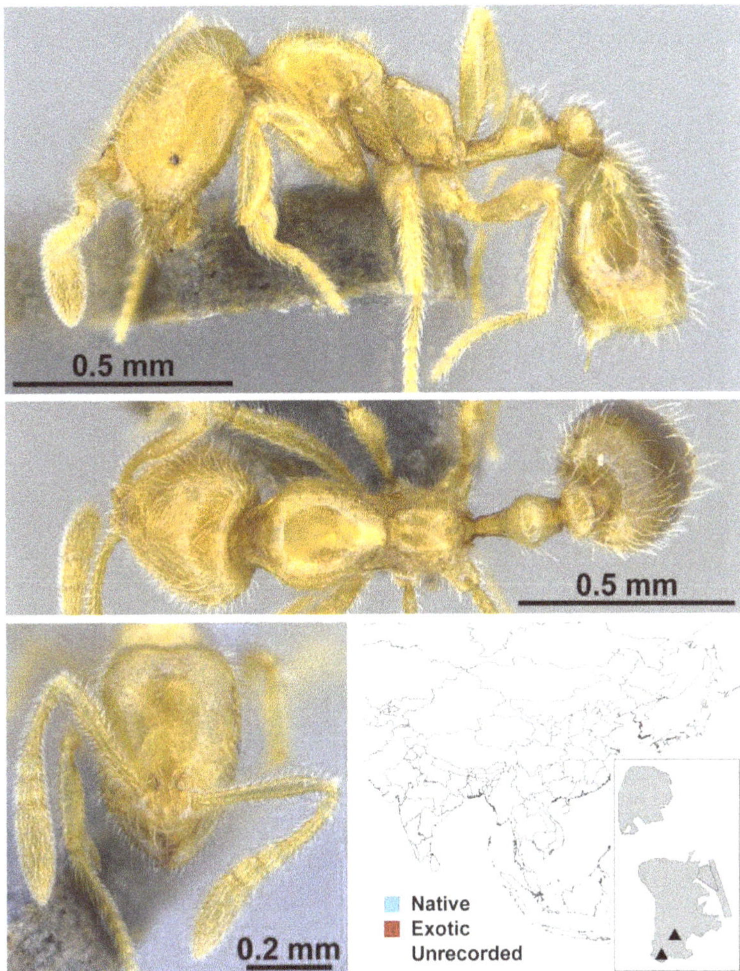

Figure A122. *Syllophopsis* sp. 1 FB worker (MAC_S18_q3_12.5_sp.4, IBBL).

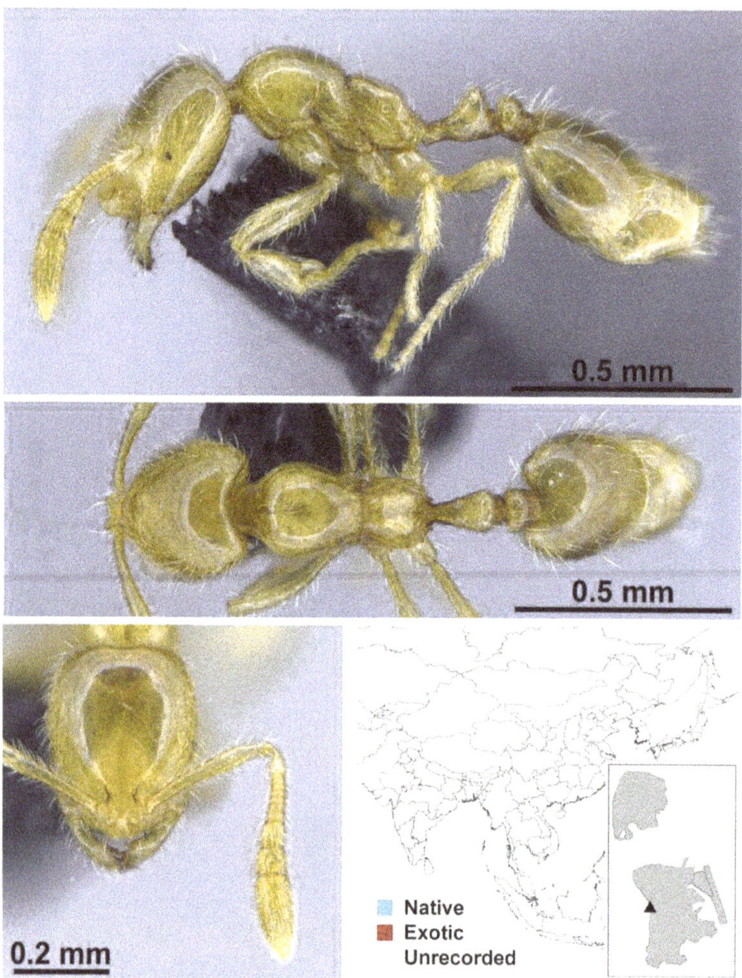

Figure A123. *Syllophopsis* sp. 2 FB worker (MAC_S20_LLSP_sp.10_top, IBBL).

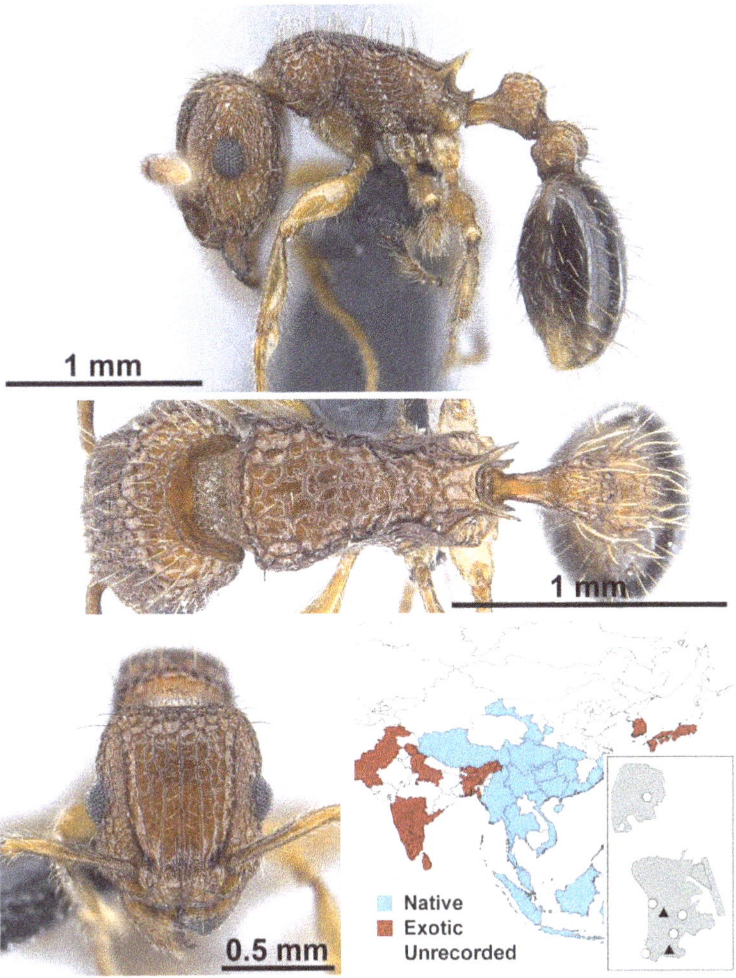

Figure A124. *Tetramorium bicarinatum* Nylander, 1846 worker (MAC_S19_LLSP_Sp.9, IBBL).

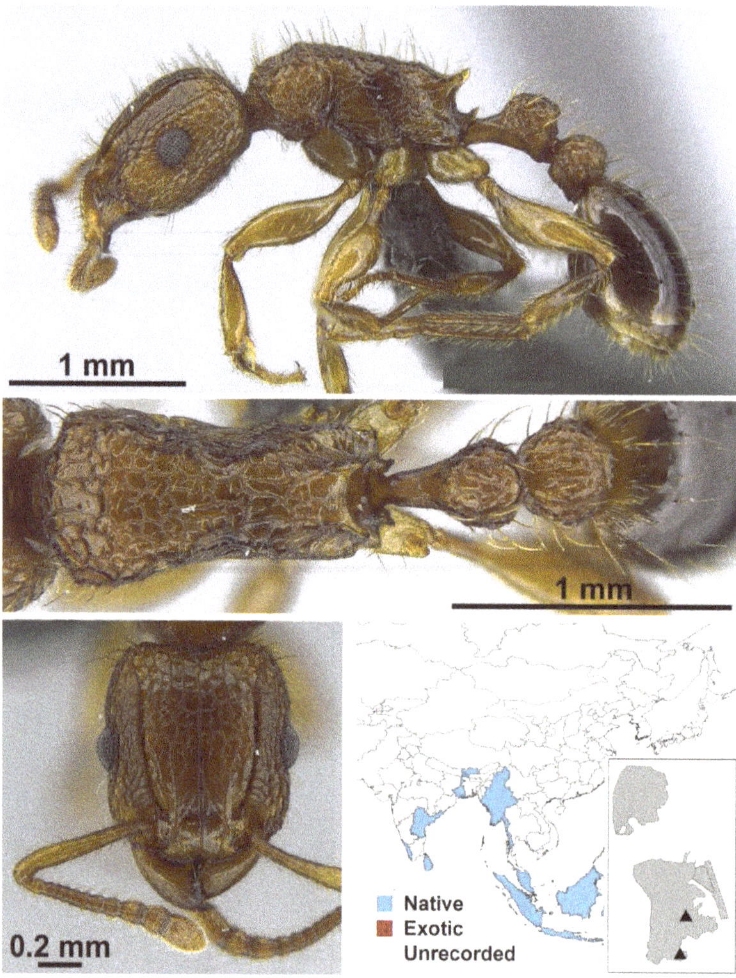

Figure A125. *Tetramorium indicum* Forel, 1913 worker (MAC_S08_T3_1m_sp.1, IBBL).

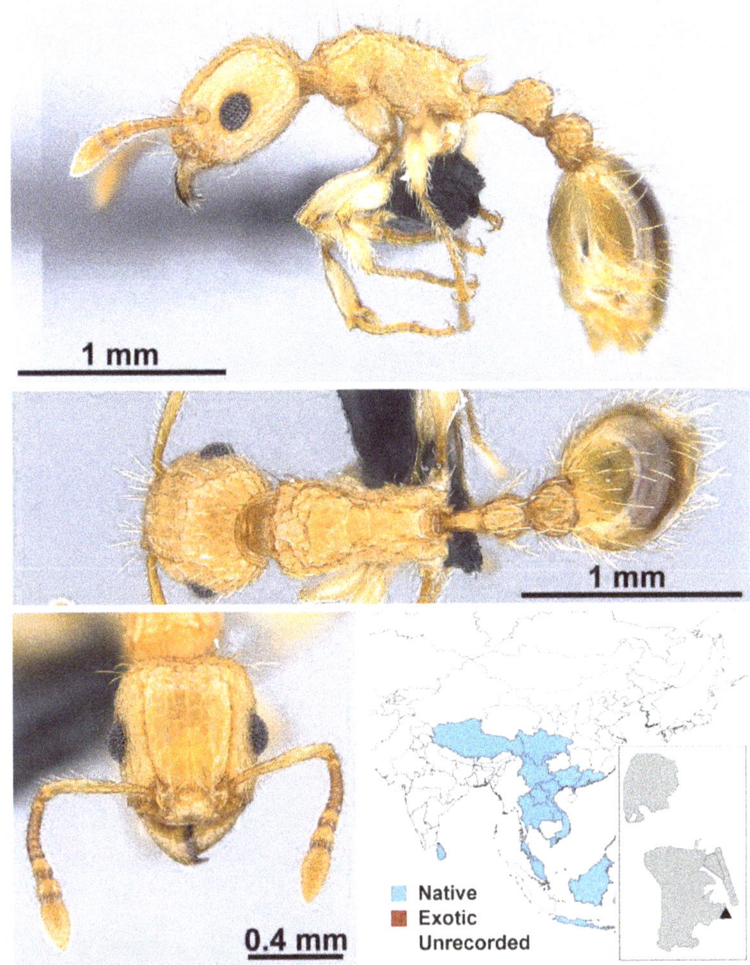

Figure A126. *Tetramorium insolens* Smith, 1861 worker (MAC_S17_LLSA_Sp.3, IBBL).

Figure A127. *Tetramorium kraepelini* Forel, 1905 worker (MAC_S8_GN1_H2_n1_sp.1, IBBL).

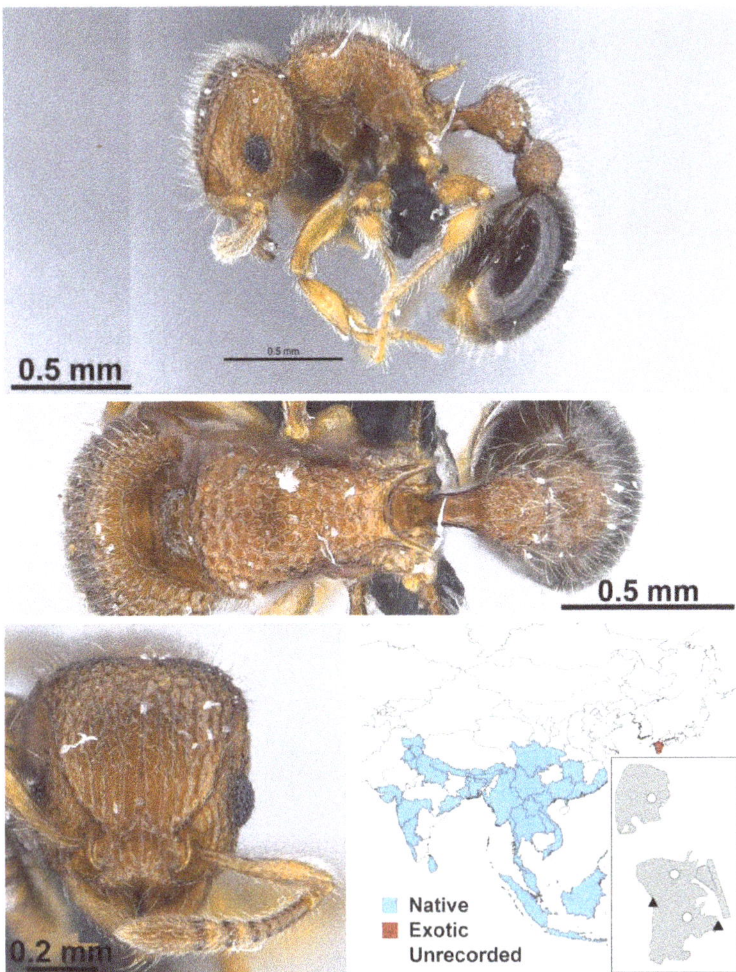

Figure A128. *Tetramorium lanuginosum* Mayr, 1870 worker (MAC_S17_LLSA_Sp.2, IBBL).

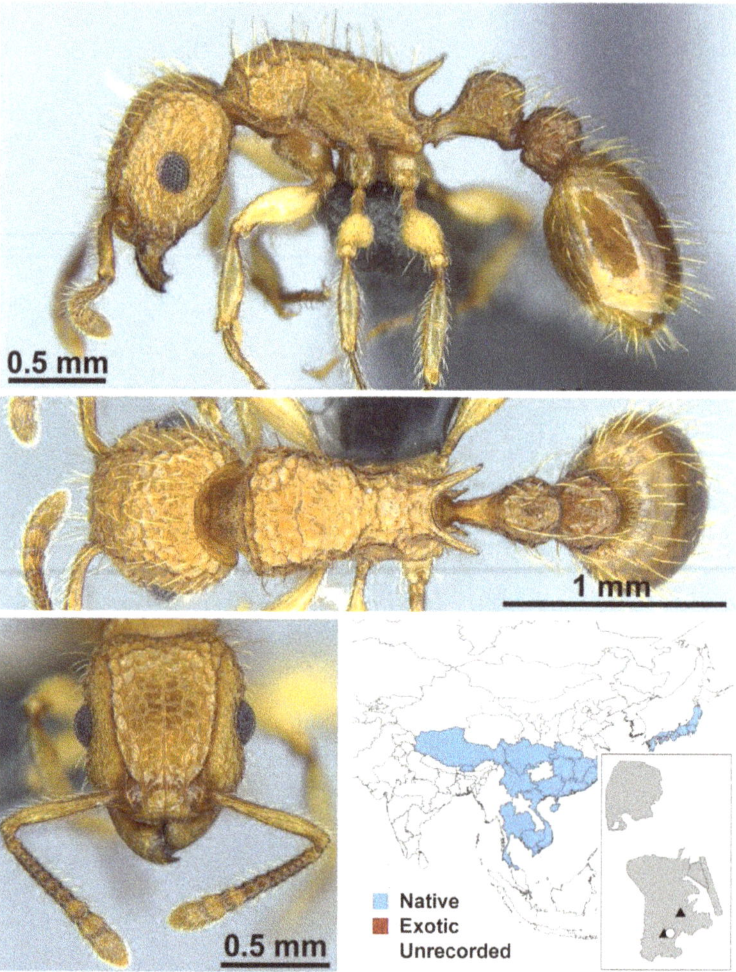

Figure A129. *Tetramorium nipponense* Wheeler, 1928 worker (MAC_S7_GN2_H4_n1, IBBL).

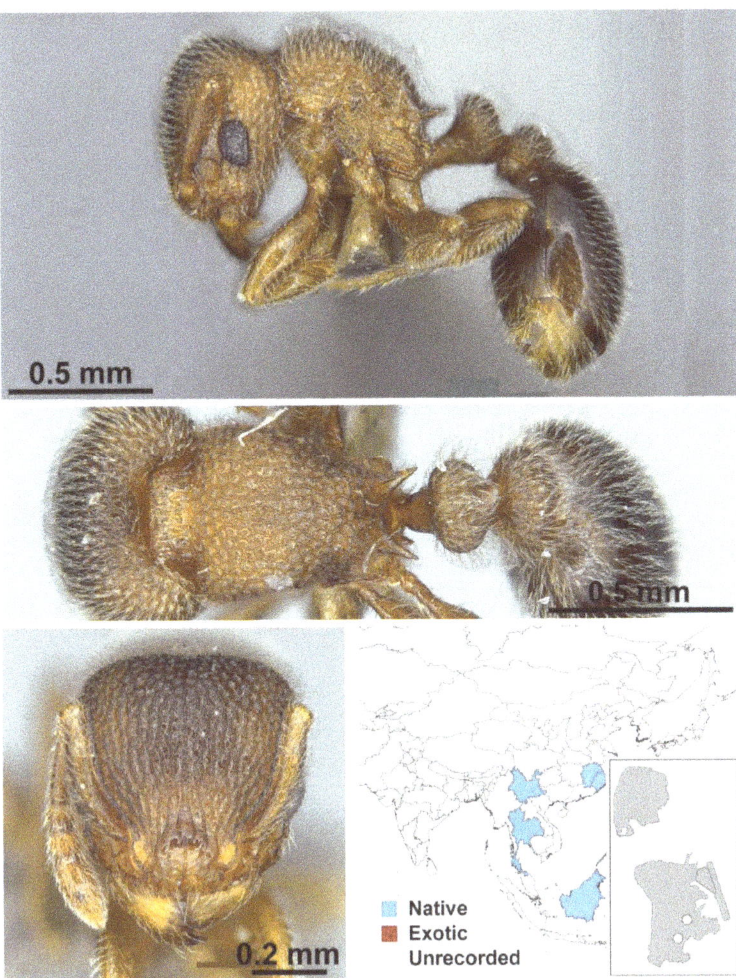

Figure A130. *Tetramorium parvispinum* Emery, 1893 worker (*Tetramorium parvispinum*, CML collection).

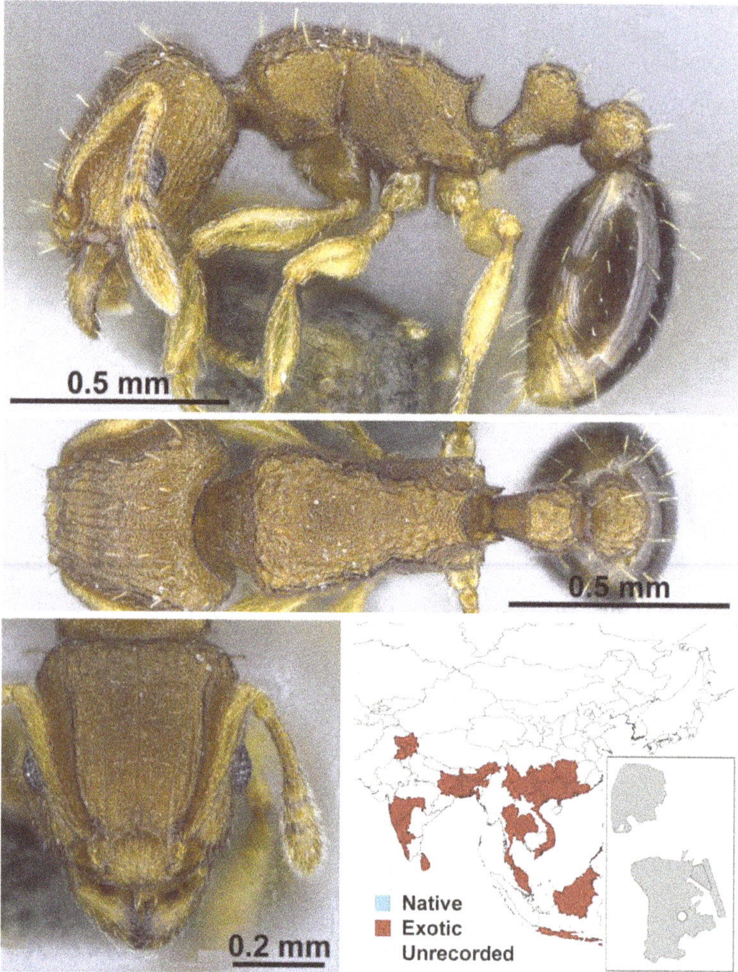

Figure A131. Tetramorium simillinum Smith, 1851 worker (Tetramorium simillinum, IBBL).

Figure A132. *Tetramorium tonganum* Mayr, 1870 worker (MAC_S10_T2_1m_sp.2, IBBL). Note that we changed the location of the map of Macao to show the localities where this species has been recorded as an exotic species in Southeast Asia.

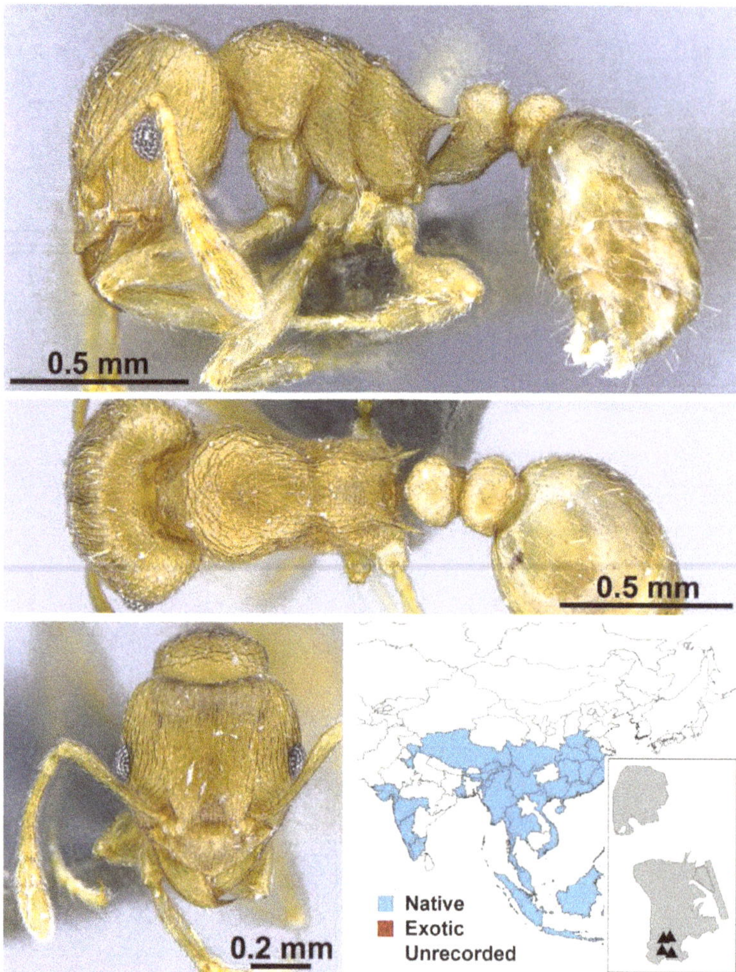

Figure A133. *Tetramorium wroughtonii* Forel, 1902 worker (MAC_S10_B03_sp.1_top, IBBL).

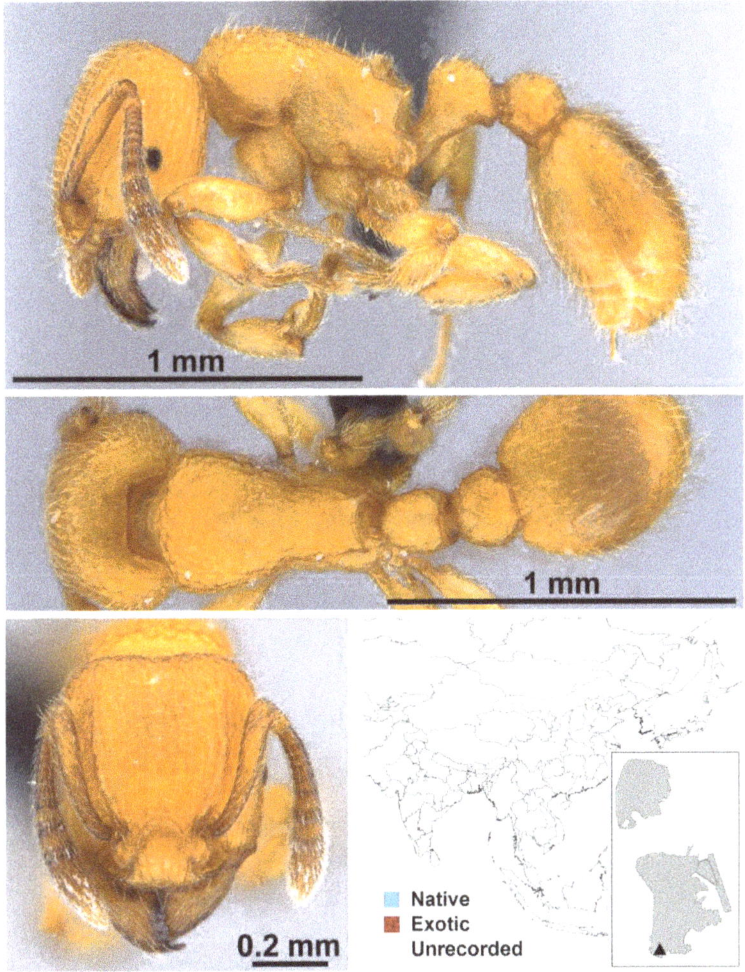

Figure A134. *Tetramorium* nr. *elisabethae* Forel, 1904 worker (MAC_S18_q1_25_Sp.2, IBBL).

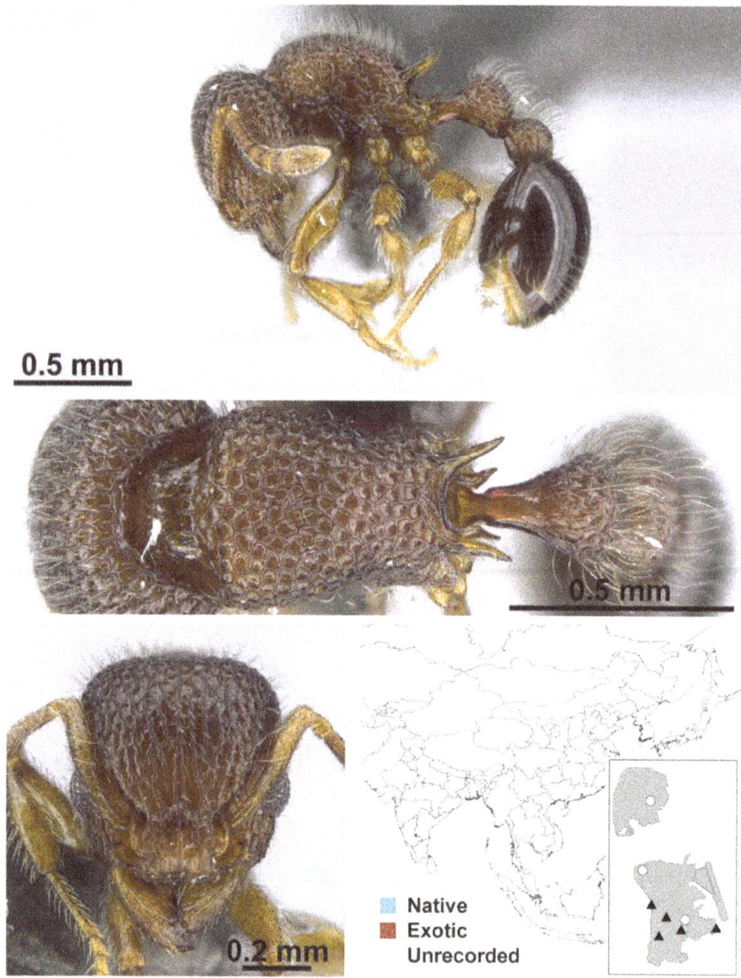

Figure A135. *Tetramorium* sp. 1 BG (*obesum* group Bolton, 1976) worker (MAC_S04_LLSP_Sp.3, IBBL).

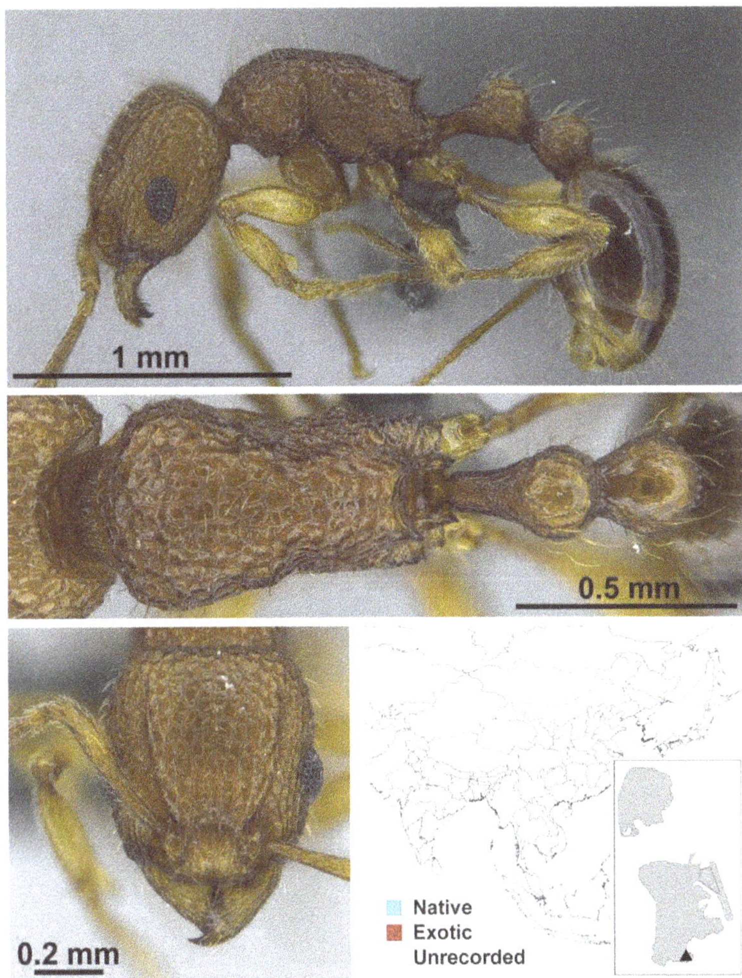

Figure A136. *Tetramorium* sp. 2 JF worker (MAC_S15_T1_3m_sp.2, IBBL).

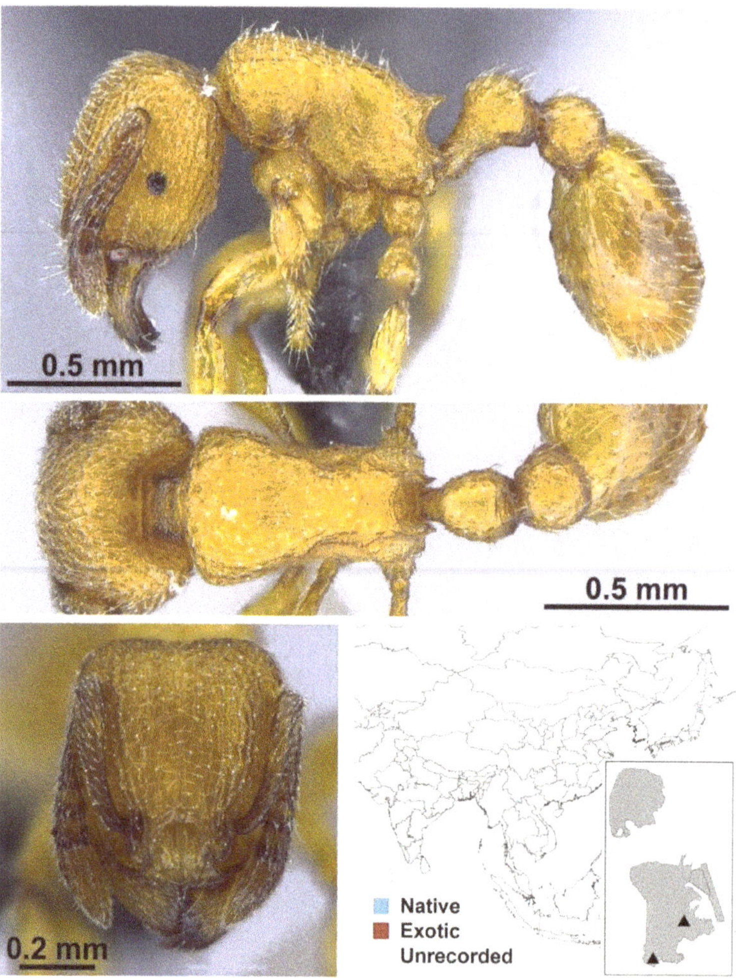

Figure A137. *Tetramorium* sp. 9 JF worker (MAC_S18_q2_25_sp.1, IBBL).

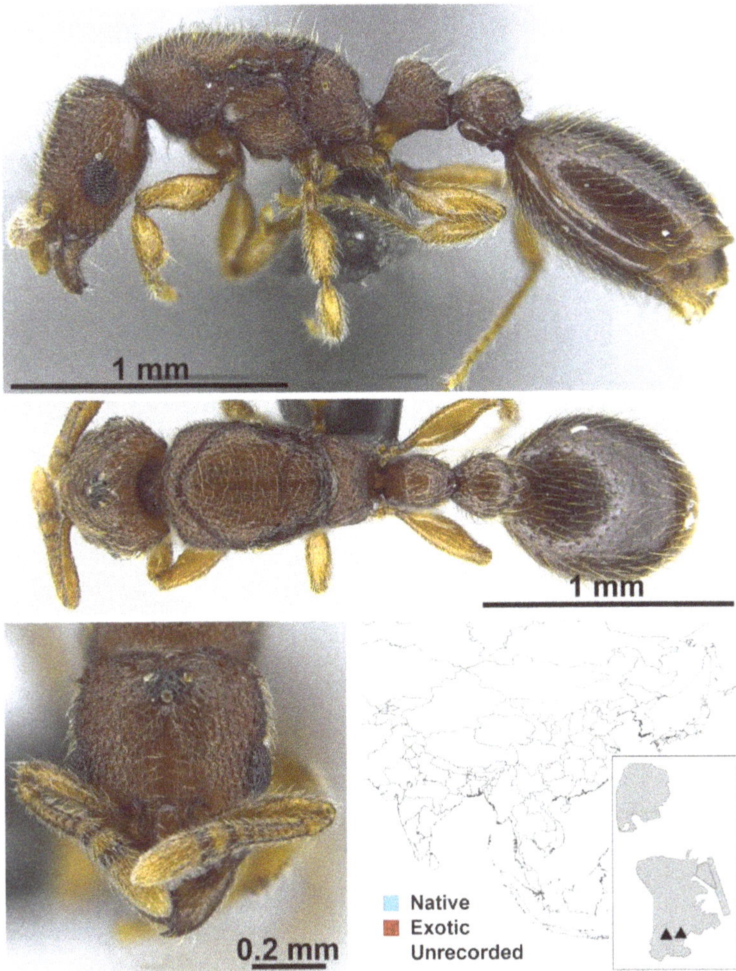

Figure A138. *Vollenhovia* sp. 1 BG queen (MAC_S06_GN3_H3_n1, IBBL).

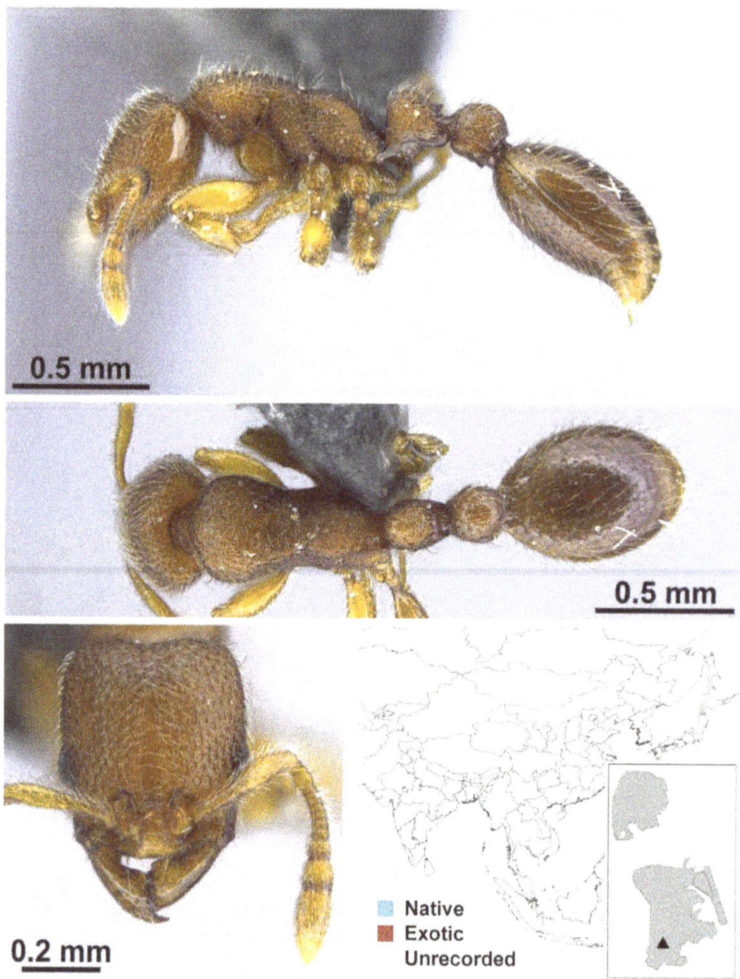

Figure A139. *Vollenhovia* sp. 2 BG worker (MAC_S02_LLSP_Sp.5, IBBL).

Appendix A.7 PONERINAE

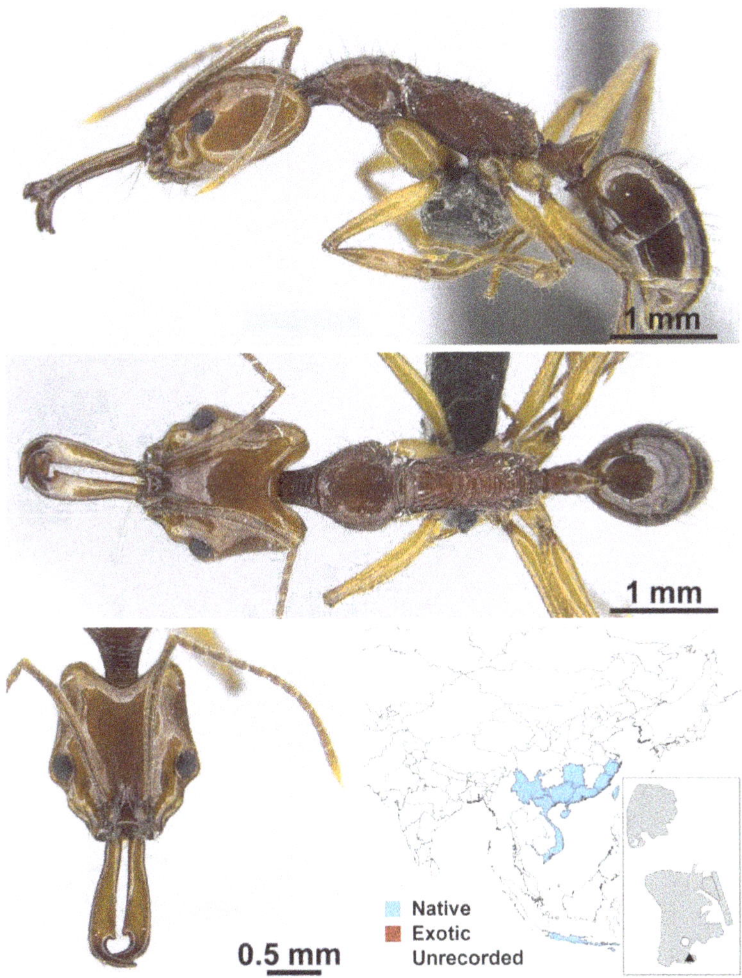

Figure A140. *Anochetus risii* Forel, 1900 worker (MAC_FB19180, IBBL).

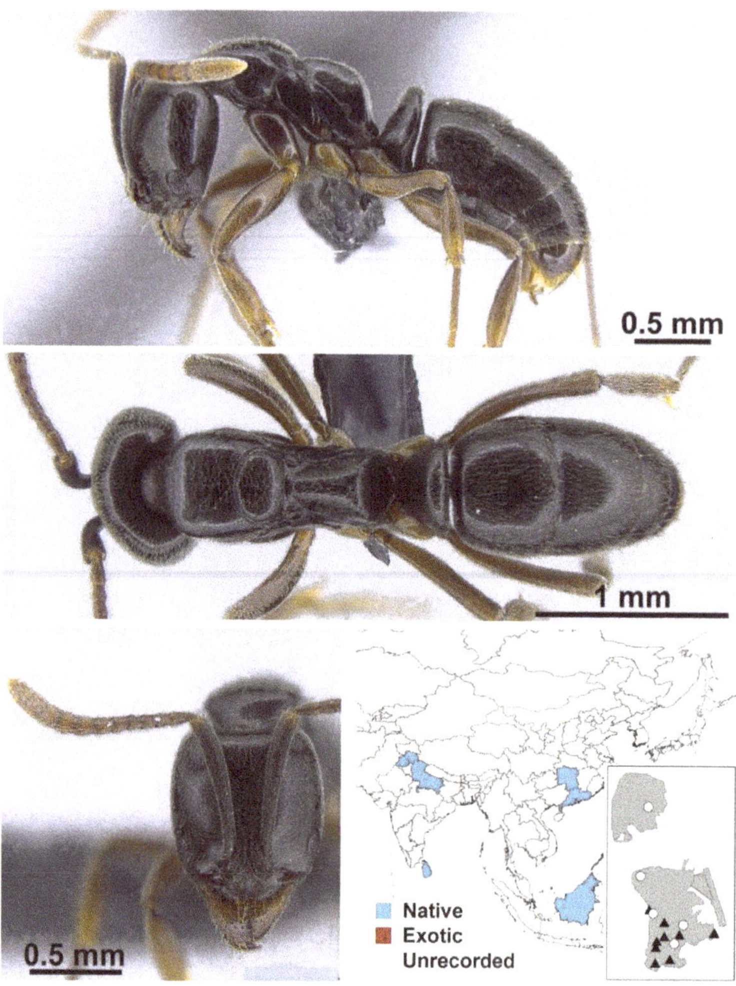

Figure A141. *Brachyponera obscurans* Mayr, 1862 worker (MAC_S19_LLSA, IBBL).

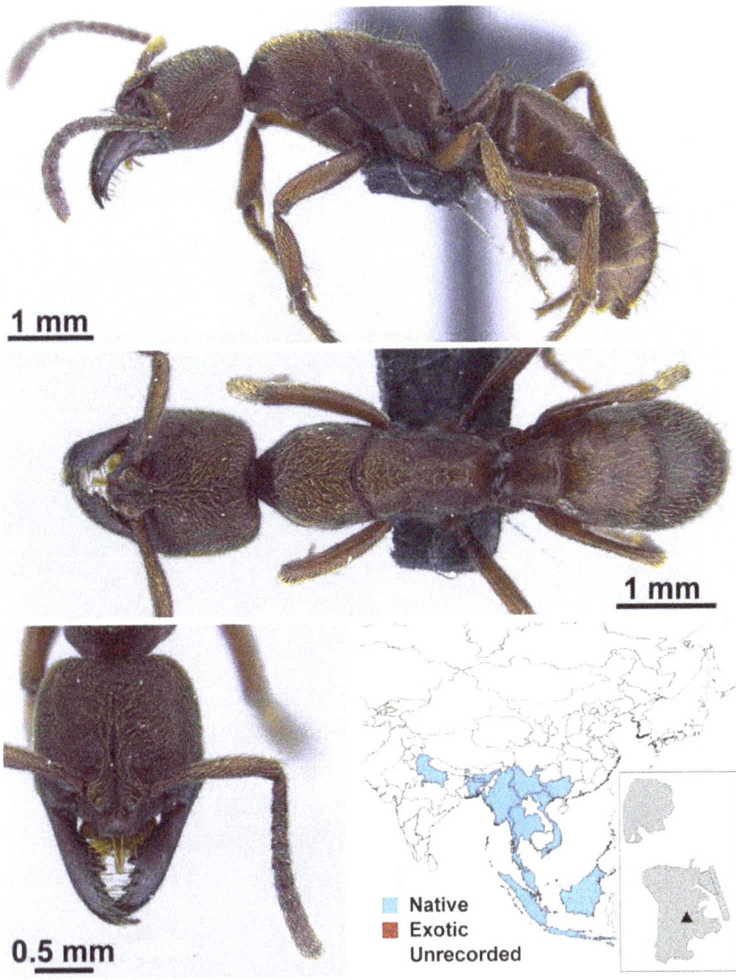

Figure A142. *Buniapone amblyops* Emery, 1887 worker (MAC_S12_q4_50_sp.2_top, IBBL).

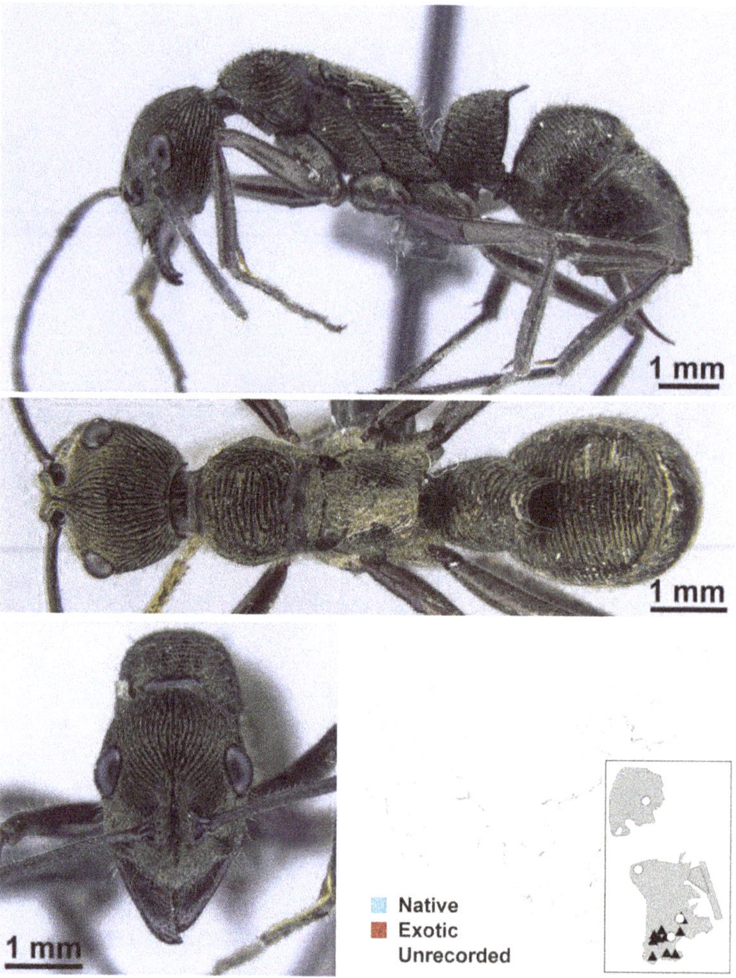

Figure A143. *Diacamma* sp. 1 worker (MAC_S15_LLSA_sp.1, IBBL).

Figure A144. *Ectomomyrmex annamitus* André, 1892 worker (MAC_S18_q2_37.5_Sp.4, IBBL).

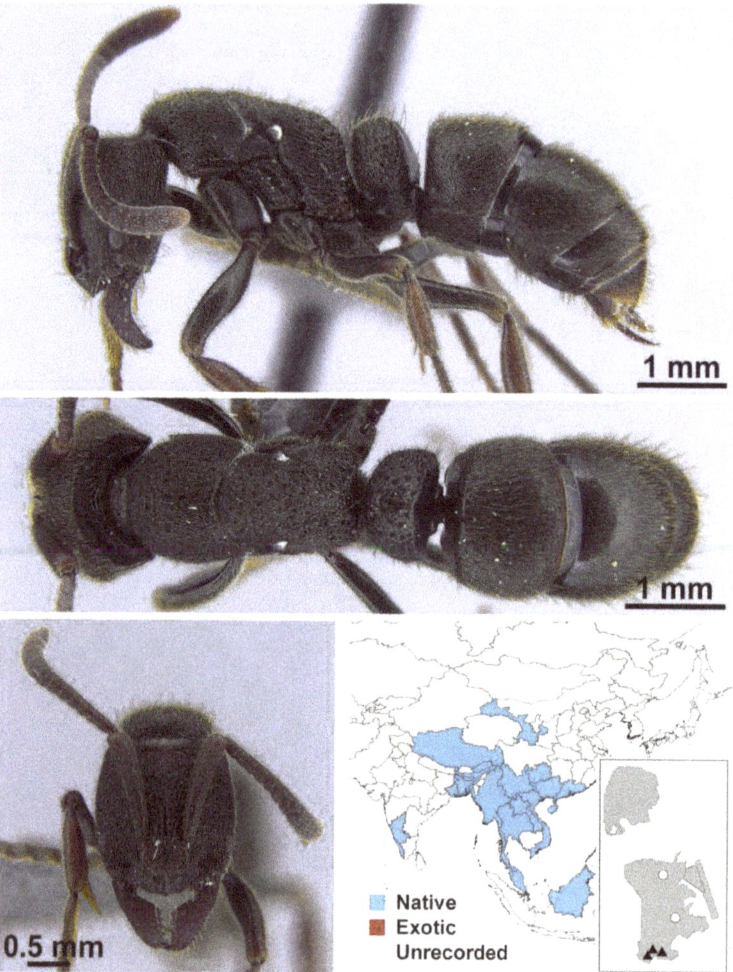

Figure A145. *Ectomomyrmex leeuwenhoecki* Forel, 1886 worker (MAC_S18_GN5_H4_n1, IBBL).

Figure A146. *Euponera pilosior* Wheeler, 1928 worker (MAC_S12_q3_50_Sp.1, IBBL).

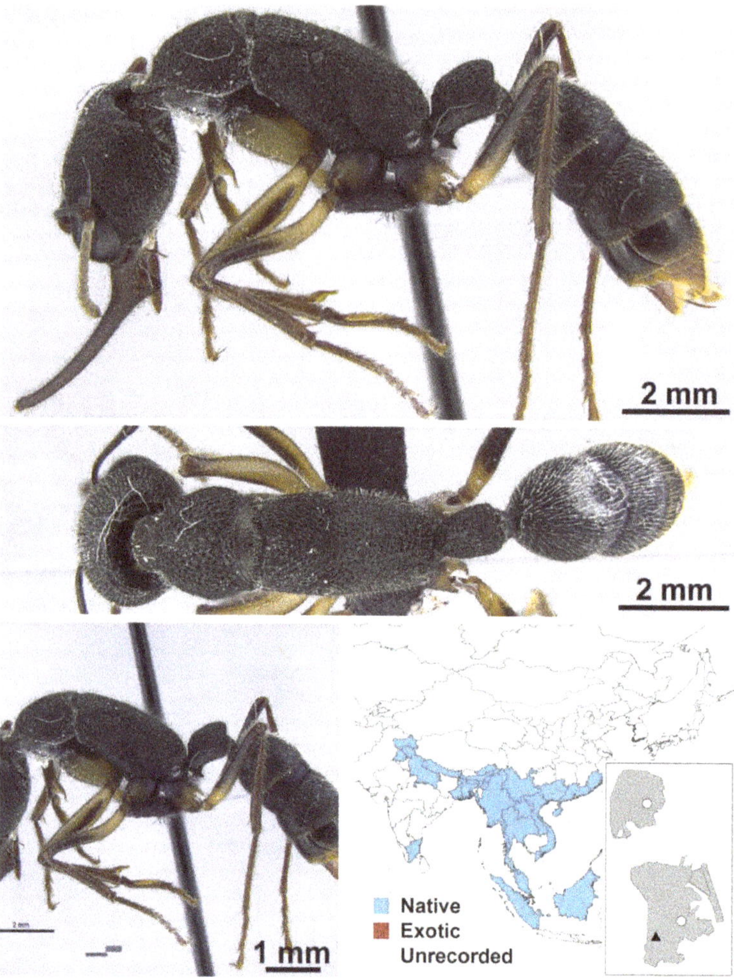

Figure A147. *Harpegnathos venator* Smith, 1858 worker (MAC_ZOO_HC07_Sp.1, IBBL).

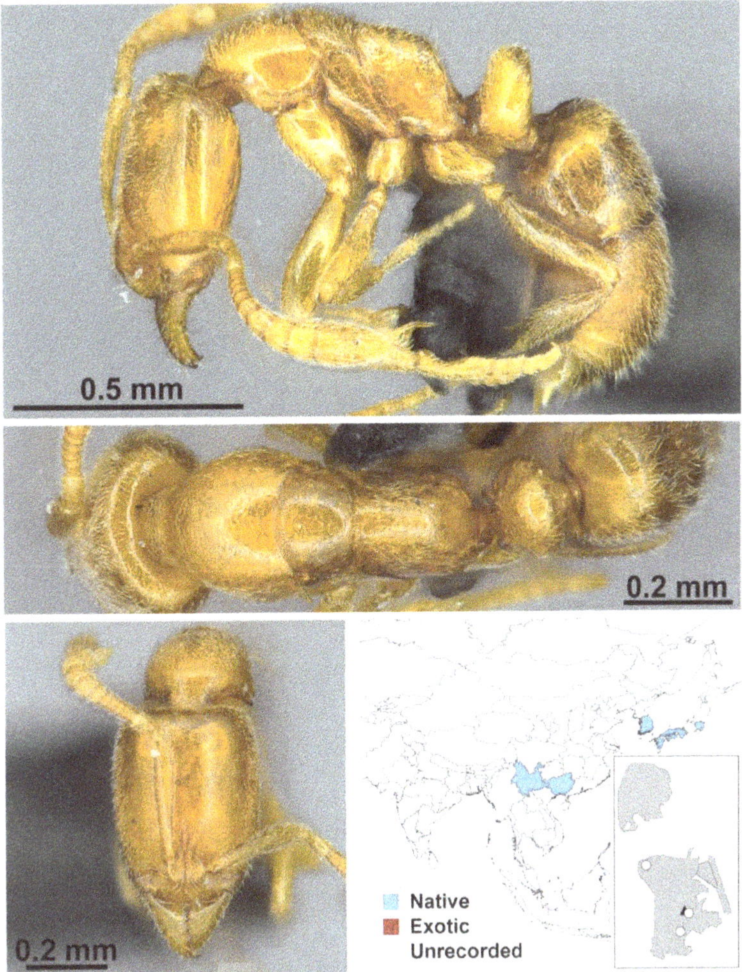

Figure A148. *Hypoponera exoecata* Wheeler, 1928 worker (*Hypoponera exoecata*, CML collection).

Figure A149. *Hypoponera* sp. psw-cn01 worker (*Hypoponera* sp. psw-cn01, CML collection).

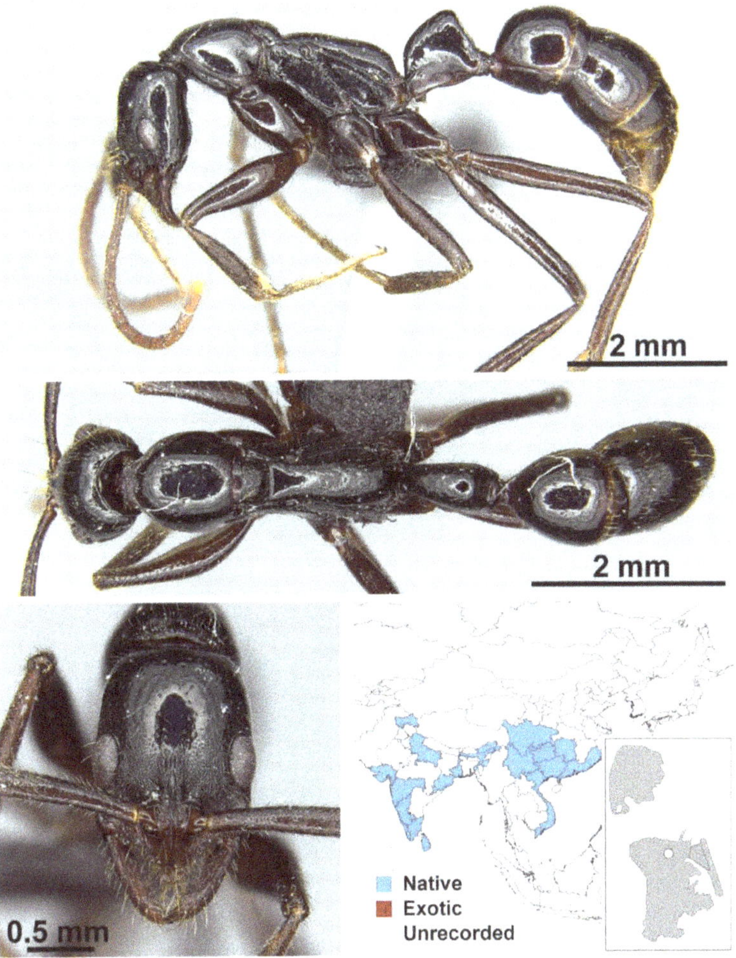

Figure A150. *Leptogenys chinensis* Mayr, 1870 worker (RHL00861, IBBL).

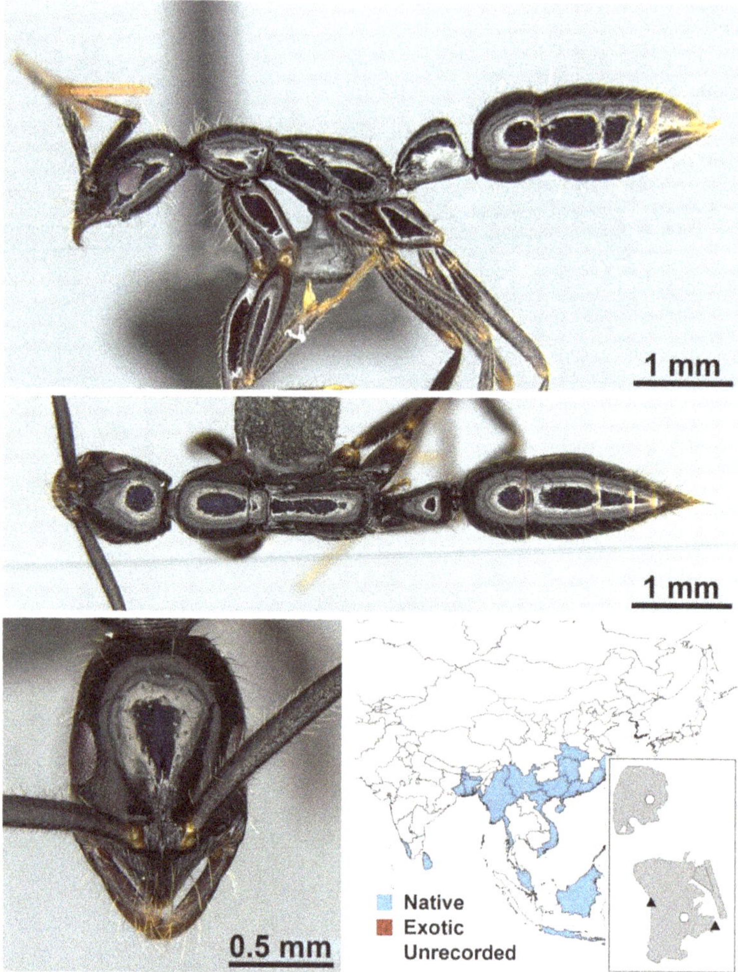

Figure A151. *Leptogenys peuqueti* André, 1887 worker (*Leptogenys peuqueti*, CML collection).

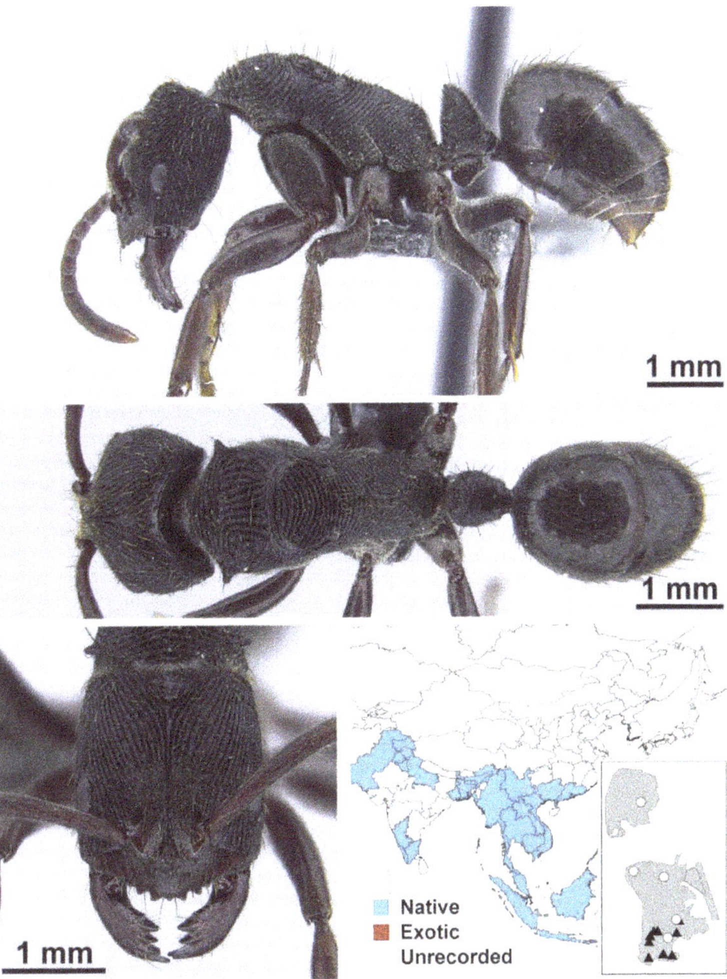

Figure A152. *Odontoponera denticulata* Smith, 1858 worker (MAC_S09_LLSA_Sp.3, IBBL).

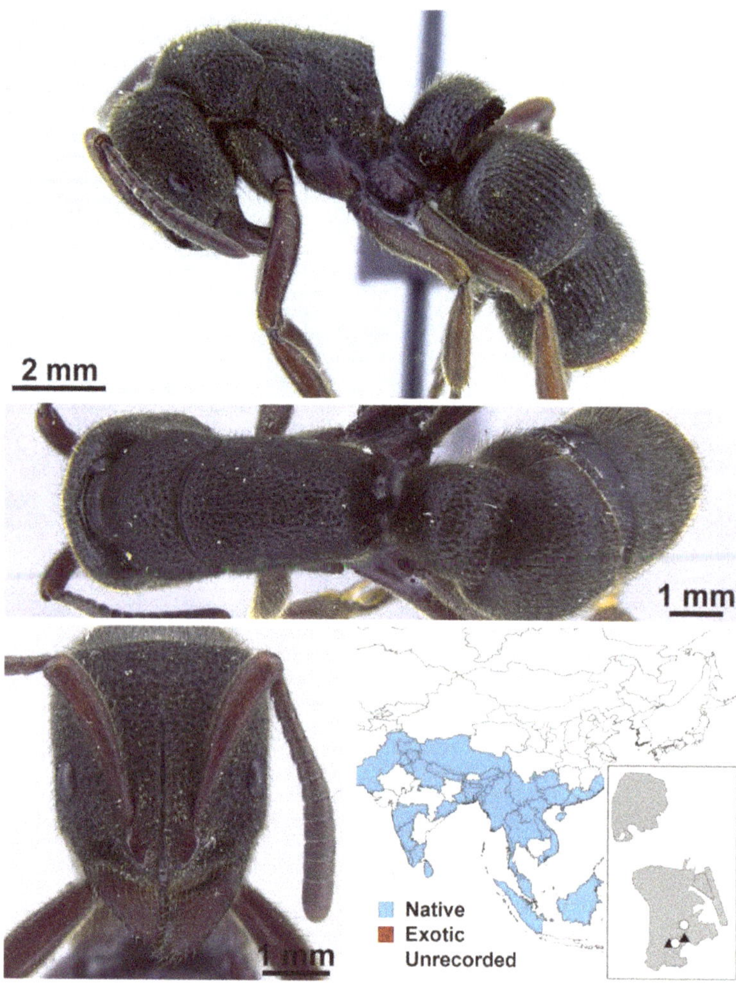

Figure A153. *Pseudoneoponera rufipes* Jerdon, 1851 worker (MAC_S03_LLSA_Sp.1, IBBL).

Appendix A.8 PROCERATIINAE

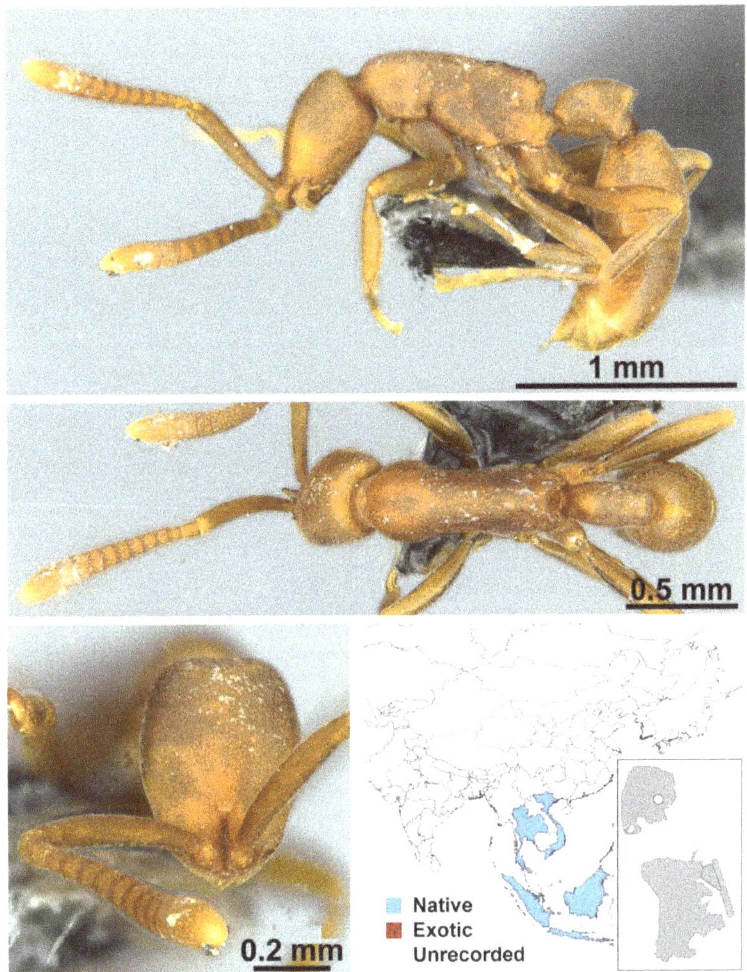

Figure A154. *Probolomyrmex dammermani* Wheeler, W. M., 1928 worker (*Probolomyrmex dammermani*, IBBL).

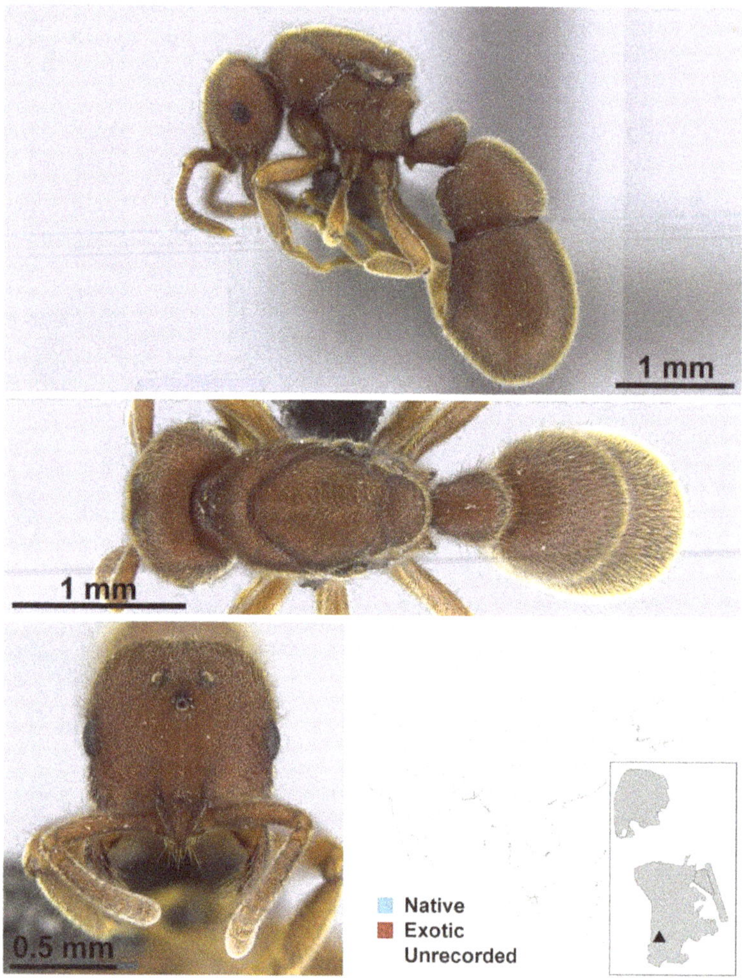

Figure A155. *Proceratium* sp. cf. *bruelheidei* Staab, Xu & Hita Garcia, 2018 queen (MAC_S05_LLSP_Sp.1, IBBL).

Appendix A.9 PSEUDOMYRMICINAE

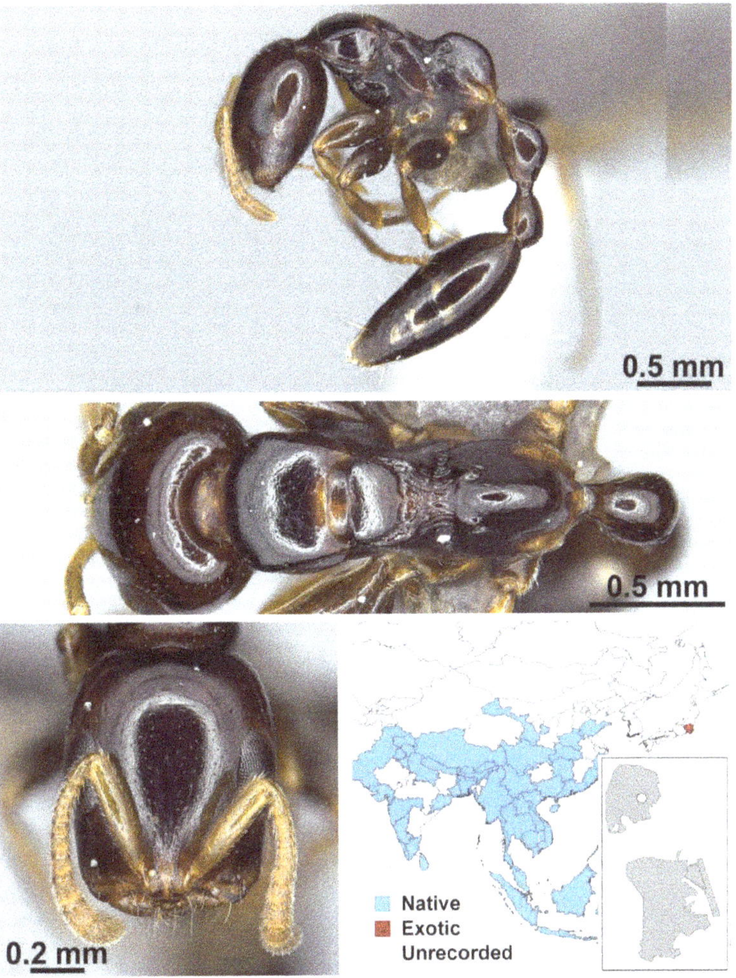

Figure A156. *Tetraponera allaborans* Walker, 1859 worker (*Tetraponera allaborans*, CML collection).

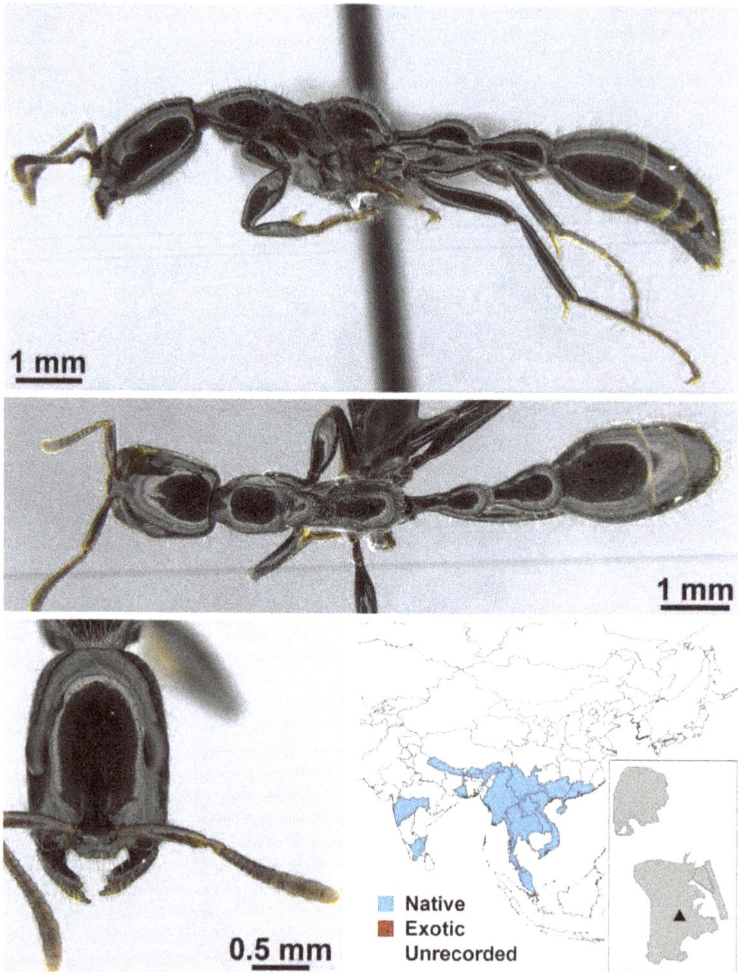

Figure A157. *Tetraponera binghami* Forel, 1902 worker (FB19140, IBBL).

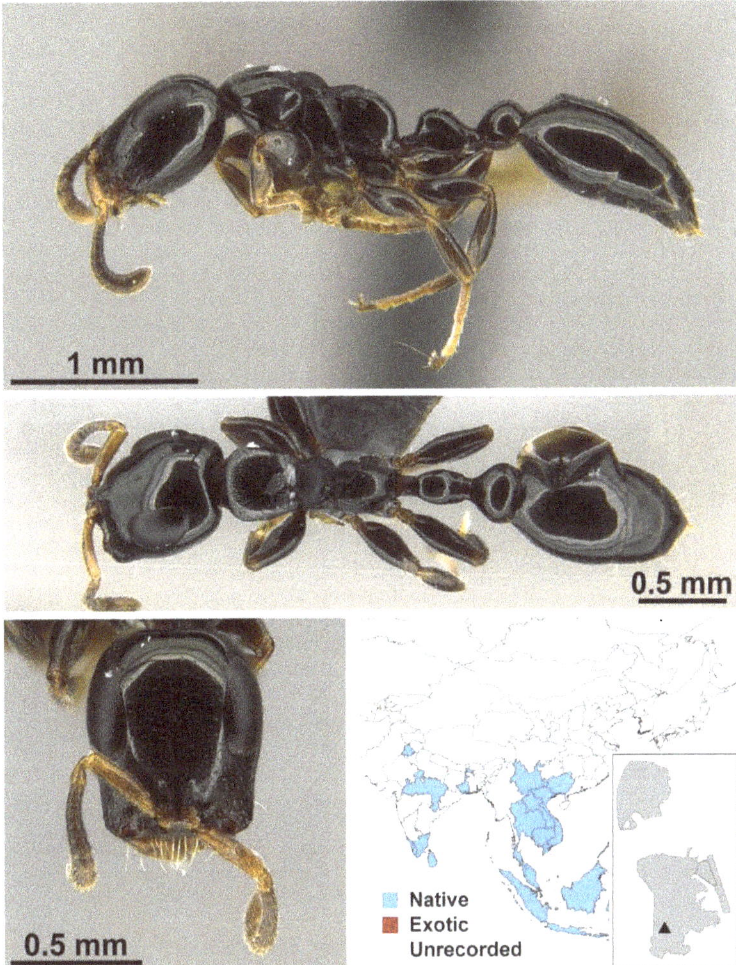

Figure A158. *Tetraponera nitida* Smith, 1860 worker (MAC_GOV_Workshop, IBBL).

References

1. United Nations. *World Urbanization Prospects 2018*, United Nations: New York, NY, USA, 2019.
2. Güneralp, B.; Seto, K.C. Futures of global urban expansion: Uncertainties and implications for biodiversity conservation. *Environ. Res. Lett.* **2013**, *8*, 1–10. [CrossRef]
3. Batty, M. The size, scale, and shape of cities. *Science (80-.)* **2008**, *319*, 769–771. [CrossRef] [PubMed]
4. McKinney, M.L. Effects of urbanization on species richness: A review of plants and animals. *Urban Ecosyst.* **2008**, *11*, 161–176. [CrossRef]
5. McKinney, M.L. Urbanization as a major cause of biotic homogenization. *Biol. Conserv.* **2006**, *127*, 247–260. [CrossRef]
6. Knop, E. Biotic homogenization of three insect groups due to urbanization. *Glob. Chang. Biol.* **2016**, *22*, 228–236. [CrossRef] [PubMed]
7. Fiera, C. Biodiversity of Collembola in urban soils and their use as bioindicators for pollution. *Pesqui. Agropecuária Bras.* **2009**, *44*, 868–873. [CrossRef]
8. Meyer, L.A.; Sullivan, S.M.P. Bright lights, big city: Influences of ecological light pollution on reciprocal stream-riparian invertebrate fluxes. *Ecol. Appl.* **2013**, *23*, 1322–1330. [CrossRef]
9. Pompeu, P.S.; Alves, C.B.M.; Callisto, M. The effects of urbanization on biodiversity and water quality in the Rio das Velhas Basin, Brazil. *Am. Fish. Soc. Symp.* **2005**, *47*, 11–22.

10. Buczkowski, G.; Richmond, D.S. The effect of urbanization on ant abundance and diversity: A temporal examination of factors affecting biodiversity. *PLoS ONE* **2012**, *7*, 1–9. [CrossRef]

11. Wang, H.F.; López-Pujol, J.; Meyerson, L.A.; Qiu, J.X.; Wang, X.K.; Ouyang, Z.Y. Biological invasions in rapidly urbanizing areas: A case study of Beijing, China. *Biodivers. Conserv.* **2011**, *20*, 2483–2509. [CrossRef]

12. Lepczyk, C.A.; Aronson, M.F.J.; Evans, K.L.; Goddard, M.A.; Lerman, S.B.; Macivor, J.S. Biodiversity in the City: Fundamental Questions for Understanding the Ecology of Urban Green Spaces for Biodiversity Conservation. *Bioscience* **2017**, *67*, 799–807. [CrossRef]

13. Ives, C.D.; Lentini, P.E.; Threlfall, C.G.; Ikin, K.; Shanahan, D.F.; Garrard, G.E.; Bekessy, S.A.; Fuller, R.A.; Mumaw, L.; Rayner, L.; et al. Cities are hotspots for threatened species. *Glob. Ecol. Biogeogr.* **2016**, *25*, 117–126. [CrossRef]

14. Kowarik, I. Novel urban ecosystems, biodiversity, and conservation. *Environ. Pollut.* **2011**, *159*, 1974–1983. [CrossRef]

15. Aronson, M.F.J.; La Sorte, F.A.; Nilon, C.H.; Katti, M.; Goddard, M.A.; Lepczyk, C.A.; Warren, P.S.; Williams, N.S.G.; Cilliers, S.; Clarkson, B.; et al. A global analysis of the impacts of urbanization on bird and plant diversity reveals key anthropogenic drivers. *Proc. R. Soc. B Biol. Sci.* **2014**, *281*, 1–8. [CrossRef]

16. Guénard, B.; Cardinal-De Casas, A.; Dunn, R.R. High diversity in an urban habitat: Are some animal assemblages resilient to long-term anthropogenic change? *Urban Ecosyst.* **2015**, *18*, 449–463. [CrossRef]

17. Donnelly, R.; Marzluff, J.M. Importance of reserve size and landscape context to urban bird conservation. *Conserv. Biol.* **2004**, *18*, 733–745. [CrossRef]

18. Evans, K.L.; Newson, S.E.; Gaston, K.J. Habitat influences on urban avian assemblages. *Ibis (Lond.)* **2009**, *151*, 19–39. [CrossRef]

19. Shwartz, A.; Muratet, A.; Simon, L.; Julliard, R. Local and management variables outweigh landscape effects in enhancing the diversity of different taxa in a big metropolis. *Biol. Conserv.* **2013**, *157*, 285–292. [CrossRef]

20. Beninde, J.; Veith, M.; Hochkirch, A. Biodiversity in cities needs space: A meta-analysis of factors determining intra-urban biodiversity variation. *Ecol. Lett.* **2015**, *18*, 581–592. [CrossRef] [PubMed]

21. Mack, R.N.; Simberloff, D.; Lonsdale, W.M.; Evans, H.; Clout, M.; Bazzaz, F.A. Biotic invasions: Causes, epidemiology, global consequences, and control. *Ecol. Appl.* **2000**, *10*, 689–710. [CrossRef]

22. Kowarik, I. On the role of alien species in urban flora and vegetation. In *Urban Ecology: An International Perspective on the Interaction Between Humans and Nature*; Marzluff, J., Shulenberger, E., Endlicher, W., Alberti, M., Bradley, G., Ryan, C., ZumBrunnen, C., Simon, U., Eds.; Springer: New York, NY, USA, 2008; p. 802. ISBN 9780387734118.

23. Ricotta, C.; La Sorte, F.A.; Pyšek, P.; Rapson, G.L.; Celesti-Grapow, L.; Thompson, K. Phyloecology of urban alien floras. *J. Ecol.* **2009**, *97*, 1243–1251. [CrossRef]

24. Costello, C.; Springborn, M.; McAusland, C.; Solow, A. Unintended biological invasions: Does risk vary by trading partner? *J. Environ. Econ. Manag.* **2007**, *54*, 262–276. [CrossRef]

25. Wittig, R. The origin and development of the urban flora of Central Europe. *Urban Ecosyst.* **2004**, *7*, 323–329. [CrossRef]

26. Von Der Lippe, M.; Kowarik, I. Do cities export biodiversity? Traffic as dispersal vector across urban-rural gradients. *Divers. Distrib.* **2008**, *14*, 18–25. [CrossRef]

27. Dawson, W.; Moser, D.; Van Kleunen, M.; Kreft, H.; Pergl, J.; Pyšek, P.; Weigelt, P.; Winter, M.; Lenzner, B.; Blackburn, T.M.; et al. Global hotspots and correlates of alien species richness across taxonomic groups. *Nat. Ecol. Evol.* **2017**, *1*, 1–7. [CrossRef]

28. Lu, J.; Li, S.P.; Wu, Y.; Jiang, L. Are Hong Kong and Taiwan stepping-stones for invasive species to the mainland of China? *Ecol. Evol.* **2018**, *8*, 1966–1973. [CrossRef] [PubMed]

29. Underwood, E.C.; Fisher, B.L. The role of ants in conservation monitoring: If, when, and how. *Biol. Conserv.* **2006**, *132*, 166–182. [CrossRef]

30. Agosti, D.; Majer, J.D.; Alonso, L.E.; Schultz, T.R. *Standard Methods for Measuring and Monitoring Biodiversity*; Smithsonian Institution Press: Washington, DC, USA, 2000; ISBN 1560988584.

31. Hölldobler, B.; Wilson, E.O. *The Ants*; Belknap Press of Harvard University Press: Cambridge, MA, USA, 1990; Volume N1, ISBN 0674040759.

32. Lach, L.; Parr, C.L.L.; Abbott, K. *Ant Ecology*, 1st ed.; Oxford University Press: Oxford, UK, 2010.

33. Andersen, A.N.; Majer, J.D. Ants show the way Down Under: Invertebrates as bioindicators in land management. *Front. Ecol. Environ.* **2004**, *2*, 291–298. [CrossRef]

34. Folgarait, P.J. Ant biodiversity and its relationship to ecosystem functioning: A review. *Biodivers. Conserv.* **1998**, *7*, 1221–1244. [CrossRef]

35. Holway, D.A.; Lach, L.; Suarez, A.V.; Tsutsui, N.D.; Case, T.J. The causes and consequences of ant invasions. *Annu. Rev. Ecol. Syst.* **2002**, *33*, 181–233. [CrossRef]

36. Holway, D.A. Edge effects of an invasive species across a natural ecological boundary. *Biol. Conserv.* **2005**, *121*, 561–567. [CrossRef]

37. Human, K.G.; Gordon, D.M. Exploitation and interference competition between the invasive Argentine ant, Linepithema humile, and native ant species. *Oecologia* **1996**, *105*, 405–412. [CrossRef]

38. Gerlach, J. Impact of the invasive crazy ant Anoplolepis gracilipes on Bird Island, Seychelles. *J. Insect Conserv.* **2004**, *8*, 15–25. [CrossRef]

39. Allen, C.R.; Birge, H.E.; Slater, J.; Wiggers, E. The invasive ant, Solenopsis invicta, reduces herpetofauna richness and abundance. *Biol. Invasions* **2017**, *19*, 713–722. [CrossRef]

40. Ness, J.H.; Bronstein, J.L. The effects of invasive ants on prospective ant mutualists. *Biol. Invasions* **2004**, *6*, 445–461. [CrossRef]

41. O'Dowd, D.J.; Green, P.T.; Lake, P.S. Invasional "meltdown" on an oceanic island. *Ecol. Lett.* **2003**, *6*, 812–817. [CrossRef]
42. Chan, K.H.; Guénard, B. Ecological and socio-economic impacts of the red import fire ant, *Solenopsis invicta* (Hymenoptera: Formicidae), on urban agricultural ecosystems. *Urban Ecosyst.* **2020**, *23*, 1–12. [CrossRef]
43. Lee, C.Y. Tropical Household Ants: Pest Status, Species Diversity, Foraging Behavior, and Baiting Studies. In Proceedings of the 4th International Conference on Urban Pests, Charleston, SC, USA, 7–10 July 2002.
44. Klotz, J.H.; Mangold, J.R.; Vail, K.M.; Davis, L.R.; Patterson, R.S. A survey of the urban pest ants (Hymenoptera: Formicidae) of Peninsular Florida. *Florida Entomol.* **1995**, *78*, 109–118. [CrossRef]
45. Fowler, H.G.; Bueno, O.C.; Sadatsune, T.; Montelli, A.C. Ants as potential vectors of pathogens in hospitals in the state of sao paulo, brazil. *Int. J. Trop. Insect Sci.* **1993**, *14*, 367–370. [CrossRef]
46. Beatson, S.H. Pharaoh's ants as pathogen vectors in hospitals. *Lancet* **1972**, *1*, 425–427. [CrossRef]
47. Brühl, C.A.; Gunsalam, G.; Linsenmair, K.E. Stratification of ants (Hymenoptera, Formicidae) in a primary rain forest in Sabah, Borneo. *J. Trop. Ecol.* **1998**, *14*, 285–297. [CrossRef]
48. Martins, M.F.d.O.; Thomazini, M.J.; Baretta, D.; Brown, G.G.; da Rosa, M.G.; Zagatto, M.R.G.; Santos, A.; Nadolny, H.S.; Cardoso, G.B.X.; Niva, C.C.; et al. Accessing the subterranean ant fauna (Hymenoptera: Formicidae) in native and modified subtropical landscapes in the neotropics. *Biota Neotrop.* **2020**, *20*, 1–16. [CrossRef]
49. Wilkie, K.T.R.; Mertl, A.L.; Traniello, J.F.A. Species diversity and distribution patterns of the ants of Amazonian ecuador. *PLoS ONE* **2010**, *5*, 1–12. [CrossRef]
50. Wong, M.K.L.; Guénard, B. Subterranean ants: Summary and perspectives on field sampling methods, with notes on diversity and ecology (Hymenoptera: Formicidae). *Myrmecol. News* **2017**, *25*, 1–16. [CrossRef]
51. Houadria, M.; Menzel, F. Digging deeper into the ecology of subterranean ants: Diversity and niche partitioning across two continents. *Diversity* **2021**, *13*, 53. [CrossRef]
52. Santos, M.N. Research on urban ants: Approaches and gaps. *Insectes Soc.* **2016**, *63*, 359–371. [CrossRef]
53. Leong, C.M.; Shiao, S.F.; Guénard, B. Ants in the city, a preliminary checklist of Formicidae (Hymenoptera) in Macau, one of the most heavily urbanized regions of the world. *Asian Myrmecol.* **2017**, *9*, 1–20. [CrossRef]
54. Brassard, F.; Leong, C.M.; Chan, H.H.; Guénard, B. A new subterranean species and an updated checklist of Strumigenys (Hymenoptera, formicidae) from Macao SAR, China, with a key to species of the greater bay area. *Zookeys* **2020**, *970*, 63–116. [CrossRef]
55. Wong, T.L.; Guénard, B. Review of ants from the genus Polyrhachis Smith (Hymenoptera:Formicidae:Formicinae) in Hong Kong and Macau, with notes on their natural history. *Asian Myrmecol.* **2021**; *13*, 1–70. [CrossRef]
56. Hui, E.C.M.; Li, X.; Chen, T.; Lang, W. Deciphering the spatial structure of China's megacity region: A new bay area—The Guangdong-Hong Kong-Macao Greater Bay Area in the making. *Cities* **2018**, *105*, 1–13. [CrossRef]
57. Statistics and Census Service Yearbook of Statistics 2019. Available online: www.dsec.gov.mo (accessed on 23 July 2021).
58. Grydehoj, A. Making Ground, Losing Space: Land Reclamation and Urban Public Space in Island Cities. *Urban Isl. Stud.* **2015**, *1*, 96–117. [CrossRef]
59. Chen, B.; Yang, Q.; Zhou, S.; Li, J.S.; Chen, G.Q. Urban economy's carbon flow through external trade: Spatial-temporal evolution for Macao. *Energy Policy* **2017**, *110*, 69–78. [CrossRef]
60. Ptak, R.; Souza, G.B. The Survival of Empire: Portuguese Trade and Society in China and the South China Sea, 1630–1754. *J. Am. Orient. Soc.* **1988**, *108*, 1630–1754. [CrossRef]
61. Leong, C.M.; Yamane, S.; Guénard, B. Lost in the city: Discovery of the rare ant genus Leptanilla (Hymenoptera: Formicidae) in Macau with description of Leptanilla macauensis sp. nov. *Asian Myrmecol.* **2018**, *10*, 1–16. [CrossRef]
62. Kouakou, L.M.M.; Yeo, K.; Kone, M.; Ouattara, K.; Kouakou, A.K.; Delsinne, T.; Dekoninck, W. Espaces verts comme une alternative de conservation de la biodiversité en villes: Le cas des fourmis (Hyménoptère: Formicidae) dans le district d'Abidjan (Côte d'Ivoire). *J. Appl. Biosci.* **2019**, *131*, 13359–13381. [CrossRef]
63. Yeo, K.; Dekoninck, W.; Kouakou, L.M.; Ouattara, K.; Konate, S. Detecting intruders: Assessment of the anthropophilic ant fauna (Hymenoptera: Formicidae) in the city of Abidjan and along access roads in Banco National Park (Côte d'Ivoire). *J. Entomol. Zool. Stud. JEZS* **2016**, *4*, 351–359.
64. Kasseney, B.D.; N'tie, T.B.; Nuto, Y.; Wouter, D.; Yeo, K.; Glitho, I.A. Diversity of ants and termites of the botanical garden of the university of lomé, Togo. *Insects* **2019**, *10*, 218. [CrossRef]
65. Taheri, A.; Wetterer, J.K.; Reyes-López, J. Tramp ants of Tangier, Morocco. *Trans. Am. Entomol. Soc.* **2017**, *143*, 267–270. [CrossRef]
66. Bernard, F. Fourmis des villes et fourmis du bled entre Rabat et Tanger. *Société Sci. Nat. Phys. Maroc* **1958**, *38*, 131–142.
67. Bernard, F. Les Fourmis des rues de Kenitra (Maroc) [Hym.]. *Bull. Société Entomol. Fr.* **1974**, *79*, 178–183.
68. Reyes-López, J.; Carpintero, S. Comparison of the exotic and native ant communities (Hymenoptera: Formicidae) in urban green areas at inland, Coastal and insular sites in Spain. *Eur. J. Entomol.* **2014**, *111*, 421–428. [CrossRef]
69. Ito, F.; Yamane, S.; Eguchi, K.; Noerdjito, W.A.; Kahono, S.; Tsuji, K.; Ohkawara, K.; Yamauchi, K.; Nishida, T.; Nakamura, K. Ant Species Diversity in the Bogor Botanic Garden, West Java, Indonesia, with Descriptions of Two New Species of the Genus Leptanilla (Hymenoptera, Formicidae). *Tropics* **2001**, *10*, 379–404. [CrossRef]
70. Rizali, A.; Bos, M.M.; Buchori, D.; Yamane, S.; Schulze, C.H. Ants in Tropical Urban Habitats: The Myrmecofauna in a Densely Populated Area of Bogor, West Java, Indonesia. *HAYATI J. Biosci.* **2008**, *15*, 77–84. [CrossRef]
71. Terayama, M. Ants from the Akasaka Imperial Gardens, Tokyo. *Mem. Natl. Sci. Museum* **2005**, *39*, 239–243.

72. Terayama, M. Ants from the Tokiwamatsu Imperial Gardens, Tokyo. *Mem. Natl. Sci. Museum Tokyo* **2005**, *39*, 245–247.
73. Terayama, M. Ants (Formicidae) from the imperial Palace, Tokyo. *Mem. Natl. Sci. Museum Tokyo* **2014**, *50*, 527–535.
74. Matsumura, S.; Yamane, S. Species composition and dominant species of ants in Jigenji Park, Kagoshima City, Japan. *Nat. Kagoshima* **2012**, *38*, 99–107.
75. Liu, K.L.; Peng, M.H.; Hung, Y.C.; Neoh, K.B. Effects of park size, peri-urban forest spillover, and environmental filtering on diversity, structure, and morphology of ant assemblages in urban park. *Urban Ecosyst.* **2019**, *22*, 643–656. [CrossRef]
76. Natuhara, Y. Ant faunae in Osaka City and three sites in Osaka Prefecture. *ARI J. Myrmecol. Soc. Jpn.* **1998**, *22*, 1–5.
77. Tan, C.K.W.; Corlett, R.T. Scavenging of dead invertebrates along an urbanisation gradient in Singapore. *Insect Conserv. Divers.* **2012**, *5*, 138–145. [CrossRef]
78. Harada, Y.; Koto, S.; Kawaguchi, N.; Sato, K.; Setoguchi, T.; Muranaga, R.; Yamashita, H.; Yo, A.; Yamane, S. Ants of Jusso, Isa City, Kagoshima Prefecture, southwestern Japan. *Bull. Biogeogr. Soc. Jpn.* **2012**, *67*, 143–152.
79. Iwata, K.; Eguchi, K.; Yamane, S. A Case Study on Urban Ant Fauna of Southern Kyusyu, Japan, with Notes on a New Monitoring Protocol (Insecta, Hymenoptera, Formicidae). *J. Asia. Pac. Entomol.* **2005**, *8*, 263–272. [CrossRef]
80. Yamaguchi, T. Influence of urbanization on ant distribution in parks of Tokyo and Chiba City, Japan I. Analysis of ant species richness. *Ecol. Res.* **2004**, *19*, 209–216. [CrossRef]
81. Park, S.H.; Hosoishi, S.; Ogata, K.; Kasuya, E. Changes of species diversity of ants over time: A case study in two urban parks. *J. Fac. Agric. Kyushu Univ.* **2014**, *59*, 71–76. [CrossRef]
82. Wang, W.; Zhang, S.; Xu, Z. Distribution patterns of ant species in Shenzhen City. *J. Southwest For. Univ.* **2012**, *32*, 69–74.
83. Khot, K.; Quadros, G.; Somani, V. Ant Diversity in an urban garden at Mumbai, Maharashtra. In Proceedings of the National Conference on Biodiversity: Status and Challenges in Conservation-FAVEO, Thane, India, 29–30 November 2013; 2013, pp. 121–125.
84. Hiroyuki, Y. List of species of Hymenoptera and Diptera in Matsuyama City, Ehime Prefecture, Shikoku, Japan. *Comm. Surv. Nat. Environ. Matsuyama City* **2012**, 167–176.
85. Harada, Y.; Yamasaki, M.; Saito, N.; Higasayama, U. Spreading of *Technomyrmex brunneus* in Shiroyama Park, Kagoshima City, southern Japan. *Nat. Kagoshima* **2021**, *47*, 275–279.
86. Miyake, K. Effect of Argentine ant invasions on Japanese ant fauna in Hiroshima Prefecture, western Japan: Preliminary report (Hymenoptera: Formicidae). *Sociobiology* **2002**, *39*, 65–474.
87. Tan, S.; Wei, H.; Liu, D. An investigation of ant species in houses and courtyards in the Chengdu area. *Sichuan J. Zool.* **2009**, *28*, 870–873.
88. Park, S.H.; Hosoishi, S.; Ogata, K. Long-term impacts of argentine ant invasion of urban parks in Hiroshima, Japan. *J. Ecol. Environ.* **2014**, *37*, 123–129. [CrossRef]
89. Touyama, Y.; Ogata, K.; Sugiyama, T. The Argentine ant, *Linepithema humile*, in Japan: Assessment of impact on species diversity of ant communities in urban environments. *Entomol. Sci.* **2003**, *6*, 57–62. [CrossRef]
90. Harada, Y.; Yamashita, M. Ants in the parks of the ruins of castles in Shikoku, western Japan. *Nat. Kagoshima* **2019**, *46*, 171–175.
91. Malozemova, L.A.; Malozemov, Y.A. Ecological Peculiarities of Ants in Urbanized Areas. *Russ. J. Ecol.* **1999**, *30*, 283–286.
92. Harada, Y. Ant fauna in Joyama Park, Hioki City, Kagoshima Prefecture, Japan, after the Technomyrmex brunneus invasion. *Nat. Kagoshima* **2020**, *47*, 173–178.
93. Roshanak, T.; Pashaei Rad, S.; Shokri, M. Faunistic investigation of ant (Hymenoptera: Formicidae) in the vicinity of Shiraz. *Environ. Sci.* **2017**, *15*, 75–91.
94. Hosoishi, S.; Rahman, M.M.; Murakami, T.; Park, S.H.; Kuboki, Y.; Ogata, K. Winter activity of ants in an urban area of western Japan. *Sociobiology* **2019**, *66*, 414–419. [CrossRef]
95. Terayama, M. Effects of the invasive ants *Linepithema humile* (Hymenoptera: Formicidae) on native ant and homopterous insect ommunities in Japan. *ARI J. Myrmecol. Soc. Jpn.* **2006**, *28*, 13–27.
96. Putyatina, T.S.; Perfilieva, K.S.; Zakalyukina, Y.V. Typification of urban habitats, with ant assemblages of Moscow city taken as an example. *Entomol. Rev.* **2017**, *97*, 1053–1062. [CrossRef]
97. Kumar, D.; Mishra, A. Ant community variation in urban and agricultural ecosystems in Vadodara District (Gujarat State), western India. *Asian Myrmecol.* **2008**, *2*, 85–93.
98. Antonov, I.A. Ant Assemblages of Two Cities with Different Ecological Conditions in Southern Cisbaikalia. *Russ. J. Ecol.* **2008**, *39*, 454–456. [CrossRef]
99. Hosaka, T.; Di, L.; Eguchi, K.; Numata, S. Ant assemblages on littered food waste and food removal rates in urban–suburban parks of Tokyo. *Basic Appl. Ecol.* **2019**, *37*, 1–9. [CrossRef]
100. Blinova, S.V. Changes in ant assemblages under conditions of a large industrial center. *Russ. J. Ecol.* **2008**, *39*, 148–150. [CrossRef]
101. Meshram, H.; Chavahan, P.; Sangode, V.; Khandwekar, V. Distribution of ant fauna in different terrestrial ecosystem in and around Nagpur city, Maharashtra India. *J. Entomol. Zool. Stud.* **2015**, *3*, 215–218.
102. Yong, G.W.J.; Wong, M.K.L.; Soh, E.J.Y. A preliminary checklist of the ant genera of Pulau Ubin, Singapore, from rapid opportunistic sampling (Hymenoptera: Formicidae). *Nat. Singap.* **2017**, *10*, 55–66.
103. Ossola, A.; Nash, M.A.; Christie, F.J.; Hahs, A.K.; Livesley, S.J. Urban habitat complexity affects species richness but not environmental filtering of morphologically-diverse ants. *PeerJ* **2015**, *3*, 1–19. [CrossRef]

104. Heterick, B.E.; Lythe, M.; Smithyman, C. Urbanisation factors impacting on ant (Hymenoptera: Formicidae) biodiversity in the Perth metropolitan area, Western Australia: Two case studies. *Urban Ecosyst.* **2013**, *16*, 145–173. [CrossRef]
105. Majer, J.D.; Brown, K.R. The effects of urbanization on the ant fauna of the Swan Coastal Plain near Perth, Western Australia. *J. R. Soc. West. Aust.* **1986**, *69*, 13–17.
106. Callan, S.K.; Mater, J.D. Impacts of an incursion of African Big-headed Ants, Pheidole megacephala (Fabricius), in urban bushland in Perth, Western Australia. *Pacific Conserv. Biol.* **2009**, *15*, 102–115. [CrossRef]
107. Heterick, B.E.; Majer, J.D. Influence of Argentine and coastal brown ant (Hymenoptera: Formicidae) invasions on ant communities in Perth gardens, Western Australia. *Urban Ecosyst.* **2000**, *4*, 277–292. [CrossRef]
108. May, J.E.; Heterick, B.E. Effects of the coastal brown ant Pheidole megacephala (Fabricius), on the ant fauna of the Perth metropolitan region, Western Australia. *Pacific Conserv. Biol.* **2000**, *6*, 81–85. [CrossRef]
109. Radchenko, A.G.; Stukalyuk, S.V.; Netsvetov, M.V. Ants (Hymenoptera, Formicidae) of Kyiv. *Entomol. Rev.* **2019**, *99*, 753–773. [CrossRef]
110. Ordóñez-Urbano, C.; Reyes López, J.; Carpintero-Ortega, S. ¿Una especie alóctona se puede ser "rara"?: El caso de Pyramica membranifera (Hymenoptera, Formicidae). *Bol. Soc. Entomológica Aragon.* **2008**, *42*, 321–323.
111. Antonova, V.; Penev, L. Change in the zoogeographical structure of ants caused by urban pressure in the Sofia region (Bulgaria). *Mymecol. News* **2006**, *8*, 271–276.
112. Penev, L.; Stoyanov, I.; Dedov, I.; Antonova, V. Patterns of Urbanisation in the City of Sofia as Shown by Carabid Beetles (Coleoptera, Carabidae), Ants (Hymenoptera, Formicidae), and Terrestrial Gastropods (Mollusca, Gastropoda Terrestria). In *Back to the Roots and Back to the Future: Towards A New Synthesis Amongst Taxonomic, Ecological and Biogeographical Approaches in Carabidology, Proceedings of the XIII European Carabidologists Meeting, Blagoevgrad, Bulgaria, 20–24 August 2007*; Pensoft: Moscow, Russia, 2008; pp. 483–509.
113. Antonova, V.; Lyubomir, P. Classification of Ant Assemblages (Hymenoptera: Formicidae) in Green Areas of Sofia. *Acta Zool. Bulg.* **2008**, *2*, 103–110. [CrossRef]
114. Dauber, J.; Eisenbeis, G. Untersuchungen zur Ameisenfauna einer urbanen Landschaft am Beispiel der Stadt Mainz. *Abh. Ber. Naturkundemus. Görlitz* **1997**, *69*, 237–244.
115. Pisarski, B.; Czechowski, W. Influence de la pression urbaine sur la myrmécofaune. *Memorab. Zool.* **1978**, *29*, 109–128.
116. Pisarski, B. Ants (Hymenoptera, Formicoidea) of Warzaw and Mazovia. *Memorab. Zool.* **1982**, *36*, 73–90.
117. Heras, P.R.; Ibáñez, M.D.M.; Sañudo, F.J.C.; Martínez, M.d.l.Á.V. Primeros datos de Formícidos (Hymenoptera, Formicidae) en parques urbanos de Madrid. *Boletín la Asoc. Española Entomol.* **2011**, *35*, 93–112.
118. Klesniaková, M.; Holecová, M.; Pavlíková, A. Interesting ant species in the urban greenery of Bratislava. *Folia Faun. Slovaca* **2016**, *21*, 235–238.
119. Trigos-Peral, G.; Rutkowski, T.; Witek, M.; Ślipiński, P.; Babik, H.; Czechowski, W. Three categories of urban green areas and the effect of their different management on the communities of ants, spiders and harvestmen. *Urban Ecosyst.* **2020**, *23*, 803–818. [CrossRef]
120. Ješovnik, A.; Bujan, J. Wooded areas promote species richness in urban parks. *Urban Ecosyst.* **2021**, 1–11. [CrossRef]
121. Behr, D.; Lippke, S.; Cölln, K. Zur Kenntnis der Ameisen von Köln (Hymenoptera, Formicidae). *Dechen. Beih.* **1996**, *35*, 215–232.
122. Ślipiński, P.; ZMihorski, M.; Czechowski, W. Species diversity and nestedness of ant assemblages in an urban environment. *Eur. J. Entomol.* **2012**, *109*, 197–206. [CrossRef]
123. Behr, D.; Cölln, K. Zur Ameisenfauna (Hymenoptera, Formicidae) von Gönnersdorf. *Dendrocopos* **1993**, *20*, 148–160.
124. Rigato, F.; Wetterer, J.K. Ants (Hymenoptera: Formicidae) of San Marino. *Nat. Hist. Sci.* **2018**, *5*. [CrossRef]
125. Espadaler, X.; López-Soria, L. Rareness of certain Mediterranean ant species: Fact or artifact? *Insectes Soc.* **1991**, *38*, 365–377. [CrossRef]
126. Vepsäläinen, K.; Ikonen, H.; Koivula, M.J. The structure of ant assemblages in an urban area of Helsinki, southern Finland. *Ann. Zool. Fennici* **2008**, *45*, 109–127. [CrossRef]
127. Trigos-Peral, G.; Reyes-López, J.L. Las hormigas como bioindicadores para la gestión de zonas verdes urbanas en el sur de la Península Ibérica. Propuesta de grupos funcionales. *Iberomyrmex* **2016**, *8*, 37–38.
128. Stukalyuk, S.V. Stratification of the ant species (Hymenoptera, Formicidae) in the urban broadleaf woodlands of the city of Kiev. *Entomol. Rev.* **2017**, *97*, 320–343. [CrossRef]
129. Reyes López, J.L.; Taheri, A. First checklist of ants (Hymenoptera, Formicidae) of urban green areas in Cádiz (Andalusia, Spain). *Boletín Asoc. Española Entomol.* **2018**, *42*, 217–223.
130. Smith, J.; Chapman, A.; Eggleton, P. Baseline biodiversity surveys of the soil macrofauna of London's green spaces. *Urban Ecosyst.* **2006**, *9*, 337–349. [CrossRef]
131. Gaspar, C.; Thirion, C. Modification des populations d'Hyménoptères sociaux dans des milieux anthropogènes. *Memorab. Zool.* **1978**, *29*, 61–77.
132. Nuhn, T.P.; Wright, C.G. An Ecological Survey of Ants (Hymenoptera: Formicidae) in a Landscaped Suburban Habitat. *Am. Midl. Nat.* **1979**, *102*, 353–362. [CrossRef]
133. Baena, M.L.; Escobar, F.; Valenzuela, J.E. Diversity snapshot of green–gray space ants in two Mexican cities. *Int. J. Trop. Insect Sci.* **2020**, *40*, 239–250. [CrossRef]

134. Menke, S.B.; Guénard, B.; Sexton, J.O.; Weiser, M.D.; Dunn, R.R.; Silverman, J. Urban areas may serve as habitat and corridors for dry-adapted, heat tolerant species; an example from ants. *Urban Ecosyst.* **2011**, *14*, 135–163. [CrossRef]
135. Miguelena, J.G.; Baker, P.B. Effects of Urbanization on the Diversity, Abundance, and Composition of Ant Assemblages in an Arid City. *Environ. Entomol.* **2019**, *48*, 836–846. [CrossRef]
136. Rocha-Ortega, M.; Castaño-Meneses, G. Effects of urbanization on the diversity of ant assemblages in tropical dry forests, Mexico. *Urban Ecosyst.* **2015**, *18*, 1373–1388. [CrossRef]
137. Toennisson, T.A.; Sanders, N.J.; Klingeman, W.E.; Vail, K.M. Influences on the structure of suburban ant (Hymenoptera: Formicidae) communities and the abundance of Tapinoma sessile. *Environ. Entomol.* **2011**, *40*, 1397–1404. [CrossRef]
138. Suarez, A.V.; Bolger, D.T.; Case, T.J. Effects of fragmentation and invasion on native ant communities in coastal southern California. *Ecology* **1998**, *79*, 2041. [CrossRef]
139. Gochnour, B.M.; Suiter, D.R.; Booher, D. Ant (Hymenoptera: Formicidae) Fauna of the Marine Port of Savannah, Garden City, Georgia (USA). *J. Entomol. Sci.* **2019**, *54*, 417–429. [CrossRef]
140. Savage, A.M.; Hackett, B.; Guénard, B.; Youngsteadt, E.K.; Dunn, R.R. Fine-scale heterogeneity across Manhattan's urban habitat mosaic is associated with variation in ant composition and richness. *Insect Conserv. Divers.* **2015**, *8*, 216–228. [CrossRef]
141. Friedrich, R.; Philpott, S.M. Nest-site limitation and nesting resources of ants (Hymenoptera: Formicidae) in urban green spaces. *Environ. Entomol.* **2009**, *38*, 600–607. [CrossRef]
142. García-Martínez, M.A.; Vanoye-Eligio, V.; Leyva-Ovalle, O.R.; Zetina-Córdoba, P.; Aguilar-Méndez, M.J.; Rosas-Mejía, M. Diversity of ants (hymenoptera: Formicidae) in a sub-montane and sub-tropical cityscape of northeastern Mexico. *Sociobiology* **2019**, *66*, 440–447. [CrossRef]
143. Fairweather, A.D.; Lewis, J.H.; Hunt, L.; McAlpine, D.F.; Smith, M.A. Ants (Hymenoptera: Formicidae) of Rockwood Park, New Brunswick: An Assessment of Species Richness and Habitat. *Northeast. Nat.* **2020**, *27*, 576–584. [CrossRef]
144. Uno, S.; Cotton, J.; Philpott, S.M. Diversity, abundance, and species composition of ants in urban green spaces. *Urban Ecosyst.* **2010**, *13*, 425–441. [CrossRef]
145. Ivanov, K.; Keiper, J. Ant (Hymenoptera: Formicidae) diversity and community composition along sharp urban forest edges. *Biodivers. Conserv.* **2010**, *19*, 3917–3933. [CrossRef]
146. Lessard, J.-P.; Christopher, M. The effects of urbanization on ant assemblages (Hymenoptera: Formicidae) associated with the Molson Nature Reserve, Quebec Jean-Philippe. *Entomol. Soc. Can.* **2005**, *137*, 215–225. [CrossRef]
147. Clarke, K.M.; Fisher, B.L.; Lebuhn, G. The influence of urban park characteristics on ant (Hymenoptera, Formicidae) communities. *Urban Ecosyst.* **2008**, *11*, 317–334. [CrossRef]
148. King, T.G.; Green, S.C. Factors affecting the distribution of pavement ants (Hymenoptera: Formicidae) in Atlantic coast urban fields. *Entomol. News* **1995**, *106*, 224–228.
149. Staubus, W.J.; Boyd, E.S.; Adams, T.A.; Spear, D.M.; Dipman, M.M.; Meyer, W.M. Ant communities in native sage scrub, non-native grassland, and suburban habitats in Los Angeles County, USA: Conservation implications. *J. Insect Conserv.* **2015**, *19*, 669–680. [CrossRef]
150. Pećarević, M.; Danoff-Burg, J.; Dunn, R.R. Biodiversity on broadway—enigmatic diversity of the societies of ants (Formicidae) on the streets of New York City. *PLoS ONE* **2010**, *5*, 1–8. [CrossRef]
151. Thompson, B.; McLachlan, S. The effects of urbanization on ant communities and myrmecochory in Manitoba, Canada. *Urban Ecosyst.* **2007**, *10*, 43–52. [CrossRef]
152. Villar, P.G.; Ríos-Casanova, L. Diversidad de hormigas (Hymenoptera: Formicidae) de La Cantera Oriente, una reserva dentro de la ciudad de México. *Dugesiana* **2020**, *27*, 29–36.
153. Stahlschmidt, Z.R.; Johnson, D. Moving targets: Determinants of nutritional preferences and habitat use in an urban ant community. *Urban Ecosyst.* **2018**, *21*, 1151–1158. [CrossRef]
154. Marussich, W.A.; Faeth, S.H. Effects of urbanization on trophic dynamics of arthropod communities on a common desert host plant. *Urban Ecosyst.* **2009**, *12*, 265–286. [CrossRef]
155. Pacheco, R.; Vasconcelos, H.L. Invertebrate conservation in urban areas: Ants in the Brazilian Cerrado. *Landsc. Urban Plan.* **2007**, *81*, 193–199. [CrossRef]
156. Santos, M.N.; Delabie, J.H.C.; Queiroz, J.M. Biodiversity conservation in urban parks: A study of ground-dwelling ants (Hymenoptera: Formicidae) in Rio de Janeiro City. *Urban Ecosyst.* **2019**, *22*, 927–942. [CrossRef]
157. Santos-Silva, L.; Vicente, R.E.; Feitosa, R.M. Ant species (Hymenoptera, Formicidae) of forest fragments and urban areas in a meridional amazonian landscape. *Check List* **2016**, *12*, 1–7. [CrossRef]
158. De Souza, D.R.; Dos Santos, S.G.; De Munhae, C.B.; De Morini, M.S.C. Diversity of epigeal ants (Hymenoptera: Formicidae) in urban areas of Alto Tietê. *Sociobiology* **2012**, *59*, 703–717. [CrossRef]
159. Lutinski, J.A.; Lopes, B.C.; de Morais, A.B.B. Diversidade de formigas urbanas (Hymenoptera: Formicidae) de dez cidades do sul do Brasil. *Biota Neotrop.* **2013**, *13*, 332–342. [CrossRef]
160. Munhae, C.d.B.; Souza-Campana, D.R.d.; Kamura, C.M.; Castro Morini, M.S.d. Ant communities (Hymenoptera: Formicidae) in urban centers of the Alto Tietê, São Paulo, Brazil. *Arq. Inst. Biol. (Sao Paulo)* **2015**, *82*, 1–5. [CrossRef]
161. Morini, M.S.d.C.; Munhae, C.d.B.; Leung, R.; Candiani, D.F.; Voltolini, J.C. Comunidades de formigas (Hymenoptera, Formicidae) em fragmentos de Mata Atlântica situados em áreas urbanizadas. *Iheringia. Série Zool.* **2007**, *97*, 246–252. [CrossRef]

162. Iop, S.; Lutinski, J.A.; Roberto, F.; Garcia, M. Formigas urbanas da cidade de Xanxerê, Santa Catarina, Brasil. *Biotemas* **2009**, *22*, 55–64. [CrossRef]

163. Matheus Caldart, V.; Iop, S.; Antonio Lutinski, J.; Roberto Mello Garcia, F. Diversidade de formigas (Hymenoptera, Formicidae) do perímetro urbano do município de Chapecó, Santa Catarina, Brasil. *Rev. Bras. Zoociências* **2012**, *14*, 81–94.

164. Ilha, C.; Maestri, R.; Antonio Lutinski, J. Assembleias de formigas (Hymenoptera: Formicidae) em áreas verdes urbanas de Chapecó, Santa Catarina. *Acta Ambient. Catarin.* **2017**, *14*, 39–50.

165. Josens, R.; Sola, F.; Lois-Milevicich, J.; Mackay, W. Urban ants of the city of Buenos Aires, Argentina: Species survey and practical control. *Int. J. Pest Manag.* **2017**, *63*, 1–11. [CrossRef]

166. Kamura, C.M.; Morini, M.S.C.; Figueiredo, C.J.; Bueno, O.C.; Campos-Farinha, A.E.C. Ant communities (Hymenoptera: Formicidae) in an urban ecosystem near the Atlantic Rainforest. *Braz. J. Biol.* **2007**, *67*, 635–641. [CrossRef]

167. Santiago, G.S.; Campos, R.B.F.; Ribas, C.R. How does landscape anthropization affect the myrmecofauna of urban forest fragments? *Sociobiology* **2018**, *65*, 441–448. [CrossRef]

168. De Souza-Campana, D.R.; Da Silva, O.G.M.; Menino, L.; Morini, M.S.D.C. Epigaeic ant (hymenoptera: Formicidae) communities in urban parks located in atlantic forest biome. *Check List* **2016**, *12*, 1967. [CrossRef]

169. Piva, A.; Campos, A.E.D.C. Ant community structure (Hymenoptera: Formicidae) in two neighborhoods with different urban profiles in the city of São Paulo, Brazil. *Psyche (Lond.)* **2012**, *2012*, 1–8. [CrossRef]

170. Alfonso Simonetti, J.; Matienzo Brito, Y.; Vázquez Moreno, L. Fauna de hormigas (Hymenoptera: Formicidae) asociadas a un sistema de producción agrícola urbano. *Fitosanidad* **2010**, *14*, 153–158.

171. Ribeiro, F.M.; Sibinel, N.; Ciocheti, G.; Campos, A.E.C. Analysis of ant communities comparing two methods for sampling ants in an urban park in the city of São Paulo, Brazil. *Sociobiology* **2012**, *59*, 971–984. [CrossRef]

172. Lutinski, J.A.; Roberto Mello Garcia, F. Análise faunística de Formicidae (Hymenoptera: Apocrita) em ecossistema degradado no município de Chapecó, Santa Catarina. *Biotemas* **2005**, *18*, 73–86.

173. Lange, D.; Vilela, A.; Erdogmus, G.; Barbosa, A. Temporal dynamic of foraging of epigeic ants in an urban forest fragment. *Bioscience* **2015**, *31*, 1501–1511. [CrossRef]

174. Starr, C.K.; Ballah, S.T. Urban Ant Fauna of Port of Spain, Trinidad, West Indies. *Living World J. Trinidad Tobago F. Nat. Club* **2017**, 49–50.

175. Soares, N.S.; Almeida, L.D.O.; Gonçalves, C.A.; Marcolino, M.T.; Bonetti, E.A.M. Survey of ants (Hymenoptera: Formicidae) in the urban area of Uberlândia, MG, Brazil. *Neotrop. Entomol.* **2006**, *35*, 324–328. [CrossRef]

176. Peel, M.C.; Finlayson, B.L.; McMahon, T.A. Updated world map of the Köppen-Geiger climate classification. *Hydrol. Earth Syst. Sci.* **2007**, *11*, 1633–1644. [CrossRef]

177. Macao Meteorological and Geophysical Bureau Macao Climate: 30-year Statistics of Some Meteorological Elements. Available online: https://www.smg.gov.mo/en/subpage/348/page/252 (accessed on 1 June 2021).

178. Governo da Região Administrativa Especial de Macau Dirreção dos Serviços de Cartografia e Cadastro Land Area Statistics. Available online: http://www.dscc.gov.mo/ENG/knowledge/geo_statistic.html (accessed on 23 June 2021).

179. IACM Macao Forest. Joint Publishing (Hong Kong). 2017; ISBN1 978-962-04-4138-7. ISBN2 962-04-4138-9.

180. Booher, D.; Macgown, J.A.; Hubbell, S.P.; Duffield, R.M. Density and Dispersion of Cavity Dwelling Ant Species in Nuts of Eastern US Forest Floors. *Trans. Am. Entomol. Soc.* **2017**, *143*, 79–93. [CrossRef]

181. R Core Team. *R: A Language and Environment for Statistical Computing*; R Foundation for Statistical Computing: Vienna, Austria, 2020.

182. Liu, C.; Guénard, B.; Garcia, F.H.; Yamane, S.; Blanchard, B.; Yang, D.R.; Economo, E. New records of ant species from Yunnan, China. *Zookeys* **2015**, *477*, 17–78. [CrossRef]

183. Guénard, B.; Weiser, M.D.; Dunn, R.R. Global models of ant diversity suggest regions where new discoveries are most likely are under disproportionate deforestation threat. *Proc. Natl. Acad. Sci. USA* **2012**, *109*, 7368–7373. [CrossRef]

184. Wickham, H. *ggplot2: Elegant Graphics for Data Analysis*; Springer: New York, NY, USA, 2016; ISBN 978-3-319-24277-4.

185. Hsieh, T.C.; Ma, K.H.; Chao, A. iNEXT: An R package for rarefaction and extrapolation of species diversity (Hill numbers). *Methods Ecol. Evol.* **2016**. [CrossRef]

186. Guénard, B.; Weiser, M.D.; Gómez, K.; Narula, N.; Economo, E.P. The Global Ant Biodiversity Informatics (GABI) database: Synthesizing data on the geographic distribution of ant species (Hymenoptera: Formicidae). *Myrmecol. News* **2017**, *24*, 83–89. [CrossRef]

187. Janicki, J.; Narula, N.; Ziegler, M.; Guénard, B.; Economo, E.P. Visualizing and interacting with large-volume biodiversity data using client-server web-mapping applications: The design and implementation of antmaps.org. *Ecol. Inform.* **2016**, *32*, 185–193. [CrossRef]

188. Wheeler, W.M. Chinese ants. *Bull. Mus. Comp. Zool.* **1921**, *64*, 529–547.

189. Wheeler, W.M. Ants collected by Professor F. Silvestri in China. *Boll. Lab. Zool. Gen. Agrar. R. Ist. Super. Agrar. Portici* **1928**, *22*, 3–38.

190. Wheeler, W.M. A list of the known Chinese ants. *Peking Nat. Hist. Bull.* **1930**, *5*, 53–81.

191. Wu, C.F. *Catalogus Insectorum Sinensium, Volume VI*; The Fan Memorial Institute of Biology, Yenching University: Beijing, China, 1941.

192. Tang, K.L.; Pierce, M.P.; Guénard, B. Review of the genus *Strumigenys* (Hymenoptera, Formicidae, Myrmicinae) in Hong Kong with the description of three new species and the addition of five native and four introduced species records. *Zookeys* **2019**, *831*, 1–48. [CrossRef]

193. Zhou, S.-Y. *Ants of Guangxi*; Guangxi Normal University Press: Guangxi, China, 2001.

194. Xu, Z. A systematic study on Chinese species of the ant genus *Oligomyrmex* Mayr (Hymenoptera: Formicidae). *Acta Zootaxonomica Sin.* **2003**, *28*, 310–322.

195. Hua, L. *List of Chinese insects Vol. IV*; Sun Yatsen University Press: Guangzhou, China, 2006; ISBN 9787306017017.

196. Eguchi, K. A revision of Northern Vietnamese species of the ant genus Pheidole (Insecta: Hymenoptera: Formicidae: Myrmicinae). *Zootaxa* **2008**, *1902*, 1–118. [CrossRef]

197. Bharti, H.; Kumar, R. Five new species of Dilobocondyla (Hymenoptera: Formicidae) with a revised key to the known species. *Asian Myrmecol.* **2013**, *5*, 29–44.

198. Dlussky, G.M.; Radchenko, A.G. The ants (Hymenoptera, Formicidae) of Vietnam. Subfamily Pseudomyrmicinae. Subfamily Myrmicinae (Tribes Calyptomyrmecini, Meranoplini, Cataulacini). Available online: GBIF.org (accessed on 21 July 2021).

199. Fisher, B.L. A new species of *Probolomyrmex* from Madagascar. In *Advances in Ant Systematics (Hymenoptera: Formicidae): Homage to E.O. Wilson—50 Years of Contributions*; American Entomological Institute: Gainesville, FL, USA, 2007.

200. Shattuck, S.O.; Gunawardene, N.R.; Heterick, B. A revision of the ant genus Probolomyrmex (Hymenoptera: Formicidae: Proceratiinae) in Australia and Melanesia. *Zootaxa* **2012**, *3444*, 40–50. [CrossRef]

201. Taylor, R.W. A monographic revision of the rare tropicopolitan ant genus Probolomyrmex Mayr (Hymenoptera: Formicidae). *Trans. R. Entomol. Soc. Lond.* **1965**, *117*, 345–365. [CrossRef]

202. Economo, E.P.; Narula, N.; Friedman, N.R.; Weiser, M.D.; Guénard, B. Macroecology and macroevolution of the latitudinal diversity gradient in ants. *Nat. Commun.* **2018**, *9*, 1–8. [CrossRef]

203. Nadkarni, N.M.; Longino, J.T. Invertebrates in Canopy and Ground Organic Matter in a Neotropical Montane Forest, Costa Rica. *Biotropica* **1990**, 286–289. [CrossRef]

204. Schmidt, F.A.; Solar, R.R.C. Hypogaeic pitfall traps: Methodological advances and remarks to improve the sampling of a hidden ant fauna. *Insectes Soc.* **2010**, *57*, 261–266. [CrossRef]

205. Ryder Wilkie, K.T.; Mertl, A.L.; Traniello, J.F.A. Biodiversity below ground: Probing the subterranean ant fauna of Amazonia. *Naturwissenschaften* **2007**, *94*, 725–731. [CrossRef]

206. Tschinkel, W.R. Back to basics: Sociometry and sociogenesis of ant societies (Hymenoptera: Formicidae). *Myrmecol. News* **2011**, *14*, 49–54.

207. Tschinkel, W.R. Insect sociometry, a field in search of data. *Insectes Soc.* **1991**, *38*, 77–82. [CrossRef]

208. Gotelli, N.J.; Ellison, A.M.; Dunn, R.R.; Sanders, N.J. Counting ants (Hymenoptera: Formicidae): Biodiversity sampling and statistical analysis for myrmecologists. *Myrmecolo. News* **2011**, *15*, 13–19.

209. Longino, J.T.; Coddington, J.; Colwell, R.K. The ant fauna of a tropical rain forest: Estimating species richness three different ways. *Ecology* **2002**, *83*, 689–702. [CrossRef]

210. Lee, R.H.; Guénard, B. Choices of sampling method bias functional components estimation and ability to discriminate assembly mechanisms. *Methods Ecol. Evol.* **2019**, *10*, 867–878. [CrossRef]

211. Yusah, K.M.; Fayle, T.M.; Harris, G.; Foster, W.A. Optimizing diversity assessment protocols for high canopy ants in tropical rain forest. *Biotropica* **2012**, *44*, 73–81. [CrossRef]

212. Kaspari, M.; Pickering, J.; Windsor, D. The reproductive flight phenology of a neotropical ant assemblage. *Ecol. Entomol.* **2001**, *50*, 382–390. [CrossRef]

213. Chong, C.S.; Li, D. Effects of forest disturbance on ant diversity in Singapore. *Dep. Biol. Sci. Natl. Univ. Singap.* **2003**, 1–4.

214. Bourguignon, T.; Dahlsjö, C.A.L.; Jacquemin, J.; Gang, L.; Wijedasa, L.S.; Evans, T.A. Ant and termite communities in isolated and continuous forest fragments in Singapore. *Insectes Soc.* **2017**, *64*, 505–514. [CrossRef]

215. MacGown, J.A.; Hill, J.G.; Deyrup, M.A. Brachymyrmex patagonicus (Hymenoptera: Formicidae), an emerging pest species in the southeastern United States. *Fla. Entomol.* **2007**, *90*, 457–464. [CrossRef]

216. Guénard, B. First record of the emerging global pest *Brachymyrmex patagonicus* Mayr 1868 (Hymenoptera: Formicidae) from continental Asia. *Asian Myrmecol.* **2018**, *10*, 1–6. [CrossRef]

217. Tschinkel, W.R. The ecological nature of the fire ant: Some aspects of colony function and some unanswered questions. In *Fire Ants and Leafcutter Ants: Biology and Management*; Lofgren, C.S., Vander Meer, R.K., Eds.; Avalon Publishing: New York, NY, USA, 1986; pp. 72–87.

218. Tschinkel, W.R. Distribution of the fire ants Solenopsis invicta and S. geminata (Hymenoptera: Formicidae) in northern Florida in relation to habitat and disturbance. *Ann. Entomol. Soc. Am.* **1988**, *81*, 76–81. [CrossRef]

219. Tschinkel, W.R.; King, J.R. The Role of Habitat in the Persistence of Fire Ant Populations. *PLoS ONE* **2013**, *8*, 1–8. [CrossRef]

220. Hoffmann, B.D.; Parr, C.L. An invasion revisited: The African big-headed ant (Pheidole megacephala) in northern Australia. *Biol. Invasions* **2008**, *10*, 1171–1181. [CrossRef]

221. Porter, S.D.; Savignano, D.A. Invasion of polygyne fire ants decimates native ants and disrupts arthropod community. *Ecology* **1990**, *71*, 2095–2106. [CrossRef]

222. Wetterer, J.K. Biology and Impacts of Pacific Island Invasive Species. 3. The African Big-Headed Ant, Pheidole megacephala (Hymenoptera: Formicidae). *Pac. Sci.* **2007**, *61*, 437–456. [CrossRef]

223. Lee, R.H.; Wang, C.L.; Guénard, B. The ecological implications of rubber-based agroforestry: Insect conservation and invasion control. *J. Appl. Ecol.* **2020**, *57*, 1605–1618. [CrossRef]
224. Liu, C.; Fischer, G.; Garcia, F.H.; Yamane, S.; Liu, Q.; Peng, Y.Q.; Economo, E.P.; Guénard, B.; Pierce, N.E. Ants of the Hengduan Mountains: A new altitudinal survey and updated checklist for Yunnan Province highlight an understudied insect biodiversity hotspot. *Zookeys* **2020**, *978*, 1–171. [CrossRef]
225. Lockwood, J.L.; Cassey, P.; Blackburn, T. The role of propagule pressure in explaining species invasions. *Trends Ecol. Evol.* **2005**, *20*, 223–228. [CrossRef]
226. Ellison, A.M. Out of Oz: Opportunities and challenges for using ants (Hymenoptera: Formicidae) as biological indicators in north-temperate cold biomes. *Myrmecol. News* **2012**, *17*, 105–119.
227. Brady, S.G.; Ward, P.S. Morphological phylogeny of army ants and other dorylomorphs (Hymenoptera: Formicidae). *Syst. Entomol.* **2005**, *30*, 593–618. [CrossRef]
228. Boswell, G.P.; Britton, N.F.; Franks, N.R. Habitat fragmentation, percolation theory and the conservation of a keystone species. *Proc. R. Soc. B Biol. Sci.* **1998**, *265*, 1921–1925. [CrossRef]
229. Schöning, C.; Kinuthia, W.; Boomsma, J.J. Does the Afrotropical army ant *Dorylus (Anomma) molestus* go extinct in fragmented forests? *J. East Afr. Nat. Hist.* **2006**, *95*. [CrossRef]

Article

Biogeography of Iberian Ants (Hymenoptera: Formicidae)

Alberto Tinaut and Francisca Ruano *

Department of Zoology, University of Granada, Campus de Fuentenueva s/n, 18071 Granada, Spain; hormiga@ugr.es
* Correspondence: fruano@ugr.es

Abstract: Ants are highly diverse in the Iberian Peninsula (IP), both in species richness (299 cited species) and in number of endemic species (72). The Iberian ant fauna is one of the richest in the broader Mediterranean region, it is similar to the Balkan Peninsula but lower than Greece or Israel, when species richness is controlled by the surface area. In this first general study on the biogeography of Iberian ants, we propose seven chorological categories for grouping thems. Moreover, we also propose eight biogeographic refugium areas, based on the criteria of "refugia-within-refugium" in the IP. We analysed species richness, occurrence and endemism in all these refugium areas, which we found to be significantly different as far as ant similarity was concerned. Finally, we collected published evidence of biological traits, molecular phylogenies, fossil deposits and geological processes to be able to infer the most probable centre of origin and dispersal routes followed for the most noteworthy ants in the IP. As a result, we have divided the Iberian myrmecofauna into four biogeographical groups: relict, Asian-IP disjunct, Baetic-Rifan and Alpine. To sum up, our results support biogeography as being a significant factor for determining the current structure of ant communities, especially in the very complex and heterogenous IP. Moreover, the taxonomic diversity and distribution patterns we describe in this study highlight the utility of Iberian ants for understanding the complex evolutionary history and biogeography of the Iberian Peninsula.

Keywords: species richness; species occurrence; endemic species; distribution ranges; dispersal routes; centre of origin; refugium areas

check for **updates**

Citation: Tinaut, A.; Ruano, F. Biogeography of Iberian Ants (Hymenoptera: Formicidae). *Diversity* **2021**, *13*, 88. https://doi.org/10.3390/d13020088

Academic Editor: Luc Legal
Received: 6 December 2020
Accepted: 12 February 2021
Published: 19 February 2021

Publisher's Note: MDPI stays neutral with regard to jurisdictional claims in published maps and institutional affiliations.

1. Introduction

Ants bring together a number of traits that make up an interesting subject for biogeographical studies: they are conspicuous and one of the most well-known taxa of the terrestrial invertebrates [1], near-ubiquitous in terrestrial ecosystems [2]. They are an ecologically dominant faunal group [3] and highly diverse but still accessible [3] and show an extreme diversity of dispersal strategies although they are sessile superorganisms [4].

Local ant species richness is strongly correlated with temperature, which is the most important factor determining the structure of ant communities [1]. Ants are highly diverse in hot and dry habitats [5] and in low-elevation, low-latitude forests [6]. The latitudinal diversity gradient [6] with an asymmetric northern hemispheric pattern has been detected for ants, which is impossible to explain by only contemporary biotic and abiotic variables [5]. None of the current factors sufficiently correlate to explain current ant diversity [2,5,7]. Moreover, modern ant assemblages suggest a climate-driven past reorganization of the Palaearctic ant fauna [7]. Many ant lineages which previously lived in a warmer Europe, were not able to survive through long-term cooling (Pliocene) and glacial cycles (Pleistocene) but some of them persisted in Indomalayan and Australian regions [7]. Thus, the high extinction pattern in the northern hemisphere appears to have conditioned its marked asymmetric latitudinal diversity gradient [5,7].

Knowledge of the ants found in the Iberian Peninsula (the IP including Andorra, Gibraltar, Portugal and Spain) is still fragmentary [8]. Moreover, there are very few publications treating data regarding their biogeography, and they are limited to some

237

genera, species or zones [8–13]. The biogeography or philogeography of other Iberian arthropods have been addressed for example for the coleopteran genera *Berberomeloe* [14], *Pimelia* [15], *Blaps* [16], *Cephalota* [17], *Hydraena* [18] and some others cited throughout this article.

The IP forms part of the Mediterranean biodiversity hotspot, unique in the Palaearctic region. It is a threatened area with elevated species richness and a high level of endemism in all the taxa [19–21]. Hewitt [22] established that the IP has an intermediate position as far as species richness and the number of paleoendemic species are concerned, compared to the other two Mediterranean peninsulas (Italy and the Balkans). This high level of diversity within the IP was attributed mainly to its environmental heterogeneity conditioned by its complex orography. The IP mountain ranges are at different latitudes and altitudes with an east-western orientation. These ranges are sometimes permeable, but others act as geographic barriers producing isolated valleys, plateaus and plains. The east-western orientation also permits the establishment of particular and markedly different microhabitats on the southern and northern slopes, which vary depending on the altitude and latitude.

On the other hand, changing paleogeography and paleoclimate in the region favoured the convergence of different lineages, sharing evolutionary history for a long period [23]. Thus, in the early Paleogene (60 mya), the small Iberian plate, fragmented during the Mesozoic (100 mya) and united with the European plate, had a semitropical climate until the end of the Miocene (23-5.3 mya) [24,25] defining similar biomes to the current African savannah. Moreover, in the southern territories a succession of separations and unifications with the African plate continued until the end of the Miocene, closing the IP with Africa during the evaporitic Messiniense [26]. These semi-tropical communities from the Miocene were composed of species of Asiatic and African origin [27]. Later, successive climatic and orographic changes in the Pliocene and Pleistocene produced migrations towards the IP [25,28–30] which led to the formation of a mosaic of fragmented populations that persisted in small refugia throughout the entire Iberian refugium (the "refugia-within-refugium" scenario) [22,31–33] which is very important for the understanding of the diversity and phylogeography of some Iberian taxa [31,32,34]. After the glacial periods the IP played an important role when the Palaearctic peninsulas acted as providers of re-colonizing species to the northern territories [35–37]. These relatively recent biogeographical processes have greatly conditioned the current faunistic composition of the IP.

Thus, to the best of our knowledge, this is the first biogeographic global study on the richness and distribution of the Iberian formicids. We have undertaken this challenge well aware that the knowledge about ant distribution and endemic species in the IP is still incomplete. Our goals have been (1) to analyse the number of ant genera and species in the IP but also to compare them with other Mediterranean countries, especially France (linking the IP with the European Palaearctic) and Morocco (with the African Palaearctic), (2) to take into account the putative refugia within the Iberian refugium, and we propose a subdivision of the IP in refugium areas and attribute a chorological category to each species, and (3) we have defined four different groups of species depending on the different dispersal patterns (centre of origin and dispersal routes) based on biological traits, molecular phylogenies, fossil deposits and geological processes, when available.

2. Materials and Methods

2.1. Species Included in the Study

In this study we have included known species in the IP (Andorra, Gibraltar, Portugal and Spain) and compared them with the species found in nearby countries such as France and Morocco. We have tried to use the most updated published lists in the four countries closely involved adding new citations such as some general reviews [38–40]. Only described and named species are included in our study (those species appearing as "sp.", species 1, 2, species under description, etc. are not included). Moreover, we have only considered mainland species, avoiding the species which only appear on the islands belonging to the four countries. In general, we have not undertaken taxonomic problems such as synonyms or misidentifications of species included in the different faunistic lists, assuming as correct the most recent reference. For the nomenclature of the social parasite species of *Temnothorax* and *Tetramorium* we have adopted the criterion of Seifert et al. [41]. With this method of managing citations our final results might show slight differences with some published lists, but this has not affected our main goals.

For the Portuguese species we focused on Salgueiro [42], taking into account previous publications of Collingwood & Prince [43] and later ones such as Boeiro et al. [44] and Gonçalves et al. [45]. We have excluded from our lists *Temnothorax caparica* Henin, Paiva and Collingwood, 2001, which resulted to be a misidentification of *Cardiocondyla mauritanica* Forel, 1890 [46].

For the Spanish species we focused on the list included in Sánchez-García et al. [47] and added some recently described species [48] and some new findings [49].

In the case of Andorra we have used the list of species of Bernadou et al. [50]. For Gibraltar there is no up-to-date catalogue and citations appear in general articles for the Iberian ants, except for some recent records that have been taken into account [51–53].

For the French species we used Casevitz-Weulersse & Galkowski [54] adding some of the variations included in Monnin et al. [55], and Seifert [56] to clarify the status and current distribution of the European *Lasius* species.

The faunistic list of the Moroccan ant species was constructed from the results of Cagniant [57] and also included the additions provided in different publications [53,58–66].

We have also used the available data on the myrmecofauna of most of the other Mediterranean countries (Table 1) permitting us to put the IP ant species into a Mediterranean context.

Table 1. List of Mediterranean countries and number of cited species.

Country/Region	N° Species	Reference
EUROPE	622	[67]
IP	299	present paper
FRANCE	211	present paper
ITALY	187	[68]
BALKANS	319	[69]
ALBANIA	79	[70]
SERBIA	116	[71]
CROATIA	140	[72]
YUGOSLAVIA (former)	171	[69]
MACEDONIA	80	[73]
GREECE	259	[74]
TURKEY	286	[75]
ISRAEL	203	[76]
EGYPT	97	[77]
TUNISIA & ALGERIA	180	[57]
MOROCCO	233	present paper

2.2. Biogeographical Analysis

We undertook three different biogeographical aspects of the Iberian biogeography: 1. Analysis of the chorology of all the species included in the Iberian list; 2. Assessment of the distribution range of these species in the IP, emphasizing the endemic, rare and relict species; 3. To analyse the colonization history of the IP and propose four different ant groups, taking into account their most probable centres of origin and dispersal routes.

2.2.1. Chorology

According to Ribera [78] we established similar chorological categories (categories one to four) but adding three more (five to seven). Specifically, the seven chorological categories we have used are: 1. southern species (S) present in northern Africa and in some areas of the IP but never extending their distribution beyond the north of the Pyrenees; 2. northern species (N), mostly distributed in the north of Europe, northern Pyrenees and some other northern areas in the IP; 3. Iberian endemic species present only in the IP (X) or extending their distribution to the northern slopes of the Pyrenees or some areas in southeastern France (XS); 4. trans-Iberian species (T) present in the northern Pyrenees, the IP and northern Africa; 5. species living only in the northern Mediterranean basin (MN); 6. species distributed over all the Mediterranean basin (M); 7. introduced species (I).

2.2.2. Distribution Range

We have also aimed to portray the distribution of all the listed species for the IP in as much detail as possible, although the ant distribution and endemic species is still incomplete and the methods and effort sampling were heterogenous (pitfall traps, visual inspection, etc.) and sometimes unknown to us, such as those taken from bibliography which only report the species occurrence. Therefore, we have defined biogeographical subdivisions following the criteria established in other studies. Arnan et al. [79] established only two subdivisions in the IP, the Atlantic and Mediterranean, but we have subdivided the two biogeographic subregions into eight refugium areas (Figure 1), following Gómez and Lunt's criteria [32] based on general biota occurrence (plants, vertebrate and invertebrates animals). We have limited the refugium areas to mountain ranges which act as borders because ants prefer low altitudes [6]. We considered an Atlantic and a Mediterranean coast refugia, which have been defined as biogeographical areas for butterflies [80]) and birds [81]. Some of these areas are supported as refugia such as the southern plateau, Atlantic and south-eastern Mediterranean for *Pimelia* species [15] or the entire Mediterranean during the Pleistocene for *Polyommatus* species [37]. Thus, we have finally defined the following refugium areas (Figure 1): 1. the Cantabric including the Cantabric and Basque ranges and a part of the eastern Galaic mountains; 2. the Pyrenean including all the southern slopes of the Pyrenean range from Navarra to Catalonia; 3. the Mediterranean, including the eastern and south-eastern coasts; 4. the Atlantic including the western Iberian coasts; 5. the northern plateau, bordered to the south by the northern slopes of the Central and Iberic ranges, and to the north by the Cantabric range; 6. the southern plateau, bordered to the north by the southern slopes of the Central and Iberic ranges and to the south by the Sierra Morena range; 7. the Guadalquivir Valley; and 8. the Ebro Valley. The geographical position of the citations has been obtained from the AntMaps.org web page [82], and confirmed with our own data and recent publications and reviews (see Material and Methods 1.1). Moreover, we have analysed the occurrence and the number of species found in every refugium area, as well as the number of refugium areas occupied by each species. We also analysed the endemic species distribution highlighting the rare species and the refugium areas which they inhabit. Finally, we tested the similarity of each of the refugium areas by mean of hierarchical clustering analysis based on Jaccard's index as the association estimate and the paired group algorithm UPGMA (Unweighted Pair Grouping method with Arithmetic Means) procedure [80,83,84] as the agglomeration criterion using PAST program V. 4.04 [85] and excluding ubiquitous species. The significant differences in similarity (Jaccard's Index) have been calculated among all the branches of the cluster [80,86,87].

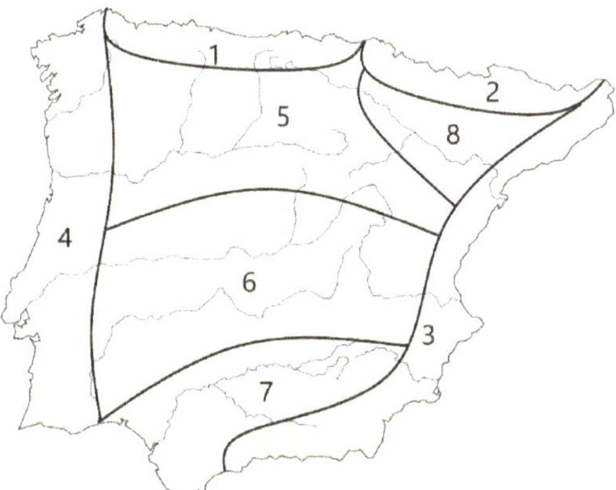

Figure 1. Aprioristic refugium areas considered in the Iberian Peninsula. 1. Cantabric; 2. Pyrenean; 3. Mediterranean coast; 4. Atlantic; 5. Northern plateau; 6. Southern plateau; 7. Guadalquivir Valley; 8. Ebro Valley.

2.2.3. Origin and Dispersal Routes

We collected biogeographical evidence from the bibliography about biological traits, dated molecular phylogenies (with different estimation methods) and fossil deposits of the different ant subfamilies, genera and species when available. We inferred from these data the origin and the most plausible dispersal routes up to the current distribution range, taking into account the most probable geology and climate distribution admitted for each epoch in the bibliography and the comparison results obtained from other taxa.

3. Results

3.1. Taxonomic Richness

The Iberian Peninsula (IP) has 299 ant species, the highest number compared to the French (211) and the Moroccan (233) ant faunistic lists (see Table S1 in Supplementary Material). The Iberian ant richness is only comparable with the 259 ant species of Greece and the 286 of Turkey. To the north the ant richness followed the latitudinal diversity gradient (Belgium 85 species [88], Norway 57 species [89]). Nevertheless, the ratio species richness/surface in the three countries revealed a similar ratio for the IP and Morocco (5.1×10^{-4} and 5.2×10^{-4} species/km^2 respectively) and much lower for France (3.2×10^{-4} species/km^2). Even in a wide view of the Mediterranean context, the ant fauna of the IP is the second in species richness, with only a few species less than the Balkan peninsula but not when the ratio of species/surface is taken into account, where the IP has a higher ratio than France, similar to Morocco, but surpassed by Greece and Israel. The Iberian ant species appear grouped in seven subfamilies (Amblyoponinae, Ponerinae, Proceratiinae, Leptanillinae, Dolichoderinae, Formicinae and Myrmicinae), but Morocco has two more subfamilies (Cerapachyinae and Dorylinae) and France one subfamily less than the IP (Amblyoponinae). The subfamilies Amblyoponinae, Cerapachyinae and Dorylinae, which are widely considered as tropical [90], are absent from France and only the subfamily Amblyoponinae is present in the IP.

The subfamily Myrmicinae had the highest species richness for the three studied areas, 114 species (54% of the total ants) in France, 171 (57%) in the IP and 147 (63%) in Morocco. The subfamily Formicinae has a lower number of species than Myrmicinae in all the three areas: 77 in France (36.5%), 95 species (33%) in the IP and 61 (26%) in Morocco. The subfamily Dolichoderinae appears far below, followed by Dorylinae and Cerapachyinae (see Supplementary material Table S1).

The most diverse genus in the IP is *Temnothorax* (47 species and 16% of the total species richness), followed by *Lasius* (25 species, 8%), *Formica* (23 sp., 8%) and *Camponotus*

(20 species, 7%). *Temnothorax* is also highly diverse in France (28 sp., 13%) and Morocco (39 sp., 17%) but represents a slightly higher percentage of the total species in the latter country. The other most abundant genera in France are *Formica* (13%), *Lasius* (12%) and *Myrmica* (10%) and in Morocco *Messor* (12%), *Aphaenogaster* (11%) and *Cataglyphis* (9%).

Although the total number of endemic species in the IP (72 sp., 24 %) is the highest in the Mediterranean countries (21 endemic ant species in Greece [74]; 25 in Israel [76]; and 56 in Morocco [57]), but again the ratio of endemic species/surface is similar to Morocco (1.25×10^{-4} endemic sp./km^2) and surpassed by Greece (1.6×10^{-4} endemic sp./km^2) and Israel (12.0×10^{-4} endemic sp./km^2). In the IP, the highest number of endemic ants belong to Myrmicinae (41 species) followed by the subfamily Formicinae (25 species, Figure 2). Amongst the Myrmicinae, the genus *Temnothorax* includes the most endemic species (16 species, 22% of the total endemic species and 30% of the full *Temnothorax* species). Nevertheless, the genus *Cataglyphis* is the most proportionally rich in endemic species, all of its 10 Iberian species are endemic. Although a positive correlation exists between the species richness and the number of endemic species within a genus (R = 0.73; $F_{(1,28)}$ = 31.81; *p* < 0.0001), some genera are especially above the expected number of endemic species, such as *Cataglyphis*, *Temnothorax* or *Goniomma*, and other genera below, such as *Lasius*, *Tetramorium* and *Formica* (Figure 3).

Moreover, the presence of an endemic genus in the IP, *Iberoformica*, with only one living species in the world, *I. subrufa*, is very remarkable. This species extends its distribution to some southern French locations. The exceptional appearance of the genus *Rossomyrmex* in the IP is also noteworthy. This genus has three different species in Asia (the Anatolian plateau, the plains of the Caspian Sea and other regions of Central Asia), which are different to the one in the IP, *R. minuchae* [30,58], the only species of this genus which exists in the full western Palaearctic.

Another interesting group of species because of their natural history are the social parasites of which 43 species appear in the IP, including temporal and permanent parasitism (22 and 21 species). Together they account for 14.4% of the total Iberian species. The number and percentage of parasite species decreased in France (20 temporal + 13 permanent which represent 15.7%), and many more in Morocco (5 temporal + 7 permanent which constitutes 5.2%).

Finally, comparing the shared species belonging to every subfamily amongst the IP, France and Morocco, (Table 2; Table S1) we obtained the highest percentage of overlapping species between the French and the Iberian ant lists (84%). Nevertheless, the IP only shared 59% of its species with the French list. Morocco shared the lowest number of species with France 21%. and the highest with the IP 37% (Table 2). A cluster based on the similarity (Jaccard's Index, Cophenetic correlation index = 0.98) amongst the three areas, grouped France and the IP together (Figure 4) without significant differences between them (Jaccard's Index = 0.52, p > 0.05) and showed significant differences between Morocco with France and the IP (Jaccard's Index = 0.12 and 0.19 respectively, p < 0.01). Moreover, the replacement of species is more frequent between Morocco and the IP, but the overlapping is mainly between France and the IP (Table S1). For instance, France has three *Bothriomyrmex* species shared with the IP, but only one of the six *Bothriomyrmex* from Morocco appear amongst the four species in the IP. Even more striking is the case of the genus *Cataglyphis*, none of the 22 species in Morocco appear amongst the 10 species in the IP, which moreover are Iberian endemic species. Something similar occurred in other genera, such as *Aphaenogaster*, *Oxyopomyrmex* and *Temnothorax*, although to a lesser extent.

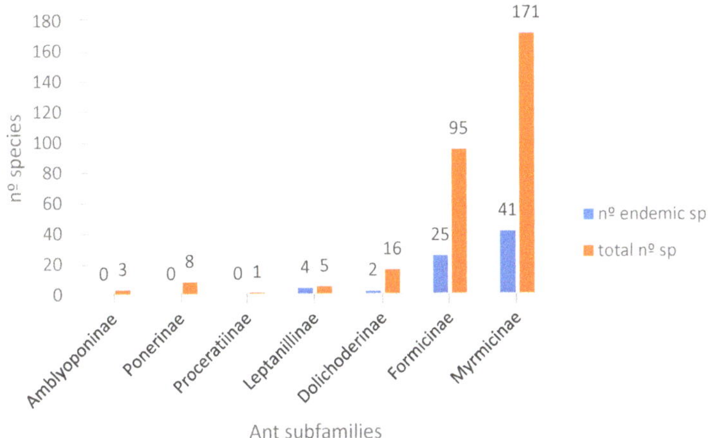

Figure 2. Number of species (orange) and endemic species (blue) belonging to each ant subfamily in the Iberian Peninsula (IP).

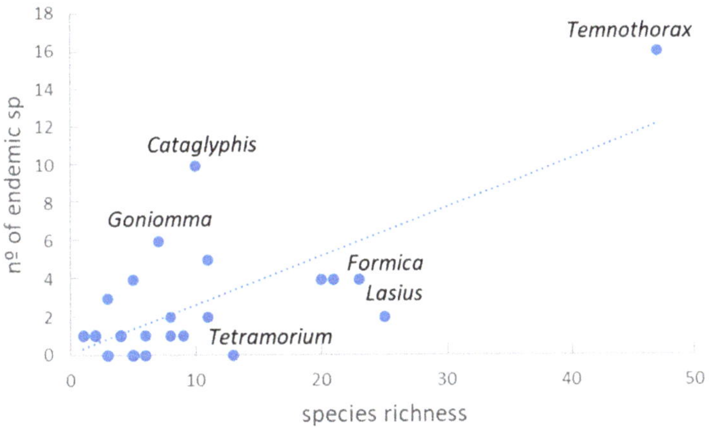

Figure 3. Positive correlation between species richness and the number of endemic species in each genera, including only native species (R = 0.73; p < 0.0001; $F_{(1,28)}$ = 33.32).

Table 2. Total shared and unshared species and percentage amongst France (Fr), Iberian Peninsula (IP) and Morocco (Mo).

Subfamily	Total Shared sp/% Fr-IP	Total Shared sp/% Fr-Mo	Total Shared sp/%IP-Fr	Total Shared sp/% IP-Mo	Total Shared sp/% Mo-Fr	Total Shared sp/% Mo/IP
Amblyoponinae	0/0%	0/0%	0/0%	2/67%	0/0%	2/67%
Cerapachyinae	0/0%	0/0%	0/0%	0/0%	0/0%	0/0%
Dorylinae	0/0%	0/0%	0/0%	0/0%	0/0%	0/0%
Ponerinae	7/87.5%	4/50%	7/87.5%	5/63%	4/80%	5/100%
Proceratiinae	1/100%	0/0%	1/100%	0/0%	0/0%	0/0%
Leptanillinae	1/100%	1/100%	1/20%	1/20%	1/33%	1/33%
Dolichoderinae	10/100%	3/30%	10/62%	4/25%	3/37%	3/37%
Formicinae	69/90%	13/17%	68/72%	19/20%	13/21%	19/31%
Myrmicinae	89/79%	28/25%	89/52%	57/33%	28/19%	45/38%
Total shared species	178/84%	49/23%	176/59%	88/29%	49/21%	87/37%
Total unshared species	33/16%	162/77%	123/41%	211/17%	184/79%	146/63%

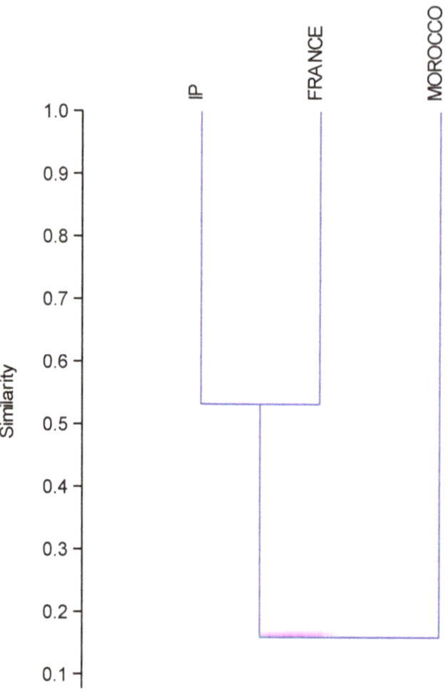

Figure 4. Cluster analysis based on the similarity (Jaccard's Index) of the ant species found in the three adjacent areas (IP, France and Morocco).

3.2. Biogeographical Analysis

3.2.1. Chorology

The most frequent chorological category in the IP was the northern species (N, 30.8%), which together with the trans-Iberian species (T, 11.7%) are 42.5% (Figure 5). On the other hand, the Mediterranean species, including the northern Mediterranean (MN, 6.7%), all the Mediterranean basin (M, 6.7%), the southern Mediterranean (S, 12%), and the Iberian endemic species (X + XS, 24.1%) represented 49.5% of the total species. The lacking percentage corresponded to the introduced species (I, 8%) (Figure 5). Amongst the ant subfamilies, the northern species (N) are clearly more abundant in Formicinae and Myrmicinae (Figure 6). In the latter subfamily the "endemic species" is the next most abundant chorological category (Figure 6). Nevertheless, the percentage of endemic species is similar for Formicinae and Myrmicinae (26% and 25%). The next most abundant category in Myrmicinae is the southern species (S), contrarily under-represented amongst the Formicinae ants (Figure 6).

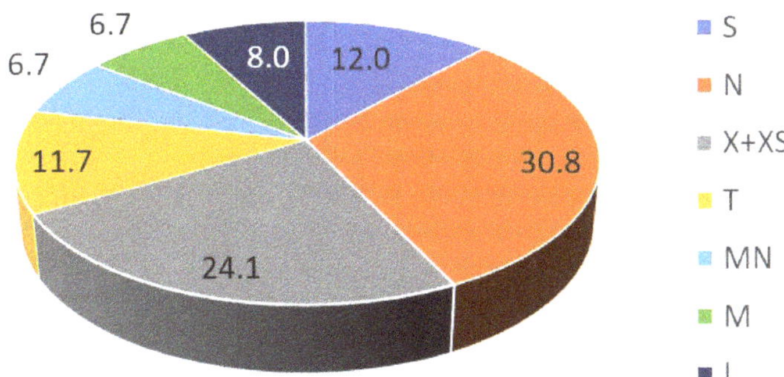

Figure 5. Percentage of species pertaining to the different chorological categories (S: Southern; N: Northern; X + XS: Endemic; T: Trans-Iberian; MN: Northern Mediterranean; M: Mediterranean; I: Introduced.

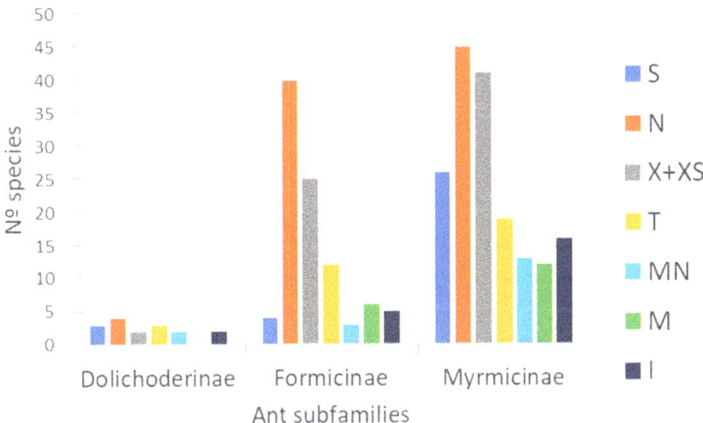

Figure 6. Chorological composition of the three more abundant subfamilies. (S: southern species, N: northern species, X + XS: endemic species, T: Trans-Iberian species, MN: north Mediterranean species, M: Mediterranean and I: introduced species).

3.2.2. Distribution Range

We have evaluated the importance of every refugium area harbouring ant species richness (see Supplementary material Table S2). We found that all the aprioristic defined refugium areas presented significant differences in similarity (Jaccard's index, $p < 0.01$) with respect to the others (Table 3). Those refugia with the high similarity index were the northern plateau, the Guadalquivir Valley and the southern plateau (Jaccard's index = 0.4; Table 3, Figure 7).

The refugium area with the highest number of ant species is the Mediterranean (144 species), followed by the Guadalquivir Valley (118 species) (Figure 8). These two refugium areas also contain the highest number of endemic species, followed by the northern plateau and the Pyrenees. The Atlantic and the southern plateau refugia contain an intermediate number of species, and the Cantabric together with the Ebro Valley present a similarly low number. The number of ubiquitous species, occurring in all the eight refugium areas was relatively low with only 46 species.

Many of the Iberian ant species appeared exclusively in one of the Iberian refugium areas (86 species, 29%) and may be considered as rare species, although they are not always endemic. Progressively a lower number of species occupied more refugium areas, with only a final increase for ubiquitous species (Table 4). The 21 endemic species which appear in only one refugium area are shown in Table 5.

Analysing the most diversified or peculiar ant genera (see Annex S2, Supplementary Material), we found the genus *Lasius* (25 species) to be the most widely distributed across all the IP and, when local, generally appear in the northern half of the IP and frequently in the Pyrenees, the Ebro Valley and the Mediterranean refugia. There are only two endemic species in this genus that are widely distributed (*L. cinereus* and *L. piliferus*). The genus *Formica* is composed of 23 species, the majority appearing in the Pyrenees (14 species) or the Mediterranean (9 species) refugia. Twelve *Formica* species of the IP belong to the *rufa* group; again, the majority of these latter species (7 of them) were present in the Pyrenean refugium and only 3 in the Guadalquivir Valley. The genus *Camponotus* is composed of 20 species, of which 16 are widely distributed in the IP. Nevertheless, the only two strictly endemic species (X chorological category) belonging to this genus (*C. haroi* and *C. amaurus*) presented a narrow distribution linked to the Mediterranean coast and the northern plateau. In the case of the genus *Cataglyphis*, 10 species are represented in the Iberian myrmecofauna, all of which are endemic (nine strictly X and one shared with France XS). Species from this genus have never been found in the most northern zones (Cantabric or Pyrenees) and principally were present in the southern areas (southern plateau, the Guadalquivir Valley and on the Mediterranean coast). Four of the endemic *Cataglyphis* species (*C. floricola, C. gadeai, C. humeya and C. tartessica*) are exclusively distributed in either the Guadalquivir Valley or the Mediterranean refugium areas. The genus *Themnothorax*, includes 47 species in the IP, 16 of which are endemic. Twenty-five percent of the *Temnothorax* species are distributed in only one refugium area, and the areas with the most *Temnothorax* species are the northern plateau, the Mediterranean coast and the Guadalquivir Valley. The refugium area with the lowest number of *Temnothorax* species is the Cantabric. Finally, the genus *Stigmatomma* is composed of only three species, with a narrow distribution, in only one or two refugium areas, thus in the Guadalquivir Valley we found the three species and one of them also on the Mediterranean and Atlantic coasts. In fact, the Guadalquivir Valley is the refugium containing the highest number of ant species linked with tropical environments, such as *Stigmatomma emeryii*, *S. impressifrons*, and the Ponerinae *Anochetus ghilianii* and *Chryptopone ochracea.*

Table 3. Jaccard's Index (black numbers) and probability associated to significant differences (blue numbers) amongst the occurrence of species in paired refugium areas of the IP.

	Cantabric	Pyrenees	Mediterranean	Atlantic	Northern Plateau	Southern Plateau	Guadalquivir Valley	Ebro Valley
Cantabric	1	<0.01	<0.01	<0.01	<0.01	<0.01	<0.01	<0.01
Pyrenees	0.28	1	<0.01	<0.01	<0.01	<0.01	<0.01	<0.01
Mediterranean	0.11	0.26	1	<0.01	<0.01	<0.01	<0.01	<0.01
Atlantic	0.12	0.25	0.33	1	<0.01	<0.01	<0.01	<0.01
Northern plateau	0.23	0.32	0.35	0.36	1	0.01	<0.01	<0.01
Southern plateau	0.1	0.22	0.3	0.37	0.4	1	<0.01	<0.01
Guadalquivir Valley	0.1	0.21	0.4	0.35	0.4	0.4	1	<0.01
Ebro Valley	0.1	0.25	0.09	0.17	0.21	0.2	0.15	1

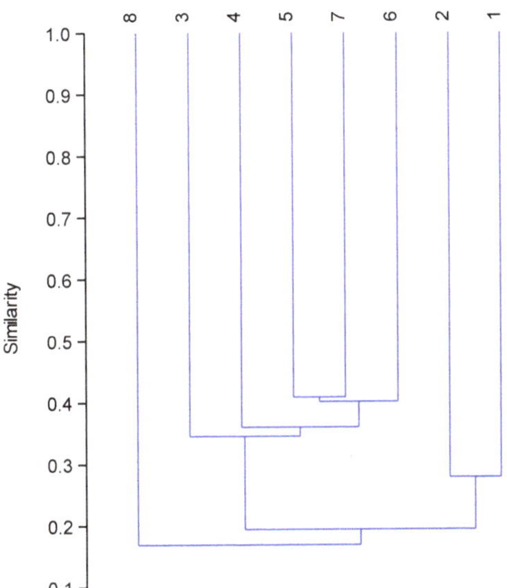

Figure 7. Cluster analysis based on the similarity of ant species occurrence (Jaccard's Index) in the eight refugium areas of the IP, all of them showing significant differences ($p \leq 0.01$; Refugium areas 1. Cantabric; 2. Pyrenean; 3. Mediterranean coast; 4. Atlantic; 5. Northern plateau; 6. Southern plateau; 7. Guadalquivir Valley; 8. Ebro Valley).

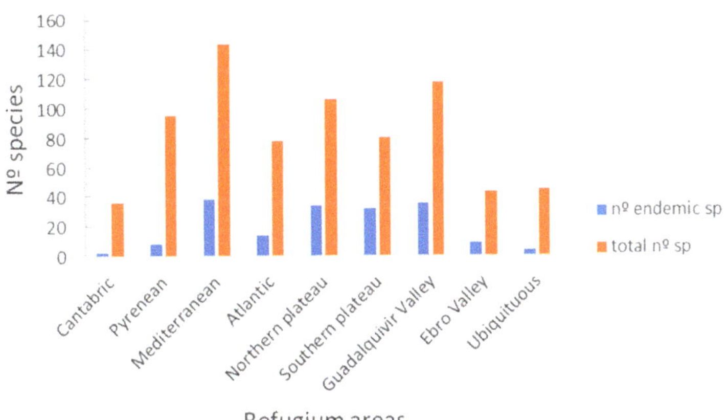

Figure 8. Number of species (orange) and endemism (blue) in the different refugium areas of the IP.

Table 4. Number of ant species occupying an increasing number of refugium areas.

N° Occupied Refugium Areas	1	2	3	4	5	6	7	8
N° species	86	46	35	32	31	16	6	46
N° endemic sp	24	12	10	10	10	2	0	5

Table 5. List of 21 endemic species appearing in only one refugium area in the IP.

Pyrenean Area	
Myrmicinae	-
Myrmica lemasnei	Bernard, 1967
Mediterranean Area	
Leptanillinae	-
Leptanilla theryi	Forel, 1903
Formicinae	-
Camponotus amaurus	Espadaler, 1997
Cataglyphis gadeai	De Haro & Collingwood, 2003
Cataglyphis humeya	Tinaut, 1991
Myrmicinae	-
Goniomma collingwoodi	Espadaler, 1997
Goniomma compressisquama	Tinaut, 1994
Messor timidus	Espadaler, 1997
Teleutomyrmex kutteri	Tinaut, 1990
Temnothorax ansei	Catarineu, Barberá & Reyes-López, 2017
Temnothorax crepuscularis	(Tinaut, 1994)
Northern Plateau Area	
Leptanillinae	-
Leptanilla charonea	López, Martínez & Barandica, 1994
Myrmicinae	-
Aphaenogaster ulibeli	Gómez & Espadaler, 2018
Myrmoxenus bernardi	(Espadaler, 1982)
Southern Plateau Area	
Leptanillinae	-
Leptanilla plutonia	López, Martínez y Barandica, 1994
Myrmicinae	-
Crematogaster fuentei	Menozzi, 1922
Oxyopomyrmex magnus	Salata & Borowiec, 2015
Guadalquivir Valley Area	
Formicinae	-
Cataglyphis floricola	Tinaut, 1993
Cataglyphis tartessica	Amor & Ortega, 2014
Myrmicinae	-
Goniomma baeticum	Reyes & Rodríguez, 1987
Ebro Valley Area	
Myrmicinae	-
Temnothorax caesari	(Espadaler, 1997)

4. Discussion

4.1. Taxonomic Richness

Ants showed a higher number of species in the IP (299 species), compared with the adjacent Mediterranean countries, with only a few species less than the Balkan Peninsula. A similar number of species is found in Greece and Turkey [74,75] (Table 1). Following the expected latitudinal diversity gradient there is a decrease in species richness to the north [88,89]. Although the IP has a higher ant species richness than Morocco [5,91,92], the last has two more subfamilies than the IP and three more than France, which follows the latitudinal diversity gradient. These results varied when the area surface is taken into account, the ratio is similar between the IP and Morocco, being higher than in France and much lower than other Mediterranean countries such as Israel.

With respect to the richness of the endemic species the same pattern occurred. The IP shows the highest number (72 endemic species), compared with other Mediterranean countries such as Greece, (21 species [74]); Israel (25 species [76]) or Morocco (56 species [57]), but when the surface areas are considered, a similar ratio between the IP and Morocco again occurred, but this was much higher in the eastern Mediterranean.

Nevertheless, not only the number of species and endemic taxa in the IP are remarkable but also the singularity of the Iberian myrmecofauna including one endemic genus (*Iberoformica*), with only one endemic species *I. subrufa* and the fossil of its probable ancestor *I. horrida* [93]. Moreover, the IP is the only place in Europe where the genus *Rossomyrmex* is located outside Asia. The presence of these two exclusive genera highlights the importance of the IP as a refugium. At least for *Rossomyrmex*, there is evidence that have been several episodes of extinction of this genus between the nearest geographical point (Turkey) and the IP. This genus has survived in the IP and not in other nearby countries, as have other emblematic animals such as the *Lynx* or the Azure-winged magpie [94].

The presence of these two ant genera as well as the low number of Iberian ant species shared with Morocco (29%) and with France (59%), while France shares with the IP 84%, and Morocco 37% with the IP, indicates that the IP harbours a unique and fairly exclusive ant fauna and functions as a "cul de sac" of ant species [94].

Several factors are responsible for the high ant biodiversity in the IP, as mentioned in the introduction, derived from the highly complex Iberian orography and convulse geological and climatic history. Another factor affecting the Iberian ant diversity is the peninsular effect, which explains the increase in diversity near the isthmus [95,96], assessed in the IP for birds [97] and butterflies [98]. Moreover, the role of the Palaearctic peninsulas during the Quaternary glaciations, as refugia for biota and centres of speciation [36], together with the Iberian complex orography producing "refugia-within-refugium" effect that has been assessed for different animal groups, contributed to increase the Iberian biodiversity [14,99].

Analysing the contribution in the IP ant diversity of the most relevant worldwide genera according to Wilson (1976 in Pie & Feitosa, [100]), only *Camponotus* make an important contribution to Iberian biodiversity, whilst *Pheidole* and *Crematogaster* had few species represented. This phenomenon is common to all Europe [69]. In the IP, the genus that we can consider "to have conquered the world" following Wilson's criteria must be *Temnothorax*, the most diverse Iberian genus (47 Iberian species) with the highest number of Iberian endemic species (16). This is an abundant genus in all the IP habitats, except in the high mountains above 2,700 m a.s.l., reflecting a high plasticity and adaptation ability. Probably its small size and its mode of mate location behaviour, by means of female pheromones, act as a speciation driver in this genus. Nevertheless, other diverse genera such as *Lasius*, *Formica* or *Camponotus* have a much lower number of endemic species than expected. Again, the mating behaviour by mass nuptial flights, which facilitates the populations mixing at a landscape level, might influence the lower rates of speciation in these genera. Moreover, other genera, such as *Cataglyphis* or *Goniomma*, showed a higher number of endemic species than expected. Other factors, probably linked with the biogeography, biotic traits and plasticity of the genera, may be responsible for this high level of endemicity.

All these data contribute to highlight the singularity of the Iberian myrmecofauna and its relevance in the Mediterranean region, placing the IP in a similar position to the Balkan Peninsula as a centre of ant diversification.

4.2. Biogeographical Analysis

4.2.1. Chorology

The chorological analysis showed two groups of ants, one widely distributed in the northern half of Iberia, sharing species with the western Palaearctic region but sometimes extending up to northern Africa, (N + T chorological categories), and the other group, mainly including species belonging to the Mediterranean basin and northern Africa (S + MN + M + X + XS chorological categories). This result is congruent with those obtained for aquatic coleopterans [78] and showed the IP acts as a transitional element between Europe (north Palaearctic) and north Africa (south Palaearctic).

This north-south distribution pattern is shown at different taxonomic levels (ant genera and species), thus we can find genera preferentially distributed in the north of the IP, such as *Formica*, *Lasius*, *Myrmica* and *Leptothorax* and others in the southern half, such as *Cataglyphis*, *Proformica*, *Stigmatomma*, *Anochetus*, *Messor*, etc. At species level, we can find the same pattern, with some typically northern species such as *Lasius fuliginosus*, most of the *Formica rufa* group, *Camponotus herculeanus*, *C. ligniperda*, *Tapinoma pygmaeum*, the three *Leptothorax* species, and more examples included in the Supplementary Material (Annex S2). In the same way, a similar number of species could be considered as southern, such as some of those included in the genus *Cataglyphis* (*C. humeya*, *C. floricola*, *C. tartessica*), *Goniomma* (*G. collingwoodi*, *G. compressiquama*), *Messor timidus*, *Aphaenogaster striativentris*, the three species of the genus *Stigmatomma*, or *Anochetus ghilianii*, amongst others (see Table S2).

4.2.2. Distribution Ranges

The complex orography of the IP produced not only a north-south division of species distribution, but even in the eight proposed refugium areas, we can find a significantly different ant fauna. This result supports the fact that similar processes affected all the biota, with common patterns emerging, although some slight differences in shape probably depend on biotic traits of the studied species such as dispersal ability [32,80,81,84,101]. Most of the proposed biogeographical subdivisions of the IP established between five (Mollusca: Pulmonata [101]) to eleven zones (birds [81]). Our proposal is closer to that of vascular plants and vertebrates [32,81,84]. The social character of ants does not appear to alter the biogeographical pattern followed for the general biota, except for their presence in the high mountains. Amongst the IP refugia the most similar for occurrence of ants (high Jaccard's diversity Index) are the northern plateau, the Guadalquivir Valley and the southern plateau. From the analysis of the species occurrence within each refugium areas, we have proved that the Mediterranean and Guadalquivir Valley refugia show the highest species richness and number of endemic species, highlighting their importance for ant conservation. In fact, Andalusia is considered the Iberian region with the highest general biodiversity [21] where both refugium areas are represented. In the Guadalquivir Valley refugium, we can find some special habitats such as the prequaternary forest with relict plant species located near the Straits of Gibraltar [102]. In this habitat we can also find the very few semitropical ant species appearing in the IP, which belong to the subfamilies Amblyoponinae together with some Ponerinae such as *Anochetus ghilianii*.

The Pyrenean and the northern plateau are the following refugium areas in biodiversity importance. The Pyrenean range is in the isthmus, concentrating the peninsular effect, which together with the high heterogeneity derived from the altitudinal variation and its west-east position, produced an important climatic variability, responsible for the species richness in this refugium. Nevertheless, the number of endemic species is not very high (10 species), pointing to the possibility of gene flow between the Pyrenean and the European populations, diluting speciation possibilities. The northern plateau is the oldest region of the IP, including the Hercynian plate; its age may be the cause of the increment in biodiversity in this refugium.

Most of the species are distributed in more than one refugium area, but about a third is found in only one IP refugium (86 species). On the other hand, only 46 species are ubiquitous in all the IP.

Similarly, from analysing the distribution of endemic species in the different refugium areas, we have been unable to find a clear pattern, finding 21 endemic species limited to only one refugium and only five endemic species being considered ubiquitous. These results are congruent with those found by different authors [78,99] pointing to different causes for the absence of clear patterns of endemicity in the IP, including sampling artifacts.

On the other hand, although mountain ranges promoted species richness and endemicity [103], this pattern is not detectable for the Iberian and Palaearctic formicids. Only one ant species is exclusive to high mountains: *Proformica longiseta* (between 1800 and

3000 m a.s.l. in the Sierra Nevada mountains [104] and the Baetic range [13]). According to these findings, the majority of the Iberian ant endemic species are related with semiarid habitats with scarce vegetation at low altitudes, such as some of the species pertaining to the genus *Temnothorax*, as *T. ansei*, *T. blascoi*, *T. caesari*, *T. crepuscularis* [105]. Another important number of endemic species are found in medium altitude mountain ranges, such as *Teleutomyrmex kutteri* (1700–2250 m a.s.l. in different Baetic ranges [106,107]), *Temnothorax gredosi* or *T. conatensis* (900–1500 m a.s.l. [108,109]).

Whereas the rare species, i.e., those found in only one or two refugium areas, include for instance the social parasite ant species, where their way of life makes them difficult to find. These are known in very few and sometimes very distant areas, such as *Rossomyrmex minuchae* [12,110], *Myrmoxenus bernardi* [111] or *Anergates atratulus* [112]. Another example of a rare species is *Aphaenogaster cardenai*, an Iberian endemic species of relatively wide distribution [113], but living in scarcely sampled habitats (the mesovoid shallow substratum, shallow caves and mines [114]).

4.2.3. Origin and Dispersal Routes

Other different causes responsible for the high ant biodiversity in the IP are related with the centre of origin and dispersal routes of the ancestors of the current ant fauna. In agreement with Blaimer et al. [115] the origin of formicids probably occurred during late Cretaceous (104–117 mya), but the rise of the modern ant fauna probably occurred during the early Cenozoic and continued until the present day. There are signs that all the genus-level taxa appeared from the first 50–60 mya are now extinct in the western Palaearctic [7]. Blaimer et al. [115] gave the estimated time of divergence for several genera from the subfamily Formicinae, thus, the clade giving place to the tribe Formicini, appeared 60 mya, and the most probable ancestral range was the Palaearctic region. These data are congruent with the estimated age of *Cataglyphoides constrictus*, a probable ancestor of the genus *Cataglyphis* (but see [116]), or *Formica horrida*, probably a species belonging to the current genus *Iberoformica* [93]. Both fossils were found in the Baltic amber (middle to late Eocene 37–42 mya) [117]. According to Blaimer et al. [115] the *Formica*/*Iberoformica* clade appearance is dated to 50 mya, the clade of *Rossomyrmex*, *Cataglyphis*/*Bajcaridris*, *Proformica* diverged in 45 mya, *Bajcaridris*/*Proformica* in 20 mya and finally *Rossomyrmex*/*Cataglyphis* in 15 mya, which are in agreement with the results of Guenard et al. [7].

The genus *Temnothorax*, which belongs to the subfamily Myrmicinae, showed its centre of origin in the Palaearctic region during the Eocene-Oligocene transition (38–33 mya) [118]. The genus *Myrmica* appeared to diversify following drastic climatic cooling in the same epoch (34 mya), and its most probable centre of origin is central or south-eastern Asia [119]. The genus *Stenamma* shared its time of origin with the previous genera but the centre of origin is estimated in the Nearctic region, and two dispersal waves to the Palaearctic were detected, the first in early Oligocene (30 mya), leaving in the Iberian myrmecofauna the species *S. striatulum*, and the second at late Pliocene (about 3 mya) with *S. debile* [120]. Undoubtedly posterior dispersals have occurred, as has been demonstrated for other insects such as the butterfly *Parnassius apollo* with a recent (late Pleistocene) colonization of most of their range [121].

With this retrospective view, we know that most of the current extant ant genera inhabited the Palaearctic region 15 mya (Middle Miocene), but the question is whether they inhabited the IP at the same time.

From the Cenozoic (60 mya), the Iberian plate was fused with the European plate [95]. Unfortunately, the ant fossil data from this time are scarce and, in some cases, too ancient to resolve the question [122,123]. We only have some scarce data from recent periods, thus, the existence of the genera *Camponotus*, *Dolichoderus*, *Iridomyrmex*, *Lasius* or *Formica*, *Liometopum* and *Messor*, in the Middle and Late Miocene of the IP has been reported (13 to 7 mya) [124,125]. The paleoclimate and paleoenvironments of the Miocene (23 mya) have been recreated more accurately through the coral reef and mammal fossils [24,29,30,126]. During this period the IP climate was semitropical and the biota was equivalent to the current African savannah [127], including important Asian and African components [27]. Nevertheless, from the Miocene until the Pliocene (5.33 mya) the biogeography of the IP was quite convulsed, especially on the southern extreme, frequently joined and separated with the African plate [26]. Moreover, the climate was cooling which finally provoked the substitution of the semitropical fauna for others better adapted to temperate or cool climates [7].

In this situation, some ant genera may have inhabited the Iberian savannah from the Miocene and later the steppes from the Pliocene and Pleistocene [7,115]. This dynamic process of sequenced colonization, extinction and diversification did not stop. Thus, during the transition Pliocene-Pleistocene, the effect of Quaternary glaciations on these Asian and African fauna dispersed to the IP, giving way to the current myrmecofauna. This view is reinforced with data on radiation of the *Formica* genus and the *Formica rufa* group species, dated in the Middle Pliocene and Pleistocene [128,129], and this was similar for other insects, as has been proved by means of molecular analyses in different coleoptera [14–18]) and in plants [130,131], pointing to a general process for all the biota of the IP.

In our analysis of the origin and dispersal routes which shaped the Iberian ant myrmecofauna, we can differentiate four groups of dispersal patterns:

Relict Species

From the middle to late Eocene (34–42 mya) [117] the climate was semi-tropical and the ant fauna known from the Baltic amber was very diverse including extant genera or their ancestors [132,133]. Some of them, such as *Formica* or *Aphaenogaster*, which nowadays have reached a high diversity, must have previously adapted to semi-tropical warm environments that were predominant when they were trapped by the amber. Later they were able to adapt to the cooling produced from the end of the Miocene and have survived until now. Nevertheless, both genera do so in different ways, one adapting to temperate or cool habitats (*Formica*, now more diversified to the north of the IP) and the other to temperate or warm habitats (*Aphaenogaster*, currently more diversified in the south of the IP). Nevertheless, some other genera were unable to adapt to the changing conditions and disappeared from the western Palaearctic [7], such as the fossils of the genera *Pachycondyla* from Denmark, preceding the Paleocene-Eocene transition (55 mya) [134]. In some rare cases, semitropical ant species such as *Anochetus* and *Stigmatomma*, found in Santo Domingo amber (16–19 mya) [117,135] and the Baltic amber [133] respectively, and currently well distributed in tropical and semitropical regions, were able to remain in some restricted locations of the IP surrounding the Straits of Gibraltar [136,137], an exceptional zone also inhabited by paleoendemic plant species [102,138]. Although Jowers et al. [9] concluded that *Anochetus ghilianii* is a recent invader of the IP, we think that the coincidence in this area of botanical paleoendemic species together with different semitropical ants (*Anochetus* and *Stigmatomma* species), it must be the result of a similar biogeographical process acting on them. Anyway, *A. ghilianii* should be considered as a relict Moroccan species and its presence in the IP is noteworthy.

Asian-IP Disjunct Species

In the Pliocene (5.33 mya) the Mediterranean Sea reopened through the Straits of Gibraltar and the climate was cooling, favouring the substitution of the existing semitropical fauna for one better adapted to temperate climates. This temperate fauna proceeded principally from Central Asia, reaching the IP as confirmed by fossil deposits [25,28–30]. The Asian colonizing waves have occurred repeatedly during the transition Pliocene-Pleistocene (2.6, 2.5 and 1.7 mya) [25] even during the late Pleistocene [139] and later. The migrant mega-fauna must have been dispersing together with invertebrates, among them obviously ants, sharing the same origin and similar ecological requirements. These Asiatic and African faunas from the later Pliocene, evolved under glacial and interglacial periods during the Pleistocene, and many of these species became extinct or remained isolated in some refugia throughout the entire Iberian refugium, setting up the current myrmecological fauna [22,31–33]. Some of these species continued isolated and now have a narrow distribution range, as is the case of the slave-making ant *R. minuchae* in Spain, and the other three species of the genus, which are dispersed, although with a wider distribution range in Asia. The genus *Rossomyrmex* can be considered as a relict in the IP. On the other hand, the species of the genus *Proformica*, the host of *Rossomyrmex*, with a probable centre of origin in Central Asia [11], were able to recolonize wide distribution ranges both before and after glacial periods, occupying all the southern Palaearctic from Central Asia to the IP. This genus, more diversified in the western Mediterranean, specifically in the IP, is not distributed in Morocco, where it has been replaced by the close genus *Bajcaridris* [11,115,140]. This fact suggests the hypothesis of a frequent dispersal route on the two sides of the Mediterranean (Figure 9) with speciation concluding in different genera (*Proformica* and *Bajcaridris*). An opposite direction of the dispersal route (from western Mediterranean to Asia) has been suggested for *Rossomyrmex* [11] as occurred, for example, with the plant genus *Odontites* [130].

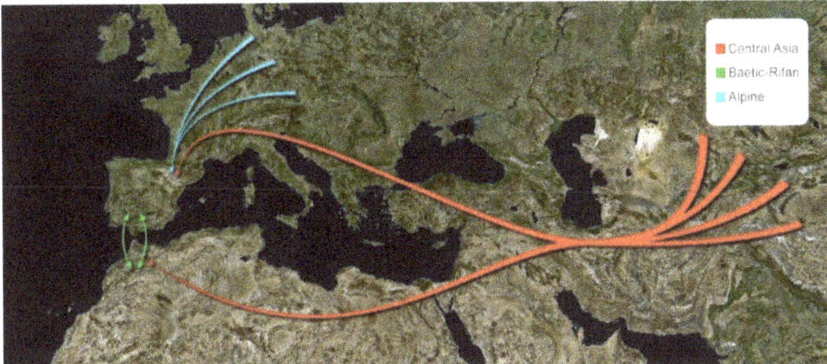

Figure 9. Suggested dispersal routes affecting the myrmecofauna of the IP. (Base map obtained from https://laboratoriorediam.cica.es/VisorRediam/ (Accessed on 2 December 2020)). Junta de Andalucía. Consejería de Agricultura, Ganadería, Pesca y Desarrollo Sostenible (ámbito de Desarrollo Sostenible).

Another genus offering interesting disjunct distributions is *Cataglyphis*, specifically the species group *altisquamis* shows two different and discontinuous distribution ranges: one group of five species distributed from Central Asia to the western Mediterranean and the other nine species distributed in the western Mediterranean (IP and Morocco) [141] (Figure 11). This second group of species is absent from Turkey to France, by the northern route, and from Egypt and Libya, by the southern route, due to extinction processes in the intermediate points. The age of the entire genus and its high diversity in Asia point to a Central Asian origin. Three different dispersal routes are possible from Asia to the western Mediterranean: 1. the simultaneous dispersal route via the two sides of the Mediterranean, 2. the dispersal by the northern Mediterranean side across to the Balkan region, and 3. the dispersal only by the southern Mediterranean side (Figure 9). The three are possible but until now only the southern route (3) has been suggested for the *Cataglyphis albicans* group [10]. Nevertheless, the three dispersal routes from Asia have also been proposed for the genus *Cephalota* (Coleoptera: Cicindelidae) [17] and may be more frequent dispersal patterns than had generally been considered. An extension of the phylogenetic study on *Cataglyphis* including middle-west distributed species should produce a wide and accurate view for the biogeography and dispersal patterns of these species.

The genera *Goniomma* and *Oxyopomyrmex*, both endemic of the Mediterranean basin, also show disjunction between the eastern and western Mediterranean distributions [142, 143] (Figure 10). Their centre of origin is unknown, although they probably appeared in the Miocene (12 mya) [90]. A similar distribution pattern is found in the coleopteran species of the genus *Pimelia* [15], *Berberomeloe* [14] and the plant genus *Odontites* [130].

Another example of current Iberian fauna showing Asian-IP disjunct distribution is the Iberian azure-winged magpie [144] or amongst the insects, the lepidopteran *Pseudochazara wiliamsi* or the coleopteran of the subgenus *Parentius* (see [145]), and together with these, we find many other species adapted to arid or semi-arid environments such as Monegros (Zaragoza) or Guadix-Baza (Granada) [146,147] or to mountain ranges [148]. Thus, all these examples reinforce the importance of Asia as a centre of origin for the Iberian fauna.

Figure 10. Disjunct distribution range of the species belonging to genus *Goniomma*. Note the presence of one species in Israel (citations from [71] are included; base map as in Figure 9).

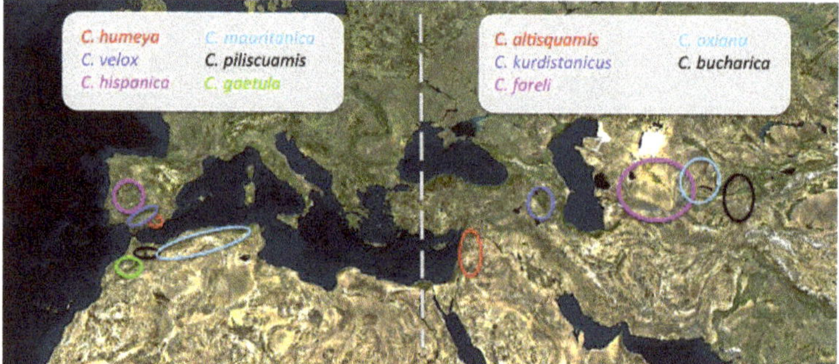

Figure 11. Disjunct distribution of the species belonging to the *Cataglyphis altisquamis* group (modified from [149]; base map as in Figure 9).

Baetic-Rifan Species

This dispersal pattern includes the shared species between the south of the IP and the north of Africa. Within this group, the ants with apterous or brachypterous females, such as *Monomorium algiricum*, *A. ghilianii*, *Stigmatomma* species or *Aphaenogaster senilis* group, gain biogeographical relevance, due to their handicap for flight dispersal across the Mediterranean Sea. One plausible explanation for this kind of distribution comes from the geological history of this territory. South-western Iberia and the north of Morocco share stratigraphic deposits, forming the Baetic-Rifan territories [150] derived from the Miocenic island of Alborán, emerged in the western Mediterranean 16 mya [126]. This island was united with Morocco until 8 mya when it again became disconnected and isolated until 5.9 mya. After this, it was again partially united with the IP until 5.3 mya. At this time, the Straits of Gibraltar reopened, dividing the territories of the Miocenic island of Alborán between the IP and Morocco. The existence of this island may explain the relatively high presence of apterous species in this zone, because apterism is one of the known island syndromes. This effect is known in the Baetic-Rifan territories not only for ants [151] but also for *Pimelia* and *Blaps* Tenebrionid [15,17] and Meloid (*Berberomeloe* group) coleoptera [14], both of them apterous. The existence of this island is more extended in geological time than the habitually invoked desiccation of the Mediterranean during the Messiniense (late Miocene) to explain the shared presence of species between the IP and northern Africa.

Alpine Species

Another important and more recent dispersal route is related with the Pleistocene glaciations, the last occurring 10,000 years ago. Many northern species survived and took refuge in the southern Palaearctic peninsulas, and during interglacial periods returned to the north and/or the populations climbed the slopes of the mountains [37]. One good example of this groupis the pair of Iberian endemic species *Formica frontalis/F. dusmeti*, very close phylogenetically to *F. truncorum*, distributed in northern Europe ([129], unpublished data). Probably the presence of *T. kutterii* in the south of the IP and *T. schneideri* in the northern IP and in the Alps is likely to have had this similar origin. Some species of *Myrmica, Temnothorax, Lasius*, etc, may belong to this group of alpine species, but this should be confirmed by phylogenetic molecular studies on the Mediterranean species.

5. Conclusions

The taxonomic diversity and distribution patterns we have presented in this study highlight the importance of Iberian ants for a better understanding of the complex evolutionary history and biogeography of the IP. Moreover, we have tried to show the importance of biogeography driving the structure of the current ant communities, ecological interactions, such as hierarchic competence and biotic factors, only explain a part of their structure [152]. Ours results on ant biogeography support the hypothesis that the centre of origin of the species and their dynamic processes (dispersal, vicariance, speciation and extinction) are the missing link to fully explain and understand the evolution and the structure of these ant communities [2,5,7]. We hope the hypothesis and proposals put forward in this study will promote new biogeographical, phylogenetic and evolutionary studies about the ant fauna of the Mediterranean Basin and produce a wide and accurate view of the centre of origin and the dispersal patterns of Iberian ants.

Supplementary Materials: The following are available online at https://www.mdpi.com/1424-281 8/13/2/88/s1, Table S1. Ant species list of France, Morocco and the IP. Table S2. List of Iberian ant species and their occurrence in the refugium areas.

Author Contributions: Conceptualization, A.T. and F.R.; methodology, A.T. and F.R.; formal analysis, A.T. and F.R.; investigation, A.T. and F.R; writing—original draft preparation, A.T. and F.R.; writing—review and editing, A.T. and F.R.; project administration, F.R.; funding acquisition, F.R. All authors have read and agreed to the published version of the manuscript.

Funding: This research was partially funded by the project RTA2015-00012-C02-02 (Ministry of Science and Innovation, INIA and FEDER funds).

Institutional Review Board Statement: Not applicable.

Informed Consent Statement: Not applicable.

Data Availability Statement: The data presented in this study are available in supplementary material.

Acknowledgments: We thank José M. Martín, Alfonso Arribas and Elvira Martín who provided some references and exchange of ideas and Pedro Sandoval for his help with the figures. Angela Tate reviewed the English edition. We are also grateful to the three anonymous reviewers who made a careful and significant improvement to the manuscript.

Conflicts of Interest: The authors declare no conflict of interest.

References

1. Jenkins, C.N.; Sanders, N.J.; Andersen, A.N.; Arnan, X.; Brühl, C.A.; Cerda, X.; Ellison, A.M.; Fisher, B.L.; Fitzpatrick, M.C.; Gotelli, N.J.; et al. Global diversity in light of climate change: The case of ants. *Divers. Distrib.* **2011**, *17*, 652–662. [CrossRef]
2. Andersen, A.N.A.N. Ant megadiversity and its origins in arid Australia. *Austral Entomol.* **2016**, *55*, 132–137. [CrossRef]
3. Economo, E.P.; Narula, N.; Friedman, N.R.; Weiser, M.D.; Guénard, B. Macroecology and macroevolution of the latitudinal diversity gradient in ants. *Nat. Commun.* **2018**, *9*, 1–8. [CrossRef] [PubMed]
4. Hakala, S.M.; Seppä, P.; Helanterä, H. Evolution of dispersal in ants (Hymenoptera: Formicidae): A review on the dispersal strategies of sessile superorganisms. *Myrmecol. News* **2019**, *29*, 35–55.
5. Dunn, R.R.; Agosti, D.; Andersen, A.N.; Arnan, X.; Bruhl, C.A.; Cerdá, X.; Ellison, A.M.; Fisher, B.L.; Fitzpatrick, M.C.; Gibb, H.; et al. Climatic drivers of hemispheric asymmetry in global patterns of ant species richness. *Ecol. Lett.* **2009**, *12*, 324–333. [CrossRef] [PubMed]
6. Fisher, B.L. Biogeography. In *Ant Ecology*; Lach, L., Parr, C.L., Abbott, K.L., Eds.; Oxford University Press: Oxford, UK, 2010; pp. 18–37.
7. Guénard, B.; Perrichot, V.; Economo, E.P. Integration of global fossil and modern biodiversity data reveals dynamism and stasis in ant macroecological patterns. *J. Biogeogr.* **2015**, *42*, 2302–2312. [CrossRef]
8. Catarineu, C.; Barberá, G.G.; Reyes-López, J.L. Zoogeography of the ants (Hymenoptera: Formicidae) of the Segura River Basin. *Sociobiology* **2018**, *65*, 383–396. [CrossRef]
9. Jowers, M.J.; Taheri, A.; Reyes-López, J. The ant Anochetus ghilianii (Hymenoptera, Formicidae), not a Tertiary relict, but an Iberian introduction from North Africa: Evidence from mtDNA analyses. *Syst. Biodivers.* **2015**, *13*, 865–874. [CrossRef]
10. Villalta, I.; Amor, F.; Galarza, J.A.; Dupont, S.; Ortega, P.; Hefetz, A.; Dahbi, A.; Cerdá, X.; Boulay, R. Origin and distribution of desert ants across the Gibraltar Straits. *Mol. Phylogenet. Evol.* **2018**, *118*, 122–134. [CrossRef] [PubMed]
11. Sanllorente, O.; Lorite, P.; Ruano, F.; Palomeque, T.; Tinaut, A. Phylogenetic relationships between the slave-making ants Rossomyrmex and their Proformica hosts in relation to other genera of the ant tribe Formicini (Hymenoptera: Formicidae). *J. Zool. Syst. Evol. Res.* **2018**, *56*, 48–60. [CrossRef]
12. Ruano, F.; Devers, S.; Sanllorente, O.; Errard, C.; Tinaut, A.; Lenoir, A. A geographical mosaic of coevolution in a slave-making host-parasite system. *J. Evol. Biol.* **2011**, *24*, 1071–1079. [CrossRef] [PubMed]
13. Sanllorente, O.; Ruano, F.; Tinaut, A. Large-scale population genetics of the mountain ant Proformica longiseta (Hymenoptera: Formicidae). *Popul. Ecol.* **2015**, *57*, 637–648. [CrossRef]
14. Sánchez-Vialas, A.; García-París, M.; Ruiz, J.L.; Recuero, E. Patterns of morphological diversification in giant Berberomeloe blister beetles (Coleoptera: Meloidae) reveal an unexpected taxonomic diversity concordant with mtDNA phylogenetic structure. *Zool. J. Linn. Soc.* **2020**, *189*, 1249–1312. [CrossRef]
15. Mas-Peinado, P.; Buckley, D.; Ruiz, J.L.; García-París, M. Recurrent diversification patterns and taxonomic complexity in morphologically conservative ancient lineages of Pimelia (Coleoptera: Tenebrionidae). *Syst. Entomol.* **2018**, *125*, 331–348. [CrossRef]
16. Condamine, F.L.; Sperling, F.A.H.; Wahlberg, N.; Rasplus, J.Y.; Kergoat, G.J. What causes latitudinal gradients in species diversity? Evolutionary processes and ecological constraints on swallowtail biodiversity. *Ecol. Lett.* **2012**, *15*, 267–277. [CrossRef]
17. Herrera-Russert, J.; López-López, A.; Serrano, J.; Matalin, A.; Galián, J. Influence of the Mediterranean basin history on the origin and evolution of the halophile tiger beetle genus Cephalota (Coleoptera: Cicindelidae). *Ann. Soc. Entomol. Fr.* **2020**, *56*, 447–454. [CrossRef]
18. Ribera, I.; Castro, A.; Díaz, J.A.; Garrido, J.; Izquierdo, A.; Jäch, M.A.; Valladares, L.F. The geography of speciation in narrow-range endemics of the "Haenydra" lineage (Coleoptera, Hydraenidae, Hydraena). *J. Biogeogr.* **2011**, *38*, 502–516. [CrossRef]
19. Myers, N.; Mittermeler, R.A.; Mittermeler, C.G.; Da Fonseca, G.A.B.; Kent, J. Biodiversity hotspots for conservation priorities. *Nature* **2000**, *403*, 853–858. [CrossRef] [PubMed]
20. Araújo, M.B.; Lobo, J.M.; Moreno, J.C. The effectiveness of Iberian protected areas in conserving terrestrial biodiversity. *Conserv. Biol.* **2007**, *21*, 1423–1432. [CrossRef] [PubMed]
21. Medail, F.; Quezel, P. Biodiversity hotspots in the Mediterranean Basin: Setting global conservation priorities. *Conserv. Biol.* **1999**, *13*, 1510–1513. [CrossRef]
22. Hewitt, G.M. Mediterranean Peninsulas: The Evolution of Hotspots. In *Biodiversity Hotspots*; Springer: Berlin/Heidelberg, Germany, 2011; pp. 123–147.
23. Martín-Piera, F.; Sanmartín, I. Biogeografía de áreas y biogeografía de artrópodos holárticos y mediterráneos. *Bol. Soc. Entomol. Aragon.* **1999**, *26*, 535–560.
24. Braga, J.C.; Martín, J.M. Los arrecifes del Mioceno de Almería. *Investig. Gest.* **1997**, *1*, 5–19.
25. Arribas, A.; Garrido, G.; Viseras, C.; Soria, J.M.; Pla, S.; Solano, J.G.; Garcés, M.; Beamud, E.; Carrión, J.S. A mammalian lost world in Southwest Europe during the late pliocene. *PLoS ONE* **2009**, *4*, e7127. [CrossRef]
26. Martín Martín, J.M.; Braga Alarcón, J.C.; Gómez Pugnaire, M.T. *Itinerarios Geológicos por Sierra Nevada*; de Andalucía, J., Ed.; Consejería de Medio Ambiente: Seville, Spain, 2008.
27. Masseti, M.; Mazza, P.P.A. Western European Quaternary lions: New working hypotheses. *Biol. J. Linn. Soc.* **2013**, *109*, 66–77. [CrossRef]

28. Arribas, A.; Riquelme, J.A.; Palmqvist, P.; Garrido, G.; Hernández, R.; Laplana, C.; Soria, J.M.; Viseras, C.; Durán, J.J.; Gumiel, P.; et al. Un nuevo yacimiento de grandes mamíferos villafranquienses en la Cuenca de Guadix-Baza (Granada): Fonelas P-1, primer registro de una fauna próxima al límite Plio-Pleistoceno en la Península Ibérica. *Bol. Geol. Min.* **2001**, *112*, 3–34.
29. Arribas, A.; Baeza, E.; Bermúdez, D.; Blanco, S.; Durán, J.J.; Garrido, G.; Gumiel, J.C.; Hernández, R.; Soria, J.M.; Viseras, C. Nuevos registros paleontológicos de grandes mamíferos en la Cuenca de Guadix-Baza (Granada): Aportaciones del Proyecto Fonelas al conocimiento sobre las faunas continentales del Plioceno-Pleistoceno europeo. *Bol. Geol. Min.* **2004**, *115*, 567–582.
30. Garrido, G. Generalidades Sobre Los Carnívoros Del Villafranquiense Superior En Relación Con El Registro Fósil De Fonelas P-1 Reflections on the Carnivores of the Upper Villafranchian Represented in the Fossil Record of the Fonelas P-1 Site. In *Vertebrados del Plioceno Superior Terminal en el Suroeste de Europa: Fonelas P-1 y el Proyecto Fonelas*; Arribas, A., Ed.; IGME: Madrid, Spain, 2008; pp. 85–146, ISBN 9788478407644.
31. Centeno-Cuadros, A.; Delibes, M.; Godoy, J.A. Phylogeography of Southern Water Vole (Arvicola sapidus): Evidence for refugia within the Iberian glacial refugium? *Mol. Ecol.* **2009**, *18*, 3652–3667. [CrossRef] [PubMed]
32. Gómez, A.; Lunt, D.H. Refugia within Refugia: Patterns of Phylogeographic Concordance in the Iberian Peninsula. In *Phylogeography of Southern European Refugia: Evolutionary Perspectives on the Origins and Conservation of European Biodiversity*; Springer: Dordrecht, The Netherlands, 2007; pp. 155–188, ISBN 9781402049040.
33. Lobo, J.M.; Moreno, J.C. Spatial and environmental determinants of vascular plant species richness distribution in the Iberian Peninsula and Balearic Islands. *Biol. J. Linn. Soc.* **2001**, *73*, 233–253. [CrossRef]
34. Miraldo, A.; Hewitt, G.M.; Paulo, O.S.; Emerson, B.C. Phylogeography and demographic history of Lacerta lepida in the Iberian Peninsula: Multiple refugia, range expansions and secondary contact zones. *BMC Evol. Biol.* **2011**, *11*, 170. [CrossRef]
35. Hewitt, G.M. Genetic consequences of climatic oscillations in the Quaternary. *Philos. Trans. R. Soc. Lond. Ser. B Biol. Sci.* **2004**, *359*, 183–195. [CrossRef]
36. Hewitt, G.M. Post-glacial re-colonization of European biota. *Biol. J. Linn. Soc.* **1999**, *68*, 87–112. [CrossRef]
37. Schmitt, T. Molecular biogeography of Europe: Pleistocene cycles and postglacial trends. *Front. Zool.* **2007**, *4*, 11. [CrossRef] [PubMed]
38. Galkowski, C. Quelques fourmis nouvelles ou intéressantes pour la faune de France (Hymenoptera, Formicidae). *Bull. Soc. Linn. Bordx. Tome* **2008**, *143*, 423–433.
39. Galkowski, C. Une liste des fourmis (Hymenoptera; Formicidae) récoltées dans la région de Grasse, avec la mention d'une nouvelle espèce de la faune de France. *Bull. Soc. linn. Provence* **2011**, *62*, 1–4.
40. Steiner, F.M.; Csősz, S.; Markó, B.; Gamisch, A.; Rinnhofer, L.; Folterbauer, C.; Hammerle, S.; Stauffer, C.; Arthofer, W.; Schlick-Steiner, B.C. Turning one into five: Integrative taxonomy uncovers complex evolution of cryptic species in the harvester ant Messor "structor". *Mol. Phylogenet. Evol.* **2018**, *127*, 387–404. [CrossRef] [PubMed]
41. Seifert, B.; Buschinger, A.; Aldawood, A.; Antonova, V.; Bharti, H.; Borowiec, L.; Dekoninck, W.; Dubovikoff, D.; Espadaler, X.; Flegr, J.; et al. Banning paraphylies and executing Linnaean taxonomy is discordant and reduces the evolutionary and semantic information content of biological nomenclature. *Insectes Soc.* **2016**, *63*, 237–242. [CrossRef]
42. Salgueiro, J. Catálogo dos formicídeos de Portugal continental e ilhas. *Bol. Soc. Entomol. Aragon.* **2002**, *31*, 145–171.
43. Collingwood, C.A.; Prince, A. A guide to ants of continental Portugal. *Bol. Soc. Port. Entomol.* **1998**, Suplement.
44. Boeiro, M.; Espadaler, X.; Azedo, A.R.; Collingwood, C.; Serrano, A.R.M. One genus and three ants species new to Portugal (Hymenoptera, Formicidae). *Bol. Soc. Entomol. Aragon.* **2009**, *45*, 515–517.
45. Gonçalves, C.; Espadaler, X.; Pereira, J.A.; Santos, S.; Patanita, M.I. Primeiros registos das espécies Strongylognathus caeciliae Forel, 1897 e Temnothorax tyndalei (Forel, 1909)(Hymenoptera, Formicidae) em Portugal Continental. *Bol. Soc. Entomol. Aragon.* **2014**, *54*, 402.
46. Henin, J.M.; Collingwood, C.; Paiva, M.R. Synonymy between Leptothorax caparica Henin, Paiva & Collingwood, 2001 and Cardiocondyla mauritanica. *Bol. Soc. Port. Entomol.* **2003**, *211*, 377–378.
47. Sánchez-García, D.; Cuesta-Segura, D.A.; Trigos-Peral, G.; Arcos, J.; Catarineu, C.; García-García, F.; Herraiz, J.A.; Espadaler, X.; Gómez, K.; Tinaut, A. *Checklist of the Iberian Myrmecofauna*, In preparation.
48. Tinaut, A.; Reyes-lópez, J. Descripción de una nueva especie para la península ibérica: Temnothorax alfacarensis n. sp. (Hymenoptera, Formicidae). *Boln. Asoc. Esp. Ent* **2020**, *44*, 359–378.
49. Casiraghi, A.; Espadaler, X.; Hidalgo, N.P.; Gómez, K. Two additions to the Iberian myrmecofauna: Crematogaster inermis Mayr, 1862, a newly established, tree-nesting species, and Trichomyrmex mayri (Forel, 1902), an emerging exotic species temporarily nesting in Spain (Hymenoptera, Formicidae). *J. Hymenopt. Res.* **2020**, *78*, 57–68. [CrossRef]
50. Bernadou, A.; Fourcassié, V.; Espadaler, X. A preliminary checklist of the ants Hymenoptera, Formicidae) of Andorra. *Zookeys* **2013**, *277*, 13–23. [CrossRef]
51. Guillem, R.; Bensusan, K. Technomyrmex vexatus (Santschi, 1919) from Gibraltar (Hymenoptera: Formicidae): A new ant species for Europe and genus for Iberia. *Myrmecol. News* **2008**, *11*, 21–23.
52. Guillem, R.; Bensusan, K. Tetramorium parvioculum sp. n. (Formicidae: Myrmicinae) a new species of the T. simillimum group from Gibraltar. *Bol. Soc. Entomol. Aragon.* **2009**, *45*, 157–161.
53. Guillem, R.; Bensusan, K. Two new species of ants (Hymenoptera: Formicidae) for Europe from southern Iberia. *Rev. Soc. Gaditana Hist. Nat.* **2019**, *13*, 5–10.

54. Casevitz-Weulersse, J.; Galkowski, C. Liste actualisée des Fourmis de France (Hymenoptera: Formicidae). *Bull. Soc. Entomol. Fr.* **2009**, *114*, 475–510.
55. Monnin, T.; Espadaler, X.; Lenoir, A.; Peeters, C. *Guide des Fourmis de France*; Belin: Paris, France, 2013.
56. Seifert, B. A taxonomic revision of the Palaearctic members of the subgenus Lasius s.str. *Soil Org.* **2020**, *92*, 15–86.
57. Cagniant, H. Liste actualisee des fourmis du Maroc (Hymenoptera: Formicidae). *Myrmecol. Nachr.* **2006**, *8*, 193–200.
58. Galkowski, C.; Cagniant, H. Contribution à la connaissance des fourmis du groupe angustulus dans le genre Temnothorax (Hymenoptera, Formicidae). *Rev. Assoc. Roussill. Entomol.* **2017**, *26*, 180–191.
59. Taheri, A.; Reyes-López, J. Exotic Ants (Hymenoptera: Formicidae) in Morocco: Checklist, Comments and New Faunistic Data. *Trans. Am. Entomol. Soc.* **2018**, *144*, 99–107. [CrossRef]
60. Taheri, A.; Reyes-López, J.L. Five new records of ants (Hymenoptera: Formicidae) from Morocco. *J. Insect Sci.* **2015**, *15*, 37. [CrossRef]
61. Delabie, J.H.C. Présence de Pheidole teneriffana Forel, 1893, au Maroc (Hym., Formicidae, Myrmicinae). *Bull. Soc. Entomol. Fr.* **2007**, *112*, 288.
62. Cagniant, H. Le Genre Cataglyphis Foerster, 1850 au Maroc (Hyménoptères Formicidae). *Orsis* **2009**, *24*, 41–71.
63. Taheri, A.; Reyes-López, J.; Espadaler, X. Citas nuevas o interesantes de hormigas (Hymenoptera, Formicidae) para Marruecos. *Bol. Soc. Entomol. Aragon.* **2010**, *47*, 299–300.
64. Taheri, A.; Reyes-López, J. Primera cita de Pyramica membranifera (Emery, 1869) (Hymenoptera, Formicidae) y listado actualizado de hormigas alóctonas para Marruecos (Norte de África). *Bol. Soc. Entomol. Aragon.* **2011**, *49*, 363.
65. Guillem, R.; Bensusan, K.; Taheri, A. First record of the ant subfamily Cerapachyinae Forel, 1893 (Hymenoptera: Formicidae) from Morocco. *Bull. l'Inst. Sci.* **2012**, *34*, 121–123.
66. Guillem, R.; Bensusan, K.; Taheri, A. New data on genus Dorylus Fabricius, 1793 (Formicidae, Dorylinae) in Morocco. *Bull. l'Inst. Sci.* **2015**, *37*, 47–51.
67. Borowiec, L. *Catalogue of ants of Europe, the Mediterranean Basin and adjacent regions (Hymenoptera: Formicidae)*; Polish Taxonomical Society: Wroclaw, Poland, 2014; Volume 25, ISBN 9788361764496.
68. Antwiki. Available online: http://antwiki.org/ (accessed on 20 November 2020).
69. Agosti, D.; Collingwood, C.A. A provisional list of the Balkan ants (Hym., Formicidae) and a key to the worker caste. I. Synonymic List. *Bull. Soc. Entomol. Fr.* **1987**, *60*, 51–62.
70. Wagner, H.C.; Seifert, B.; Borovsky, R.; Paill, W. Insight into the ant diversity of the Vjosa valley, Albania (Hymenoptera: Formicidae). *Acta ZooBot Austria* **2018**, *155*, 315–321.
71. Petrov, I.Z. A list of currently known ant species (Formicidae, Hymenoptera) of Serbia. *Arch. Biol. Sci.* **2004**, *56*, 121–125. [CrossRef]
72. Bracko, G. Review of the ant fauna (Hymenoptera: Formicidae) of Croatia. *Acta Entomol. Slov.* **2006**, *14*, 131–156.
73. Bracko, G.; Wagner, H.C.; Schulz, A.; Gioahin, E.; Maticic, J.; Tratnik, A. New investigation and a cheklist of the ants (Hymenoptera: Formicidae) of the republic of Macedonia. *North West. J. Zool.* **2014**, *10*, 10–24.
74. Borowiec, L.; Salata, S. Ants of Greece-additions and corrections (Hymenoptera: Formicidae). *Genus* **2012**, *24*, 335–401.
75. Kiran, K.; Karaman, C. First annotated cheklist of the ant fauna of Turkey (Hymenoptera. Formicidae). *Zootaxa* **2012**, *3548*, 1–38. [CrossRef]
76. Vonshak, M.; Ionescu-Hirsch, A. A checklist of the ants of Israel (Hymenoptera: Formicidae). *Isr. J. Entomol.* **2009**, *39*, 33–55.
77. Mohamed, S.; Zalat, S.; Fadi, H.; Gadalla, S.; Sharaf, M. Taxonomy of ant species (Hymenoptera: Formicidae) collected by pitfall traps from Sinai and Delta region, Egypt. *Egypt. J. Nat. Hist.* **2001**, *3*, 40–61. [CrossRef]
78. Ribera, I. Biogeography and conservation of Iberian water beetles. *Biol. Conserv.* **2000**, *92*, 131–150. [CrossRef]
79. Arnan, X.; Cerdá, X.; Retana, J. Relationships among taxonomic, functional, and phylogenetic ant diversity across the biogeographic regions of Europe. *Ecography* **2017**, *40*, 448–457. [CrossRef]
80. Romo, H.; García-Barros, E. Biogeographic regions of the Iberian Peninsula: Butterflies as biogeographical indicators. *J. Zool.* **2010**, *282*, 180–190. [CrossRef]
81. Carrascal, L.M.; Lobo, J.M. Respuestas a viejas preguntas con nuevos datos: Estudio de los patrones de distribución de la avifauna española y consecuencias para su conservación. In *Atlas de las Aves Reproductoras de España*; Martí, R., Del Moral, J.C., Eds.; Dirección General de la Conservación de la Naturaleza y Sociedad española de Ornitología: Madrid, Spain, 2003; pp. 651–668.
82. antmaps.org. Available online: https://antmaps.org/? (accessed on 2 December 2020).
83. Sokal, R.R.; Rohlf, F.J. *Biometry*, 3rd ed.; Freeman and Company: New York, NY, USA, 1995; ISBN 0-7167-2411-1.
84. Moreno Saiz, J.C.; Donato, M.; Katinas, L.; Crisci, J.V.; Posadas, P. New insights into the biogeography of south-western Europe: Spatial patterns from vascular plants using cluster analysis and parsimony. *J. Biogeogr.* **2013**, *40*, 90–104. [CrossRef]
85. Hammer, D.A.T.; Ryan, P.D.; Hammer, Ø.; Harper, D.A.T. Past: Paleontological Statistics Software Package for Education and Data Analysis. *Palaeontol. Electron.* **2001**, *4*, 178.
86. Real, R. Tables of significant values of Jaccard's index of similarity. *Miscel·lania Zool.* **1999**, *22*, 29–40.
87. Baroni-Urbani, C.; Buser, M.W. Similarity of binary data. *Syst. Zool.* **1976**, *25*, 251–259. [CrossRef]
88. Brosens, D.; Vankerhoven, F.; Ignace, D.; Wegnez, P.; Noé, N.; Heughebaert, A.; Borteis, J.; Dekoninck, W. FORMIDABEL: The Belgian Ants Database. *Zookeys* **2013**, *306*, 59–70.
89. Ødegaard, F. New and little known ants (Hymenoptera, Formicidae) in Norway. *Nor. J. Entomol.* **2013**, *60*, 172–175.

90. Moreau, C.S.; Bell, C.D. Testing the museum versus cradle tropical biological diversity hypothesis: Phylogeny, diversification, and ancestral biogeographical range evolution of the ants. *Evolution* **2013**, *67*, 2240–2257. [CrossRef]
91. Mittelbach, G.G.; Schemske, D.W.; Cornell, H.V.; Allen, A.P.; Brown, J.M.; Bush, M.B.; Harrison, S.P.; Hurlbert, A.H.; Knowlton, N.; Lessios, H.A.; et al. Evolution and the latitudinal diversity gradient: Speciation, extinction and biogeography. *Ecol. Lett.* **2007**, *10*, 315–331. [CrossRef]
92. Willig, M.R.; Kaufman, D.M.; Stevens, R.D. Latitudinal Gradients of Biodiversity: Pattern, Process, Scale, and Synthesis. *Annu. Rev. Ecol. Evol. Syst.* **2003**, *34*, 273–309. [CrossRef]
93. Gómez, K.; Lorite, P.; García, F.; Tinaut, A.; Espadaler, X.; Palomeque, T.; Sanllorente, O.; Trager, J. Differentiating Iberoformica and Formica (Serviformica) with description of the sexual castes of Formica (Serviformica) gerardi Bondroit, 1917 stat. rev. *Sociobiology* **2018**, *65*, 463–470. [CrossRef]
94. O'Regan, H.J. The Iberian Peninsula—Corridor or cul-de-sac? Mammalian faunal change and possible routes of dispersal in the last 2 million years. *Quat. Sci. Rev.* **2008**, *27*, 2136–2144. [CrossRef]
95. Lomolino, M.V.; Riddle, B.R.; Whittaker, R.J. *Biogeography: Biological Diversity across Space and Time*; Sinauer Associates: Sunderland, MA, USA, 2017; ISBN 9781605354729.
96. Jenkins, D.G.; Rinne, D. Red herring or low illumination? The peninsula effect revisited. *J. Biogeogr.* **2008**, *35*, 2128–2137. [CrossRef]
97. González-Taboada, F.; Nores, C.; Álvarez, M.Á. Breeding bird species richness in Spain: Assessing diversity hypothesis at various scales. *Ecography* **2007**, *30*, 241–250. [CrossRef]
98. Martin, J.; Gurrea, P. The Peninsular Effect in Iberian Butterflies (Lepidoptera: Papilionoidea and Hesperioidea). *J. Biogeogr.* **1990**, 85–96. [CrossRef]
99. Abellán, P.; Svenning, J.-C. Refugia within refugia-patterns in endemism and genetic divergence are linked to Late Quaternary climate stability in the Iberian Peninsula. *Biol. J. Linn. Soc.* **2014**, *113*, 13–28. [CrossRef]
100. Pie, M.R.; Feitosa, R.M. Relictual ant lineages (Hymenoptera: Formicidae) and their evolutionary implications. *Myrmecol. News* **2016**, *22*, 55–58.
101. Puente, A.I.; Altonaga, K.; Prieto, C.E.; Rallo, A. Delimitation of biogeographical areas in the Iberian Peninsula on the basis of Helicoidea species (Pulmonata: Stylommatophora). *Glob. Ecol. Biogeogr. Lett.* **1998**, *7*, 97–113. [CrossRef]
102. Molina-Venegas, R.; Aparicio, A.; Lavergne, S.; Arroyo, J. Climatic and topographical correlates of plant palaeo- and neoendemism in a Mediterranean biodiversity hotspot. *Ann. Bot.* **2017**, *119*, 229–238. [CrossRef]
103. Rahbek, C.; Borregaard, M.K.; Antonelli, A.; Colwell, R.K.; Holt, B.G.; Nogues-Bravo, D.; Rasmussen, C.M.Ø.; Richardson, K.; Rosing, M.T.; Whittaker, R.J.; et al. Building mountain biodiversity: Geological and evolutionary processes. *Science* **2019**, *365*, 1114–1119. [CrossRef]
104. Fernández Escudero, I.; Tinaut, A.; Ruano, F. Ovarian Maturation under Cold Winter Conditions in a High-Mountain Ant (Hymenoptera: Formicidae). *Environ. Entomol.* **1997**, *26*, 1373–1377. [CrossRef]
105. Catarineu, C.; Barberá, G.G.; Reyes-López, J.L. A new ant species, Temnothorax ansei sp.n. (Hymenoptera: Formicidae) from the arid environments of South-Eastern Spain. *Sociobiology* **2017**, *64*, 138–145. [CrossRef]
106. Tinaut, A. Teleutomyrmex kutteri, spec. nov. A new species from Sierra Nevada (Granada, Spain). *Spixiana* **1990**, *13*, 201–208.
107. Reyes Lopez, J.; Benavente Martínez, A. Nueva cita de Teleutomyrmex kutteri Tinaut, 1990 (Hym. Formicidae) para la Península Ibérica. *Bol. Soc. Entomol. Aragon.* **2011**, *49*, 206.
108. Lebas, C.; Galkowski, C.; Wegnez, P.; Espadaler, X.; Blatrix, R. The exceptional diversity of ants on mount Coronat (Pyrénées-Orientales), and Temnothorax gredosi (Hymenoptera, Formicidae) new to France. *Rev. Assoc. Roussill. Entomol.* **2015**, *24*, 24–33.
109. García García, F.; Espadaler Gelabert, X.; Cuesta-Segura, A.D.; Sánchez-García, D. Primera cita ibérica para Temnothorax conatensis Galkowski & Lebas, 2016, y actualización de la distribución para Temnothorax grouvellei (bondroit, 1918) (Hymenoptera: Formicidae). *Iberomyrmex* **2018**, *10*, 22–27.
110. Azcárate, F.M.; Rota, C.; Hevia, V.; Silvestre, M.; Tinaut, A.; Ruano, F.; Seoane, J.; Martín Azcárate, F.; Rota, C.; Hevia, V.; et al. Primera cita de Rossomyrmex minuchae Tinaut, 1981 (Hymenoptera, Formicidae) en el Sistema Central (España). *Bol. Asoc. Esp. Entomol.* **2016**, *40*, 535–537.
111. Reyes-López, J.L.; Obregón Romero, R.; López Tirado, J. Nuevo registro de Myrmoxenus bernardi (Espadaler, 1982) (Hymenoptera: Formicidae) para la península ibérica. *Bol. Assoc. Esp. Entomol.* **2012**, *36*, 427–432.
112. Tinaut, A.; Ruano, F.; Martínez, M.D. Biology, distribution and taxonomic status of the parasitic ants of the Iberian Peninsula (Hymenoptera: Formicidae, Myrmicinae). *Sociobiology* **2005**, *46*, 1–42.
113. Ortuño, V.M.; Gilgado, J.D.; Tinaut, A. Subterranean ants: The case of Aphaenogaster cardenai (Hymenoptera: Formicidae). *J. Insect Sci.* **2014**, *14*, 212. [CrossRef]
114. Tinaut, A.; Barranco, P.; Fernández, D. Nuevas localidades para Aphaenogaster cardenai Espadaler, 1981 (Hymenoptera, Formicidae). *Bol. Asoc. Esp. Entomol.* **2015**, *39*, 421–424.
115. Blaimer, B.B.; Brady, S.G.; Schultz, T.R.; Lloyd, M.W.; Fisher, B.L.; Ward, P.S. Phylogenomic methods outperform traditional multi-locus approaches in resolving deep evolutionary history: A case study of formicine ants. *BMC Evol. Biol.* **2015**, *15*, 1–14. [CrossRef]

116. Radchenko, A.G.; Khomych, M.R. Ants of the extinct genus Cataglyphoides Dlussky, 2008 (Hymenoptera: Formicidae: Formicinae) from the late Eocene European ambers. *Invertebr. Zool.* **2020**, *17*, 154–161. [CrossRef]

117. La Polla, J.S.; Dlussky, G.M.; Perrichot, V. Ants and the fossil record. *Annu. Rev. Entomol.* **2013**, *58*, 609–630. [CrossRef]

118. Prebus, M. Insights into the evolution, biogeography and natural history of the acorn ants, genus Temnothorax Mayr (hymenoptera: Formicidae). *BMC Evol. Biol.* **2017**, *17*, 250. [CrossRef]

119. Jansen, G.; Savolainen, R.; Vepsäläinen, K. Phylogeny, divergence-time estimation, biogeography and social parasite-host relationships of the Holarctic ant genus Myrmica (Hymenoptera: Formicidae). *Mol. Phylogenet. Evol.* **2010**, *56*, 294–304. [CrossRef] [PubMed]

120. Branstetter, M.G. Origin and diversification of the cryptic ant genus Stenamma Westwood (Hymenoptera: Formicidae), inferred from multilocus molecular data, biogeography and natural history. *Syst. Entomol.* **2012**, *37*, 478–496. [CrossRef]

121. Todisco, V.; Gratton, P.; Cesaroni, D.; Sbordoni, V. Phylogeography of Parnassius apollo: Hints on taxonomy and conservation of a vulnerable glacial butterfly invader. *Biol. J. Linn. Soc.* **2010**, *101*, 169–183. [CrossRef]

122. Alonso, J. El yacimiento de ámbar cretácico de Peñacerrada (Álava). Una extraordinaria puerta de acceso al Aptiense (112-114 m.a.). *Rev. PH* **1999**, *29*, 142–147. [CrossRef]

123. Peñalver, E.; Martínez Delclòs, X. Importancia Patrimonial del Arroyo de la Pascueta, un yacimiento de ámbar Cretácico con insectos fósiles en Rubielos de Mora. In *El patrimonio paleontológico de Teruel*; Meléndez Hevia, G., Mollá Peñalver, E., Eds.; Instituto de Estudios Turolenses: Teruel, Spain, 2002; pp. 201–208.

124. Peñalver, E.; Martínez-Delclòs, X.; Arillo, A. Yacimientos con insectos fósiles en España. *Rev. Esp. Paleontol.* **1999**, *14*, 231–245.

125. Peñalver, E.; Martínez Delclòs, X. Insectos del Mioceno inferior de Ribesalbes (Castellón, España). Hymenoptera. *Treb. Mus. Geol. Barc.* **2000**, *9*, 97–153.

126. Braga, J.C.; Martín, J.M.; Quesada, C. Patterns and average rates of late Neogene- Recent uplift of the Betic Cordillera, SE Spain. *Geomorphology* **2003**, *50*, 3–26. [CrossRef]

127. Agustí, J.; Antón, M. *Memoria de la tierra: Vertebrados fósiles de la Península Ibérica*; Ediciones del Serbal: Barcelona, Spain, 1997; ISBN 978-84-7628-195-6.

128. Goropashnaya, A.V.; Fedorov, V.B.; Seifert, B.; Pamilo, P. Phylogenetic relationships of Palaearctic Formica species (hymenoptera, Formicidae) based on mitochondrial cytochrome b sequences. *PLoS ONE* **2012**, *7*, e41697. [CrossRef] [PubMed]

129. Goropashnaya, A.V.; Fedorov, V.B.; Pamilo, P. Recent speciation in the Formica rufa group ants (Hymenoptera, Formicidae): Inference from mitochondrial DNA phylogeny. *Mol. Phylogenet. Evol.* **2004**, *32*, 198–206. [CrossRef]

130. Gaudeul, M.; Véla, E.; Rouhan, G. Eastward colonization of the Mediterranean Basin by two geographically structured clades: The case of Odontites Ludw. (Orobanchaceae). *Mol. Phylogenet. Evol.* **2016**, *96*, 140–149. [CrossRef] [PubMed]

131. Inda, L.A.; Sanmartí, I.; Buerki, S.; Catal, P. Mediterranean origin and Miocene–Holocene Old World diversification of meadow fescues and ryegrasses (Festuca subgenus Schedonorus and Lolium). *J. Biogeogr.* **2014**, *41*, 600–614. [CrossRef]

132. Wheeler, W.M. *The Ants of the Baltic Amber*; Teubner, B.G., Ed.; BG Teubner: Berlin, Germany, 1915.

133. Dlussky, G.M. Genera of ants (Hymenoptera: Formicidae) from Baltic amber (in russian). *Paleontol. J.* **1997**, *6*, 50–62.

134. Rust, J.; Andersen, N.M. Giant ants from the Paleogene of Denmark with a discussion of the fossil history and early evolution of ants (Hymenoptera: Formicidae). *Zool. J. Linn. Soc.* **1999**, *125*, 331–348. [CrossRef]

135. de Andrade, M.L. *Fossil Odontomachiti Ants from the Dominican republic (Amber Collection Stuttgart: Hymenoptera, Formicidae. VII: Odontomachiti)*; Staatliches Museum für Naturkunde: Berlin, Germany, 1994.

136. Tinaut, A. El género Amblyopone Erichson en la península ibérica (Hymenoptera, Formicidae). *Miscel.lania Zool.* **1988**, *12*, 189–193.

137. Tinaut, A.; Bensusan, K. Second record of Amblyopone impressifrons (Emery, 1869) (Formicidae, Amblyoponinae) for Iberia, with some coomments on the genus. *Bol. Assoc. Esp. Entomol.* **2011**, *35*, 509–514.

138. Rodríguez-Sánchez, F.; Pérez-Barrales, R.; Ojeda, F.; Vargas, P.; Arroyo, J. The Strait of Gibraltar as a melting pot for plant biodiversity. *Quat. Sci. Rev.* **2008**, *27*, 2100–2117. [CrossRef]

139. Ros-Montoya, S. El Padul (Granada): Presencia de Mamut Lanudo (Mammuthus primigenius Blumenbach) en el Sur de España. *Misc. Paleontológica* **2005**, *6*, 275–292.

140. Ward, P.S.; Brady, S.G.; Fisher, B.L.; Schultz, T.R. Phylogeny and biogeography of dolichoderine ants: Effects of data partitioning and relict taxa on historical inference. *Syst. Biol.* **2010**, *59*, 342–362. [CrossRef]

141. Tinaut, A. Situación taxonómica del género Cataglyphis Förster, 1850 en la Península Ibérica III. El grupo de C. velox Santschi, 1929 y descripción de Cataglyphis humeya, sp. n. (Hymenoptera, Formicidae). *Eos* **1991**, *66*, 215–227.

142. Espadaler, X. Les formigues granivores de la mediterránia Occidental. *Treb. Inst. Catalana Hist. Nat.* **1981**, *9*, 39–44.

143. Salata, S.; Borowiec, L. A taxonomic revision of the genus Oxyopomyrmex André, 1881 (Hymenoptera: Formicidae). *Zootaxa* **2015**, *4025*, 1–66. [CrossRef] [PubMed]

144. Kryukov, A.; Iwasa, M.A.; Kakizawa, R.; Suzuki, H.; Pinsker, W.; Haring, E. Synchronic east-west divergence in azure-winged magpies (Cyanopica cyanus) and magpies (Pica pica). *J. Zool. Syst. Evol. Res.* **2004**, *42*, 342–351. [CrossRef]

145. Tinaut, A. A new species of the genus *Rossomyrmex* Arnoldi, 1928 from Turkey (Hymenoptera. Formicidae). *Graellsia* **2007**, *63*, 135–142. [CrossRef]

146. Ribera, I.; Blasco-Zumeta, J. Biogeographical links between steppe insects in the Monegros region (Aragon, NE Spain), the eastern Mediterranean, and central Asia. *J. Biogeogr.* **1998**, *25*, 969–986. [CrossRef]

147. Sánchez Piñero, F.; Tinaut, A.; Aguirre-Segura, A.; Miñano, J.; Lencina, J.L.; Ortiz-Sánchez, F.J.; Pérez-López, F.J. Terrestrial arthropod fauna of arid areas of SE Spain: Diversity, biogeography, and conservation | Elsevier Enhanced Reader. *J. Arid Environ.* **2011**, *75*, 1321–1332. [CrossRef]
148. Olivares, F.J.; Barea-Azcón, J.M.; Pérez-López, F.J.; Tinaut, A.; Henares, I. *Las mariposas diurnas de Sierra Nevada*; de Andalucía, J., Ed.; Consejería de Medio Ambiente: Seville, Spain, 2011.
149. Tinaut, A. Taxonomic situation of the genus Cataglyphis Förster, 1850 in the Iberian Peninsula II. New position for C. viatica (Fabricius, 1787) and redescription of C. velox Santschi, 1929 stat. n.(Hymenoptera, Formicidae). *Eos* **1990**, *66*, 49–59.
150. Martín, J.M.; Puga-Bernabéu, Á.; Aguirre, J.; Braga, J.C. Miocene Atlantic-Mediterranean seaways in the Betic cordillera (Southern Spain). *Rev. Soc. Geol. Esp.* **2014**, *27*, 175–186.
151. Tinaut, A.; Heinze, J. Wing reduction in ant queens from arid habitats. *Naturwissenschaften* **1992**, *79*, 84–85. [CrossRef]
152. Cerdá, X.; Arnan, X.; Retana, J. Is competition a significant hallmark of ant (Hymenoptera: Formicidae) ecology? *Myrmecol. News* **2013**, *18*, 131–147.

Article

The Coupled Influence of Thermal Physiology and Biotic Interactions on the Distribution and Density of Ant Species along an Elevational Gradient

Lacy D. Chick [1,2,*], Jean-Philippe Lessard [3], Robert R. Dunn [4,5] and Nathan J. Sanders [6]

[1] Hawken School, Gates Mills, OH 44040, USA
[2] Department of Ecology and Evolutionary Biology, University of Tennessee, Knoxville, TN 37996, USA
[3] Department of Biology, Concordia University, Montreal, QC H4B-1R6, Canada; jp.lessard@concordia.ca
[4] Department of Applied Ecology, North Carolina State University, Raleigh, NC 27695, USA; rrdunn@ncsu.edu
[5] Natural History Museum of Denmark, University of Copenhagen, 1350 Copenhagen, Denmark
[6] Department of Ecology and Evolutionary Biology, University of Michigan, Ann Arbor, MI 48109, USA;
    njsander@umich.edu
* Correspondence: lacy.chick@hawken.edu; Tel.: +1-843-450-9854

Received: 31 October 2020; Accepted: 26 November 2020; Published: 30 November 2020

Abstract: A fundamental tenet of biogeography is that abiotic and biotic factors interact to shape the distributions of species and the organization of communities, with interactions being more important in benign environments, and environmental filtering more important in stressful environments. This pattern is often inferred using large databases or phylogenetic signal, but physiological mechanisms underlying such patterns are rarely examined. We focused on 18 ant species at 29 sites along an extensive elevational gradient, coupling experimental data on critical thermal limits, null model analyses, and observational data of density and abundance to elucidate factors governing species' elevational range limits. Thermal tolerance data showed that environmental conditions were likely to be more important in colder, more stressful environments, where physiology was the most important constraint on the distribution and density of ant species. Conversely, the evidence for species interactions was strongest in warmer, more benign conditions, as indicated by our observational data and null model analyses. Our results provide a strong test that biotic interactions drive the distributions and density of species in warm climates, but that environmental filtering predominates at colder, high-elevation sites. Such a pattern suggests that the responses of species to climate change are likely to be context-dependent and more specifically, geographically-dependent.

Keywords: ants; community structure; physiology; interactions; temperature

1. Introduction

One of the most striking patterns in nature is that the number of species varies, often systematically, along environmental gradients. Explaining this variation has attracted the attention of ecologists and biogeographers for decades [1], if not longer [2], and has inspired empirical studies in fields ranging from physiological ecology to macroecology and global change biology [3]. However, why does the number of species that coexist in a particular assemblage vary? One possibility is that, broadly speaking, species differ in how they respond to biotic and abiotic factors along environmental gradients, and these differences among species, in turn, influence abundance, distribution, community composition, and broad-scale patterns of diversity. For instance, temperature tends to decrease systematically with elevation and latitude [4,5] and as a result, the abiotic environment at high-elevation and high-latitude sites might be more physiologically stressful for potential colonizers than at low-elevation and low-latitude sites. In such a model, temperature acts as a filter, permitting the occurrence of only those species with traits that allow them to persist at low temperatures [6–8].

Of course, multiple factors can and do simultaneously operate to shape communities, and different factors might be more important in different locations [9–12]. Wallace (1878), Dobzhansky (1950), and Fischer (1960) all suggested that negative interspecific interactions (competition, predation, parasitism) might be more intense in benign, stable environments [13–15]. Indeed, a growing number of investigators have explored the geography of biotic interactions [16] with some studies suggesting that negative interactions might limit the distributions of species and pose a cap to the number of species that can coexist in benign environments [11,17]. Two studies along elevational gradients hint at such a scenario: in hummingbird assemblages in the Andes [18] and in ant assemblages in the U.S. and Europe [19], there is evidence that interspecific interactions shape community membership at low elevations, but that more stressful environmental conditions (e.g., cold temperatures) shape communities at high elevations. Such studies are important because they suggest a mechanism, but do so based on community phylogenetic approaches, which rely on numerous underlying assumptions and can give misleading answers about the processes that actually structure communities [9,20].

More compelling evidence for geographic variation in the relative influence of climate and biotic interactions on the species in assemblages might come from field-based measurements of physiological tolerances (e.g., [8,21–23]) and/or detailed studies of the outcomes of interactions among species, i.e., actual measurements of individual-level functional traits and observations of interactions in the field [24–27]. Such studies, however, are rare due to logistical constraints and many traits that are often measured do not directly relate to tolerance of the abiotic environment.

Like other ectotherms, ants exhibit thermal sensitivity, and species differ in their thermal tolerances (i.e., the ability to tolerate either extreme temperatures or a broad range of temperatures; [28–31]. Thermal tolerance in ants may be related to total abundance and range size [32–34], foraging activity [35,36], and broad-scale patterns of diversity [8,32,34,37–39]. If a species occurs at all locations with suitable environmental conditions, then the environment alone would be the sole driver of its distribution. However, if the observed range of a species is smaller than its expected range based on environmental tolerance alone, then some other factor, such as competitive interactions or dispersal limitation, acts to shape the distribution of species among local communities along the gradient [40,41]. If the same suite of factors affects the distribution of many species, then such factors are expected to also influence the distribution of diversity, as diversity is simply a collective property.

Competitive interactions are widely thought to influence the structure and dynamics of some local assemblages, and might shape broad-scale patterns in the distribution of species as well. Competition likely structures local ant assemblages [42], yet its effects are mediated by temperature altering interactions between dominant and subordinate species [28,35,43] and the activities of particular species [27,35]. Here, we examine co-occurrence patterns among ant species along an environmental gradient to assess whether species occur less than expected by chance as would be expected if competition structures communities [44]. We note that considerable debate still exists about whether co-occurrence patterns are evidence of ecological interactions (e.g., [45]). However, we use them here as only one several lines of evidence to ask a series of inter-related questions about the factors that govern the distribution, abundance, and density of ant species along an extensive elevational gradient. Specifically, we ask (1) Are abundance and species density related to environmental temperature? (2) Does thermal tolerance predict elevational range size and species density, as would be the case if temperature were the sole driver of species distributions? and (3) Do species co-occur less among assemblages than would be expected if temperature alone limits membership? Based on the suspicions of early biogeographic pioneers (e.g., [13–15]), we predicted that physiological constraints would limit community membership at high-elevation sites, filtering species that have the physiological capacity to withstand more extreme temperatures, but that interspecific interactions shape assemblages at lower elevation sites that are more environmentally benign. We further predicted that thermal tolerance would be the best predictor of the occurrence and number of species in high elevation communities where environmental filtering predominates, but not at low elevation where biotic interactions are most frequent and intense.

2. Materials and Methods

2.1. Sampling

This work was done in Great Smoky Mountains National Park, TN, USA at sites that were situated in mixed hardwood forests and were located in areas away from roads, heavily visited trails, or other recent human disturbances. All sites were highly suitable for a variety of ant species, with similar vegetative cover and habitat to account for differences in habitat specificity of the ant species in the area. We systematically sampled 29 sites (from 379 to 1828 m) in June to August 2004–2007. These sites had a total temperature range of −8.0 to 29.7 °C (mean annual temperatures ranged from 7.7 to 13.3 °C) and ranged 1308–1928 mm in annual precipitation, respectively. Winkler samplers were used to extract ants from the leaf litter in 16 1-m^2 quadrats at each site in a haphazardly placed 50 × 50 m quadrat. At each site, species density is the observed number of species collected in the 50 × 50 m quadrat, and abundance is the number of 1-m^2 quadrats in which any species was detected. This estimate of abundance, which is actually "occurrence" [9,46–49] is preferable to a count of worker number because ants are social, and because counts of colonies is challenging when species have multiple nests per colony and occur in the leaf litter. We differentiate abundance from occurrence because our measure of abundance combines all species, whereas occurrence implies the presence of only a single species. At eight of the sites, ants were also collected using an array of ten pitfall traps over two years [50]. The number of species collected by pitfall traps did not differ from the number collected by the Winkler samplers (paired t = 1.88, n = 8, p = 0.11). Similarly, the fauna sampled by the pitfall traps was similar to the fauna sampled by the Winkler samplers [50]. At most of the sites, an asymptotic species richness estimator (Chao2 in this case) plateaued, suggesting that sampling within sites approached completeness [51]. Moreover, a Chao2 estimate of richness among all sites suggests, at least using these sampling techniques at similar sites, that there would be approximately 45 species in total, and we captured 38 species in our systematic sampling (Table 1). Therefore, we suggest that these communities are adequately sampled, while acknowledging that we undoubtedly missed some subterranean species, species that might be partially arboreal, or other exceedingly rare species in these habitats. Nevertheless, we systematically sampled each site.

In July 2011 and 2012, 31 sites were visited (from 375 to 1825 m; 17 of which were in the previously sampled sites in 2004–2007) in order to collect live individual ants for physiological tolerance estimates. At each of these 31 sites, the same Winkler extraction methods as in the previous sampling were used to extract ants from the leaf litter. However, litter was collected from only 10 1-m^2 quadrats per site instead of 16, and live ants were extracted from the leaf litter by sifting through the litter in the field rather than returning them to the lab to use Winkler extractors, as is typically the case with Winkler sampling methods. These modifications were made because we were not aiming to sample the entire community and because live specimens were needed. Finally, we also baited for ants by placing laminated index cards stocked with ~5 g of tuna in oil and individuals were hand-collected at the site. For any species detected either at the bait, the Winkler extraction method, or in general hand collecting, ten live individuals were obtained and returned to the lab (1–2 h from the field site) to conduct thermal tolerance assays.

2.2. Assessing Thermal Tolerance

Critical thermal minima (CT$_{min}$) and maxima (CT$_{max}$) were used to examine the physiological constraints imposed on species across the environmental gradient. For each species collected at each site, thermal tolerance assays were performed on 5 individuals for CT$_{min}$ and 5 individuals for CT$_{max}$, which were estimated by documenting the temperature at which individuals lost the ability of righting response. Loss of righting response is measured as the point in which an organism is flipped on its dorsum and can no longer independently right itself. This measure is considered an ecologically relevant endpoint for physiological tolerance because as an organism becomes incapacitated, it can no longer forage or escape predation [52,53].

We used methods described in Warren and Chick (2013) to estimate thermal tolerance for each species at each of the 31 sampled sites [33]. Individuals were transferred to 16 mm glass test tubes, plugged with cotton to reduce thermal refuges, and were placed in an Ac-150-A40 refrigerated water bath (NesLab, ThermoScientific, IL, USA). Water bath temperatures were raised or lowered at a rate of $1\,^\circ\text{C min}^{-1}$ until thermal tolerance was reached. Thermal tolerance was characterized as the highest and lowest temperatures at which an individual could no longer retain locomotor ability, respectively. One vial contained only a copper-constantan Type-T thermocouple (Model HH200A, Omega, CT, USA) and was used to monitor temperature inside the tubes and to ensure accurate readings. All tolerance tests were performed within 5 h of field collection to reduce potential acclimation to the lab thermal environment; however, a subsequent common garden experiment indicated no effects of acclimation on thermal limits. A mean temperature of the loss of righting response served as the index for thermal tolerance for each species at each site. All ants were preserved individually in 2.0-mL vials containing 95% ethanol, and placed in NJS's private collection at the University of Tennessee.

2.3. Are Abundance and Species Density Related to Environmental Temperature?

Current mean temperatures (1950–2000) were extrapolated for each site from the WorldClim database [54] at a resolution of 30 arc-seconds (758 m distance at a latitude of 35° N). Previous work in this region [4] modeled climate based on empirical data collected from a 120-sensor temperature logger network. While data from the 120-sensor network are more fine-scaled, they were not used in this study because they were collected for a shorter time period (2005–2006) and because temperature measured in the data loggers is related to elevation in much the same way as WorldClim data. Similarly, data from weather stations arrayed in the region indicate that temperature declines in a manner comparable to the model used by WorldClim. For these reasons, the WorldClim dataset was used here so that our findings may be more comparable to studies along other gradients where fine-scale resolution may not exist.

Total abundance (the total number of 1-m² quadrats in which a species was detected, combined for all species) and species density (the number of species in the 50 × 50 m quadrat) were plotted against mean annual temperature (MAT; we note that MAT was strongly correlated with both January minimum and July maximum). Least squares regressions were used to examine the relationship between temperature and total abundance as well as the relationship between temperature and species density. If temperature is an important determinant of species density and abundance, and colder temperatures filter species from the regional species pool, we would expect to find a linear relationship in which both species density and abundance declined with decreasing temperature.

2.4. Does Thermal Tolerance Predict Elevational Range Size and Species Density as Would Be the Case If Temperature Were the Sole Driver of Species Occurrence?

To test whether physiological tolerance of environmental temperature influences spatial variation in species density, the relationship between the thermal ranges of species (i.e., $\text{CT}_{\max}$–$\text{CT}_{\min}$) and the environmental conditions across the gradient were examined. We first asked whether species with broader thermal tolerances had broader elevational ranges and higher elevational midpoints. For each species, we combined the sampling data and plotted the highest elevation at which it was collected minus the lowest elevation at which it was collected and determined the elevational range of each species. To calculate elevational midpoints, we calculated the mean of the highest elevation and lowest elevation at which each species was collected [55]. These values were then compared to the thermal range of each species. We predicted that if temperature were an important determinant of the range sizes of species, then species at higher elevations that are able to withstand colder temperatures would have broader thermal ranges than species at lower elevations that may be confined by their physiological temperature tolerances. Additionally, species with broader thermal ranges typically have broader geographic ranges and thereby also have higher elevational midpoints [56].

Many species likely overlap in the range of temperatures at which they can occur based on their physiological thermal ranges. Yet, if species do not occupy the same environmental conditions as would be predicted by their thermal tolerances alone, then some other factor accounts for at least some of the variation in species density and occurrence. To determine whether thermal tolerance influenced species density, we asked whether the species occurring in a particular community were simply the collection of those species whose thermal tolerances overlapped the annual range of temperatures of that particular place. To do this, we extracted the annual range of temperatures (maximum temperature of the warmest month–minimum temperature of the coldest month) for each of the 27 of the 29 sites for which we had estimates of species density and calculated the mean maximum and minimum thermal tolerances of each of the 18 species across the 27 sites for which we had thermal tolerance data (two sites were omitted because species found at these sites did not have thermal tolerance data and therefore we could not estimate expected densities). The extent of overlap between physiological ranges of the ants and environmental temperatures of the sites (henceforth, thermal overlap) was then calculated. For any given species × site combination, this is simply the range of shared temperatures for both the species and the site. These values were then used to estimate a probability of occurrence for each species at each site using logistic regression models.

In the logistic regression models, the probability of occurrence of one species was determined based on its thermal overlap, as well as the thermal overlaps and recorded presences of the other 17 species in the regional species pool. This approach allowed us to determine a probability of occurrence for each species at each site based on overlapping physiological and environmental conditions (thereby incorporating variation in physiological thermal ranges and environmental thermal ranges between sites), as well as actual occurrences of other species (thereby incorporating the possibility for species co-occurrences). So as not to bias the models, presence data for the focal species were not included when estimating the probability of occurrence of that species, as including the actual occurrence of a species would inherently increase its probability of occurring in a given area.

Finally, to estimate expected species density based on thermal overlap alone, the independent probabilities of occurrence for each species were simply summed at each site. This expected species density would then be the number of species that could occur at a particular place along the gradient if temperature and temperature alone limited community membership. We then plotted observed species density against the expected species density based on thermal limits alone. If the slope of the line of expected species density plotted against observed species density equals 1, then temperature would be the sole predictor of species density. Both presences and absences of species are evident in the site × species matrix. One possibility is that the absences were not true absences. So, as a test whether the potential pseudo-absences in the site × species matrix could influence the result, we filled in the matrix so that all sites between the highest and lowest elevation at which a species was recorded were counted as presences. We then compared the expected species density if each species occurred at every site within its range to the predicted range based on thermal tolerance alone.

2.5. Do Species Co-Occur Less among Assemblages than Would Be Expected If Temperature Alone Limits Membership?

Null model analyses were used to ask whether species co-occur non-randomly (some species pair combinations being less frequent than expected by chance alone) among sites, as would be predicted if competitive interactions influenced the distribution of ants. In particular, the C-score of Stone and Roberts (1990) was used to quantify co-occurrence patterns (for methods on C-score analysis, see Appendix A) [57]. C-scores that are not significantly larger than expected by chance indicate random species distributions among sites (i.e., segregation), and C-scores that are smaller than expected by chance indicate species non-random patterns of occurrence (i.e., aggregation).

This analysis was conducted for all 29 sites to determine a general pattern, and then for the 12 communities at high (>1000 m) and 17 communities at low (<1000 m) elevations separately to examine whether the signature of competition varied along the environmental gradient.

3. Results

3.1. Are Abundance and Species Density Related to Environmental Temperature?

Species density (the number of species per site) ranged from 1 to 22 (mean = 9.44), and abundance (the total number of occurrences) per site ranged from 2 to 140 (mean = 49.5). Abundance ($r^2 = 0.34$, $p < 0.001$; Figure 1a) and species density ($r^2 = 0.47$, $p < 0.0001$; Figure 1b) both declined as mean annual temperature (MAT) declined. Species had a high degree of overlap at the lower elevations compared to higher elevations in the 2004–2007 sampling (Table 1).

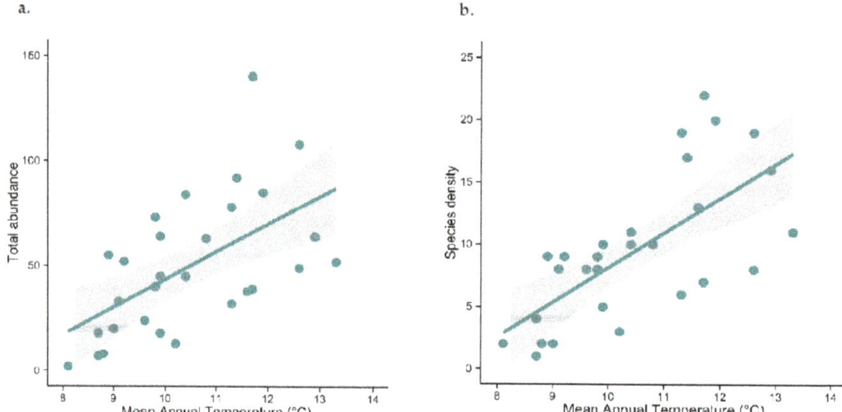

Figure 1. The relationship between (**a**) total abundance and temperature and (**b**) species density and temperature. Temperatures are current (1950–2000) mean temperatures for each site extrapolated from the WorldClim database [54] at a resolution of 30 arc-seconds. The line in each figure is the best-fit linear regression.

Table 1. Species occurrence matrix for the 29 elevations sampled during 2004–2007. Filled boxes indicate species were observed at the given elevation (i.e., presence) and empty boxes indicate species were not observed at that elevation [58].

Species	Elevation (m)																												
	379	388	403	440	462	511	520	625	656	688	719	770	786	900	941	941	988	1087	1231	1300	1342	1419	1456	1530	1651	1658	1707	1742	1828
Aphaenogaster rudis	■	■	■	■	■	■	■	■	■	■	■	■	■	■	■	■	■	■	■	■	■	■	■	■	■				■
Myrmecina americana	■	■	■	■	■	■	■	■	■	■	■	■	■	■	■	■	■	■	■	■	■		■	■	■				
Lasius alienus	■	■	■	■	■	■	■	■	■	■	■	■	■	■	■	■	■	■	■		■		■	■	■				
Stigmatomma pallipes	■		■	■	■	■			■	■	■		■	■		■	■	■	■	■	■		■						
Ponera pennsylvanica	■	■	■	■	■	■	■	■	■	■	■	■	■		■	■		■	■	■	■								
Stenamma diecki	■	■	■	■	■			■	■	■	■		■	■				■	■		■	■	■						
Stenamma meridionalis	■	■	■			■	■		■	■	■			■				■	■	■		■	■		■		■		
Myrmica punctiventris		■	■		■	■	■		■	■	■			■		■		■	■		■			■					
Stenamma schmitti	■	■	■	■	■	■	■		■	■		■						■	■		■			■					
Nylanderia faisonensis	■	■	■		■	■	■	■		■		■	■		■			■	■		■								
Temnothorax curvispinosus	■	■		■		■	■	■		■			■		■	■		■											
Strumigenys ohioensis	■	■	■		■	■		■	■	■			■	■		■													
Stenamma brevicorne		■				■		■	■		■		■	■		■							■		■				■
Stenamma impar			■		■		■			■	■			■		■													■
Brachymyrmex depilis	■				■	■	■	■	■	■			■			■		■											
Proceratium silaceum					■	■	■	■	■				■			■													
Camponotus chromaiodes					■	■	■	■			■			■				■											
Prenolepis imparis						■	■	■		■				■				■				■							
Strumigenys clypeata				■		■		■		■			■	■		■				■									
Strumigenys rostratum	■			■	■			■		■	■		■					■											
Temnothorax longispinosus				■			■	■				■			■												■		
Aphaenogster fulva								■		■	■		■	■															
Formica subsericea						■			■		■				■									■		■			
Stenamma sp. A										■					■								■						
Aphaenogaster carolonensis							■	■	■	■																			
Camponotus americanus	■					■				■																			
Camponotus subbarbatus							■			■														■					
Myrmica pinetorum							■			■																			
Lasius umbratus								■			■										■								
Monomorium minimum						■				■																			
Myrmica spatulata					■		■																						
Strumigenys ornata						■				■																			
Tapinoma sessile										■																			
Camponotus pennsylvanicus																				■									
Crematogaster minutissima			■																										
Myrmica latifrons					■																								
Pheidole bicarinata									■																				
Solenopsis molesta				■																									

3.2. Does Thermal Tolerance Predict Range Size and Species Density, as Would Be the Case If Temperature Were the Sole Driver of Species Occurrence?

We first asked whether species with broader thermal ranges had broader elevational ranges and higher elevational midpoints, as would be predicted if temperature were an important factor determining range sizes of species. There was a positive relationship between elevational ranges and thermal ranges ($r^2 = 0.48$, $p = 0.001$, Figure 2a) as well as a positive relationship between elevational midpoints and thermal ranges ($r^2 = 0.38$, $p = 0.004$, Figure 2b). Species with broader thermal ranges occurred at more elevations and tended to have higher elevational midpoints. We stress that these were lab-measured thermal tolerances and not simply the temperatures of sites at which species were collected. While not much data exist on critical thermal minima for these species, our critical thermal maxima data are similar to those found in other North American ant species [29,59].

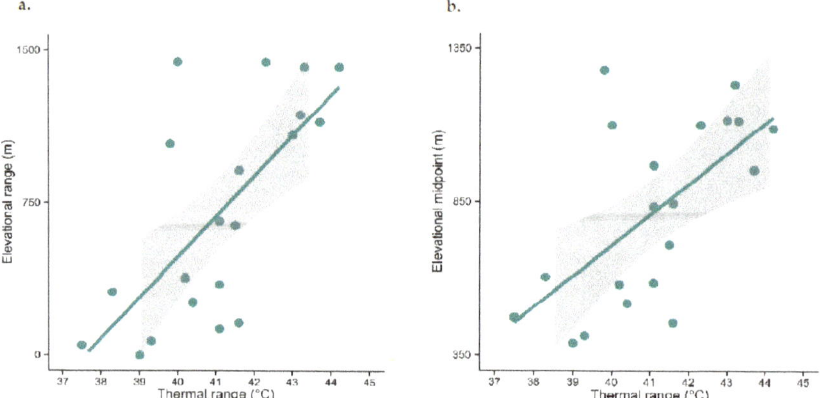

Figure 2. Thermal ranges (CT_{max}–CT_{min}) show a positive relationship with (**a**) elevational ranges ($r^2 = 0.48$, $p = 0.001$) and (**b**) elevational midpoints ($r^2 = 0.38$, $p = 0.004$) of 20 species for which we obtained both physiological and distributional data. Elevational ranges were calculated as the highest elevation at which a species was recorded minus the lowest elevation at which a species was recorded.

To determine if physiological thermal tolerances alone could predict species density, we examined the relationship between the environmental conditions at each site and the composite thermal ranges of species found at that site. In comparing sites, the thermal limits of species within sites declined with mean annual temperature ($CT_{max} = r^2 = 0.66$, $p < 0.0001$, Figure 3a; $CT_{min} = r^2 = 0.67$, $p < 0.0001$, Figure 3b). So, all species occurring at the warmest sites had, on average, the highest CT_{max} values (Figure 3a) and all species occurring at the coldest sites had, on average, the lowest CT_{min} values (Figure 3b). Furthermore, CT_{max} (Figure 3c) and CT_{min} (Figure 3d) values within and among species also follow this pattern, exhibiting both intra- and interspecific variation along the elevational gradient.

It is common to interpolate the sites at which a species could occur based on its upper and lower elevations. Here, we do something similar; we interpolate the sites at which a species could occur based on its thermal tolerance. We assume a species can occur at all sites along the gradient within its physiological thermal range (where MAT of the site is higher than its CT_{min} but lower than its CT_{max}). When we performed this interpolation, we found that at lower temperatures, observed richness more closely matched expected richness based on thermal constraints alone; however, at warmer temperatures, there was more deviation in observed richness from the null expectation (Figure 4), indicating that at low elevations, temperature is not the sole driver of species density.

3.3. Do Species Co-Occur Less among Assemblages than Would Be Expected If Temperature Alone Limits Membership?

When all 29 sites along the gradient were considered, species co-occurred much less than expected by chance (i.e., they were strongly segregated; observed C-score = 12.66; simulated C-score = 10.92; SES = 7.81; $p < 0.0001$). However, when co-occurrence patterns were examined separately, species in the warm, low-elevation sites (<1000 m) were significantly segregated among assemblages (SES = 2.06; $p = 0.02$), but species in cold, high-elevation sites (>1000 m) showed no significant deviation from randomness with respect to one another (SES = 0.94; $p = 0.17$).

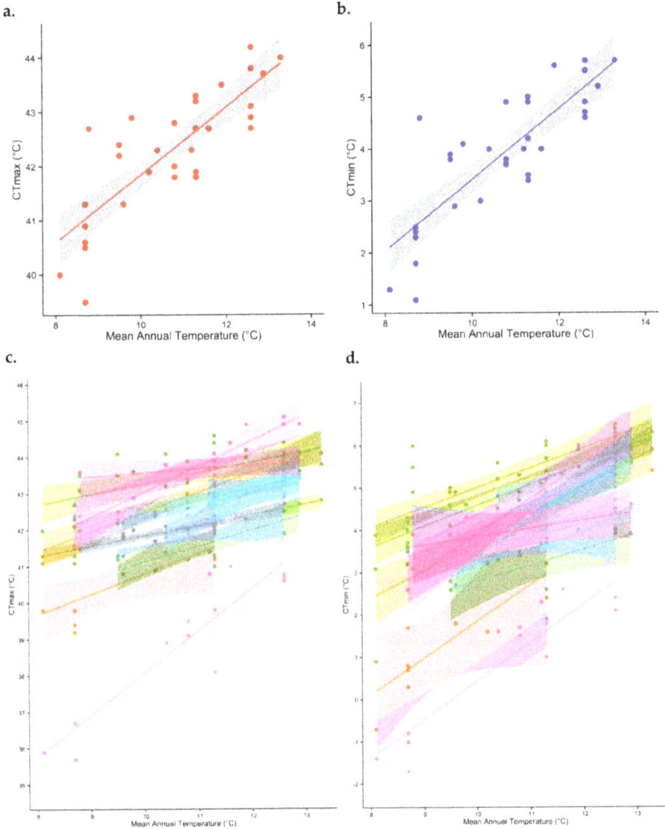

Figure 3. Species (**a**,**b**) and populations (**c**,**d**) that occur at the warmest sites have, on average, the highest CT_{max} values (**a**,**c**), and those that occur at the coldest sites have, on average, the lowest CT_{min} values (**b**,**d**). Each point is the mean of the thermal limits for all species pooled at each site (**a**,**b**) and for each population of a species (**c**,**d**) averaged for each site. Each color in (**c**,**d**) corresponds to different ant species collected across multiple populations along the elevational gradient, and each point within the color corresponds to the pooled individual CT_{max} (**c**) and CT_{min} (**d**) for an ant species at each site. Temperatures were extrapolated from WorldClim [54] at a resolution of 30 arc-seconds and represent mean temperatures from 1950–2000.

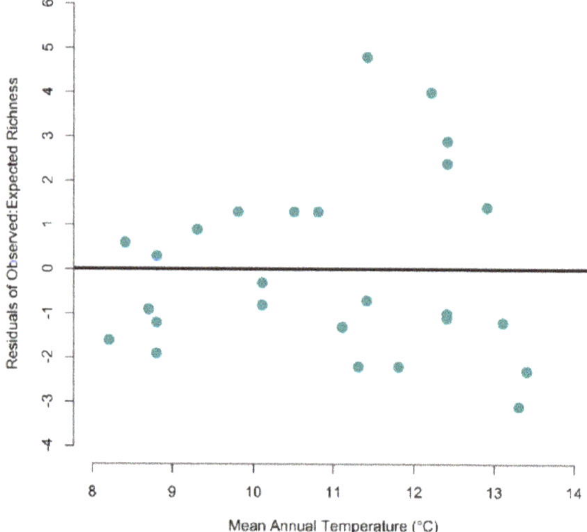

Figure 4. Residuals of the observed and expected species density based on thermal overlap in physiological limits and environmental temperatures. Black line indicates the 1:1 line to illustrate deviation from the null expectation.

4. Discussion

One of the most striking patterns in nature is that species density, abundance, and diversity vary, often systematically, along environmental gradients. Importantly, the influence of biotic interactions relative to abiotic factors shifts with elevation and environmental conditions. Such a finding supports the notion that the processes shaping community structure are context-dependent, and that both biotic and abiotic factors interact to determine the distribution and density of species among assemblages.

Temperature is related to total abundance and species density, and temperature (especially cold temperature) likely limits the ranges of species as well [8,34,38]. Species with broad elevational ranges also have broad thermal ranges. However, we need to elucidate why temperature matters. In this case, we can rule out some temperature-dependent mechanisms as an influence on patterns of diversity. One such influence that we can rule out is the metabolic theory of ecology, which depends on temperature-dependent activation energies [60], as previous work with this system [9] and others [61,62] have demonstrated. Similarly, the pattern of ant diversity here probably does not arise because of variation in in situ temperature-dependent speciation rates, since none of the studied species are endemic to the study region. In addition, temperature and net primary productivity (NPP) are not correlated spatially in this system, and NPP is weakly and negatively, rather than positively, related to ant diversity [9]. Finally, although it has been suggested that one effect of temperature on ant diversity is via the effects of temperature on ant foraging and access to resources, experimental manipulation in this same ant study system found that changes in temperature did not limit access to resources by ants [63].

One possibility is that low temperatures limit abundance, and might alter community evenness and species-abundance distributions and in turn, species density at high-elevation sites [64], but see [65]. Specifically, underlying physiological constraints exert a filter on community membership by allowing only certain cold-tolerant taxa to establish and persist at high elevations, as low temperatures limit both overwintering success as well as slow the rate of brood development. Previous work in this system [19] found that assemblages at high-elevation sites are characterized by the presence of fewer and clustered lineages, as might be expected if only the species of the restricted subset lineages with the ability to tolerate cold climates persist at high elevations. Our measurements of physiological tolerance lend support to the conclusions from previous community phylogenetic approaches. On average, populations at high-elevation sites tended to have lower CT_{min} values than did populations at warmer low-elevation sites. In fact, thermal breadth also increased with elevation, suggesting that communities at higher elevations consist of individuals that can withstand a wider range of environmental temperatures than low-elevation species.

We found that at warmer sites, observed species density varied more from the densities predicted by physiological–environmental matching alone, but observed densities in colder and more stressful conditions approximated expected densities. Populations at higher elevations (and latitudes) often persist in areas that are colder than would be expected based on their CT_{min} values [55] and low-elevation (and low-latitude) populations can persist in regions that are much warmer than their CT_{max} would suggest. Here, we found that populations in high elevation assemblages occur at temperatures that are colder than would be expected given their thermal tolerances; this has been referred to as "overfilling" the thermal niche space [66]. In contrast, populations at low elevations do not occur in all of the places they might, based on their thermal tolerances alone, which has been dubbed "underfilling" of thermal niche space. It has been suggested that overfilling of niche space is due to winter survival mechanisms of physiological cold tolerance and behavioral avoidance strategies (e.g., diapause) [29]. This seems likely in our case, though we did not specifically examine any potential overwintering mechanisms.

Underfilling of thermal niche space in warm sites might be due to biotic interactions, as has often been suggested in the literature [66]. Here, our null model approach lends support to the idea that interspecific interactions, especially in warm sites, limit community membership. The idea that interactions structure ant assemblages is not new [25,67,68]. In fact, the strongest evidence for the effects of competition on ant assemblages comes from the collapse of many ant assemblages in the face of competitively dominant invasive species [69], non-random patterns of co-occurrence among assemblages [70,71], temporal, spatial, and resource partitioning within assemblages [24,72,73] and the influence of competitively dominant native species [25,67,68]. Here, we focused on the co-occurrence of native species, and use observed interactions as well as the C-score of Stone and Roberts [57] and null model analyses to show evidence of the role of biotic interactions within communities. When we examined the warmer sites (below <1000 m elevation) and the colder sites (above >1000 m in elevation) separately, we found evidence for the signature of interspecific interaction in low-elevation sites, but not high-elevation sites. That is, species co-occurred less than expected at low-elevation sites, as would be predicted if competitive exclusion structured communities [74] at high-elevation sites, species co-occurred randomly with respect to one another. These null models alone do not directly implicate interactions, but they agree with other independent lines of evidence. First, community phylogenetic evidence points to the role of interspecific competition in shaping low-elevation but not high-elevation sites [19]. Second, there is more deviation from the null expectation in low-elevation assemblages based on physiology-environment associations alone.

5. Conclusions

One of the fundamental tenets of biogeography is that abiotic and biotic factors interact to shape the distributions of species and the organization of communities, with interactions being more important in benign environments, and environmental filtering more important in physiologically stressful environments [75,76]. Null models and community phylogenetic studies at large spatial scales, and manipulative experiments at small spatial scales, have hinted at such a scenario [18]. Taken together, our results, using a combination of observational data, null models, and physiological measurements, provide a strong test that interspecific interactions drive the distributions and density of species in warm climates, but that physiologically-driven environmental filtering predominates at high-elevation sites, at least in temperate forests of the US.

Our results also have implications for predicting the responses of biodiversity to ongoing climate change, a topic at the forefront of ecology and diversity science [77]. Forecasts of biodiversity change in response to climate change, often rely on matching the thermal tolerances of species to thermal environments in the future, and most show that species in the warmest places are the most susceptible to ongoing warming because species are operating close to their thermal maxima, so any increase in temperature essentially pushes these species over the thermal edge [66,78]. While some studies have pointed out that organisms in the warmest places can modulate their behavior to escape stressfully high temperatures, they have generally overlooked the fact that these warm places are also where organisms are likely to face the most negative consequences of interspecific interactions. For instance, our thermal constraints models showed that diversity varied most from our expectation in the warmest places, because that is also where biotic interactions among species are the most important in limiting community membership. So, while positive interactions among species might buffer species in the face of climate change, negative interactions such as competition might exacerbate the effects of climate change on biodiversity in warm environments. Models that focus on the future of biodiversity in warm environments, where most of biodiversity is, should also examine the combined and relative effects of biotic interactions and abiotic constraints and how these processes scale up to influence patterns of diversity.

Author Contributions: Conceptualization, L.D.C., J.-P.L., R.R.D., and N.J.S.; methodology, L.D.C., J.-P.L., N.J.S.; validation, L.D.C. and N.J.S.; formal analysis, L.D.C., J.-P.L., R.R.D., and N.J.S.; investigation, L.D.C. and J.-P.L.; resources, L.D.C., J.-P.L. and N.J.S.; data curation, L.D.C. and N.J.S.; writing—original draft preparation, L.D.C., R.R.D. and N.J.S.; writing—review and editing, L.D.C., J.-P.L., R.R.D. and N.J.S.; visualization, L.D.C. and N.J.S.; supervision, R.R.D. and N.J.S.; project administration, L.D.C., J.-P.L., and N.J.S.; funding acquisition, N.J.S. and R.R.D. All authors have read and agreed to the published version of the manuscript.

Funding: This research was funded by the U.S. Department of Energy PER award (DEFG02-08ER64510) and a National Science Foundation Dimensions of Biodiversity grant (NSF-1136703). Additional support was granted to J.-P.L. and L.D.C. from the Department of Ecology & Evolutionary Biology at the University of Tennessee.

Acknowledgments: We are grateful to Quentin Read, Nicholas Gotelli, Carsten Rahbek, and Michael Borregaard for their comments and computational expertise on the null models. We thank Kathryn Stuble, Courtney Patterson, and David Fowler for field assistance. We also thank anonymous reviewers for comments on previous versions of this manuscript. We acknowledge the National Park Service for collection permits for Great Smoky Mountains National Park (GRSM-2011-SCI-0050, GRSM-2013-SCI-1071).

Conflicts of Interest: The authors declare no conflict of interest. The funders had no role in the design of the study; in the collection, analyses, or interpretation of data; in the writing of the manuscript, or in the decision to publish the results.

Appendix A

The C-score quantifies the number of "checkerboard units" for each species pair, where a checkerboard unit is a 2 × 2 submatrix of the form 01 10 or 10 01. For each species pair, the number of checkerboard units is $(R_i-S)(R_j-S)$, where R_i is the number of occurrences (equal to the row total) for species i, R_j is the number of occurrences for species j, and S is the number of sample plots in which both species occur. The C-score is the average number of checkerboard units for each unique species pair. If this index is unusually large compared with a null distribution, there is less pairwise species co-occurrence (segregation) than expected by chance. If the index is unusually small, there is more species co-occurrence (aggregation) than expected. We compared the observed C-scores to those generated from 5000 randomly constructed assemblages (using null models in EcoSim version 7.72) [79].

A fixed-fixed null model [80] was used for which both row totals and column totals are fixed within sites and among species, which maintains differences in species density among sites and total occurrences among species. Gotelli [80] suggests that SIM9 is appropriate for analyzing co-occurrence patterns of species from "island lists" and has a low probability of Type I errors.

References

1. MacArthur, R.H.; Diamond, J.M.; Karr, J.R. Density compensation in island faunas. *Ecology* **1972**, *53*, 330–342. [CrossRef]
2. von Humboldt, A.; Sabine, E.J. *Aspects of Nature, in Different Lands and Different Climates with Scientific Elucidations*, 3rd ed.; J. Murray: London, UK, 1849.
3. Diez, J.; Ibáñez, I.; AJ, M.-R.; Mazer, S.; Crimmins, T.; Crimmins, M.; Bertelsen, C.; Inouye, D. Forecasting phenology: From species variability to community patterns. *Ecology* **2012**, *15*, 545–553. [CrossRef] [PubMed]
4. Fridley, J.D. Downscaling Climate over Complex Terrain: High Finescale (<1000 m) Spatial Variation of Near-Ground Temperatures in a Montane Forested Landscape (Great Smoky Mountains). *J. Appl. Meteorol. Climatol.* **2009**, *48*, 1033–1049. [CrossRef]
5. McCain, C.; Colwell, R. Assessing the threat to montane biodiversity from discordant shifts in temperature and precipitation in a changing climate. *Ecol. Lett.* **2011**, *14*, 1236–1245. [CrossRef]
6. Addo-Bediako, A.; Chown, S.L.; Gaston, K.J. Thermal tolerance, climatic variability and latitude. *Proc. R. Soc. Lond. B* **2000**, *267*, 739–745. [CrossRef]
7. Sunday, J.M.; Bates, A.E.; Dulvy, N.K. Global analysis of thermal tolerance and latitude in ectotherms. *Proc. R. Soc. B* **2011**, *278*, 1823–1830. [CrossRef]
8. Bishop, T.R.; Robertson, M.P.; Van Rensburg, B.J.; Parr, C.L. Coping with the cold: Minimum temperatures and thermal tolerances dominate the ecology of mountain ants: Thermal tolerances of mountain ants. *Ecol. Entomol.* **2017**, *42*, 105–114. [CrossRef]
9. Sundqvist, M.K.; Sanders, N.J.; Wardle, D.A. Community and Ecosystem Responses to Elevational Gradients: Processes, Mechanisms, and Insights for Global Change. *Annu. Rev. Ecol. Evol. Syst.* **2013**, *44*, 261–280. [CrossRef]
10. Peters, M.K.; Hemp, A.; Appelhans, T.; Behler, C.; Classen, A.; Detsch, F.; Ensslin, A.; Ferger, S.W.; Frederiksen, S.B.; Gebert, F.; et al. Predictors of elevational biodiversity gradients change from single taxa to the multi-taxa community level. *Nat. Commun.* **2016**, *7*, 13736. [CrossRef]
11. Jankowski, J.; Robinson, S.; Levey, D. Squeezed at the top: Interspecific aggression may constrain elevational ranges in tropical birds. *Ecology* **2010**, *91*, 1877–1884. [CrossRef]
12. Galbreath, K.E.; Hafner, D.J.; Zamudio, K.R. When cold is better: Climate-driven elevation shifts yield complex patterns of diversification and demography in an Alpine specialist (American Pika, *Ochotona princeps*). *Evolution* **2009**, *63*, 2848–2863. [CrossRef] [PubMed]
13. Wallace, A.R. Tropical nature and other essays. *Nature* **1878**, *18*, 140–141. [CrossRef]
14. Dobzhansky, T. Evolution in the tropics. *Am. Sci.* **1950**, *38*, 209–221.
15. Fischer, A.G. Latitudinal Variations in Organic Diversity. *Evolution* **1960**, *14*, 64–81. [CrossRef]
16. Schemske, D.; Mittelbach, G.; Cornell, H.; Sobel, J.; Roy, K. Is there a latitudinal gradient in the importance of biotic interactions? *Annu. Rev. Ecol. Evol. Syst.* **2009**, *40*, 245–269. [CrossRef]
17. Kozak, K.; Wiens, J. Phylogeny, ecology, and the origins of climate–richness relationships. *Ecology* **2012**, *93*, S167–S181. [CrossRef]
18. Graham, C.; Parra, J.; Rahbek, C. Phylogenetic structure in tropical hummingbird communities. *Proc. Natl. Acad. Sci. USA* **2009**, *106*, 19673–19678. [CrossRef] [PubMed]
19. Machac, A.; Janda, M.; Dunn, R.; Sanders, N. Elevational gradients in phylogenetic structure of ant communities reveal the interplay of biotic and abiotic constraints on diversity. *Ecography* **2011**, *34*, 364–371. [CrossRef]
20. Losos, J. Phylogenetic niche conservatism, phylogenetic signal and the relationship between phylogenetic relatedness and ecological similarity among species. *Ecol. Lett.* **2008**, *11*, 995–1003. [CrossRef]
21. Helmuth, B. Climate Change and Latitudinal Patterns of Intertidal Thermal Stress. *Science* **2002**, *298*, 1015–1017. [CrossRef]
22. Sinclair, B.; Scott, M.; Klok, C. Determinants of terrestrial arthropod community composition at Cape Hallett, Antarctica. *Antarct. Sci.* **2006**, *18*, 303–312. [CrossRef]
23. Buckley, L.; Rodda, G.; Jetz, W. Thermal and energetic constraints on ectotherm abundance: A global test using lizards. *Ecology* **2008**, *89*, 48–55. [CrossRef] [PubMed]
24. Albrecht, M.; Gotelli, N. Spatial and temporal niche partitioning in grassland ants. *Oecologia* **2001**, *126*, 134–141. [CrossRef] [PubMed]

25. Parr, C. Dominant ants can control assemblage species richness in a South African savanna. *J. Anim. Ecol.* **2008**, *77*, 1191–1198. [CrossRef]
26. Violle, C.; Enquist, B.J.; McGill, B.J.; Jiang, L.; Albert, C.H.; Hulshof, C.; Jung, V.; Messier, J. The return of the variance: Intraspecific variability in community ecology. *TREE* **2012**, *27*, 244–252. [CrossRef]
27. Stuble, K.L.; Rodriguez-Cabal, M.A.; McCormick, G.L.; Jurić, I.; Dunn, R.R.; Sanders, N.J. Tradeoffs, competition, and coexistence in eastern deciduous forest ant communities. *Oecologia* **2013**, *171*, 981–992. [CrossRef]
28. Cerda, X.; Retana, J.; Cros, S. Thermal Disruption of Transitive Hierarchies in Mediterranean Ant Communities. *J. Anim. Ecol.* **1997**, *66*, 363. [CrossRef]
29. Diamond, S.E.; Sorger, D.M.; Hulcr, J.; Pelini, S.L.; Toro, I.D.; Hirsch, C.; Oberg, E.; Dunn, R.R. Who likes it hot? A global analysis of the climatic, ecological, and evolutionary determinants of warming tolerance in ants. *Glob. Chang. Biol.* **2012**, *18*, 448–456. [CrossRef]
30. Kaspari, M.; Clay, N.A.; Lucas, J.; Yanoviak, S.P.; Kay, A. Thermal adaptation generates a diversity of thermal limits in a rainforest ant community. *Glob. Chang. Biol.* **2015**, *21*, 1092–1102. [CrossRef]
31. Chick, L.D.; Perez, A.; Diamond, S.E. Social dimensions of physiological responses to global climate change: What we can learn from ants (Hymenoptera: Formicidae). *Myrmecol. News* **2017**, *25*, 29–40.
32. Geraghty, M.J.; Dunn, R.R.; Sanders, N.J. Body size, colony size, and range size in ants (Hymenoptera: Formicidae): Are patterns along elevational and latitudinal gradients consistent with Bergmann's rule. *Myrmecol. News* **2007**, *10*, 51–58.
33. Warren, R.J., II; Chick, L. Upward ant distribution shift corresponds with minimum, not maximum, temperature tolerance. *Glob. Chang. Biol.* **2013**, *19*, 2082–2088. [CrossRef] [PubMed]
34. Diamond, S.E.; Chick, L.D. Thermal specialist ant species have restricted, equatorial geographic ranges: Implications for climate change vulnerability and risk of extinction. *Ecography* **2018**. [CrossRef]
35. Lessard, J.; Fordyce, J.; Gotelli, N.; Sanders, N. Invasive ants alter the phylogenetic structure of ant communities. *Ecology* **2009**, *90*, 2664–2669. [CrossRef]
36. Stuble, K.L.; Pelini, S.L.; Diamond, S.E.; Fowler, D.A.; Dunn, R.R.; Sanders, N.J. Foraging by forest ants under experimental climatic warming: A test at two sites. *Ecol. Evol.* **2013**, *3*, 482–491. [CrossRef]
37. Kaspari, M. A primer on ant ecology. In *Ants—Standard Methods for Measuring and Monitoring Biodiversity*; Smithsonian Institution Press: Washington, DC, USA; London, UK, 2000; pp. 9–24.
38. Bujan, J.; Roeder, K.A.; de Beurs, K.; Weiser, M.D.; Kaspari, M. Thermal diversity of North American ant communities: Cold tolerance but not heat tolerance tracks ecosystem temperature. *Glob. Ecol. Biogeogr.* **2020**, *29*, 1486–1494. [CrossRef]
39. Dunn, R.R.; Guenard, B.S.; Weiser, M.D.; Sanders, N.J. Geographic gradients. In *Ant Ecology*; Lach, L., Parr, C.L., Abbott, K.L., Eds.; Oxford University Press: New York, NY, USA, 2010; pp. 38–58.
40. Guisan, A.; Rahbek, C. SESAM—A new framework integrating macroecological and species distribution models for predicting spatio-temporal patterns of species assemblages. *J. Biogeogr.* **2011**, *38*, 1433–1444. [CrossRef]
41. Fordham, D.; Mellin, C.; Russell, B.; Akçakaya, R.; Bradshaw, C.; ME, A.-L.; Shepherd, S.; Brook, B. Population dynamics can be more important than physiological limits for determining range shifts under climate change. *Glob. Chang. Biol.* **2013**, *19*, 3224–3237. [CrossRef]
42. Cerdá, X.; Arnan, X.; Retana, J. Is competition a significant hallmark of ant (Hymenoptera: Formicidae) ecology? *Myrmecol. News* **2013**, *18*, 131–147.
43. Bestelmeyer, B.T. The trade-off between thermal tolerance and behavioural dominance in a subtropical South American ant community. *J. Anim. Ecol.* **2000**, *69*, 998–1009. [CrossRef]
44. Gotelli, N.J.; Mccabe, D.J. Species Co-occurence: A Meta-analysis of J.M. Diamond's Assembly Rules Model. *Ecology* **2002**, *83*, 6. [CrossRef]
45. Blanchet, F.G.; Cazelles, K.; Gravel, D. Co-occurrence is not evidence of ecological interactions. *Ecol. Lett.* **2020**, *23*, 1050–1063. [CrossRef] [PubMed]
46. Kaspari, M.; O'Donnell, S.; Kercher, J. Energy, density, and constraints to species richness: Ant assemblages along a productivity gradient. *Am. Nat.* **2000**, *155*, 280–293. [CrossRef] [PubMed]
47. Longino, J.; Coddington, J.; Colwell, R.K. The ant fauna of a tropical rain forest: Estimating species richness three different ways. *Ecology* **2002**, *83*, 689–702. [CrossRef]

48. Gotelli, N.J.; Ellison, A.M.; Dunn, R.R.; Sanders, N.J. Counting ants (Hymenoptera: Formicidae): Biodiversity sampling and statistical analysis for myrmecologists. *Myrmecol. News* **2011**, *15*, 13–19.

49. Gotelli, N.J.; Colwell, R.K. Estimating species richness. *Biol. Divers. Front. Meas. Assess.* **2011**, *12*, 39–54.

50. Lessard, J.; Dunn, R.; Parker, C.; Sanders, N. Rarity and diversity in forest ant assemblages of Great Smoky Mountains National Park. *Southeast. Nat.* **2007**, *6*, 215–228. [CrossRef]

51. Sanders, N.J.; Lessard, J.-P.; Fitzpatrick, M.C.; Dunn, R.R. Temperature, but not productivity or geometry, predicts elevational diversity gradients in ants across spatial grains. *Glob. Ecol. Biogeogr.* **2007**, *16*, 640–649. [CrossRef]

52. Lutterschmidt, W.; Hutchison, V. The critical thermal maximum: Data to support the onset of spasms as the definitive end point. *Can. J. Zool.* **1997**, *75*, 1553–1560. [CrossRef]

53. Lighton, J.R.; Turner, R.J. Thermolimit respirometry: An objective assessment of critical thermal maxima in two sympatric desert harvester ants, Pogonomyrmex rugosus and P. californicus. *J. Exp. Biol.* **2004**, *207*, 1903–1913. [CrossRef]

54. Hijmans, R.; Cameron, S.; Parra, J.; Jones, P.; Jarvis, A. Very high resolution interpolated climate surfaces for global land areas. *Int. J. Climatol. A J. R. Meteorol. Soc.* **2005**, *25*, 1965–1978. [CrossRef]

55. Rohde, K.; Heap, M.; Heap, D. Rapoport's Rule Does Not Apply to Marine Teleosts and Cannot Explain Latitudinal Gradients in Species Richness. *Am. Nat.* **1993**, *142*, 1–16. [CrossRef]

56. Sanders, N. Elevational gradients in ant species richness: Area, geometry, and Rapoport's rule. *Ecography* **2002**, *25*, 25–32. [CrossRef]

57. Stone, L.; Roberts, A. The checkerboard score and species distributions. *Oecologia* **1990**, *85*, 74–79. [CrossRef] [PubMed]

58. Sanders, N.J.; Lessard, J.-P.; Dunn, R.R. *Great Smoky Mountain Ant Community Composition 2020, 91888 Bytes*; Zenodo: Genève, Switzerland, 2020.

59. Kaspari, M.; Bujan, J.; Roeder, K.A. Species energy and Thermal Performance Theory predict 20-yr changes in ant community abundance and richness. *Ecology* **2019**, *100*, 7. [CrossRef] [PubMed]

60. Brown, J.H.; Gillooly, J.F.; Allen, A.P.; Savage, V.M.; West, G.B. Toward a metabolic theory of ecology. *Ecology* **2004**, *85*, 1771–1789. [CrossRef]

61. Hawkins, B.; Albuquerque, F.; Araújo, M.; Beck, J.; Bini, L.; Cabrero-Sañudo, F.J.; Castro-Parga, I.; Diniz-Filho, J.A.F.; Ferrer-Castán, D.; Field, R.; et al. A global evaluation of metabolic theory as an explanation for terrestrial species richness gradients. *Ecology* **2007**, *88*, 1877–1888. [CrossRef]

62. McCain, C.M.; Sanders, N.J. Metabolic theory and elevational diversity of vertebrate ectotherms. *Ecology* **2010**, *91*, 601–609. [CrossRef]

63. Lessard, J.-P.; Sackett, T.E.; Reynolds, W.N.; Fowler, D.A.; Sanders, N.J. Determinants of the detrital arthropod community structure: The effects of temperature and resources along an environmental gradient. *Oikos* **2011**, *120*, 333–343. [CrossRef]

64. McGlinn, D.J.; Engel, T.; Blowes, S.A.; Gotelli, N.J.; Knight, T.M.; McGill, B.J.; Sanders, N.; Chase, J.M. A multiscale framework for disentangling the roles of evenness, density, and aggregation on diversity gradients. *Ecology* **2020**, e03233. [CrossRef]

65. Gifford, M.E.; Kozak, K.H. Islands in the sky or squeezed at the top? Ecological causes of elevational range limits in montane salamanders. *Ecography* **2012**, *35*, 193–203. [CrossRef]

66. Sunday, J.M.; Bates, A.E.; Dulvy, N.K. Thermal tolerance and the global redistribution of animals. *Nat. Clim. Chang.* **2012**, *2*, 686–690. [CrossRef]

67. Andersen, A. Regulation of "momentary" diversity by dominant species in exceptionally rich ant communities of the Australian seasonal tropics. *Am. Nat.* **1992**, *140*, 401–420. [CrossRef] [PubMed]

68. Parr, C.L.; Sinclair, B.J.; Andersen, A.N.; Gaston, K.J.; Chown, S.L. Constraint and competition in assemblages: A cross-continental and modeling approach for ants. *Am. Nat.* **2005**, *165*, 481–494. [CrossRef] [PubMed]

69. Holway, D.A.; Lach, L.; Suarez, A.V.; Tsutsui, N.D.; Case, T.J. The Causes and Consequences of Ant Invasions. *Annu. Rev. Ecol. Syst.* **2002**, *33*, 181–233. [CrossRef]

70. Gotelli, N.; Arnett, A. Biogeographic effects of red fire ant invasion. *Ecol. Lett.* **2000**, *3*, 257–261. [CrossRef]

71. Sanders, N.; Gotelli, N.; Heller, N.; Gordon, D. Community disassembly by an invasive species. *Proc. Natl. Acad. Sci. USA* **2003**, *100*, 2474–2477. [CrossRef]

72. Cros, S.; Cerdá, X.; Retana, J. Spatial and temporal variations in the activity patterns of Mediterranean ant communities. *Ecoscience* **1997**, *4*, 269–278. [CrossRef]

73. Sanders, N.; Gordon, D. Resource-dependent interactions and the organization of desert ant communities. *Ecology* **2003**, *84*, 1024–1031. [CrossRef]

74. Gotelli, N.J.; Graves, G.R.; Rahbek, C. Macroecological signals of species interactions in the Danish avifauna. *Proc. Natl. Acad. Sci. USA* **2010**, *107*, 5030–5035. [CrossRef]

75. MacArthur, R.; Levins, R. The limiting similarity, convergence, and divergence of coexisting species. *Am. Nat.* **1967**, *101*, 377–385. [CrossRef]

76. Weiher, E.; Keddy, P. Assembly rules, null models, and trait dispersion: New questions from old patterns. *Oikos* **1995**, *74*, 159–164. [CrossRef]

77. Urban, M.C. Accelerating extinction rish from climate change. *Science* **2015**, *348*, 571–573. [CrossRef] [PubMed]

78. Deutsch, C.A.; Tewksbury, J.J.; Huey, R.B.; Sheldon, K.S.; Ghalambor, C.K.; Haak, D.C.; Martin, P.R. Impacts of climate warming on terrestrial ectotherms across latitude. *Proc. Natl. Acad. Sci. USA* **2008**, *105*, 6668–6672. [CrossRef] [PubMed]

79. Gotelli, N.; Entsminger, G. EcoSim: Null Models Software for Ecology. Available online: https://zenodo.org/record/16504#.X8Sjqdr7RPY (accessed on 1 October 2020).

80. Gotelli, N. Null model analysis of species co-occurrence patterns. *Ecology* **2000**, *81*, 2606–2621. [CrossRef]

Publisher's Note: MDPI stays neutral with regard to jurisdictional claims in published maps and institutional affiliations.

Article

The Evolution and Biogeography of *Wolbachia* in Ants (Hymenoptera: Formicidae)

Manuela O. Ramalho [1,*] and Corrie S. Moreau [1,2]

[1]   Department of Entomology, Cornell University, Ithaca, NY 14853, USA; corrie.moreau@cornell.edu
[2]   Department of Ecology and Evolutionary Biology, Cornell University, Ithaca, NY 14853, USA
*   Correspondence: manu.ramalho@cornell.edu

Received: 13 October 2020; Accepted: 10 November 2020; Published: 12 November 2020

Abstract: *Wolbachia* bacteria are widely distributed across invertebrate taxa, including ants, but several aspects of this host-associated interaction are still poorly explored, especially with regard to the ancestral state association, origin, and dispersion patterns of this bacterium. Therefore, in this study, we explored the association of *Wolbachia* with Formicidae in an evolutionary context. Our data suggest that supergroup F is the ancestral character state for *Wolbachia* infection in ants, and there is only one transition to supergroup A, and once ants acquired infection with supergroup A, there have been no other strains introduced. Our data also reveal that the origin of *Wolbachia* in ants likely originated in Asia and spread to the Americas, and then back to Asia. Understanding the processes and mechanisms of dispersion of these bacteria in Formicidae is a crucial step to advance the knowledge of this symbiosis and their implications in an evolutionary context.

Keywords: endosymbiont; ant; vertical transmission; biogeography; ancestral state reconstruction; phylogeny

1. Introduction

There are many examples of insect–microbe interactions providing benefits to all players involved [1–5], and one of the most well-documented bacteria associated with insects is *Wolbachia* [6,7]. This bacterium is known for modifying the host's reproduction for its own benefit, including the induction of cytoplasmic incompatibility, parthenogenesis, and male-killing or feminization [8]. However, it is not known whether these functions are related to a particular *Wolbachia* strain or host [9]. As this bacterium has been found in association with hundreds of hosts and encompasses an immense diversity, the classification of strains into supergroups has been proposed and about 17 supergroups, to date, have been reported through genotyping a single gene, several genes, or even genomic approaches [10]. These supergroups are called "A to S", with G and R no longer considered separate supergroups [11,12]. Of these, supergroups found exclusively in arthropods belong to the A, B, E, H, I, K, M, N, O, P, Q, and S supergroups [13–17], with supergroup F being common for nematodes and arthropods [18,19]. Although we know that *Wolbachia* is associated with several arthropods hosts, we focused on ants, a highly diverse group with more than 13,000 described species and a global distribution [20]. In addition, it is one of the groups of insects best studied in terms of association with *Wolbachia*. This offers us an excellent opportunity to explore the evolution and biogeography of this association.

Wolbachia has been identified from several ant genera, but few studies have succeeded in identifying the implications for these ant-associated interactions. *Monomorium pharaonis* appears to have an accelerated colony life cycle [21] and *Tapinoma melanocephalum* has a nutritional upgrade of vitamin B [22] in the presence of *Wolbachia*. However, whether these are specific cases within all ant diversity is still unknown. Despite the fact that these bacteria have been identified in several ant genera, little is known about the diversity and evolutionary history of this symbiosis; see [23–35].

Historically, studying these interactions was conducted by targeted PCR amplification and sequencing of the *Wolbachia* surface protein (*wsp*) gene [36]. However, the use of this gene alone has been shown to not be appropriate for diversity studies since it has been reported to have a high rate of recombination [36]. Therefore, the Multilocus Sequence Typing (MLST) approach [37] has been shown to be a more reliable alternative, targeting five different *Wolbachia* genes, *coxA*, *fbpA*, *ftsZ*, *gatB*, and *hcpA*, instead of just one [38]. These five genes are spread across the *Wolbachia* genome and evolve under purifying selection [38], making them more appropriate for phylogenetic analysis. Once a unique sequence is found with the combination of all of these genes, a sequence type (ST) is assigned and is stored in the *Wolbachia* MLST database (https://pubmlst.org/wolbachia/).

Many studies have reliably used the MLST approach [39–42], including some studies of ants [9,24,28,32,33]. However, no studies, to date, in ants have leveraged ancestral state reconstruction and biogeographic range evolution to understand the origin and biogeography of this symbiosis. As a robust phylogenetic tree is necessary to investigate lifestyle transitions in *Wolbachia*'s evolutionary history, this study aimed to (1) infer the evolutionary history of *Wolbachia* associated with Formicidae through the MLST approach (five genes), controlling for their different evolutionary models; (2) reconstruct the ancestral state of the different *Wolbachia* supergroups; (3) investigate the dispersion patterns of these bacteria associated with Formicidae through biogeographic range evolution analyses.

2. Material and Methods

2.1. Phylogenetic Reconstruction

Wolbachia sequences isolated from 70 Formicidae hosts (2079 bp from the five genes) were downloaded from the *Wolbachia* MLST database (https://pubmlst.org/wolbachia/), as well as metadata information. ST78 from *Opistophthalmus chaperi* (supergroup F) was added as an outgroup because it has not been found to be associated with ants but shows high similarity with strains associated with ants (Table 1). The sequences were aligned using ClustalW [43] of the BioEdit software [44] and later were used for partitioned phylogenetic reconstruction. PartitionFinder2 (2.1.1) [45] was used to choose the best model of molecular evolution and returned four partitions: charset Subset1 = 1–402 pb; charset Subset2 = 403–831 pb; charset Subset3 = 832–1266 pb and 1636–2079 pb; charset Subset4 = 1267–1635 pb, with GTR + G for Subset1, GTR + I + G for Subset2, GTR + I + G for Subset3, and TIM + I + G for Subset4. Bayesian inference was implemented using MrBayes (3.2.6) [46] on the Cipres Science Gateway [47,48] for phylogenetic reconstruction with the Markov chain Monte Carlo analysis for 1,000,000 generations with sampling every 1000 generations, and discarding the first 25% of trees as burnin. We used the chronos function with the correlated model available in the Ape package [49] with R software [50] to reconstruct a chronogram with a relative time scale [51–53].

Table 1. *Wolbachia* sequence samples associated with Formicidae included in the present study.

id	Supergroup	Host Genus	Host Species	Locality	Submitter	Sequence Type (ST)
2	A	*Solenopsis*	*invicta*	Argentina	Laura Baldo	29
4	A	*Camponotus*	*pennsylvanicus*	USA	Laura Baldo	33
103	A	*Formica*	*occulta*	USA	Jacob Russell	43
104	A	*Pseudomyrmex*	*apache*	USA	Jacob Russell	44
105	A	*Stenamma*	*snellingi*	USA	Jacob Russell	45
106	A	*Azteca*	spp.	Ecuador	Jacob Russell	46
107	A	*Wasmannia*	spp.	Peru	Jacob Russell	47

Table 1. *Cont.*

id	Supergroup	Host Genus	Host Species	Locality	Submitter	Sequence Type (ST)
108	A	*Metapone*	*madagascaria*	Madagascar	Jacob Russell	48
109	A	*Myrmica*	*incompleta*	USA	Jacob Russell	49
110	A	*Polyergus*	*breviceps*	USA	Jacob Russell	50
111	A	*Technomyrmex*	*albipes*	Philippines	Jacob Russell	19
112	A	*Polyrhachis*	*vindex*	Philippines	Jacob Russell	51
113	A	*Anoplolepis*	*gracillipes*	Philippines	Jacob Russell	52
114	A	*Notoncus*	spp.	Australia	Jacob Russell	53
115	A	*Leptomyrmex*	spp.	Australia	Jacob Russell	19
116	A	*Myrmecorhynchus*	spp.	Australia	Jacob Russell	54
117	A	*Pheidole*	*minutula*	French Guiana	Jacob Russell	55
119	A	*Lophomyrmex*	spp.	Thailand	Jacob Russell	52
120	A	*Camponotus*	*leonardi*	Thailand	Jacob Russell	57
121	A	*Pheidole*	*vallicola*	USA	Jacob Russell	58
122	A	*Rhytidoponera*	*metallica*	Australia	Jacob Russell	59
124	A	*Pheidole*	*plagiara*	Thailand	Jacob Russell	19
125	A	*Pheidole*	*sauberi*	Thailand	Jacob Russell	19
126	A	*Pheidole*	*gatesi*	Vietnam	Jacob Russell	60
127	A	*Pheidole*	spp.	Thailand	Jacob Russell	61
129	A	*Dorymyrmex*	*elegans*	USA	Jacob Russell	63
134	A	*Odontomachus*	*clarus*	USA	Jacob Russell	111
135	A	*Ochetellus*	*glaber*	Australia	Jacob Russell	112
137	A	*Pheidole*	*coloradensis*	USA	Jacob Russell	114
138	A	*Pheidole*	*micula*	USA	Jacob Russell	115
139	A	*Pheidole*	*vistana*	Mexico	Jacob Russell	116
140	A	*Pheidole*	*obtusospinosa*	USA	Jacob Russell	117
141	A	*Pheidole*	spp.	Indonesia	Jacob Russell	118
143	A	*Aenictus*	spp.	Thailand	Jacob Russell	120
144	A	*Crematogaster*	spp.	Thailand	Jacob Russell	121
145	A	*Solenopsis*	spp.	Thailand	Jacob Russell	122
146	A	*Leptogenys*	spp.	Thailand	Jacob Russell	19
147	A	*Pheidole*	*planifrons*	Thailand	Jacob Russell	19
148	A	*Monomorium*	*chinense*	Thailand	Jacob Russell	123
149	F	*Ocymyrmex*	*picardi*	Congo (DRC]	Jacob Russell	124
558	A	*Camponotus*	*textor*	Brazil	Manuela Ramalho	347
1827	A	*Paratrechina*	*longicornis*	Taiwan	Tseng ShuPing	19
1828	F	*Paratrechina*	*longicornis*	Taiwan	Tseng ShuPing	471
1868	A	*Cephalotes*	*atratus*	Brazil	Madeleine Kelly	494
1869	A	*Cephalotes*	*atratus*	French Guiana	Madeleine Kelly	495
1870	A	*Cephalotes*	*atratus*	Brazil	Madeleine Kelly	496

Table 1. *Cont.*

id	Supergroup	Host Genus	Host Species	Locality	Submitter	Sequence Type (ST)
1871	A	*Cephalotes*	*atratus*	Guyana	Madeleine Kelly	497
1872	A	*Cephalotes*	*atratus*	Brazil	Madeleine Kelly	498
1873	A	*Cephalotes*	*atratus*	Peru	Madeleine Kelly	499
1989	A	*Cardiocondyla*	spp.	India	Manisha Gupta	550
1996	F	*Paratrechina*	spp.	India	Manisha Gupta	557
2010	A	*Pheidole*	spp.	India	Manisha Gupta	571
2017	A	*Anoplolepis*	*gracillipes*	Malaysia	Tseng ShuPing	52
2022	A	*Camponotus*	spp.	Malaysia	Tseng ShuPing	576
2023	A	*Camponotus*	spp.	Malaysia	Tseng ShuPing	577
Not defined	A	*Solenopsis*	spp.	Brazil	Cintia Martins	314
Not defined	A	*Solenopsis*	spp.	Brazil	Cintia Martins	315
Not defined	A	*Solenopsis*	spp.	Brazil	Cintia Martins	316
Not defined	A	*Solenopsis*	spp.	Brazil	Cintia Martins	317
Not defined	A	*Solenopsis*	spp.	Brazil	Cintia Martins	318
Not defined	A	*Solenopsis*	spp.	Brazil	Cintia Martins	319
Not defined	A	*Solenopsis*	spp.	Brazil	Cintia Martins	320
Not defined	A	*Solenopsis*	spp.	Brazil	Cintia Martins	321
Not defined	A	*Solenopsis*	spp.	Brazil	Cintia Martins	322
Not defined	A	*Solenopsis*	spp.	Brazil	Cintia Martins	323
Not defined	A	*Solenopsis*	spp.	Brazil	Cintia Martins	324
Not defined	A	*Solenopsis*	spp.	Brazil	Cintia Martins	325
Not defined	A	*Solenopsis*	spp.	Brazil	Cintia Martins	326
Not defined	A	*Solenopsis*	spp.	Brazil	Cintia Martins	327
Not defined	A	*Solenopsis*	spp.	Brazil	Cintia Martins	328
59	F	*Opistophthalmus*	*chaperi*	South Africa	Laura Baldo	78

These data were retrieved from the *Wolbachia* Multilocus Sequence Typing (MLST) database (https://pubmlst.org/wolbachia/). Included are identification (id), provided for each submission, *Wolbachia* supergroup, host genus, host species, locality, submitter (researcher responsible for submitting the sequences), and the sequence type (ST) found.

2.2. Reconstruction of the Ancestral State of Wolbachia

To reconstruct the ancestral state of *Wolbachia* diversity associated with Formicidae, we assigned each supergroup A or F to each tip in our topology. Although there are only a few (three) observations from the supergroup F associated with ants, we included them in the analysis. However, several other studies have already reported that supergroup A is more common in ants [9,28,33], which gives us support that our data are a real representation of what we find in nature. As we only have one observation of *Wolbachia* belonging to the supergroup B associated with Formicidae (see [9]), we decided to remove this sample from the subsequent analyses. The 'equal rates' (ER) model and the 'all rates different' (ARD) model were compared to determine which model best explains our data using the likelihood ratio test (LRT) [54,55]. The Ape and Phytools packages [49,56] from the R software [50] were used to reconstruct ancestral character states. We used the MCMC approach to sample character histories from the probability distribution using stochastic character mapping [57]. This method samples character histories in direct proportion to their posterior probability under a model. Using the SIMMAP function [58], we generated 100 stochastic character maps from our dataset. To summarize the set of stochastic maps in a more meaningful way, we estimated the number of changes of each type, the proportion of time spent in each state, and the later probabilities that each internal node is in each state, under the best model.

2.3. Biogeographic Range Evolution Analyses

We used the inferred phylogeny and noted where, geographically, the *Wolbachia* ST was recovered (Asia, Africa, North America, South America, or Oceania) to implement biogeographic range evolution analyses using the R package BioGeoBEARS v1.1.1 [59] and estimated the ancestral range of *Wolbachia* associated with Formicidae. We followed the recommendations and parameters available on BioGeoBEARS PhyloWiki (http://phylo.wikidot.com/biogeobears) to test whether the observed biogeographic distribution of *Wolbachia* associated with Formicidae is best explained with a model that allows for vicariance and long-distance dispersal (DEC + J model) [60] versus a model that allows for only vicariance (DEC model) [61]. In addition, as *Wolbachia* is widely spread across these five biogeographic areas, we set max_range_size = 5. We used the likelihood ratio test (LRT) and the Akaike information criterion (AIC) to see which model best fits the data.

3. Results

Our *Wolbachia* phylogeny appears to be fairly robust when examining the results of posterior probability (PP); however, it grouped taxonomically unrelated ants, indicating that there is a lack of specificity of the host. Samples from 35 ant genera were included in this analysis, and as an example, 11 strains of *Camponotus* recovered from different locations around the world (Brazil, USA, Malaysia, India, and Thailand) were distributed across the phylogeny. This suggests the lack of codivergence of the bacteria and ant host. However, there was a clear and robust distinction between *Wolbachia* supergroups A and F associated with Formicidae (Figure 1).

Based on the likelihood ratio test, the ARD (all rates different) model (LTR= −6.975) was the best fit referring to the transition rates between each state to estimate the ancestral states of the *Wolbachia* supergroup associated with Formicidae when compared with ER (equal rates) (LTR= −8.530). Our ancestral state reconstruction (ASR) results with all 70 strains of *Wolbachia* associated with Formicidae show that the ancestor's supergroup was F and that was consistent across all 100 replicates, which suggests that supergroup F was the ancestor character state (Figure 2A). In addition, the transition from supergroup ancestor F to supergroup A occurred only once within the Formicidae family (Figure 2B). Therefore, once *Wolbachia* supergroup A is acquired, there is no transition to another type of *Wolbachia*.

For the biogeographic analysis, the DEC + J model was returned as the best fit to our data (AIC weight ratio model = 1.28×10^{13}), which allows for vicariance and long-distance dispersal of *Wolbachia* in Formicidae (see Table 2). According to this scenario, the ancestral origin of *Wolbachia* associated with Formicidae remains ambiguous; however, it is most probable that it originated in Asia. Then, this supergroup F, rarely found among ants, seems to have expanded to Africa. More information about *Wolbachia* associated with Formicidae in the African region could confirm this trend. Another lineage of *Wolbachia*, supergroup A, is the most common among Formicidae and appears to have likely expanded to South America. One clade then expanded into North America. In the other major clade, there was a second expansion of *Wolbachia* back to Asia from South America, introducing supergroup A into this region. Our survey revealed only a few representatives of *Wolbachia* associated with Formicidae in Oceania and Africa (Madagascar), but our results show that infections in these regions are more recent (Figure 3). Our survey did not recover any data from Europe or Central America.

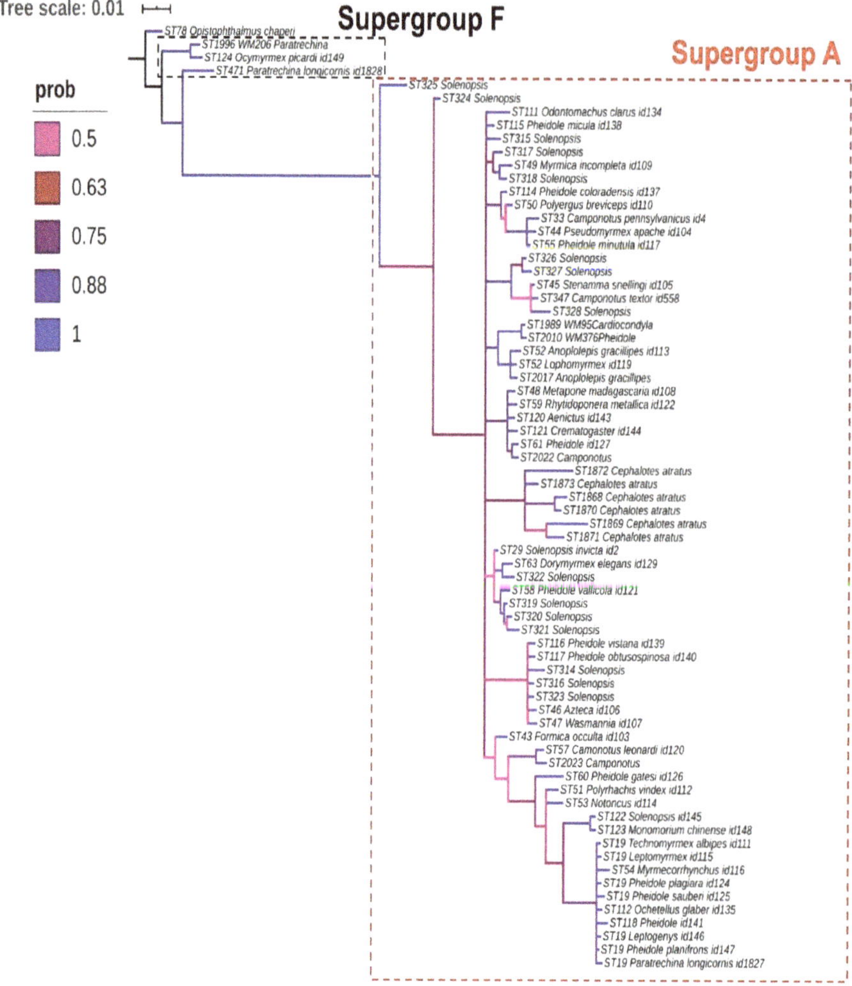

Figure 1. Bayesian majority-rule consensus tree (MrBayes) from the concatenated dataset (*coxA*, *fbpA*, *ftsZ*, *gatB*, and *hcpA* genes) of *Wolbachia* strains in Formicidae. Note that most strains belong to supergroup A, highlighted by a dotted box red, with the exceptions of three sequence types (STs), which were classified in supergroup F, highlighted by a dotted box black, and ST78 as outgroup.

Table 2. Likelihood ratio test (LRT) and the Akaike information criterion (AIC) of the DEC model and the DEC + J model to select the optimal model of ancestral areas of *Wolbachia* associated with Formicidae.

Model	Likelihood Ratio Test (LRT)	Akaike Information Criterion (AIC)
DEC + J Model	**−77.355**	**1.28 × 10^{13}**
DEC Model	−108.541	7.77× 10^{-14}

Note that the DEC + J model is highlighted in bold because vicariance and long-distance dispersal better explained our data.

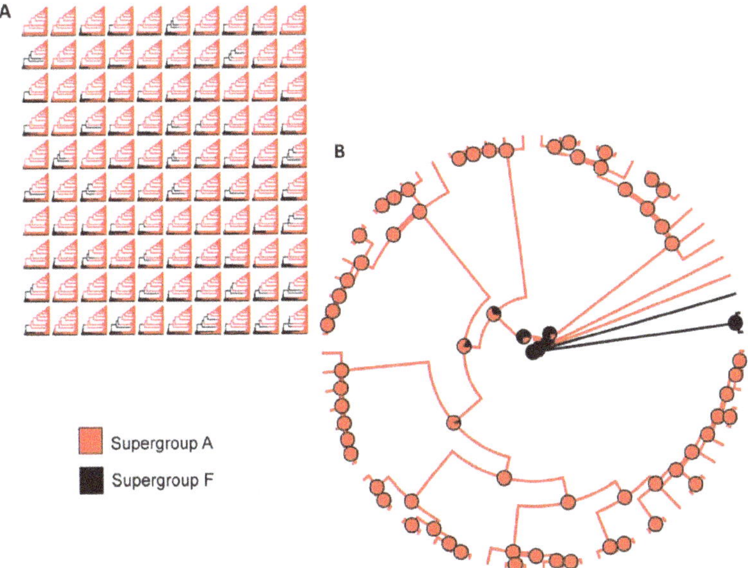

Figure 2. Ancestral state reconstruction of *Wolbachia* associated with Formicidae. (**A**) One hundred stochastic character maps from our dataset. (**B**) Summary of all stochastic character maps for ancestral state reconstruction. Model = all rates different (ARD).

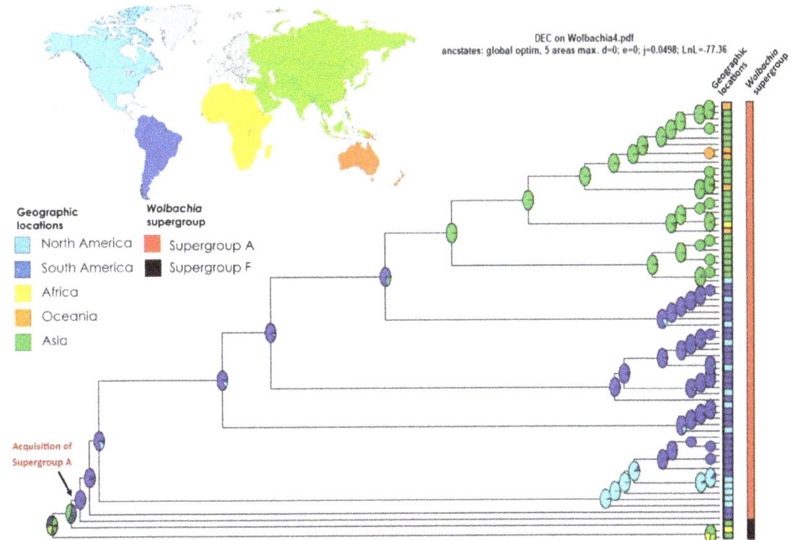

Figure 3. Ancestral range estimation of *Wolbachia* associated with Formicidae (MLST approach) with the chronogram using BIOGEOBEARS (DEC + J model). The pies are color-coded with the highest maximum likelihood probability of locations on the continents, and the boxes in the corner are colored according to the geographic locations and *Wolbachia* supergroup of the tips. The arrow indicates when *Wolbachia* supergroup A was acquired.

4. Discussion

This is the first study that has sought to investigate the ancestral state of supergroups, as well as to understand the ancestral range of origin in the evolutionary context of *Wolbachia* associated with ants. In the past, identifying *Wolbachia* infection was done by genotyping a single gene, *wsp* only. In general, with this approach, the *Wolbachia* strains associated with Formicidae belonged to supergroups A and B. With the MLST approach, now with five genes, it was found that the vast majority of Formicidae strains belong to supergroup A, followed by some representatives from supergroup F. There is only a single observation of supergroup B associated with *Pheidole sciophila* found in Mexico [9] and, therefore, it was not included in the present study. In order to understand if supergroup B, in nature, is atypical as a symbiont of Formicidae or if this is mainly a sampling bias, further studies are needed. In an extensive study, Russell and colleagues [9] used the MLST approach to understand the evolution of this interaction between *Wolbachia* and its hosts. For this, the authors focused their studies on two insect groups: ants and lycaenid butterflies. In addition to showing the presence of these bacteria in these hosts, the authors also concluded that the *Wolbachia* bacteria found in each of these groups of insects are different and highly specialized. Furthermore, they concluded that phylogenetic and geographic barriers can influence the evolutionary divergence of these bacteria. The authors found about 41 STs (sequence types) of *Wolbachia* associated with Formicidae, with few sample sequences included from South America, a hotspot for biodiversity [62,63]. Thus, in the present study, we were able to expand and include samples from this region and add more STs included in the *Wolbachia* MLST database since 2009 (Table 1).

Including *Wolbachia* associated with 35 different ant genera, our results indicate a lack of codivergence of *Wolbachia* with their ant hosts, corroborating what has been found by other studies [9,28,33,64]. The evidence for this conclusion is supported since taxonomically unrelated ants were grouped together in our *Wolbachia* phylogeny. Surprisingly, supergroup F was the ancestral character state for *Wolbachia* present in Formicidae, despite being less frequent than supergroup A. As already mentioned above, this supergroup was originally described in nematodes, but with the MLST approach, it has also been found in ants, although less frequently than supergroup A [9,32]. In addition, our results agree with the findings of Comandatore et al. [65] which suggest that supergroup F is a basal branching lineage for all *Wolbachia*.

This symbiont has already been reported to be transmitted vertically from queens to eggs in several host insects [40,66–68], including ants [29,69], and horizontal transmission of *Wolbachia* in ants is not common but may occasionally occur in related hosts [9,32,70]. In addition, Tseng et al. [32] showed that not all strains of ant *Wolbachia* have the same potential for being horizontally transferred. Another interesting aspect about these two supergroups being present in ants is that the evolutionary histories seem to be different, with supergroup A being transmitted vertically (maternal) and supergroup F being acquired horizontally [32].

Regarding the potential origin of the ancestral range of *Wolbachia* in Formicidae, our results seem to explain the dispersions by across continents. Other studies also show that geography can impact *Wolbachia* diversity on different ant host [9,28,33] and also in other insects, such as butterflies and moths [71]; however, few studies have focused on the origin and pattern of dispersion of these bacteria in different hosts until recently [72]. Our data suggest that the origin of *Wolbachia* in Formicidae happened in Asia. The supergroup F, although with few samples in ants, seems to have expanded to Africa and Asia. With the evolution of supergroup A associated with ants, *Wolbachia* then appeared in South America, with one major clade remaining in South America with a single introduction into North America and other South American clades resulting in a jump back to Asia with introductions from Asia into Oceania and Africa (Madagascar). Russell et al. [9] found evidence that *Wolbachia* strains were grouped according to samples from the Old and New Worlds, and our results also found this grouping. Our work highlights that to fully understand the evolutionary history and mechanisms of *Wolbachia* dispersion in ants or any other host group, we need much broader sampling of host species and geographic locations. However, our work is shedding light on the evolution and biogeographic history of this symbiotic interaction.

Author Contributions: M.O.R. and C.S.M. contributed equally to this study. All authors have read and agreed to the published version of the manuscript.

Funding: This work was supported in part by funding to C.S.M. from the National Science Foundation (NSF DEB 1900357).

Acknowledgments: This manuscript was initiated as part of a class project in the ModelBased Phylogenetics and Hypothesis Testing course in the Department of Entomology at Cornell University. We would like to thank Augusto Santos Rampasso for help in some aspects of the analysis of the data in this study.

Conflicts of Interest: The authors declare no conflict of interest.

References

1. Kikuchi, Y.; Hosokawa, T.; Fukatsu, T. Insect-microbe mutualism without vertical transmission: A stinkbug acquires a beneficial gut symbiont from the environment every generation. *Appl. Environ. Microbiol.* **2007**, *73*, 4308–4316. [CrossRef]

2. Buchner, P. *Endosymbiosis of Animals with Plant Microorganisms.1965.-Google Acadêmico*; Buchner, P., Ed.; Interscience Publishers: New York, NY, USA, 1965.

3. Bourtzis, K.; Miller, T. *Insect Symbiosis*; CRC Press: Boca Ranton, FL, USA, 2006; Volume 2.

4. Hu, Y.; Sanders, J.G.; Łukasik, P.; D'Amelio, C.L.; Millar, J.S.; Vann, D.R.; Lan, Y.; Newton, J.A.; Schotanus, M.; Kronauer, D.J.C.; et al. Herbivorous turtle ants obtain essential nutrients from a conserved nitrogen-recycling gut microbiome. *Nat. Commun.* **2018**, *9*, 964. [CrossRef]

5. Saraithong, P.; Li, Y.; Saenphet, K.; Chen, Z.; Chantawannakul, P. Midgut bacterial communities in the giant Asian honeybee (*Apis dorsata*) across 4 developmental stages: A comparative study. *Insect Sci.* **2017**, *24*, 81–92. [CrossRef]

6. Russell, J.A.; Funaro, C.F.; Giraldo, Y.M.; Goldman-Huertas, B.; Suh, D.; Kronauer, D.J.C.; Moreau, C.S.; Pierce, N.E. A Veritable Menagerie of Heritable Bacteria from Ants, Butterflies, and Beyond: Broad Molecular Surveys and a Systematic Review. *PLoS ONE* **2012**, *7*, e51027. [CrossRef]

7. Hilgenboecker, K.; Hammerstein, P.; Schlattmann, P.; Telschow, A.; Werren, J.H. How many species are infected with Wolbachia? A statistical analysis of current data. *FEMS Microbiol. Lett.* **2008**, *281*, 215–220. [CrossRef]

8. Werren, J.H.; Baldo, L.; Clark, M.E. Wolbachia: Master manipulators of invertebrate biology. *Nat. Rev. Microbiol.* **2008**, *6*, 741–751. [CrossRef]

9. Russell, J.A.; Goldman-Huertas, B.; Moreau, C.S.; Baldo, L.; Stahlhut, J.K.; Werren, J.H.; Pierce, N.E. Specialization and geographic isolation among Wolbachia symbionts from ants and lycaenid butterflies. *Evolution* **2009**, *63*, 624–640. [CrossRef]

10. Lefoulon, E.; Clark, T.; Borveto, F.; Perriat-Sanguinet, M.; Moulia, C.; Slatko, B.E.; Gavotte, L. Pseudoscorpion Wolbachia symbionts: Diversity and evidence for a new supergroup S. *BMC Microbiol.* **2020**, *20*, 188. [CrossRef]

11. Baldo, L.; Werren, J.H. Revisiting Wolbachia supergroup typing based on WSP: Spurious lineages and discordance with MLST. *Curr. Microbiol.* **2007**, *55*, 81–87. [CrossRef]

12. Gerth, M. Classification of Wolbachia (Alphaproteobacteria, Rickettsiales): No evidence for a distinct supergroup in cave spiders. *Infect. Genet. Evol.* **2016**, *43*, 378–380. [CrossRef]

13. Bordenstein, S.; Rosengaus, R.B. Discovery of a novel Wolbachia supergroup in isoptera. *Curr. Microbiol.* **2005**, *51*, 393–398. [CrossRef] [PubMed]

14. Glowska, E.; Dragun-Damian, A.; Dabert, M.; Gerth, M. New Wolbachia supergroups detected in quill mites (Acari: Syringophilidae). *Infect. Genet. Evol.* **2015**, *30*, 140–146. [CrossRef] [PubMed]

15. Lo, N.; Casiraghi, M.; Salati, E.; Bazzocchi, C.; Bandi, C. How Many Wolbachia Supergroups Exist? *Mol. Biol. Evol.* **2002**, *19*, 341–346. [CrossRef] [PubMed]

16. Lo, N.; Paraskevopoulos, C.; Bourtzis, K.; O'Neill, S.L.; Werren, J.H.; Bordenstein, S.R.; Bandi, C. Taxonomic status of the intracellular bacterium Wolbachia pipientis. *Int. J. Syst. Evol. Microbiol.* **2007**, *57*, 654–657. [CrossRef] [PubMed]

17. Ros, V.I.D.; Fleming, V.M.; Feil, E.J.; Breeuwer, J.A.J. How diverse is the genus Wolbachia? Multiple-gene sequencing reveals a putatively new Wolbachia supergroup recovered from spider mites (Acari: Tetranychidae). *Appl. Environ. Microbiol.* **2009**, *75*, 1036–1043. [CrossRef]

18. Ferri, E.; Bain, O.; Barbuto, M.; Martin, C.; Lo, N.; Uni, S.; Landmann, F.; Baccei, S.G.; Guerrero, R.; de Souza Lima, S.; et al. New Insights into the Evolution of Wolbachia Infections in Filarial Nematodes Inferred from a Large Range of Screened Species. *PLoS ONE* **2011**, *6*, e20843. [CrossRef]

19. Lefoulon, E.; Gavotte, L.; Junker, K.; Barbuto, M.; Uni, S.; Landmann, F.; Laaksonen, S.; Saari, S.; Nikander, S.; de Souza Lima, S.; et al. A new type F Wolbachia from Splendidofilariinae (Onchocercidae) supports the recent emergence of this supergroup. *Int. J. Parasitol.* **2012**, *42*, 1025–1036. [CrossRef]

20. Bolton, B. An Online Catalog of the Ants of the World. Available online: http://www.antcat.org/ (accessed on 20 October 2016).

21. Singh, R.; Linksvayer, T.A. Wolbachia-infected ant colonies have increased reproductive investment and an accelerated life cycle. *J. Exp. Biol.* **2020**, *223*. [CrossRef]

22. Cheng, D.; Chen, S.; Huang, Y.; Pierce, N.E.; Riegler, M.; Yang, F.; Zeng, L.; Lu, Y.; Liang, G.; Xu, Y. Symbiotic microbiota may reflect host adaptation by resident to invasive ant species. *PLoS Pathog.* **2019**, *15*, e1007942. [CrossRef]

23. Fernando de Souza, R.; Daivison Silva Ramalho, J.; Santina de Castro Morini, M.; Wolff, J.L.C.; Araújo, R.C.; Mascara, D. Identification and Characterization of Wolbachia in Solenopsis saevissima Fire Ants (Hymenoptera: Formicidae) in Southeastern Brazil. *Curr. Microbiol.* **2009**, *58*, 189–194. [CrossRef]

24. Frost, C.L.; Fernandez-Marin, H.; Smith, J.E.; Hughes, W.O.H. Multiple gains and losses of Wolbachia symbionts across a tribe of fungus-growing ants. *Mol. Ecol.* **2010**, *19*, 4077–4085. [CrossRef]

25. Frost, C.L.; Pollock, S.W.; Smith, J.E.; Hughes, W.O.H. Wolbachia in the flesh: Symbiont intensities in germ-line and somatic tissues challenge the conventional view of Wolbachia transmission routes. *PLoS ONE* **2014**, *9*, e95122. [CrossRef]

26. Martins, C.; Souza, R.F.; Bueno, O.C. Presence and distribution of the endosymbiont Wolbachia among Solenopsis spp. (Hymenoptera: Formicidae) from Brazil and its evolutionary history. *J. Invertebr. Pathol.* **2012**, *109*, 287–296. [CrossRef]

27. Kautz, S.; Rubin, B.E.R.; Moreau, C.S. Bacterial Infections across the Ants: Frequency and Prevalence of *Wolbachia, Spiroplasma*, and *Asaia. Psyche A J. Entomol.* **2013**, *2013*, 1–11. [CrossRef]

28. Ramalho, M.O.; Martins, C.; Silva, L.M.R.; Martins, V.G.; Bueno, O.C. Intracellular symbiotic bacteria of Camponotus textor, Forel (Hymenoptera, Formicidae). *Curr. Microbiol.* **2017**, *74*, 589–597. [CrossRef]

29. Ramalho, M.O.; Vieira, A.S.; Pereira, M.C.; Moreau, C.S.; Bueno, O.C. Transovarian Transmission of Blochmannia and Wolbachia Endosymbionts in the Neotropical Weaver Ant Camponotus textor (Hymenoptera, Formicidae). *Curr. Microbiol.* **2018**, *75*, 866–873. [CrossRef]

30. Ramalho, M.O.; Bueno, O.C.; Moreau, C.S. Species-specific signatures of the microbiome from Camponotus and Colobopsis ants across developmental stages. *PLoS ONE* **2017**, *12*, e0187461. [CrossRef]

31. Ramalho, M.O.; Bueno, O.C.; Moreau, C.S. Microbial composition of spiny ants (Hymenoptera: Formicidae: Polyrhachis) across their geographic range. *BMC Evol. Biol.* **2017**, *17*, 96. [CrossRef]

32. Tseng, S.P.; Wetterer, J.K.; Suarez, A.V.; Lee, C.Y.; Yoshimura, T.; Shoemaker, D.W.; Yang, C.C.S. Genetic Diversity and Wolbachia Infection Patterns in a Globally Distributed Invasive Ant. *Front. Genet.* **2019**, *10*, 838. [CrossRef]

33. Kelly, M.; Price, S.L.; de Oliveira Ramalho, M.; Moreau, C.S. Diversity of Wolbachia Associated with the Giant Turtle Ant, Cephalotes atratus. *Curr. Microbiol.* **2019**, *76*, 1330–1337. [CrossRef]

34. Ramalho, M.O.; Moreau, C.S.; Bueno, O.C. The Potential Role of Environment in Structuring the Microbiota of Camponotus across Parts of the Body. *Adv. Entomol.* **2019**, *7*, 47–70. [CrossRef]

35. Reeves, D.D.; Price, S.L.; Ramalho, M.O.; Moreau, C.S. The Diversity and Distribution of Wolbachia, Rhizobiales, and Ophiocordyceps Within the Widespread Neotropical Turtle Ant, Cephalotes atratus (Hymenoptera: Formicidae). *Neotrop. Entomol.* **2020**, *49*, 52–60. [CrossRef] [PubMed]

36. Baldo, L.; Lo, N.; Werren, J.H. Mosaic nature of the wolbachia surface protein. *J. Bacteriol.* **2005**, *187*, 5406–5418. [CrossRef] [PubMed]

37. Paraskevopoulos, C.; Bordenstein, S.; Wernegreen, J.; Werren, J.; Bourtzis, K. Toward a Wolbachia multilocus sequence typing system: Discrimination of Wolbachia strains present in Drosophila species. *Curr. Microbiol.* **2006**, *53*, 388–395. [CrossRef]

38. Baldo, L.; Dunning Hotopp, J.C.; Jolley, K.A.; Bordenstein, S.R.; Biber, S.A.; Choudhury, R.R.; Hayashi, C.; Maiden, M.C.J.; Tettelin, H.; Werren, J.H. Multilocus sequence typing system for the endosymbiont Wolbachia pipientis. *Appl. Environ. Microbiol.* **2006**, *72*, 7098–7110. [CrossRef]

39. Ali, H.; Muhammad, A.; Hou, Y. Infection Density Dynamics and Phylogeny of Wolbachia Associated with Coconut Hispine Beetle, Brontispa longissima (Gestro) (Coleoptera: Chrysomelidae), by Multilocus Sequence Type. *Artic. J. Microbiol. Biotechnol.* **2018**, *28*, 796–808. [CrossRef]

40. Ali, H.; Muhammad, A.; Sanda Bala, N.; Hou, Y. The Endosymbiotic Wolbachia and Host COI Gene Enables to Distinguish Between Two Invasive Palm Pests; Coconut Leaf Beetle, Brontispa longissima and Hispid Leaf Beetle, Octodonta nipae. *J. Econ. Entomol.* **2018**, *111*, 2894–2902. [CrossRef]

41. Betelman, K.; Caspi-Fluger, A.; Shamir, M.; Chiel, E. Identification and characterization of bacterial symbionts in three species of filth fly parasitoids. *FEMS Microbiol. Ecol.* **2017**, *93*. [CrossRef]

42. Prezotto, L.F.; Perondini, A.L.P.; Hernández-Ortiz, V.; Marino, C.L.; Selivon, D. Wolbachia strains in cryptic species of the Anastrepha fraterculus complex (Diptera, Tephritidae) along the Neotropical Region. *Syst. Appl. Microbiol.* **2017**, *40*, 59–67. [CrossRef]

43. Higgins, D.; Bleasby, A.; Fuchs, R. CLUSTAL V: Improved software for multiple sequence alignment. *Comput. Appl.* **1992**, *8*, 189–191.

44. Hall, T. BioEdit: A user-friendly biological sequence alignment editor and analysis program for Windows 95/98/NT. *Nucleic Acids Symp. Ser.* **1999**, *41*, 95–98.

45. Lanfear, R.; Frandsen, P.B.; Wright, A.M.; Senfeld, T.; Calcott, B. PartitionFinder 2: New Methods for Selecting Partitioned Models of Evolution for Molecular and Morphological Phylogenetic Analyses. *Mol. Biol. Evol.* **2016**, *34*, msw260. [CrossRef]

46. Huelsenbeck, J.; Ronquist, F. MRBAYES: Bayesian inference of phylogenetic trees. *Bioinformatics* **2001**, *17*, 754–755. [CrossRef] [PubMed]

47. Miller, M.A.; Pfeiffer, W.; Schwartz, T. Creating the CIPRES Science Gateway for inference of large phylogenetic trees. In Proceedings of the 2010 Gateway Computing Environments Workshop (GCE), New Orleans, LA, USA, 14 November 2010; pp. 1–8.

48. Miller, M.A.; Pfeiffer, W.; Schwartz, T. The CIPRES Portals. Available online: http://www.phylo.org/ (accessed on 20 September 2020).

49. Paradis, E.; Claude, J.; Strimmer, K. APE: Analyses of Phylogenetics and Evolution in R language. *Bioinformatics* **2004**, *20*, 289–290. [CrossRef] [PubMed]

50. R Core Team. *R: A Language and Environment for Statistical Computing*; R Foundation for Statistical Computing: Vienna, Austria, 2020; Available online: http://www.R-project.org/ (accessed on 20 September 2020).

51. Kim, J.; Sanderson, M.J. Penalized Likelihood Phylogenetic Inference: Bridging the Parsimony-Likelihood Gap. *Syst. Biol.* **2008**, *57*, 665–674. [CrossRef] [PubMed]

52. Paradis, E. Molecular dating of phylogenies by likelihood methods: A comparison of models and a new information criterion. *Mol. Phylogenet. Evol.* **2013**, *67*, 436–444. [CrossRef]

53. Sanderson, M.J. Estimating Absolute Rates of Molecular Evolution and Divergence Times: A Penalized Likelihood Approach. *Mol. Biol. Evol.* **2002**, *19*, 101–109. [CrossRef]

54. Pagel, M. Detecting correlated evolution on phylogenies: A general method for the comparative analysis of discrete characters. *Proc. R. Soc. London. Ser. B Biol. Sci.* **1994**, *255*, 37–45. [CrossRef]

55. Schluter, D.; Price, T.; Mooers, A.Ø.; Ludwig, D. Likelihood of ancestor states in adaptive radiation. *Evolution* **1997**, *51*, 1699–1711. [CrossRef]

56. Maintainer, L.J.R.; Revell, L.J. *Package "Phytools" Title Phylogenetic Tools for Comparative Biology (and Other Things)*; R Package. 2020. Available online: https://besjournals.onlinelibrary.wiley.com/doi/full/10.1111/j.2041-210X.2011.00169.x (accessed on 12 November 2020).

57. Huelsenbeck, J.P.; Nielsen, R.; Bollback, J.P. Stochastic Mapping of Morphological Characters. *Syst. Biol.* **2003**, *52*, 131–158. [CrossRef]

58. Bollback, J.P. SIMMAP: Stochastic character mapping of discrete traits on phylogenies. *BMC Bioinf.* **2006**, *7*, 88. [CrossRef] [PubMed]

59. Matzke, N.J. BioGeoBEARS: An R package for model testing and ancestral state estimation in historical biogeography. *Methods Ecol. Evol.* **2014**.

60. Matzke, N.J. Model selection in historical biogeography reveals that founder-event speciation is a crucial process in island clades. *Syst. Biol.* **2014**, *63*, 951–970. [CrossRef] [PubMed]

61. Ree, R.H.; Smith, S.A. Maximum Likelihood Inference of Geographic Range Evolution by Dispersal, Local Extinction, and Cladogenesis. *Syst. Biol.* **2008**, *57*, 4–14. [CrossRef] [PubMed]
62. Fuentes-Castillo, T.; Hernández, H.J.; Pliscoff, P. Hotspots and ecoregion vulnerability driven by climate change velocity in Southern South America. *Reg. Environ. Chang.* **2020**, *20*, 1–15. [CrossRef]
63. Cantidio, L.S.; Souza, A.F. Aridity, soil and biome stability influence plant ecoregions in the Atlantic Forest, a biodiversity hotspot in South America. *Ecography* **2019**, *42*, 1887–1898. [CrossRef]
64. Martins, C.; Correa Bueno, O. *Determinação de Cepas de Wolbachia em Populações Naturais de Solenopsis spp. (Hymenoptera: Formicidae) Analisadas via Multilocus Sequence Typing (MLST): Diversidade Genética, Coevolução e Recombinação*; Universidade Estadual Paulista (UNESP): Rio Claro, Brazil, 2014.
65. Comandatore, F.; Sassera, D.; Montagna, M.; Kumar, S.; Koutsovoulos, G.; Thomas, G.; Repton, C.; Babayan, S.A.; Gray, N.; Cordaux, R.; et al. Phylogenomics and Analysis of Shared Genes Suggest a Single Transition to Mutualism in Wolbachia of Nematodes. *Genome Biol. Evol.* **2013**, *5*, 1668–1674. [CrossRef]
66. Narita, S.; Shimajiri, Y.; Nomura, M. Strong cytoplasmic incompatibility and high vertical transmission rate can explain the high frequencies of Wolbachia infection in Japanese populations of Colias erate poliographus (Lepidoptera: Pieridae). *Bull. Entomol. Res.* **2009**, *99*, 385–391. [CrossRef]
67. Gerth, M.; Röthe, J.; Bleidorn, C. Tracing horizontal *Wolbachia* movements among bees (Anthophila): A combined approach using multilocus sequence typing data and host phylogeny. *Mol. Ecol.* **2013**, *22*, 6149–6162. [CrossRef]
68. Duplouy, A.; Couchoux, C.; Hanski, I.; van Nouhuys, S. Wolbachia Infection in a Natural Parasitoid Wasp Population. *PLoS ONE* **2015**, *10*, e0134843. [CrossRef]
69. Bouwma, A.M.; Shoemaker, D. *Wolbachia* wSinvictaA Infections in Natural Populations of the Fire Ant *Solenopsis invicta*: Testing for Phenotypic Effects. *J. Insect Sci.* **2011**, *11*, 1–19. [CrossRef] [PubMed]
70. Tolley, S.J.A.; Nonacs, P.; Sapountzis, P. Wolbachia Horizontal Transmission Events in Ants: What Do We Know and What Can We Learn? *Front. Microbiol.* **2019**, *10*, 296. [CrossRef] [PubMed]
71. Ahmed, M.Z.; Araujo-Jnr, E.V.; Welch, J.J.; Kawahara, A.Y. Wolbachia in butterflies and moths: Geographic structure in infection frequency. *Front. Zool.* **2015**, *12*, 16. [CrossRef] [PubMed]
72. Ahmed, M.Z.; Breinholt, J.W.; Kawahara, A.Y. Evidence for common horizontal transmission of Wolbachia among butterflies and moths. *BMC Evol. Biol.* **2016**, *16*, 1–16. [CrossRef] [PubMed]

Publisher's Note: MDPI stays neutral with regard to jurisdictional claims in published maps and institutional affiliations.

Article

Digging Deeper into the Ecology of Subterranean Ants: Diversity and Niche Partitioning across Two Continents

Mickal Houadria * and Florian Menzel

Institute of Organismic and Molecular Evolution, Johannes-Gutenberg-University Mainz, Hanns-Dieter-Hüsch-Weg 15, 55128 Mainz, Germany; menzelf@uni-mainz.de

* Correspondence: mickal.houadria@free.fr

Abstract: Soil fauna is generally understudied compared to above-ground arthropods, and ants are no exception. Here, we compared a primary and a secondary forest each on two continents using four different sampling methods. Winkler sampling, pitfalls, and four types of above- and below-ground baits (dead, crushed insects; melezitose; living termites; living mealworms/grasshoppers) were applied on four plots (4 × 4 grid points) on each site. Although less diverse than Winkler samples and pitfalls, subterranean baits provided a remarkable ant community. Our baiting system provided a large dataset to systematically quantify strata and dietary specialisation in tropical rainforest ants. Compared to above-ground baits, 10–28% of the species at subterranean baits were overall more common (or unique to) below ground, indicating a fauna that was truly specialised to this stratum. Species turnover was particularly high in the primary forests, both concerning above-ground and subterranean baits and between grid points within a site. This suggests that secondary forests are more impoverished, especially concerning their subterranean fauna. Although subterranean ants rarely displayed specific preferences for a bait type, they were in general more specialised than above-ground ants; this was true for entire communities, but also for the same species if they foraged in both strata.

Keywords: soil arthropods; pitfall; bait; turnover; food specialisation; stratification; sampling methods; hypogaeic

Citation: Houadria, M.; Menzel, F. Digging Deeper into the Ecology of Subterranean Ants: Diversity and Niche Partitioning across Two Continents. *Diversity* **2021**, *13*, 53. https://doi.org/10.3390/d13020053

Academic Editors: Alan N. Andersen and Luc Legal

Received: 4 November 2020

Accepted: 26 January 2021

Published: 29 January 2021

Publisher's Note: MDPI stays neutral with regard to jurisdictional claims in published maps and institutional affiliations.

1. Introduction

From the tip of the leaves in the canopy of a forest, to the epiphytes living on the branches all the way down the bark, the leaf litter, and even within the first metre of soil, ants are present everywhere. Be they above-ground or hypogaeic foragers, ants contribute significantly to ecosystem functions. In addition to being important in trophic functions as predators [1], their digging activity aerates the soil and promotes nutrient cycling [2]. They often display vertical stratification, with many species being specialised in certain microhabitats. Such stratification has been studied mostly above ground to date, with a focus on canopy vs. understory vs. leaf litter dwellers [3–5]. However, we know even less about the subterranean ants that live only a few centimetres away from our feet—within the soil. We adopt the terminology of [6], with 'subterranean' referring to all ants (and baits) from below ground, whereas 'hypogaeic' refers to species that predominantly live and forage below ground, thus excluding above-ground species that might nest in the soil but do not forage there.

A recent study from the Amazon rainforest [7] captured nine hypogaeic species (out of 47) with subterranean baits that could not be sampled using other methods. This demonstrates that conventional ant sampling methods neglect this fauna almost completely [8–10]; this is also because published protocols on recommended sampling methods mostly include above-ground baits, pitfalls, and Winkler sampling [11], which misses out hypogaeic species. However, hypogaeic ant communities include important lineages which could shed light on the early development of ancestral ants. They include phylogenetically ancient and conserved ant lineages. For example, the recently discovered *Martialis heureka* represents the sister lineage to all other ants. It is a hypogaeic ant, with the first two specimen being collected in soil core samples [12]. Within the clade that is sister to *Martialis*, the blind, hypogaeic subfamily Leptanillinae is sister to all other ant taxa, which implies that cryptobiotic, blind ants evolved and diversified early in ant evolution [12]. Thus, hypogaeic ants may become key to our understanding of the ecology and evolution of early ants. Many specialised and cryptobiotic ant genera, which often include blind species, are collected almost exclusively using subterranean baiting or soil/leaf-litter sampling (such as *Anillomyrma*, *Prionopelta, Simopelta, Leptanilla*, and *Aenictus*) [6]. However, a severe obstacle to research on subterranean ants is that exhaustive and reliable sampling remains challenging.

The difficult sampling also limits our understanding of ecosystem processes in which these ants are involved. For instance, the environmental importance of army ants in tropical ecosystems has been demonstrated through the study of above-ground species such as *Eciton* [1,13]. However, as the majority of army ant genera are hypogaeic, we probably underestimate the ecological impact of army ants [14]. Here, two important aspects to understand ecological niches are stratification, i.e., the degree of specialisation in subterranean habitats, and dietary specialisation. A good way to determine these is offering different baits [15] in a single grid-point location but on both strata.

Most studies with subterranean baits have used mixtures of processed food resources [1,8,13,15–17]. This is a highly effective way to attract the maximum possible range of ants but does not yield information on the dietary specialisation of subterranean ants. Al-though baits that mimic natural resources necessarily sample only a subset of the ant fauna [16,18], they provide valuable information on the dietary preferences of these species [15], which are not only important aspects of their ecology, but will also improve how to sample certain species specifically. On a similar note, it is not known if ants forage on the same food sources above and below ground. For instance, certain ant species that tend aphids above ground could be preying on termites below ground (or vice versa).

Like other cryptobiotic ants, hypogaeic species might be especially sensitive to disturbance [19]. Therefore, it is crucial to analyse the ant diversity of subterranean communities across primary and disturbed habitats. To facilitate conservation efforts, it is also important to know if different strata respond to habitat disturbance in a similar fashion and if this differs between tropical regions.

In this study, we investigated the diversity and ecology of subterranean ants at rainforest sites in the Neotropics and Paleotropics. Using a complex sampling design, with 64 spatial replicates per site, above-ground baits were placed at the identical location to the subterranean baits in order to separate ants that forage in both strata to those unique to one of them. To compare baiting efficiency to more conventional methods, pitfalls and Winkler extractions were also performed in each plot (Figure S1). This standardised design was completed in four tropical rainforest sites, providing an extraordinarily large data base to study the ecology and distribution of subterranean ants. The focal research areas of our study are supported by a recent review on subterranean ants [6], which pointed out a lack of publications in the aspects we have addressed here.

First, we compared subterranean baiting to the other three methods regarding taxonomic yield and complementarity, and identified the species only caught with this method. Second, we compared ant assemblages attracted to aboveground and subterranean baits, and calculated stratum specialisation for each species. Third, we analysed α and β diversity above and below ground to determine whether patterns of biodiversity were consistent across sites. Finally, based on the ant communities found at the four different bait types, we analysed trophic specialisation in aboveground and subterranean ant communities.

2. Material and Methods

2.1. Study Site

Ant sampling was done in two neotropical and two paleotropical sites in 2012 (French Guiana, France) and 2013 (Sabah, Borneo, Malaysia), respectively. On each continent, we sampled one primary and one secondary forest, henceforth abbreviated 'NPF' (neotropical primary forest), 'PSF' (paleotropical secondary forest), and so on. In the neotropics, the Les Nouragues Nature Reserve was studied as a primary forest (NPF, 4°05′ N, 52°41′ W, 90 m a.s.l.), which covers >100,000 ha of pristine forest. The secondary forest site was a 16-ha forest fragment, bordered by urban grass and surrounded by residential areas on the Campus Agronomique in Kourou (NSF, 5°09′ N, 52°39′ W, 8 m a.s.l.). In the paleotropics, primary forest was sampled in the Danum Valley Conservation Area (PPF, Sabah, Malaysian Borneo; 4°55′ N, 117°40′ E, 296 m a.s.l.). The site is part of a 438-km^2 primary forest dominated by Dipterocarpaceae trees. As a secondary forest, the Malua Forest Reserve was chosen (PSF, 4°24′ N, 118°14′ E, 6 m a.s.l.). It comprises 35 km^2 of production forest, which was selectively logged in the 1980s.

2.2. Sampling Design

In each site, we established four plots of 4 × 4 grid points each, i.e., a total number of 64 grid points. Within a plot, the distance between grid points was 10 m (Figure S1); the four plots were separated by at least 150 m. At each grid point, we placed pitfalls, above-ground baits, and subterranean traps. Winkler sampling was conducted for all areas within four grid points, resulting in four Winkler samples per plot and 16 per site (Figure 1). Data from the above-ground baits and pitfalls were partly analysed and published previously [2,15,20–22].

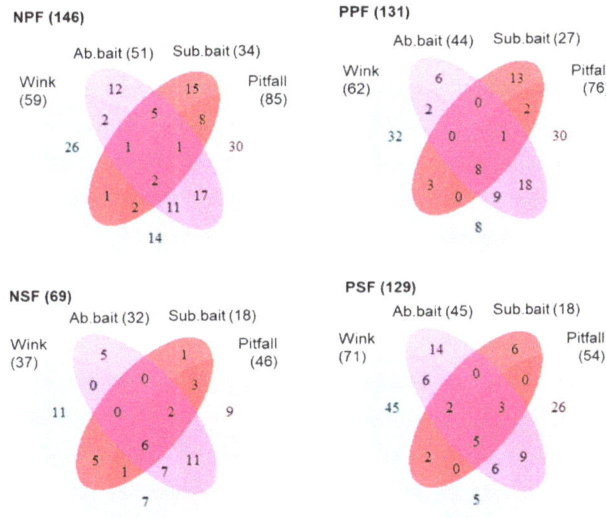

Figure 1. Venn diagrams for each site, comparing the ant species sampled by the four different sampling methods. Wink, Ab.bait, and Sub.bait stand for Winkler samples, above-ground baiting, and subterranean baiting, respectively. The numbers on overlapping sections are the number of species common to the sampling methods concerned, otherwise they are unique to one sampling method. Overall species richness for each method and site is written in brackets.

2.2.1. Above-Ground Baits

The baits were presented in small plastic boxes with slits at the sides for 90 min and then quickly retrieved. We presented four bait types that reflected natural resources available to ants in a rainforest: melezitose (reflecting trophobiont honeydew); dead, crushed insects (carrion); living termites (small prey); and living grasshoppers (large prey). The four different above- and below-ground baits were separately presented at a given time, i.e., there were never two baits displayed on a grid point simultaneously. Melezitose was presented in a 20% m/v solution on paper tissue; approx. 10 living termites were kept free on the paper towel and usually stayed on a small piece of humid termite mound which was added. The living grasshopper was tethered to the bait box to prevent escape. At each grid point, all above-ground baits were presented separately during the day (between 10:00 to 15:00) and night (between 20:00 to 23:00). For the present study, the two time periods were pooled per grid point and bait type, in order to be comparable to the subterranean baits left for the whole 24-h cycle. Additional above-ground baits that have been presented but were not considered for this study include sucrose, bird faeces, and seeds [2,15,20–22]. Therefore, numbers of species or occurrences may differ from our previous studies.

2.2.2. Subterranean Baits

Subterranean traps consisted of closed Eppendorf cups (1.5 mL) that contained a food resource. The cups had four holes drilled in the upper half to allow access by ants, as described by Andersen and Brault [10] (see also Figure S1). Based on previous studies [7,18,23], the highest species richness is obtained between 10 and 20 cm and most species are sampled within a 24 h timeframe [7,10]. Therefore, after litter removal, traps were buried 10 cm deep into the soil, and carried a wire which stuck out from the soil for quick retrieval. For practical reasons (weather and preparation), one bait type was displayed for 24 h simultaneously on all 64 grid points within a site. We used the same four baits as above (melezitose, termites, crushed insects), but replaced live grasshoppers by live 1.5-cm mealworms as it seemed more suitable as a large subterranean prey [18]. To present melezitose, a cotton ball was compacted in the concave lids, which was then saturated with 2 mL of 20% m/v solution. The crushed insect paste was compressed in a similar way. To keep the mealworms in place and prevent them falling in the killing solution, the mealworm was pinned to the lid using the tip of a needle, minimizing the wound for the mealworm to be still capable of wriggling. For the last bait, three to four termites were glued alive to a 1-cm piece of toothpick which was then inserted through the lid in a similar fashion as the mealworm. Preliminary tests showed that mealworms and termites survive for over 20 h in this state.

2.2.3. Pitfalls and Winkler Extractions

Sampling methods were performed at separate times to avoid interference between them. Pitfalls were left open for 12 h during the day (6:00–18:00) or night (18:00–6:00); this was done three times consecutively before retrieval. For the present study, day and night pitfall data were pooled for each grid point. Leaf litter was sampled at four places in each plot. We sampled litter and superficial soil from 1 m^2 quadrats in the middle of each of the four grid points. Samples were put in Winkler extraction devices for 48 h; ants were collected in an ethylene-glycol solution.

2.2.4. Species Identification

All ants were identified to genera using Bolton [24] and using a binocular microscope (Leica S8AP0). Reference collections with voucher specimens of all ant species were prepared. Morphospecies numbers were assigned consecutively in the order of detection in samples. The collections are deposited at the Institute of Organismic and Molecular Evolution, University of Mainz, Germany. Several morphospecies, including the most common ones, were identified to species using specialised literature [25–27] or the help of specialists (Tables S1 and S2). In particular, several *Pheidole* and *Crematogaster* species were kindly identified by John Longino and Bonnie Blaimer, respectively. Notably, the identification of taxonomically challenging dimorphic genera such as *Pheidole* and *Carebara* is easier with bait samples than with other sampling methods because major and minor workers are often collected together. This way, more conspecific individuals are present, especially in the case of bait monopolization, which is helpful especially in order to identify species with dimorphic workers (pers. comm. J. Longino 2013).

2.3. Statistical Analysis

2.3.1. Sampling and Baiting Complementarity

The exhaustiveness of Winkler samples, both baiting systems (with bait types summed) and pitfalls was estimated by calculating Cole's rarefaction curves. We also estimated the expected species richness with the Chao2 species richness estimator with EstimateS 9.0 [28]. All other statistics were calculated using R version 3.6.2 (R Core Team 2018, Vienna, Austria; https://www.R-project.org/). The complementarity of species captured was visualised using Venn diagrams, R (VennDiagram, Venneuler packages). For the subsequent analyses, we only considered above-ground and subterranean baiting data, disregarding pitfalls and Winkler samples.

2.3.2. Diversity Measures above and below Ground

Species richness per grid point (α diversity) was calculated including singletons, separately above and below ground. These species numbers were compared using mixed-effects models with species richness as the dependent variable, site and stratum as fixed variables and grid points as random effects. Species turnover (β diversity) was calculated as the proportion of unique species, viz. $(b + c)/(a + b + c)$, with a being the species shared between two grid points, and b and c being the species unique to the two samples of each comparison. For vertical turnover, we compared these values between above-ground and subterranean assemblages of the same grid point. The values were arcsine-square root transformed and compared between sites using linear models (LM); pairwise comparisons were based on model summaries.

Horizontal turnover was calculated as the proportion of unique species in pairwise comparisons between all grid points within the same 4×4 plot. This was done separately for above-ground and subterranean baits. Horizontal turnover values were then transformed as above and compared between sites and strata using a linear model with stratum and site and their interaction as fixed factors. Models were analysed using Anova (package car) in R.

2.3.3. Trophic and Strata Specialisation

Both aboveground and subterranean baits are active sampling methods, which capture only foraging ants that are attracted. Thus, it makes most sense to compare specifically these two sampling methods to assess trophic and strata specialization. Here, we included only species with at least five occurrences at above-ground or subterranean baits. This way, there were enough occurrences that the species might have foraged at all four resources. For strata specialisation, we calculated the number of occurrences above and below ground for each species. These were compared using χ^2 tests, followed by correction for false discovery rate.

Next to strata specialisation, food preferences were calculated using a randomization algorithm based on the number of occurrences per bait. A null model randomly swapped occurrences over 1000 permutations; the bait was defined as 'preferred' if realised occurrences were higher than the 95% confidence interval of the random expectation [15]. This was done separately for both strata.

Furthermore, we assessed the overall degree of food specialisation (*fs*). A common species is more likely to encounter different food types by chance; therefore we rarefied down to five occurrences and calculated *fs*; this was done 1000 times per species. 'Food specialisation' *fs* was calculated similarly to Simpson's diversity index as $fs = \Sigma p_i^2$, where p_i is the number of occurrences at bait type i divided by the total number of occurrences for the given species. *fs* considers the relative specialisation of a species on any resource and ranges from 0 to 1. In contrast, 'food preferences' provide information about the actual resources that are foraged. A species specialised in one resource will prefer it, but a species with several preferences may or may not be highly specialised overall [15]. Food specialisation was calculated separately for above- and below-ground baits and compared using *t*-tests. In a second analysis, we compared food specialisation for only those species which foraged in both strata using a paired *t*-test. Food specialisation of the overall species community was further compared between strata using the $H_{2'}$ index [29] using the R package bipartite, which ranges from 0 (maximum generalisation) to 1 (maximum specialisation of the entire community).

3. Results

3.1. Species Richness for Different Sampling Techniques and Strata

Subterranean baits yielded the lowest richness in all four sites (18 in both secondary forests and up to 34 in NPF), but always had between one (PSF) and 15 (NPF) unique species which were not captured with any other method (Figure 1). In contrast, Winkler samples (followed by pitfalls) yielded the highest species numbers, giving between 37 (NSF) and 71 (PSF) species per site (Figure 1). Furthermore, overall expected richness (Chao 2 estimators) was highest for Winkler and pitfall samples compared to the two baiting methods (Figure 2). In above-ground baits, ants were attracted onto 62 to 64 out of 64 grid points per site. In contrast, subterranean baits attracted ants onto only 49 (PPF) to 57 (PSF) out of 64 grid points.

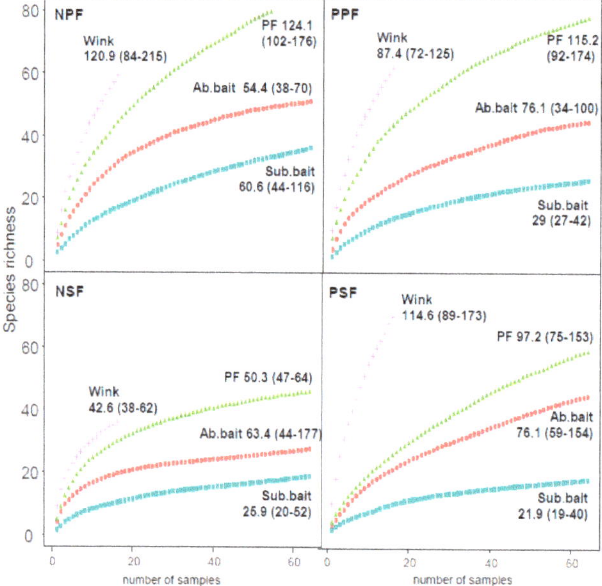

Figure 2. Rarefaction curves for all four different sampling methods. For both subterranean and above-ground baiting, the four different baits were pooled according to grid points before rarefaction. Numbers in bold represent Chao 2 estimates of total species richness for each sampling method, with confidence intervals in brackets. Wink, Ab. bait, Sub.bait, and PF stand for Winkler samples, above-ground baiting, subterranean baiting, and pitfalls, respectively.

Species richness per grid point was always higher above ground than below ground ($F_1 = 418.0$, $p < 0.0001$; Figure 3A). Above-ground richness differed substantially between sites, whereas below-ground richness did not (site × stratum interaction: $F_3 = 12.8$, $p < 0.0001$; site: $F_3 = 13.7$, $p < 0.0001$).

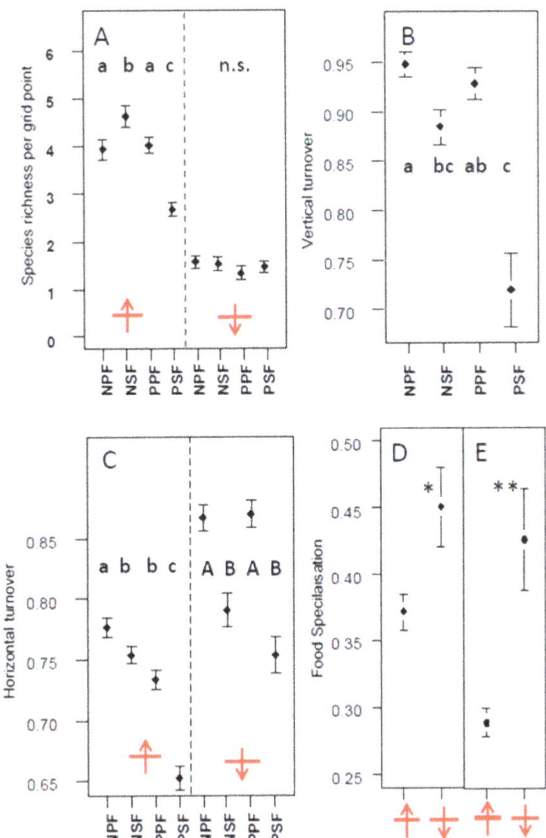

Figure 3. Diversity in above-ground and subterranean baits (arrows pointing up and down, respectively). (**A**) Species richness per grid point for above-ground and subterranean baiting. (**B**) Vertical turnover, given as the proportion of species unique to either above-ground or subterranean baits (for each grid point separately). (**C**) Horizontal turnover, given as the proportion of species unique to a grid point compared to any other grid point of the same 4 × 4 plot (separately for above-ground and subterranean baits). Sites with the same letters are not significantly different. (**D,E**) Species-specific food specialisation (*fs*) in assemblages above or below ground, with higher values indicating higher specialisation. (**D**) depicts all species with at least five occurrences; (**E**) depicts only those species that occurred in both strata with at least five occurrences. Significant differences are denoted with asterisks (* $p < 0.05$, ** $p < 0.01$).

3.2. Taxonomic Composition and Stratum Specialisation at Different Baits

Subfamily and genus composition differed between above-ground and subterranean baits. Dolichoderinae were absent from subterranean baits, but occurred above ground in all sites (Figure 4). Among the Formicinae, certain genera were only found above ground (*Euprenolepis, Camponotus, Dinomyrmex, Polyrhachis*), whereas others were only found in the subterranean baits (*Acropyga, Brachymyrmex, Pseudolasius*). Only *Nylanderia* was found in both strata. In the Ectatomminae, *Ectatomma* was only found in above-ground baits, whereas *Gnamptogenys* only occurred below ground. Among the Ponerinae, *Hypoponera* and *Parvaponera* were exclusive to subterranean baits, whereas *Mayaponera, Leptogenys*, and *Mesoponera* were found in both baited strata. Among the Myrmicinae, several genera had species occurring in one or the other strata (*Pheidole, Solenopsis, Carebara, Lophomyrmex, Crematogaster*), with several remarkable genera exclusive to the subterranean strata (*Acromyrmex*, *Hylomyrma*, *Octostruma*, and *Epelysidris*). In the Dorylinae, *Labidus* was found in both strata, whereas *Dorylus*, *Aenictus*, and *Acanthostichus* were exclusive to subterranean baits.

In total, 12–28% of the species per site were significantly more common at subterranean baits than at aboveground baits below ground (considering only species with more than four occurrences, Figure 5). Except for *Solenopsis* sp. 4 in NPF, all of these species were rare or entirely absent from pitfalls and Winkler samples and thus may be considered truly hypogaeic. Most of these species were Myrmicinae, with only one Ponerinae and three Formicinae species (Table S3). In turn, 60–76% of the species were more common above ground, and 5–24% occurred in both strata in similar frequencies.

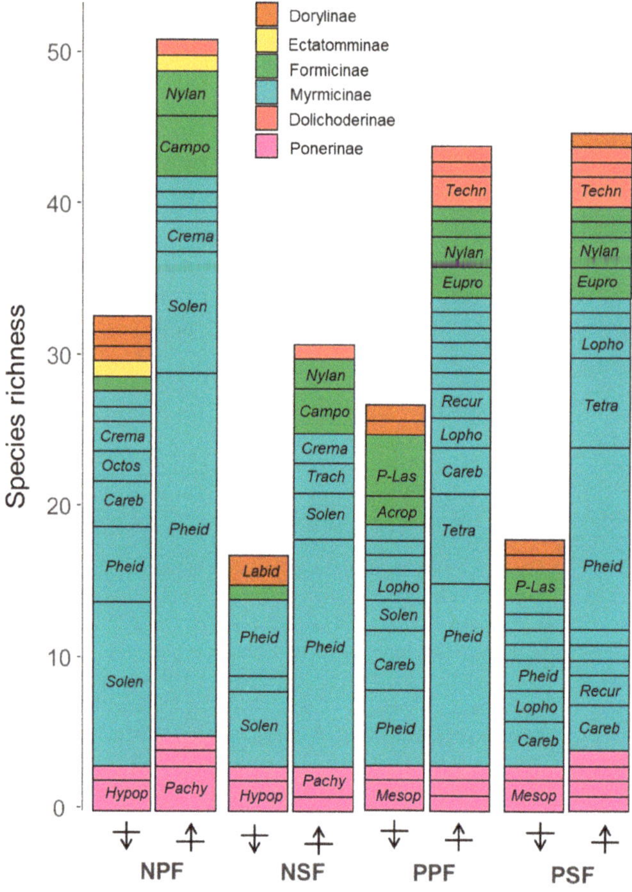

Figure 4. Species richness per subfamily and site for both subterranean and above-ground baiting. Each rectangle represents a different genus, proportional in size to the number of species. Abbreviations are denoted when at least two species were found within the genus. P-Las: *Pseudolasius*; Acrop: *Acropyga*; Lopho: *Lophomyrmex*; Solen: *Solenopsis*; Careb: *Carebara*; Mesop: *Mesoponera*; Techn: *Technomyrmex*; Hypop: *Hypoponera*; Nylan: *Nylanderia*; Eupro: *Eupronolepis*; Recur: *Recurvidris*; Tetra: *Tetramorium*; Pheid: *Pheidole*; Crema: *Crematogaster*; Octos: *Octostruma*; Pachy: *Pachycondyla*; Campo: *Camponotus*; Labid: *Labidus*; Trach: *Trachymyrmex*.

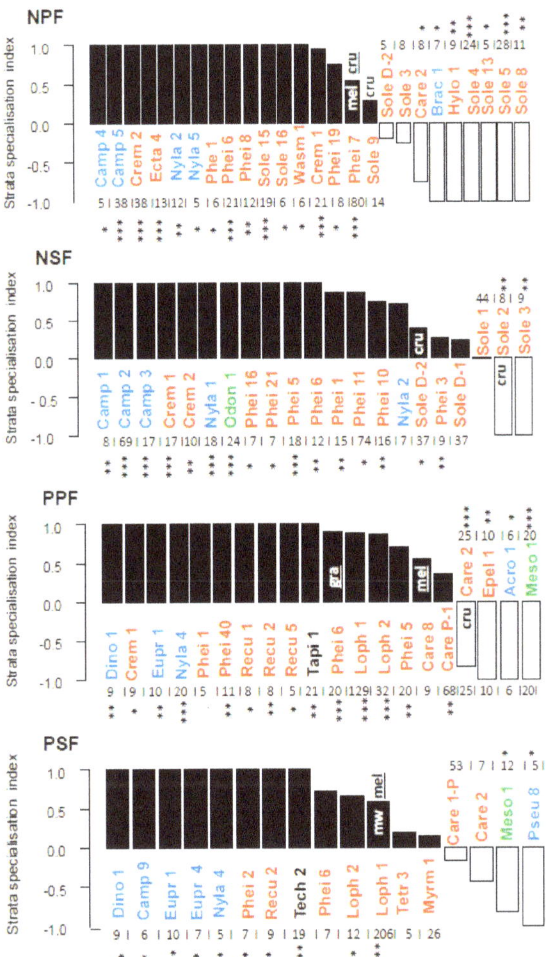

Figure 5. Strata specialisation for each species with five or more occurrences. The values range from 1 (exclusively above-ground) to −1 (exclusively subterranean), based on the number of occurrences. Significant differences between strata (χ^2 tests) are denoted with asterisks (* $p < 0.05$, ** $p < 0.01$, *** $p < 0.001$); numbers indicate the number of occurrences per strata. Food preferences are denoted when significant: *cru*, crushed insects, mel: melezitose, gra: grasshopper, mw: mealworm; if the abbreviation is underlined it means the preference was found above ground. Blue, orange, black, and green respectively represent the subfamilies Formicinae, Myrmicinae, Dolichoderinae, and Ponerinae. Genera are abbreviated as follows: Camp: *Camponotus*, Crem: *Crematogaster*, Ect: *Ectatomma*, Nyl: *Nylanderia*, Phei: *Pheidole*, Sole: *Solenopsis*; Wasm: *Wasmannia*; Care: *Carebara*, Brac: *Brachymyrmex*, Hylo: *Hylomyrma*, Odon: *Odontomachus*; Dino: *Dinomyrmex*; Eupr: *Euprenolepis*; Recu: *Recurvidris*; Tapi: *Tapinoma*; Loph: *Lophomyrmex*; Epel: *Epelysidris*; Acro: *Acropyga*, Meso: *Mesoponera*, Techn: *Technomyrmex*; Tetr: *Tetramorium*; Myrm: *Myrmicaria*. The following morphospecies were identified to species: (neotropical primary forest (NPF) and neotropical secondary forest (NSF)) Camp5: *Camponotus femoratus*; Crem2: *Crematogaster levior*; Phei7: *Pheidole* cf. *nitella*; Ecta4: *Ectatomma tuberculatum*; Camp2: *Camponotus* cf. *melanoticus*; Crem1: *Crematogaster limata*; Odon1: *Odontomachus haematodus*; Phei3: *Pheidole subarmata*; Phei10: *Pheidole zeteki*; Phei11: *Pheidole pugnax*; (paleotropical primary forest (PPF) and paleotropical secondary forest (PSF)) Dino1: *Dinomyrmex gigas*; Loph1: *Lophomyrmex bedoti*; Loph2: *Lophomyrmex longicornis*; Epel1: *Epelysidris brocha*. See also Tables S1 and S2.

3.3. Vertical and Horizontal Turnover

On both continents, the two secondary forests had a lower vertical species turnover than the two primary forests (LM; factor site: $F_3 = 18.7$, $p < 0.0001$) (Figure 3B). It was lowest in PSF, probably because the two numerically dominant species, *Carebara* sp.1-Phg (formerly *Pheidologeton*) and *Lophomyrmex bedoti*, occurred in both strata in high frequencies (Figure 5). In contrast, the most common species on above-ground baits in NPF (*Pheidole* cf. *nitella*, *Crematogaster levior*, *Camponotus femoratus*) and NSF (*Pheidole subarmata*, *Pheidole pugnax*, *Camponotus* cf. *melanoticus*) were all strata specialists, being significantly more common above ground.

Within the same stratum, horizontal turnover between grid points of the same 4×4 plot ranged from 65% to 90% (Figure 3C). It was significantly lower above ground ($F_1 = 311.3$, $p < 0.0001$). For both strata, horizontal turnover differed between sites ($F_3 = 40.9$, $p < 0.0001$); this effect was particularly high below ground (site:strata: $F_3 = 7.3$, $p < 0.0001$) (Figure 3C).

3.4. Dietary Characteristics for Species and Whole Communities

In both neotropical sites, 'crushed insects' was the most preferred bait with three subterranean and two above-ground species, among which one also preferred melezitose (Figure 5). Melezitose was preferred by one *Pheidole* species in NPF, one *Carebara* species in PPF and by *Lophomyrmex bedoti* in PSF. Concerning the large prey items, mealworms were preferred in subterranean baits only by *Lophomyrmex bedoti* in PSF, and grasshoppers (above ground) were only preferred by *Pheidole* sp. 6 in PPF. The different above-ground bait types often yielded additional species in above-ground baits, but this was not always the case for the hypogaeic baits (Figure S2). In general, food specialisation was higher below-ground than above-ground. This was evident from species-specific *fs* values calculated for all species with at least five occurrences above or below ground (t-test: $t_{41.8} = 2.42$, $p = 0.020$, Figure 3D), but also when we only considered species found in both strata (paired t-test: $t_{11} = 3.30$, $p = 0.0071$, Figure 3E). Moreover, below-ground communities were more specialised for all sites except NSF, as indicated by the H_2' index. For NPF, PPF, and PSF, it ranged from 0.070 to 0.092 above ground and from 0.19 to 0.32 below ground, whereas these values were 0.064 and 0.060 for NSF, respectively.

4. Discussion

Although less diverse than the three other sampling methods used in this study, subterranean baiting provided a remarkable ant community in itself (see Text S1 for a comparison of all sampling methods). In each site, 12–20% of the species were significantly more common below ground or even confined to this stratum (Figure 3). For these species, we were able to assess food preferences and to demonstrate trophic differences depending on whether they were foraging above ground or underground.

4.1. Species Richness in Above-Ground and Subterranean Baits

Richness was always lower below ground for all four sites. This aspect was consistent at the grid point level (α diversity), the plot level (data not shown), and the site level (Chao 2 estimators, Figure 2). However, these results are in accord with only one of the few studies [30] that have performed above- and below-ground baiting. Only in the exceptionally rich NPF, Chao 2 estimates were higher for subterranean than for above-ground baits. This would be more consistent with a previous study, which found that depending on the habitat, the hypogaeic species richness could equal that of above-ground baits [31]. From a taxonomical perspective, all subterranean ant studies are useful in providing valuable geographical and methodological information on this still mostly-unknown stratum. However, in terms of comparing ecological aspects and stratification, we believe that having subterranean baits over a 24-h cycle-compared to baited cards only presented during day time—neglects the numbers of above-ground nocturnal ants which could also come to baits [21]. This could potentially explain why these studies [30,31] found lower species richness differences between above-ground and hypogaeic ant species. One study [32] found that honey baits attracted fewer species below ground, but sardine baits attracted similar amounts below and above ground. Our two comparable baits would be melezitose and crushed insects, as sources of carbohydrates and protein, respectively, for which we consistently found lower occurrences in the subterranean strata.

4.2. Taxonomic Composition above and below Ground

Subfamilies and genera differed in their occupancy of the two strata. For instance, Dolichoderinae were never found below ground in any site, although they were present on the above-ground baits. To our knowledge, this is consistent with only one study [30]. In other studies, mainly from South and Central America, *Linepithema*, *Azteca*, and *Dorymyrmex* were uncommon but occasionally present in subterranean baits [31–35]. In the Ponerinae, *Hypoponera* was only sampled in subterranean baits, although only half of the mentioned studies found this rather common genus. To our knowledge, *Parvaponera* has never been sampled using hypogaeic baiting. Within the Dorylinae, an interesting discovery was *Dorylus*, which has never been baited except in traditional oil soil soaking [14,36,37]; all other Dorylinae genera were reported previously in at least one other study. Among the Myrmicinae, several genera have never been reported before from subterranean baits, such as *Acromyrmex*, *Hylomyrma*, and *Epelysidris*. This is also the only subfamily in which multiple species (13) occurred often enough in both strata to analyse dietary preferences and specialisation. In the Formicinae, *Brachymyrmex* was regularly found in subterranean baits, except for in two studies [30,31]. *Nylanderia* was found frequently in only one other study [34]. To our knowledge, two other genera common in our subterranean baits, *Acropyga* and *Pseudolasius*, have never been reported from subterranean baits before. Finally, the genus *Camponotus* was absent from our subterranean samples, as expected from their ecology. Other studies found *Camponotus* below ground [30,33,35], but the rare occurrence of these and other usually above-ground ants could also be due to contamination [10,34]. For instance, poor compaction of the soil covering the subterranean bait may facilitate access for above-ground fauna (MH pers. obs.). Although our subterranean sampling was thorough (1024 baits per site), certain subfamilies which we were expecting [6,38], such as the Amblyoponinae (*Prionopelta*) and Leptanillinae (*Leptanilla*), even when present in pitfalls and Winkler samples in NSF, were absent from our subterranean samples. Potentially, the baits used were not potent enough in smell (compared to sardine baits) or were simply not adapted to these species' diets (geophilomorph centipedes [6]). Note that although some species of *Pseudolasius*, *Gnamptogenys* and *Acromyrmex* were recorded only in subterranean baits here, these genera are generally not hypogaeic.

4.3. Vertical and Horizontal Turnover

Although more species were above-ground specialists than subterranean specialists, 12–20% of all species found at baits were subterranean specialists, which is similar to a recent study [34]. Thus, subterranean ants represent a community of their own. In addition to the higher horizontal turnover, both primary forests also had a higher vertical species turnover, i.e., more species were unique to each stratum. Moreover, both primary forests had more hypogaeic specialist species (with at least five occurrences) than the secondary forests. Above ground, grid points shared more species than below ground (lower horizontal species turnover), which confirms an earlier study reporting high local turnover among subterranean ants [39]. Possibly, hypogaeic species are more limited in foraging distance or in the ways they detect food resources [1]. This might explain a finer-grained community, and thus a higher horizontal species turnover below ground. Furthermore, in both strata and tropical regions, horizontal species turnover was lowest in the secondary forests. This scales up to a lower richness of subterranean ants on a site level for secondary forests, as suggested by Chao 2 estimates.

The higher diversity in primary forests suggests that hypogaeic ants, like other cryptobiotic species, are particularly vulnerable to disturbance [19]. This explains their lower species richness, but also their lower spatial turnover (horizontal and vertical) in secondary forests. This effect was consistent across the two tropical regions, although it is notable that the neotropics were more diverse in our study, and at the site level the two paleotropical sites were similar in species richness. However, considerably more studies are required before any global assertions can be confirmed statistically.

4.4. Dietary Preferences and Food Specialisation

Subterranean communities were more specialised than above-ground assemblages, demonstrating that they show a higher degree of specificity in what they forage for. Notably, crushed insects were the most preferred resource in both tropical regions and both strata. This was even more pronounced below ground, where crushed insect was the only preferred bait. This makes sense as nitrogen is among the most limited resources in tropical ecosystems [40]. Interestingly, *Pheidole* sp. 7 (cf. *nitella*) in NPF preferred crushed insects in the above-ground strata and melezitose in the subterranean strata; in PSF *Lophomyrmex* sp.1 (*L. bedotti*) preferred melezitose above ground and mealworms in subterranean baits. This supports the idea that certain species may display niche variation depending on the ecological context [2,41]. Based on the ecology of certain subterranean species, we were expecting melezitose to be preferred by species which tend subterranean mealybugs, such as *Pseudolasius* and *Acropyga* [42]. However, it would make also sense that these species are not limited in carbohydrates as they tend coccids in their own nests and rather go foraging for complementary resources. Interestingly, some unexpected species were found foraging melezitose, such as *Labidus* sp. 2, an army ant that had two out of three occurrences on melezitose, and *Mesoponera* sp. 1, a ponerine, was found eight times out of 33 on melezitose. Based on their taxonomy and morphology, these species would be considered specialised predators [43], but their foraging of melezitose suggests otherwise. This calls into question the categorization of trophic traits based on morphological criteria [44]. Concerning predation, more ants were found at mealworms than at grasshoppers, but in most cases this involved species which seemed to be more opportunistic than specialised predators (*Pheidole* and *Lophomyrmex*). We believe that the partial immobilization of the prey in the subterranean Eppendorf and baited boxes did not always prevent smaller opportunist ants from foraging them (M.H pers. obs.). This, however, does not explain why no ants were found to prefer termites, although termites should be an important prey for hypogaeic tropical ants [45].

5. Conclusions and Outlook

Our study showed that subterranean baits provide a reliable means of investigating hypogaeic and other ant species, which are less frequently (or not at all) sampled by pitfalls, above-ground baits, and Winkler sampling. By parallel sampling with multiple baits above and below ground, which has rarely been done before, we assessed strata and trophic specialisation in the subterranean ants, and showed that species were more specialised below ground. Moreover, species widely considered to be specialist predators were actually more generalist. This highlights the fact that the ecology of subterranean ants is still largely unknown and awaits further study, particularly in terms of foraging ecology. For example, in a habitat where smell should be a prevalent feature of food detection (instead of eyesight), we were not expecting a non-odorous sugary bait to be as attractive as strongly smelling baits such as decaying insects. More research on vertical stratification within the soil is urgently needed in order to understand underground food webs especially since subterranean habitats are among the highly threatened, but least studied ecosystems [46].

On a community level, we showed that species turnover between above-ground and subterranean strata was considerably higher in primary forests, regardless of the tropical region. Together with our data on horizontal species turnover, this tentatively suggests that the subterranean ant fauna community may vary more between primary and secondary forests than the above ground community. Hence, subterranean ant communities might be more sensitive to the ecological state of a forest. Especially given the lower time effort needed, subterranean baits should hence become part of standard sampling protocols that aim for complete inventories. A further research avenue concerns temporal asynchrony within this little-known community, i.e., whether circadian rhythms affect ant communities in a stratum with no light.

Supplementary Materials: The following materials are available online at https://www.mdpi.com/1424-2818/13/2/53/s1, Figure S1: The four different sampling methods used; Figure S2: Venn Diagrams comparing the ant species attracted to the four different baits. Table S1: Ant species from NPF and NSF; Table S2: Ant species from PPF and PSF; Table S3: Hypogaeic species. Text S1: Comparing Sampling Methods.

Author Contributions: Conceptualization and methodology: F.M. and M.H.; field work and species identification: M.H.; statistical analysis and writing: M.H. and F.M.; funding acquisition: F.M. All authors have read and agreed to the published version of the manuscript.

Funding: This research was funded by the German Research Foundation (DFG), grant numbers ME 3842/1-1 and 3842/6-1.

Institutional Review Board Statement: Not applicable.

Informed Consent Statement: Not applicable.

Data Availability Statement: All data are available from the authors upon request.

Acknowledgments: We thank the Sabah Biodiversity Council (SaBC), Danum Valley Management Committee (DVMC), and Southeast Asian Rainforest Research Programme (SEARRP) for research permissions in Danum and Malua. In particular, Glen Reynolds (Danum) and our collaborator Arthur Y.C. Chung (FRC, Sandakan) were of great help. In addition to this, thanks to the great Danum and Malua staff for making our stay a pleasurable experience. Furthermore, we thank Jérôme Chave for permission to work at Les Nouragues, and Jérôme Orivel for help in Kourou. Thanks also to Heike Stypa for the whole field work logistics and Mona-Isabel Schmitt, Johanna Arndt, and Eric Schneider for help in the field, and to Jack Longino and Bonnie Blaimer for help with ant identification. This research was made possible by the grants ME 3842/1-1 and ME 3842/6-1 of the Deutsche Forschungsgemeinschaft (DFG).

Conflicts of Interest: The authors declare no conflict of interest.

References

1. Berghoff, S.; Weissflog, A.; Linsenmair, K.E.; Hashim, R.; Maschwitz, U. Foraging of a hypogaeic army ant: A long neglected majority. *Insectes Sociaux* **2002**, *49*, 133–141. [CrossRef]
2. Houadria, M.; Menzel, F. Temporal and dietary niche is context-dependent in tropical ants. *Ecol. Entomol.* **2020**, *45*. [CrossRef]
3. Tanaka, H.O.; Yamane, S.; Itioka, T. Within-tree distribution of nest sites and foraging areas of ants on canopy trees in a tropical rainforest in Borneo. *Popul. Ecol.* **2010**, *52*, 147–157. [CrossRef]
4. Floren, A.; Biun, A.; Linsenmair, K.E. Arboreal ants as key predators in tropical lowland rainforest trees. *Oecologia* **2002**, *131*, 137–144. [CrossRef] [PubMed]
5. Bruhl, C.A.; Gunsalam, G.; Linsenmair, K.E. Stratification of ants (Hymenoptera, Formicidae) in a primary rain forest in Sabah, Borneo. *J. Trop. Ecol.* **1998**, *14*, 285–297. [CrossRef]
6. Wong, M.K.L.; Guénard, B. Subterranean ants: Summary and perspectives on field sampling methods, with notes on diversity and ecology (Hymenoptera: Formicidae). *Myrmecol. News* **2017**, *25*, 1–16.
7. Ryder Wilkie, K.T.; Mertl, A.L.; Traniello, J.F.A. Biodiversity below ground: Probing the subterranean ant fauna of Amazonia. *Naturwissenschaften* **2007**, *94*, 725–731. [CrossRef]
8. Longino, J.T.; Colwell, R.K. Biodiversity assessment using structured inventory: Capturing the ant fauna of a tropical rain forest. *Ecol. Appl.* **1997**, *7*, 1263–1277. [CrossRef]
9. Agosti, D.; Alonso, L. The ALL Protocol. In *Ants: Standard Methods for Measuring and Monitoring Biodiversity*; Agosti, D., Majer, J., Alonso, E., Schultz, T., Eds.; Smithsonian Institution Press: Washington, DC, USA, 2000; p. 280, ISBN 1-56098-858-4.
10. Andersen, A.N.; Brault, A. Exploring a new biodiversity frontier: Subterranean ants in northern Australia. *Biodivers. Conserv.* **2010**, *19*, 2741–2750. [CrossRef]
11. Agosti, D.; Majer, J.D.; Alonso, L.E.; Schultz, T.R. *Standard Methods for Measuring and Monitoring Biodiversity*, 2000th ed.; Duke, J., Ed.; Smithsonian Institution Press: Washington, DC, USA, 2000; Volume 233, ISBN 1560988584.
12. Rabeling, C.; Brown, J.M.; Verhaagh, M. Newly discovered sister lineage sheds light on early ant evolution. *Proc. Natl. Acad. Sci. USA* **2008**, *105*, 14913–14917. [CrossRef]
13. Brown, B.; Feener, D.H.J. Parasitic phorid flies (diptera: Phoridae) associated with army ants (hymenoptera: Formicidae: Ecitoninae, dorylinae) and their conservation biology. *Biotropica* **1998**, *30*, 482–487. [CrossRef]

14. Berghoff, S.; Maschwitz, U.; Linsenmair, K.E. Influence of the hypogaeic army ant *Dorylus (Dichthadia) laevigatus* on tropical arthropod Communities. *Int. Assoc. Ecol.* **2003**, *135*, 149–157. [CrossRef] [PubMed]
15. Houadria, M.; Salas-Lopez, A.; Orivel, J.; Blüthgen, N.; Menzel, F. Dietary and temporal niche differentiation in tropical ants-Can they explain local ant coexistence? *Biotropica* **2015**, *47*, 208–217. [CrossRef]
16. Fowler, H.G.; Delabie, J.H.C. Resource partitioning among epigaeic and hypogaeic ants (Hymenoptera: Formicidae) of a Brazilian cocoa plantation. *Ecol. Austral.* **1995**, *5*, 117–124.
17. Díaz, J.A.; Cabezas-Díaz, S. Seasonal variation in the contribution of different behavioural mechanisms to lizard thermoregulation. *Funct. Ecol.* **2004**, *18*, 867–875. [CrossRef]
18. Yamaguchi, T.; Hasegawa, M. An experiment on ant predation in soil using a new bait trap method. *Ecol. Res.* **1996**, *11*, 11–16. [CrossRef]
19. Andersen, A.N.; Penman, T.D.; Debas, N.; Houadria, M. Ant community responses to experimental fire and logging in a eucalypt forest of south-eastern Australia. *For. Ecol. Manag.* **2009**, *258*, 188–197. [CrossRef]
20. Houadria, M.; Menzel, F. What determines the importance of a species for ecosystem processes? Insights from tropical ant assemblages. *Oecologia* **2017**, *184*. [CrossRef]
21. Houadria, M.; Blüthgen, N.; Salas-Lopez, A.; Schmitt, M.I.; Arndt, J.; Schneider, E.; Orivel, J.; Menzel, F. The relation between circadian asynchrony, functional redundancy, and trophic performance in tropical ant communities. *Ecology* **2016**, *97*, 225–235. [CrossRef]
22. Grevé, M.E.; Houadria, M.; Andersen, A.N.; Menzel, F. Niche differentiation in rainforest ant communities across three continents. *Ecol. Evol.* **2019**, 1–15. [CrossRef]
23. Eguchi, K.; Bui, T.V. Exploration of subterranean ant fauna by underground bait-trapping—A case study in Binh Chau-Phuoc Buu Nature Reserve, Vietnam (Insecta: Hymenoptera: Formicidae). *ARI* **2009**, *32*, 20–24.
24. Bolton, B. *Identification Guide to the Ant Genera of the World*; Harvard University Press: Cambridge, MA, USA, 1997; ISBN 0-674-44280-6.
25. Schmidt, C.; Shattuck, S. The higher classification of the ant subfamily Ponerinae (Hymenoptera: Formicidae), with a review of ponerine ecology and behavior. *Zootaxa* **2014**, *3817*, 1–242. [CrossRef] [PubMed]
26. Fayle, T.M. Key to the Workers of the 100 Ant Genera and 12 Ant Subfamilies of Borneo in English and Malay. Available online: http://www.tomfayle.com/Key%20to%20the%20ant%20genera%20of%20Borneo%20v1%20(English-Malay).pdf (accessed on 21 October 2020).
27. Rigato, F. Revision of the myrmicine ant genus Lophomyrmex, with a review of its taxonomic position (Hymenoptera: Formicidae). *Syst. Entomol.* **1994**, *19*, 47–60. [CrossRef]
28. Colwell, R.K. *EstimateS: Statistical Estimation of Species Richness and Shared Species from Samples, Version 9*; User's Guide and application; Department of Ecology & Evolutionary Biology, University of Connecticut: Storrs, CT, USA, 2013; Available online: http://purl.oclc.org/estimates (accessed on 1 April 2020).
29. Blüthgen, N.; Menzel, F.; Blüthgen, N. Measuring specialization in species interaction networks. *BMC Ecol.* **2006**, *67*, 9.
30. Ryder Wilkie, K.T.; Mertl, A.L.; Traniello, J.F.A. Species diversity and distribution patterns of the ants of Amazonian ecuador. *PLoS ONE* **2010**, *5*. [CrossRef]
31. Schmidt, F.A.; Diehl, E. What is the effect of soil use on ant communities? *Neotrop. Entomol.* **2008**, *37*, 381–388. [CrossRef]
32. De Souza, D.R.; Stingel, E.; De Almeida, L.C.; Lazarini, M.A.; Munhae, C.D.B.; Bueno, O.C.; Archangelo, C.R.; Morini, M.S.D.C. Field methods for the study of ants in sugarcane plantations in Southeastern Brazil. *Sci. Agric.* **2010**, *67*, 651–657. [CrossRef]
33. Schmidt, F.A.; Solar, R.R.C. Hypogaeic pitfall traps: Methodological advances and remarks to improve the sampling of a hidden ant fauna. *Insectes Sociaux* **2010**, *57*, 261–266. [CrossRef]
34. Torres, M.T.; Souza, J.L.P.; Baccaro, F.B. Distribution of epigeic and hypogeic ants (Hymenoptera: Formicidae) in ombrophilous forests in the Brazilian Amazon. *Sociobiology* **2020**, *67*, 186–200. [CrossRef]
35. Pacheco, R.; Vasconcelos, H.L. Subterranean pitfall traps: Is it worth including them in your ant sampling protocol? *Psyche (London)* **2012**, *2012*, 20–23. [CrossRef]
36. Sheikh, A.H.; Ganaie, G.A.; Thomas, M.; Bhandari, R.; Rather, Y.A. Ant pitfall trap sampling: An overview. *J. Entomol. Res.* **2018**, *42*, 421–436. [CrossRef]
37. Weissflog, A.; Sternheim, E.; Dorow, W.H.O.; Berghoff, S.; Maschwitz, U. How to study subterranean army ants: A novel method for locating and monitoring field populations of the South East Asian army ant Dorylus (Dichthadia) laevigatus Smith, 1857 (Formicidae, Dorylinae) with observations on their ecology. *Insectes Sociaux* **2000**, *47*, 317–324. [CrossRef]
38. Wong, M.K.L.; Guénard, B. *Leptanilla hypodracos* sp. n., a new species of the cryptic ant genus *Leptanilla* (Hymenoptera, Formicidae) from Singapore, with new distribution data and an updated key to Oriental *Leptanilla* species. *Zookeys* **2016**, *551*, 129–144. [CrossRef]
39. Jacquemin, J.; Drouet, T.; Delsinne, T.; Roisin, Y.; Leponce, M. Soil properties only weakly affect subterranean ant distribution at small spatial scales. *Appl. Soil Ecol.* **2012**, *62*, 163–169. [CrossRef]
40. Davidson, D.W. Resource discovery versus resource domination in ants: A functional mechanism for breaking the trade-off. *Ecol. Entomol.* **1998**, 484–490. [CrossRef]
41. Wong, M.K.L.; Guénard, B.; Lewis, O.T. Trait-based ecology of terrestrial arthropods. *Biol. Rev.* **2019**, *94*, 999–1022. [CrossRef]
42. Delabie, J.H.C. Trophobiosis between Formicidae and Hemiptera (Sternorrhyncha and Auchenorrhyncha): An Overview. *Neotrop. Entomol.* **2001**, *30*, 501–516. [CrossRef]

43. Rogerio, R.S.; Brandao, C.R.F. Morphological patterns and community organization in leaf-litter ant assemblages. *Ecol. Monogr.* **2010**, *80*, 107–124.
44. Gibb, H.; Stoklosa, J.; Warton, D.I.; Brown, A.M.; Andrew, N.R.; Cunningham, S.A. Does morphology predict trophic position and habitat use of ant species and assemblages? *Oecologia* **2015**, *177*, 519–531. [CrossRef]
45. Tuma, J.; Eggleton, P.; Fayle, T.M. Ant-termite interactions: An important but under-explored ecological linkage. *Biol. Rev.* **2019**, *95*. [CrossRef]
46. Mammola, S.; Cardoso, P.; Culver, D.C.; Deharveng, L.; Ferreira, R.L.; Fišer, C.; Galassi, D.M.P.; Griebler, C.; Halse, S.; Humphreys, W.F.; et al. Scientists' warning on the conservation of subterranean ecosystems. *Bioscience* **2019**, *69*, 641–650. [CrossRef]

 *diversity*

Article

Do Dominant Ants Affect Secondary Productivity, Behavior and Diversity in a Guild of Woodland Ants?

Jean-Philippe Lessard [1,2,*], **Katharine L. Stuble** [1,3] **and Nathan J. Sanders** [1,4]

[1] Department of Ecology and Evolutionary Biology, University of Tennessee, Knoxville, TN 37996, USA;
    kstuble@holdenfg.org (K.L.S.); njsander@umich.edu (N.J.S.)
[2] Department of Biology, Concordia University, Montreal, QC H4B-1R6, Canada
[3] The Holden Arboretum, 9500 Sperry Rd, Kirtland, OH 44094, USA
[4] Department of Ecology and Evolutionary Biology, University of Michigan, Ann Arbor, MI 48109, USA
* Correspondence: jp.lessard@concordia.ca; Tel.: +1-514-848-2424 (ext. 5184)

Received: 23 October 2020; Accepted: 29 November 2020; Published: 2 December 2020

Abstract: The degree to which competition by dominant species shapes ecological communities remains a largely unresolved debate. In ants, unimodal dominance–richness relationships are common and suggest that dominant species, when very abundant, competitively exclude non-dominant species. However, few studies have investigated the underlying mechanisms by which dominant ants might affect coexistence and the maintenance of species richness. In this study, we first examined the relationship between the richness of non-dominant ant species and the abundance of a dominant ant species, *Formica subsericea*, among forest ant assemblages in the eastern US. This relationship was hump-shaped or not significant depending on the inclusion or exclusion of an influential observation. Moreover, we found only limited evidence that *F. subsericea* negatively affects the productivity or behavior of non-dominant ant species. For example, at the colony-level, the size and productivity of colonies of non-dominant ant species were not different when they were in close proximity to dominant ant nests than when they were away and, in fact, was associated with increased productivity in one species. Additionally, the number of foraging workers of only one non-dominant ant species was lower at food sources near than far from dominant *F. subsericea* nests, while the number of foragers of other species was not negatively affected. However, foraging activity of the non-dominant ant species was greater at night when *F. subsericea* was inactive, suggesting a potential mechanism by which some non-dominant species avoid interactions with competitively superior species. Gaining a mechanistic understanding of how patterns of community structure arise requires linking processes from colonies to communities. Our study suggests the negative effects of dominant ant species on non-dominant species may be offset by mechanisms promoting coexistence.

Keywords: behavioral interactions; coexistence; co-occurrence; competitive exclusion; dominance; Formicidae; scale

---

1. Introduction

Dominant species are those species that excel at exploiting and sequestering resources [1] thereby affecting the behavior and population dynamics of other species and potentially the structure of communities. Janzen [2] and Connell [3] famously argued that if populations of dominant species were not kept in check by predators or parasites, they would become overabundant and exclude other species, leading to a decrease in species richness. In relation to this hypothesis, several studies have documented a unimodal relationship between the richness of species in a local community and the abundance of dominant species [4,5]. The most common explanation for this unimodal relationship is that at low levels of dominance, the abundance of both dominant and non-dominant ants increases as environmental stress decreases. Then, as dominant species become more abundant, they competitively exclude non-dominant species leading to the descending portion of the hump-shaped curve [4,5]. An alternative explanation is that the abundance of dominant and non-dominant species peaks at a different point along the abiotic stress gradient. One approach for teasing apart these alternative explanations is to test predictions regarding the influence of competition with a dominant species across levels of organization, from individuals to entire communities.

In animals, dominant species are those with high abundance and biomass, which may be attained through exploitative and interference competition. Dominance can be attained via exploitative competition or interference competition. In ants, dominant species often display aggressive behavior enabling them to defend territories and monopolize food resources. Aside from the unimodal relationship between species richness and the abundance of dominant species [4,5], evidence that dominant species have community-wide effects on other species includes the effects of competitively dominant exotic species on the diversity [6–9], spatial arrangement [10] and phylogenetic structure of native ant communities [11]. Additional evidence stems from the effects dominant ants have on patterns of co-occurrence in arboreal assemblages of the tropics [12–17] and the dominance-species richness relationship [4,5,18,19]. However, a recent global study suggests that exotic dominant ants are more likely to exert community-wide effects than are native dominant ants [20]. Moreover, experimental removals of dominant species have yielded mixed results [21–23]. In sum, whether or not native dominant ants affect other members of the community deserves further exploration.

Several studies have documented the effects of dominant ant species on the behavior, resource use and fitness of non-dominant ant colonies [24–28], while others have documented their effects on community structure [4,18,29–33]. Examining the outcome of interspecific interactions across organization levels is key to elucidating the mechanism by which competition might shape community structure [34]. However, few studies [22,23,33] have linked the effects of a single dominant ant species on individual colonies, populations or communities of competitively inferior ants. In ants, community-level effects, relate to the richness or other multi-species metrics of diversity (including composition). Population-level effects relate to the abundance of a species at a site (i.e., number of colonies or workers). Individual effects relate to characteristics of the colonies of a given species. Worker-level effects relate to the behavior of individual ant workers when confronted with workers of another species at a given food source.

Community-level effects of dominant ants on non-dominant ant species are well documented but much less is known regarding the individual-level effects (i.e., colony-level effects) of dominant ants on non-dominant ants. It is well established that dominant ants often interfere with resource exploitation by non-dominant ant workers via negative behavioral effects [8,29,35–38]. Moreover, it is often assumed that by interfering with the foraging activities of non-dominant species, dominant ants also affect rates of resource acquisition and sequestration at the individual colony-level (but see [27]). If that were the case, the effect of dominant ants should be observed at the individual-level by decreasing the productivity, size and/or fitness of colonies of non-dominant ants [26,31]. However, whether the negative effect of dominant species on colony success of non-dominant ants is the rule rather than the exception, and whether colony-level effects always translate into community-level effects, is unclear [39].

Effects of dominant species in the genus *Formica* (*rufa* group) on non-dominant ant species have been well documented in a series of seminal studies in ant assemblages in Finland [26–28,32,33]. *Formica* ants in the rufa group form large colonies and are aggressive but are uncommon in temperate forests of North America. However, *Formica* ants in the *fusca* group, which form smaller colonies and are less aggressive, are ubiquitous. Here, we examine how one of the most numerically and behaviorally dominant species in a low-elevation temperate forest in the eastern US, *Formica subsericea* (*fusca* group) Say, affects non-dominant ant species. We first investigate community-level effects by testing whether the relationship between the richness of non-dominant species and the abundance of *F. subsericea* best fits a unimodal or linear model [4,5,18]. Then, to elucidate whether community-level effects of *F. subsericea* concords with individual-level and behavioral effects, we asked the following questions: (i) the colony density and/or abundance of non-dominant ant species decreases with increasing abundance of *F. subsericea*, (ii) the colony size and/or productivity of colonies of non-dominant ant species is lower near than far from the nest of *F. subsericea*, (iii) resource use by non-dominant ants is lower near than far from *F. subsericea* nests and (iv) non-dominant ant foraging activity is lower when *F. subsericea* are active than when inactive.

2. Materials and Methods

2.1. Study Site

We conducted this study in a mid-elevation (~740 m elevation) forest in the southern Appalachian mountain range, within Great Smoky Mountains National Park (35°38′41″ N, 83°35′06″ W) in June–August 2008 and 2009. Dominant overstory vegetation included *Liriodendron tulipifera*, *Halesia tetraptera var. monticola*, *Tilia americana var. heterophylla*, *Acer rubrum*, *Magnolia acuminata* and *Fraxinus americana*. Dominant understory vegetation included *Acer pensylvanicum*, *Calycanthus floridus* and *Rhododendron maximum*. *Formica subsericea* Say is common in open and partially open woodlands and forest ecotones of eastern North American temperate forests [40]. Nests of *F. subsericea* are usually deep and the nest entrance is often in a dead log or adjacent to large stones. In the absence of a log or rock, *F. subsericea* often forms small mounds covered by leaf-litter around the nest entrance.

In temperate forests of eastern North America, *F. subsericea* is one of several locally dominant species [29,41]. Although colony size and rates of aggression by *F. subsericea* are low relative to dominant species found in arid and tropical systems. In this particular part of the National Park, the density of *F. susbericea* colonies is higher than anywhere else in the region [42–44] and therefore, its presence likely affects other species. Workers of *F. subsericea* do not maintain exclusive foraging territories (i.e., absolute territories) but do aggressively defend their nest and food. Typically, workers of *F. subsericea* rapidly recruit to food resources, displace other species and monopolize the resource. Workers of *Camponotus pennsylvanicus* sometimes manage to displace *F. subsericea* from food resources but this depends on the distance of the food resource from the *F. subsericea* nest and the size of the *F. subsericea* colony (J.-P. Lessard pers. obs.). However, because *C. pennsylvanicus* was not common at the study plots we worked in, we did not consider it dominant [45]. In our study plots, the mean distance between a focal nest of *F. subsericea* and the nearest *F. subsericea* neighbor nest was 10.25 ± 3.15 m, which roughly corresponds to the average radius of a *F. subsericea* foraging territory in the studied sites (foraging territories are usually asymmetrical in shape).

2.2. Sampling

Though the density of *F. subsericea* is high in this region [42–44], there was considerable among-site variation. We took advantage of this natural variation in the density of *F. subsericea* to examine the potential effects of this species on community and population-level effects, as well as on behavior. Though this is a commonly used approach in studies of dominance on ant community structure [4,5,18,46], we acknowledge that it is correlative only and a long-term removal experiment would provide additional insights. Here, we located 10 sites separated by at least 50 m. Each site was 300 m² and had 12 sampling stations arranged in a 3 × 4 grid and separated by 10 m. At each site we surveyed ants using baits and pitfall traps during peak ant activity (June–July) in both 2008 and 2009. Within these 10 sites we also located 16 nests of *F. subsericea*, which were later used for colony-level tests of dominance effects. To assess the colony-level effects of *F. subsericea* on non-dominant species, we compared behavioral and colony traits near and far from *F. subsericea* nests.

2.2.1. Community and Population-Level Effects

To estimate the relative abundance of ant species in our system, we placed pitfall traps at each of the 12 sampling stations at the 10 sites within 1 meter of the bait station. Pitfall traps are commonly used to estimate the relative abundance of ants [4,5,18,35,47]. We set up pitfall traps 24 h after the last baiting trial. Each pitfall trap (55 mm diameter, 75 mm deep) was partially filled with propylene glycol (low toxicity antifreeze), buried flush with the ground and left in place for 72 h. All ants were counted, identified to species and deposited in NJ Sanders's personal collection.

Here, we used the total number of pitfall traps in which a species occurred (i.e., incidence) as an estimate of its abundance for that site since worker incidence and the number of workers were strongly correlated ($R^2 = 0.95$) [48–50]. Pitfall trap incidence is a more conservative approach to estimate the abundance of ant colonies and has been used in previous studies with temperate forests ant assemblages [48,51].

2.2.2. Individual-Level Effects

By setting up observation grids around *F. subsericea* nests we found that activity levels of *F. subsericea* workers is 3× higher within 1 m of the nest entrance than anywhere else within a 10 m radius (unpublished data). Thus, to test whether *F. subsericea* affected the size and productivity of colonies of non-dominant ants, we collected entire colonies of the two most abundant non-dominant species on baits—*Aphaenogaster rudis* and *Nylanderia faisonensis*—near (≤1 m) and far (5–10 m) from 16 *F. subsericea* nests. Colonies of *A. rudis* in this system are typically monodomous (i.e., occupy only one nest) and monogynous (i.e., only one queen is present in the colony), although in some instances 2–3 queens were found in a colony [52]. Colonies of *N. faisonensis* in our system appeared to all be monodomous but others have found polydomous colonies of that species [53]. All colonies of *N. faisonensis* were monogynous, which is consistent with previous records for that species [53].

Both *A. rudis* and *N. faisonensis* frequently interacted with *F. subsericea* on baits and their colonies are conspicuous and relatively easy to collect. Another non-dominant ant species, *Myrmica punctiventris*, was also abundant on baits and frequently interacted with *F. subsericea* but we were unable to locate enough colonies of *M. punctiventris* for statistical analyses. For each of these species, we attempted to collect one *A. rudis* and one *N. faisonensis* colony near and one colony of each species far from each of the 16 *F. subsericea* nests. Colonies were typically found in dead branches and leaf litter. Dead branches were moved directly into a large plastic container with edges coated in Teflon, while making sure to capture any workers/queens attempting to escape. When in the leaf litter, a shovel was used to collect the leaf litter as well as 15–20 cm of soil underneath it, while making sure to capture any escaping individuals and visually inspecting surrounding leaf litter. We were able to locate 24 *A. rudis* colonies (near = 12, far = 12) and 14 *N. faisonensis* colonies (near = 8, far = 6). Colonies were collected during the last 2 weeks of July 2008. For each colony, we counted the number of brood, workers and queens. Due to the size and quantity of *N. faisonensis* eggs, it was impossible to accurately estimate the number of eggs in the colonies sampled. Colony size was calculated as the total number of workers in a colony and colony productivity was calculated as the proportion of brood to workers in a colony [54,55].

2.2.3. Worker-Level Effects

We placed bait stations near (≤1 m) and far (5–10 m) from 12 focal *F. subsericea* nests. Baiting stations consisted of 4 laminated index cards arranged in a square, with 10 cm between each card. We used cotton balls dipped into a sugar-water solution on two of the baits and cat food on the other two. We operated the baits both between 1:00 p.m. and 5:00 p.m., and, on the same day, between 9:00 p.m. and 1:00 a.m. At night, we used a red-light headlamp to avoid interfering with ant foraging activities [56–58]. We visited baits every hour for three hours. Baiting was not conducted on days with heavy precipitation. At each bait station we tallied (1) the number of workers from the three most common non-dominant ants on baits (i.e., *A. rudis*, *N. faisonensis* and *M. punctiventris*), (2) the total abundance of non-dominant workers (i.e., all ants other than *F. subsericea*) and (3) the total species richness of non-dominant species. For each *F. subsericea* nest, we pooled observations from the 4 baits and all three observational visits to get a single value for each of the parameters listed above both near and far from each *F. subsericea* colony.

2.3. Estimating Dominance

We used baits to quantify the outcome of behavioral interactions and the relative abundance of ant species at food resources, a common technique in studies of ant ecology [28,29,31,35,47,59,60]. Baits consisted of 5 g of cat food (11% protein, 4% fat, 78% water) on a white laminated index cards positioned on the ground. Cat food is a useful resource for baiting for ants because it can be retrieved both in liquid and solid form and contains a mix of protein, lipids and salts. It has been used in several community-level studies with ants to assess behavioral dominance and species richness [4,61,62]. In our study system, all of the ant species that were recorded on sugar-based baits were also recorded on cat food baits. At each of the 10 sites, we placed baits at the 12 sampling stations and recorded the number of workers of each species as well as any interspecific behavioral interactions. We visited each bait station for 1 min every 15 min for 3 h and visited each of the 10 sites once (for a total of 1440 censuses). One of us (JPL) conducted all of the baiting trials on sunny or partially sunny days between 1:00 p.m. and 5:00 p.m., during peak ant activity [41].

To determine which ant species were dominants, previous studies have used a variety of approaches, from mainly qualitative [63] to a mix of dominance metrics [18,29,47,60]. Because different approaches to estimating dominance can yield different results [60] and because dominance in ants has several components, we used three different approaches for estimating dominance: behavioral dominance, numerical dominance and ecological dominance.

To estimate behavioral dominance, we tallied the number of wins and losses in encounters for each species observed at baits. A win consisted of a species attacking another species (i.e., aggressive behavior) causing the losing ant to leave the bait station (i.e., submissive behavior); the ant that left the bait after such an encounter would in turn get a loss. At any given bait, the outcome of the interaction between two given species was counted only once to insure independence among observations. We then used the Colley ranking method [64] to rank each species from most dominant to most submissive as in Reference [47]. The Colley method estimates the dominance hierarchy based on both the proportion of wins for a species out of the total number of interactions for that species, as well as the relative strength of the opponents in inter-specific interactions. Thus, winning an interaction against a dominant species is worth more than winning against a submissive species. The Colley method was designed to rank intercollegiate American football teams and does not require that every species interact with every other species in order to obtain an accurate ranking of the species in the hierarchy.

We also calculated numerical dominance by estimating the abundance of each species in pitfall traps across all 10 sites [65]. We used the total number of pitfall traps in which a species occurred as an estimate of numerical dominance for that species. We estimated ecological dominance by extrapolating total biomass for each ant species based on known body size and abundance. We used a modified version of the scaling equation presented in Reference [66] so that we could estimate individual worker mass from body size data

$$M = (4.7297 \times 10^{-4})\ WL^{3.179.} \tag{1}$$

in this Equation (1), WL is Weber's Length, which is a commonly used proxy for body size and which we derived from published work [67]. We then estimated the average biomass of ant colonies for each species we caught in pitfall traps by multiplying the predicted mass (M) of workers by the by colony size, where colony size was an average extracted from the literature [67]. Finally, to estimate the total biomass of an ant species at our site, we multiplied the predicted colony biomass by the incidence of that species across all pitfall traps (Table 1).

We ranked each species that was recorded at baits and in pitfall traps according to the three dominance indices described above. We determined that *F. subsericea* was the most dominant because it was the only species to be ranked no lower than 3rd in any dominance hierarchy for any metric for all of the dominance indices (Table 2). *Formica subsericea* also came out as the most dominant ant species when, for each species, we summed the rank occupied in each of the three hierarchies

Table 1. Metrics used to estimate behavioral dominance, numerical dominance and ecological dominance.

Species	Worker Count	Abundance	Worker Body Size	Predicted Worker Mass	Colony Size
	No. of Workers	No. of Incidence	Weber's Length (mm)	(g)	No. of Workers
Aphaenogaster fulva Roger	10	3	1.48	1.64	281
Aphaenogaster rudis Enzmann	712	123	1.41	1.41	303
Camponotus americanus Mayr	3	3	2.59	9.74	3560
Camponotus pennsylvanicus (DeGeer)	20	18	2.52	8.93	2222
***Formica subsericea* Say**	**181**	**57**	**2.33**	**6.96**	**8919**
Lasius alienus (Förster)	87	19	1.42	1.44	3000
Myrmica punctiventris Roger	107	36	1.53	1.83	86
Nylanderia faisonensis (Forel)	25	18	0.61	0.10	268
Prenolepis imparis (Say)	8	3	0.91	0.35	3370
Temnothorax longispinosus (Roger)	5	4	0.62	0.10	47

Worker count is the sum of all workers and abundance is the sum of all incidences of a given species recorded across all sites (120 pitfall traps). Body size and colony size data were extracted from Reference [67]. Weber's length of mesosoma is a proxy for body size commonly used in ant studies. It is measured as the length separating the anterior edge of pronotum from the posterior corner of metapleuron. Predicted worker mass was calculated based on worker body size estimates and the equation provided in Reference [66]. The dominant species is indicated in bold.

Table 2. List of species recorded both in pitfall trap and on baits in Great Smoky Mountains National Park.

Species	Behavioral Dominance	Numerical Dominance	Total Predicted Biomass (kg)
	Colley Ranking	No. of Pitfall Traps	No. Pitfall Traps $\times$ Colony Size $\times$ Worker Biomass
Aphaenogaster fulva Roger	0.71 (2)	3 (8)	1.39 (8)
Aphaenogaster rudis Enzmann	0.40 (7)	123 (1)	52.55 (5)
Camponotus americanus Mayr	0.68 (4)	3 (9)	104. 06 (3)
Camponotus pennsylvanicus (DeGeer)	0.93 (1)	18 (5)	357.19 (2)
***Formica subsericea* Say**	**0.70 (3)**	**57 (2)**	**3538.74 (1)**
Lasius alienus (Förster)	0.68 (5)	19 (4)	82.19 (4)
Myrmica punctiventris Roger	0.40 (8)	36 (3)	5.66(6)
Nylanderia faisonensis (Forel)	0.08 (10)	18 (6)	0.47 (9)
Prenolepis imparis (Say)	0.58 (6)	3 (10)	3.54 (7)
Temnothorax longispinosus (Roger)	0.20 (9)	4 (7)	0.02 (10)

Listed are scores for behavioral dominance, numerical dominance and total biomass. For each dominance category, ranks are indicated in parentheses. The most dominant species is indicated in bold.

2.4. Statistical Analyses

We examined the relationship between ant species richness and the incidence of *F. subsericea*. At each site, we estimated the abundance (i.e., incidence) of *F. subsericea* as the number of pitfall traps containing at least one *F. subsericea* worker. Species richness was calculated as the total number of ant species recorded at a site across all pitfall traps (excluding *F. subsericea*). One data point was a highly influential observation, so we conducted this part of the analyses with and without the outlier. Because it has been suggested that the species richness of non-dominants might be negatively related to the abundance of dominant ants only at high levels of dominance [4,5], we considered whether the relationship between the species richness of non-dominants and abundance of *F. subsericea* was best described by either a linear least squares regression or a polynomial regression. We compared the fit of the models by comparing the adjusted r^2 and Akaike information criterion (AIC) values for each fit.

2.4.1. Is the Pooled Abundance of Non-Dominant Species Negatively Related to the Abundance of the Dominant Species?

We estimated the abundance of non-dominant ant species at each of the 10 sites by counting the number of pitfall traps in which each non-dominant ant species was recorded and then pooling the abundance of all non-dominant ants. We then related the abundance of non-dominant species to that of the dominant species using linear regression. We also tested whether the effect of *F. subsericea* on non-dominant ants differed between common and rare species by pooling the incidence of common (*A. rudis*, *N. faisonensis*, *M. punctiventris*) and less-common species (all other species) separately.

2.4.2. Is the Colony Size and/or Productivity of Non-Dominant Ant Species Negatively Influenced by Proximity to a *F. subsericea* Colony?

We used paired *t*-tests to determine whether colony size and productivity of *A. rudis* differed between colonies near and far from a focal *F. subsericea* nests. Because *N. faisonensis* colonies could not always be paired, we used one-sample *t*-tests to assess whether mean colony size and productivity was higher far from *F. subsericea* nests relative to near. The data were not always normally distributed, thus we used both parametric (i.e., one-sample, two-sample or paired *t*-tests) and non-parametric (i.e., Wilcoxon or Wilcoxon signed rank tests) tests for all colony-level analyses.

2.4.3. Is Resource Use by Non-Dominant Ants Negatively Influenced by Proximity to a *F. subsericea* Colony?

We tested whether the abundance and species richness of non-dominant workers at baits depended on distance from the focal nest using a paired *t*-test and a Wilcoxon signed rank test. We performed analyses separately for day and night baiting sessions. We did not combine the distance and time variables into a single model because the observations were paired.

2.4.4. Are Temporal Patterns of Foraging Activity in Non-Dominant Species Are Negatively Related to Those of *F. subsericea*?

It is possible that non-dominant species might take advantage of lower foraging activities of *F. subsericea* at night to exploit resources and avoid interference competition. Thus, we assessed whether foraging activity by non-dominant ants varied between day and night. We estimated differences between diurnal and nocturnal foraging patterns by subtracting the number of workers of non-dominant species recorded during the night from the number recorded during the day. In addition, we estimated temporal variability in the richness of foraging ant species by subtracting the number of species recorded at baits during the night from the number recorded during the day. Depending on the normality of the data, we used one-sample *t*-tests or one-sample Wilcoxon signed rank tests to ask whether the difference between the (1) total abundance of workers and (2) total number of species of non-dominant ants foraging during day and night differed from zero.

3. Results

The incidence of the dominant *F. subsericea* varied from 1 to 8 (of 12) pitfall traps per site. The total abundance of non-dominant species (the sum of incidences across 12 pitfall traps at a given site) varied from 24 to 45 and total richness of non-dominant species varied from 4 to 12 species per site.

3.1. Species Richness and Dominance

When considering all sampled sites, there was only a marginally significant polynomial relationship between the abundance of *F. subsericea* and species richness ($r^2 = 0.53$, $r^2_{adjusted} = 0.39$, $n = 10$, $p = 0.07$, AIC = 17.90 for quadratic fit vs. $r^2 = -0.26$, $r^2_{adjusted} = -0.12$, $n = 10$, $p = 0.95$, AIC = 23.40 for linear fit; Figure 1). The relationship was significant only if one observation was removed ($r^2 = 0.63$, $r^2_{adjusted} = 0.51$, $n = 9$, $p = 0.05$, AIC = 10.36 for quadratic fit vs. $r^2 = 0.25$, $r^2_{adjusted} = 0.15$, $n = 9$, $p = 0.16$, AIC = 14.64 for linear fit). Note that adjusted r^2 values can be negative when the model fits the data poorly [68].

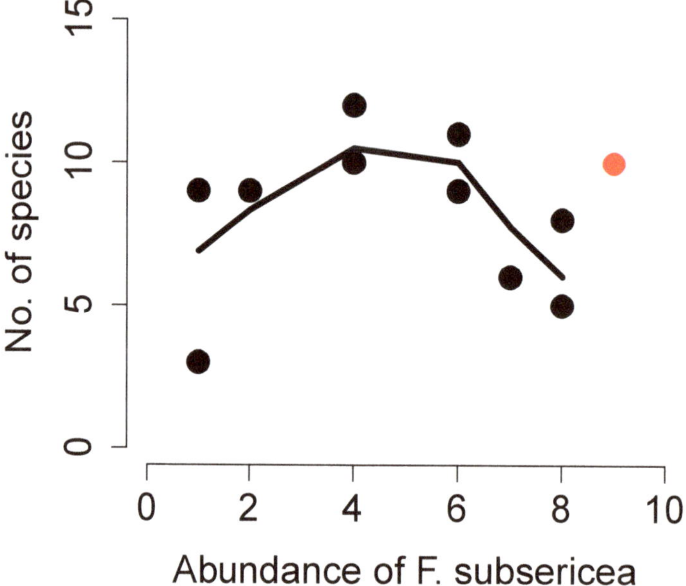

Figure 1. The unimodal relationship between the richness of non-dominant ant species recorded in pitfall traps at a site and the number of pitfall traps in which a dominant *F. subsericea* was recorded. The relationship is statistically significant only if the outlying point in red is excluded.

3.2. Is the Pooled Abundance of Non-Dominant Species Negatively Related to the Abundance of the Dominant Species?

There was no relationship between the abundance of *F. subsericea* and the total abundance (incidences) of non-dominant species ($r^2 = 0.28$, $r^2_{adjusted} = 0.07$, $n = 10$, $p = 0.32$, AIC = 39.77 for quadratic fit vs. $r^2 = 0.04$, $r^2_{adjusted} = -0.07$, $n = 10$, $p = 0.57$, AIC = 39.00 for linear fit). This relationship was not different when the statistical outlier was removed ($r^2 = 0.17$, $r^2_{adjusted} = -0.10$, $n = 9$, $p = 0.57$, AIC = 30.80 for quadratic fit vs. $r^2 = 0.08$, $r^2_{adjusted} = -0.05$, $n = 9$, $p = 0.47$, AIC = 29.76 for linear fit). The relationship between the abundance of non-dominant ants and dominant ants did not depend on whether the non-dominant species were common ($r^2 = 0.30$, $r^2_{adjusted} = 0.10$, $n = 10$, $p = 0.29$, AIC = 28.87 for quadratic fit vs. $r^2 = 0.30$, $r^2_{adjusted} = 0.21$, $n = 10$, $p = 0.10$, AIC = 26.91 for linear fit) or rare ($r^2 = 0.37$,

$r^2_{\text{adjusted}} = 0.19$, $n = 10$, $p = 0.20$, AIC = 27.65 for quadratic fit vs. $r^2 = 0.00$, $r^2_{\text{adjusted}} = -0.12$, $n = 10$, $p = 0.94$, AIC = 30.30 for linear fit).

3.3. Is Colony Size and/or Productivity of Non-Dominant Ant Species Negatively Influenced by Proximity to a Dominant Ant Colony?

Neither colony size (Figure 2a) nor colony productivity (Figure 2b) of *A. rudis* depended on distance from *F. subsericea* nests (Table 3). Likewise, the colony size of *N. faisonensis* did not depend on distance from *F. subsericea* nests (Figure 2a), whereas productivity was, on average, 2× higher near than far from *F. subsericea* nests (Figure 2b), though this result was marginally significant (Table 3).

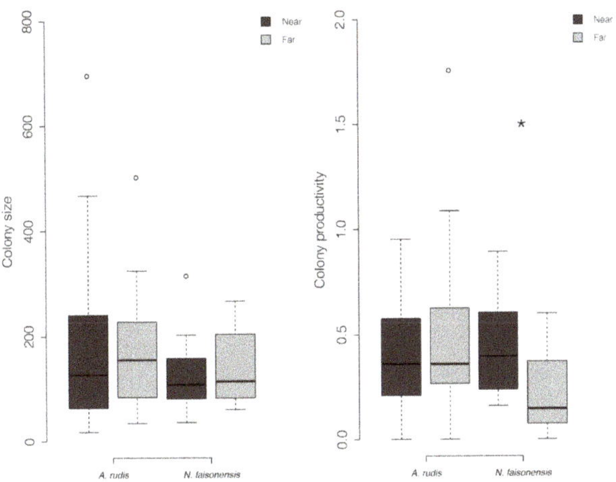

Figure 2. Mean (**a**) colony size (number of workers) and (**b**) colony productivity (number of broods/number of workers) of non-dominant ant species *A. rudis* and *N. faisonensis* near and far from focal nests of dominant *F. subsericea*. Asterisks indicate significant differences ($p < 0.05$).

Table 3. Variation in colony size and productivity of colonies of non-dominant ant species (Ar and Pf) near and far from *F. subsericea* nests.

Response Variable	Taxa	Test	n	Test Statistic	p
Colony size	*A. rudis*	Wilcoxon Rank Sums	10	Z = 0.12	0.91
Colony size	*N. faisonensis*	Two-sample *t*-test	11.53	t-Ratio = −0.21	0.42
Colony productivity	*A. rudis*	Wilcoxon Rank Sums	10	Z = −0.46	0.64
Colony productivity	*N. faisonensis*	Wilcoxon Rank Sums	11.61	Z = −1.74	0.08

3.4. Is Resource Exploitation by Non-Dominant Ants Negatively Influenced by Proximity to a Dominant Ant Colony?

The mean number of *F. subsericea* workers recorded at baits was 3× higher near (≤1 m) than far (≥5 m) from the focal nest (Table 4, Figure 3a). During the day, the number of workers of non-dominant species was 50% lower at baits near than far from focal *F. subsericea* nests but the richness of non-dominant species did not differ between near and far baits (Table 4). There were 4× fewer *A. rudis* workers on baits near than far from *F. subsericea* nests during the day (Table 4, Figure 3a). However, there was no difference in the number of *N. faisonensis* (Table 4, Figure 3a) or *M. punctiventris* (Table 4, Figure 3a) workers on baits near and far from *F. subsericea* nests during the day.

Table 4. Difference in resource use of dominant (Fs) and non-dominant ant species near and far from nests of *F. subsericea*.

Source of Variation	Response Variable	Taxa	Test	Test Statistic	*p*
Distance (day)	Species richness	all	Paired *t*-test	t-Ratio = −1.20	0.13
Distance (day)	Worker abundance	all	Wilcoxon Sign-Rank	Z = 23.00	0.04 *
Distance (day)	Worker abundance	Ar	Wilcoxon Sign-Rank	Z = 27.50	0.01 *
Distance (day)	Worker abundance	Fs	Wilcoxon Sign-Rank	Z = −39.00	<0.0001 **
Distance (day)	Worker abundance	M	Wilcoxon Sign-Rank	Z = −9.00	0.13
Distance (day)	Worker abundance	Nf	Wilcoxon Sign-Rank	Z = −2.00	0.41
Distance (night)	Species richness	all	Paired *t*-test	t-Ratio = 0.00	1.00
Distance (night)	Worker abundance	all	Wilcoxon Sign Rank	Z = −2.00	0.46
Distance (night)	Worker abundance	Ar	Wilcoxon Sign-Rank	Z = −2.00	0.55
Distance (night)	Worker abundance	Fs	Wilcoxon Sign-Rank	NA	NA
Distance (night)	Worker abundance	M	Wilcoxon Sign-Rank	Z = −1.50	0.41
Distance (night)	Worker abundance	Nf	Paired *t*-test	t-Ratio = −2.36	0.02 *

Differences are shown separately for day and night baiting trials. The table shows significant differences for parametric or non-parametric statistical tests, depending on the distribution of the data (* $p < 0.05$, ** $p < 0.01$). Ar = *A. rudis*; Nf = *N. faisonensis*; Mp = *M. punctiventris*.

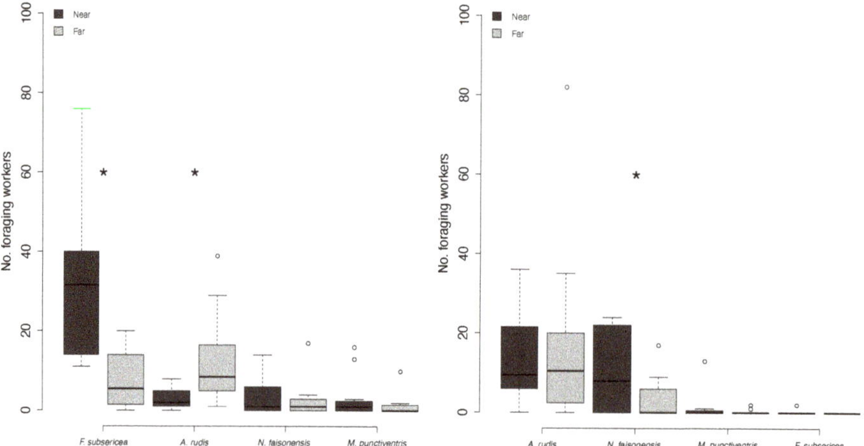

Figure 3. Mean number of workers of *A. rudis*, *M. punctiventris* and *N. faisonensis* recorded on baits near and far from focal nests of dominant *F. subsericea* (**a**) during the day and (**b**) at night. Asterisks indicate significant differences between near and far baits ($p < 0.05$).

In only one instance did we find *F. subsericea* workers foraging at night (i.e., one worker at one bait), suggesting that *F. subsericea* is largely diurnal, at least in this system. At night, there was no difference in the number of workers of non-dominant species (Table 4, Figure 3b), nor in the number of non-dominant species (Table 4, Figure 3b) on baits near and far from focal *F. subsericea* nests. Likewise, there was no difference in the number of *A. rudis* (Table 4, Figure 3b) or *M. punctiventris* workers (Table 4, Figure 3b) on baits near and far from *F. subsericea* nests. There were, however, 2× more *N. faisonensis* workers on baits near than far from *F. subsericea* nests at night (Table 4, Figure 3b).

3.5. Are Temporal Patterns of Foraging Activity in Non-dominant Species Negatively Related to Those of the Dominant Species?

There were higher numbers of workers of non-dominant species foraging on baits at night than during the day (mean difference = −28.5 ± 9.60; Table 4; Figure 4a). Additionally, there were fewer species of non-dominant ants recorded on baits at night than during the day (mean difference = 1.33 ± 0.36; Table 4; Figure 4b). There were 2× more *A. rudis* workers recorded on baits at night than during the day

(mean difference = −14.92 ± 5.83, Tables 5 and 6). The same trend was seen in N. faisonensis, with 2×
more workers on baits at night than during the day (mean difference = −8.50 ± 3.29, Tables 5 and 6).
There were 2× fewer M. punctiventris workers recorded on baits at night than during the day (mean
difference = −2.92 ± 1.19, Tables 5 and 6).

Table 5. Difference in the number of species and worker of non-dominant ants recorded on baits
between the day and night (day–night). The table shows significant differences for parametric or
non-parametric statistical tests, depending on the distribution of the data (* $p < 0.05$, ** $p < 0.01$).

Response Variable	Taxa	Test	Test Statistic	p
Species richness	all	*t*-test	t = 3.75	0.002 **
Worker abundance	all	*t*-test	t = −2.94	0.007 **
Worker abundance	Ar	*t*-test	t = −2.56	0.01 *
Worker abundance	Fs	*t*-test	t = 5.57	<0.0001 **
Worker abundance	M	Wilcoxon Signed-Rank	Z = 22.5	0.002 **
Worker abundance	Nf	*t*-test	t = −2.58	0.01 *

Table 6. Mean number of workers (±SE) recorded at baits during the day and at night.

Species	Day		Night	
	Mean	SE	Mean	SE
Aphaenogaster rudis Enzmann	7.67	1.95	15.13	3.6
Camponotus americanus Mayr	1	0.73	0.83	0.5
Camponotus pennsylvanicus (DeGeer)	0.38	0.25	1.38	0.76
Formica subsericea Say	20.25	3.97	0.08	0.08
Lasius alienus (Förster)	2.29	2.25	2.33	2.16
Myrmica punctiventris Roger	2.21	0.88	0.75	0.54
Nylanderia faisonensis (Forel)	3.08	0.97	7.33	1.88
Prenolepis imparis (Say)	0	0	4.25	4.25
Temnothorax curvispinosus (Mayr)	0.63	0.22	0	0
Temnothorax longispinosus (Roger)	0.63	0.28	0	0

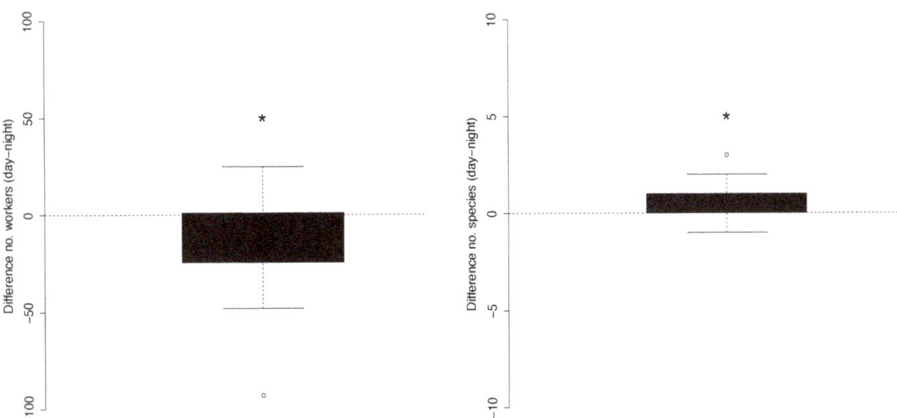

Figure 4. Mean difference in the (**a**) number of workers and (**b**) species richness of non-dominant ant
species recorded on baits during day and night. Asterisk indicates a significant difference in foraging
activity between day and night ($p < 0.01$).

4. Discussion

Increases in the abundance of one species are often associated with decreases in abundance and
richness of other species it competes with [2,3]. If these patterns resulted from competitive exclusion,

one would expect negative impacts of the dominant species on the subordinate species to trickle up and down organization levels. Here, we found only weak support that the abundance of dominant *F. subsericea* ants was related to the richness of non-dominant ant species across sites [4,5,18,19,44]. However, the shape of that relationship was sensitive to the inclusion/exclusion of a highly influential observation. Moreover, except for a few exceptions, population-, individual- and worker level variables such as abundance, productivity and foraging behavior did not match the predictions of competitive exclusion and its impact on diversity maintenance. It might thus be that competitive interactions do not lead to the exclusion of non-dominant ants [69,70].

One mechanism by which dominant ant species might exert community-level influence on non-dominant species is by reducing the amount of resources available in the system (e.g., dominance-impoverishment-rule [71] but see Reference [19]. Although in our system dominant ant species do not appear to affect population-level impact on the abundance of non-dominant ants, they could nevertheless have individual-level effects. Specifically, by interfering with resource use and/or resource acquisition by non-dominant species [29,32], dominant species may negatively affect the size, productivity and fitness of colonies of non-dominant ants [26]. For example, in Finnish ant assemblages colonies of non-dominant ants found inside the territory of dominant *Formica* tend to be smaller than those found outside dominant *Formica* territories [26]. However, proximity to a *F. subsericea* nest, the dominant ant species at our study sites, did not affect colony size in two of the most common non-dominant ant species (*A. rudis* and *N. faisonensis*). Gibb and Hochuli (2004) experimentally excluded dominant *Iridomyrmex purpureus* and found only congeneric species to benefit from competitive release, suggesting that dominant species might have greater effects on closely related species than they do on those that are distantly related [11]. In our system, *F. subsericea* was the only representative of the genus and the principal competitors are quite smaller in size suggesting perhaps lower overlap in resource use under natural conditions.

While we did not observe any effect of dominant ants on colony size, we found evidence of positive effects on another individual-level variable. Colony productivity of *N. fasionensis* tended to be higher in colonies near *F. subsericea* nests relative to colonies that were far from these nests (note that $p = 0.06$). Thus, contrary to our expectations, dominant ants did not negatively affect the size and productivity of subordinate colonies of these common species and may actually be associated with enhanced productivity in one species—*N. fasionensis*. Such positive effects of a large-bodied dominant species on small-bodied non-dominant species can arise when the indirect effects of competition outweigh the direct effects [72]. Davidson [73] who was working with harvester ants in desert systems showed that despite dietary overlap between a large species (*Pogonomyrmex rugosus*) and a small species (*Pheidole xerophila*), the large species facilitated the small species indirectly by suppressing populations of an intermediate-size species (*Pogonomyrmex desertorum*). It may thus be the case that in our system *F. subsericea* (large size) suppress populations of *A. rudis* (medium size), thereby facilitating *N. fasionensis* (small). Apparent facilitation might thus be at play [73].

In sum, we did not find strong community, population or individual-level evidence that dominant *F. subsericea* negatively affects non-dominant ants. We did, however, limit our analyses to a subsample of the most common species in the assemblage. One possibility is that dominant ants do not affect the resource use of all of the non-dominant species equally. For example, although *F. subsericea* interferes with foraging and resource use by *A. rudis*, two non-dominant ant species (*N. faisonensis* and *M. punctiventris*) did not exhibit lower resource use near *F. subsericea* nests relative to far. Thus, while *F. subsericea* was dominant over *N. faisonensis* and *M. punctiventris* in aggressive encounters, it did not decrease the resource use of these species at transient food items. However, we did not estimate the amount of time spent at resources or the rate at which resources were removed, which might be more accurate estimates of resource use by ants [74]. Nevertheless, the lower foraging among *A. rudis* observed near than far from *F. subsericea* nests did not translate to reduced colony size or productivity. Herbers [30] noted that temperate forest ant assemblages can be limited by the availability of nest sites, rather than food resources. Thus, the benefit of nesting in a suitable patch might outweigh the cost incurred by increased competition for food resources when nesting near a dominant ant nest.

The seemingly weak influence of competitive interactions with dominant ants on non-dominant ants at the community, population and individual level suggests a need for further examination of the life strategies that allow these species to coexist. Previous work in southeastern temperate forests of USA showed that the structure of these ant communities are highly influenced by temperature filtering; from regional to micro-scale [43,75,76]. Moreover, in these ant communities, trade-off between thermal tolerance and competitive ability provide a mechanism by which species prevent competitive exclusion. Non-dominant species indeed appear to forage at lower temperatures exploiting shaded parts of the forest floors while dominant ants appear to dominate open canopy habitats where sunlight reach the forest floor [61]. Results from the present study further demonstrate that non-dominant ant species can forage during both the night and the day whereas dominant *Formica* only forage during the day, which suggests a strategy to avoid interference competition [59,77,78]. In our system, interspecific overlap in seasonal activity is high, with most species reaching peak foraging activity in the warmest months of the year [79]. However, our results suggest that on a daily basis, variation in activity levels may be important for coexistence. While there were more species active during the day when *F. subsericea* forages than at night when *F. subsericea* is not active, there were more workers of non-dominant ants foraging at night. Thus, colonies of some non-dominant ants might send more workers out to forage at night than during the day to avoid aggressive encounters with dominant species, a phenomenon that has been previously documented with dominant *Formica* in Finnish forest ant assemblages [28].

Our results suggest that, although dominant ants might sometimes play an important role in structuring ant communities, we find very little evidence of the negative effects of the dominant species in this system across levels of organizations. Note however that whereas much of the previous work on dominant species focused on extremely abundant and aggressive ant species, the dominance of our focal species here is more relative and less absolute. Moreover, our study, as well as most studies that have examined richness-dominance relationship in ants, are observational, which limits our ability to infer causality. Nevertheless, we conclude that although competition evidently occurs in ant communities, other coexistence mechanisms in place may be sufficient to prevent local extinctions [69]. Moreover, there are multiple explanations for the commonly observed hump-shaped relationship between the abundance of dominant ants and species richness among ant communities. Variation in the abiotic environment alone could be driving this widespread pattern if the fitness of non-dominant and dominant species peak at different places along micro-environmental gradients. Further, our results indicate that some non-dominant species may actually benefit from nesting in the vicinity of dominant ant species [80], as was evidenced by higher productivity and nocturnal foraging in a non-dominant species near than far from dominant ant nests. Disentangling the direct and indirect effects of dominant ants on non-dominant ants likely requires further field experiments that manipulate the density of dominant species (rather than just presence/absence) to assess colony-level as well as population- and community-level responses of non-dominant ants (e.g., References [22,23,81–83]).

Author Contributions: Conceptualization, J.-P.L., K.L.S. and N.J.S.; methodology, J-P.L. and N.J.S.; validation, J.-P.L., K.L.S. and N.J.S.; formal analysis, J.-P.L.; investigation, J.-P.L.; resources, J.-P.L. and N.J.S.; data curation, J.-P.L.; writing—original draft preparation, J.-P.L. and N.J.S.; writing—review and editing, J.-P.L., K.L.S. and N.J.S.; visualization, J.-P.L.; supervision, N.J.S.; project administration, J.-P.L. and N.J.S.; funding acquisition, N.J.S. All authors have read and agreed to the published version of the manuscript.

Funding: This research was funded by the U.S. Department of Energy PER award (DEFG02-08ER64510) and a National Science Foundation Dimensions of Biodiversity grant (NSF-1136703). Additional support was granted to J.-P.L. from the Department of Ecology and Evolutionary Biology at the University of Tennessee, an NSERC-PGS doctoral scholarship and a NSF-DDIG.

Acknowledgments: We thank Alan Andersen, Tara Sackett, Mariano Rodriguez-Cabal, David Fowler and three anonymous reviewers for making comments that greatly improved the manuscript, Noa Davidai and Benoit Guenard for help in the field and Gabriel Munoz for help with the visuals. This study was permitted by and enhanced through collaboration with the Great Smoky Mountains National Park.

References

1. Grime, J.P. *Colonization, Succession and Stability*; Blackwell: Oxford, UK, 1984.

2. Janzen, D.H. Herbivores and the Number of Tree Species in Tropical Forests. *Am. Nat.* **1970**, *104*, 501–528. [CrossRef]

3. Connell, J.H. On the role of natural enemies in preventing competitive exclusion in some marine animals and in rain forest trees. In *Dynamics of Population*; Den Boer, P.J., Gradwell, G.R., Eds.; Pudoc: Wageningen, The Netherlands, 1970.

4. Parr, C.L. Dominant ants can control assemblage species richness in a South African savanna. *J. Anim. Ecol.* **2008**, *77*, 1191–1198. [CrossRef] [PubMed]

5. Parr, C.L.; Sinclair, B.J.; Andersen, A.N.; Gaston, K.J.; Chown, S.L. Constraint and competition in assemblages: A cross-continental and modeling approach for ants. *Am. Nat.* **2005**, *165*, 481–494. [CrossRef] [PubMed]

6. Gotelli, N.J.; Arnett, A.E. Biogeographic effects of red fire ant invasion. *Ecol. Lett.* **2000**, *3*, 257–261. [CrossRef]

7. Holway, D.A. Effect of Argentina ant invasions on ground-dwelling arthropods in northern California riparian woodlands. *Oecologia* **1998**, *116*, 252–258. [CrossRef] [PubMed]

8. Human, K.G.; Gordon, D.M. Exploitation and interference competition between the invasive Argentine ant, Linepithema humile, and native ant species. *Oecologia* **1996**, *105*, 405–412. [CrossRef] [PubMed]

9. Porter, S.; Savignano, D. Invasion of Polygyne Fire Ants Decimates Native Ants and Disrupts Arthropod Community. *Ecology* **1990**, *71*, 2095–2106. [CrossRef]

10. Sanders, N.J.; Gotelli, N.J.; Heller, N.E.; Gordon, D.M. Community disassembly by an invasive species. *Proc. Natl. Acad. Sci. USA* **2003**, *100*, 2474–2477. [CrossRef]

11. Lessard, J.-P.; Fordyce, J.A.; Gotelli, N.J.; Sanders, N.J. Invasive ants alter the phylogenetic structure of ant communities. *Ecology* **2009**, *90*, 2664–2669. [CrossRef]

12. Sanders, N.J.; Crutsinger, G.M.; Dunn, R.R.; Majer, J.D.; Delabie, J.H.C. An ant mosaic revisited: Dominant ant species disassemble arboreal ant communities but co-occur randomly. *Biotropica* **2007**, *39*, 422–427. [CrossRef]

13. Davidson, D.W.; Lessard, J.P.; Bernau, C.R.; Cook, S.C. The tropical ant mosaic in a primary Bornean rain forest. *Biotropica* **2007**, *39*, 468–475. [CrossRef]

14. Jackson, D.A. Ant Distribution Patterns in a Cameroonian Cocoa Plantation—Investigation of the Ant Mosaic Hypothesis. *Oecologia* **1984**, *62*, 318–324. [CrossRef] [PubMed]

15. Leston, D. Neotropical Ant Mosaic. *Ann. Entomol. Soc. Am.* **1978**, *71*, 649–653. [CrossRef]

16. Majer, J.D.; Delabie, J.H.C.; Smith, M.R.B. Arboreal Ant Community Patterns in Brazilian Cocoa Farms. *Biotropica* **1994**, *26*, 73–83. [CrossRef]

17. Adams, E.S. Territory Defense by the Ant Azteca-Trigona—Maintenance of an Arboreal Ant Mosaic. *Oecologia* **1994**, *97*, 202–208. [CrossRef]

18. Andersen, A.N. Regulation of Momentary Diversity by Dominant Species in Exceptionally Rich Ant Communities of the Australian Seasonal Tropics. *Am. Nat.* **1992**, *140*, 401–420. [CrossRef]

19. Sheard, J.K.; Nelson, A.S.; Berggreen, J.D.; Boulay, R.; Dunn, R.R.; Sanders, N.J. Testing trade-offs and the dominance–impoverishment rule among ant communities. *J. Biogeogr.* **2020**, *47*, 1899–1909. [CrossRef]

20. Arnan, X.; Andersen, A.N.; Gibb, H.; Parr, C.L.; Sanders, N.J.; Dunn, R.R.; Angulo, E.; Baccaro, F.B.; Bishop, T.R.; Boulay, R.; et al. Dominance–diversity relationships in ant communities differ with invasion. *Glob. Chang. Biol.* **2018**, *24*, 4614–4625. [CrossRef]

21. Gibb, H. Dominant meat ants affect only their specialist predator in an epigaeic arthropod community. *Oecologia* **2003**, *136*, 609–615. [CrossRef]

22. Gibb, H.; Hochuli, D.F. Removal experiment reveals limited effects of a behaviorally dominant species on ant assemblages. *Ecology* **2004**, *85*, 648–657. [CrossRef]

23. Gibb, H.; Johansson, T. Field tests of interspecific competition in ant assemblages: Revisiting the dominant red wood ants. *J. Anim. Ecol.* **2011**, *80*, 548–557. [CrossRef] [PubMed]

24. Deslippe, R.J.; Savolainen, R. Mechanisms of Competition in a Guild of Formicine Ants. *Oikos* **1995**, *72*, 67–73. [CrossRef]

25. Herbers, J.M.; Banschbach, V.S. Food supply and reproductive allocation in forest ants: Repeated experiments give different results. *Oikos* **1998**, *83*, 145–151. [CrossRef]

26. Savolainen, R. Colony Success of the Submissive Ant Formica-Fusca within Territories of the Dominant Formica-Polyctena. *Ecol. Entomol.* **1990**, *15*, 79–85. [CrossRef]

27. Savolainen, R. Interference by Wood Ant Influences Size Selection and Retrieval Rate of Prey by Formica-Fusca. *Behav. Ecol. Sociobiol.* **1991**, *28*, 1–7. [CrossRef]

28. Vepsalainen, K.; Savolainen, R. The Effect of Interference by Formicine Ants on the Foraging of Myrmica. *J. Anim. Ecol.* **1990**, *59*, 643–654. [CrossRef]

29. Fellers, J.H. Interference and Exploitation in a Guild of Woodland Ants. *Ecology* **1987**, *68*, 1466–1478. [CrossRef]

30. Herbers, J.M. Community Structure in North Temperate Ants—Temporal and Spatial Variation. *Oecologia* **1989**, *81*, 201–211. [CrossRef]

31. Savolainen, R.; Vepsalainen, K. A Competition Hierarchy among Boreal Ants—Impact on Resource Partitioning and Community Structure. *Oikos* **1988**, *51*, 135–155. [CrossRef]

32. Savolainen, R.; Vepsalainen, K. Niche Differentiation of Ant Species within Territories of the Wood Ant Formica Polyctena. *Oikos* **1989**, *56*, 3–16. [CrossRef]

33. Savolainen, R.; Vepsalainen, K.; Wuorenrinne, H. Ant Assemblages in the Taiga Biome—Testing the Role of Territorial Wood Ants. *Oecologia* **1989**, *81*, 481–486. [CrossRef] [PubMed]

34. Tilman, D. The Importance of the Mechanisms of Interspecific Competition. *Am. Nat.* **1987**, *129*, 769–774. [CrossRef]

35. Cerdá, X.; Retana, J.; Manzaneda, A. The role of competition by dominants and temperature in the foraging of subordinate species in Mediterranean ant communities. *Oecologia* **1998**, *117*, 404–412. [CrossRef]

36. Holway, D.A. Competitive mechanisms underlying the displacement of native ants by the invasive Argentine ant. *Ecology* **1999**, *80*, 238–251. [CrossRef]

37. Erős, K.; Maák, I.; Markó, B.; Babik, H.; Ślipiński, P.; Nicoară, R.; Czechowski, W. Competitive pressure by territorials promotes the utilization of unusual food source by subordinate ants in temperate European woodlands. *Ethol. Ecol. Evol.* **2020**, *32*, 457–465. [CrossRef]

38. Law, S.J.; Parr, C. Numerically dominant species drive patterns in resource use along a vertical gradient in tropical ant assemblages. *Biotropica* **2020**, *52*, 101–112. [CrossRef]

39. Andersen, A.N. Not enough niches: Non-equilibrial processes promoting species coexistence in diverse ant communities. *Austral. Ecol.* **2008**, *33*, 211–220. [CrossRef]

40. Francoeur, A. Révision taxonomique des espèces néarctiques du groupe fusca, genre Formica (Formicidae, Hymenoptera). *Mem. Soc. Ent. Québec* **1973**, *3*, 1–316.

41. Fellers, J.H. Daily and Seasonal Activity in Woodland Ants. *Oecologia* **1989**, *78*, 69–76. [CrossRef]

42. Lessard, J.-P.; Dunn, R.R.; Parker, C.R.; Sanders, N.J. Rarity and Diversity in Forest Ant Assemblages of Great Smoky Mountains National Park. *Southeast. Nat.* **2007**, *6*, 215–228. [CrossRef]

43. Lessard, J.-P.; Sackett, T.E.; Reynolds, W.N.; Fowler, D.A.; Sanders, N.J. Determinants of the detrital arthropod community structure: The effects of temperature and resources along an environmental gradient. *Oikos* **2011**, *120*, 333–343. [CrossRef]

44. Sanders, N.J.; Lessard, J.-P.; Fitzpatrick, M.C.; Dunn, R.R. Temperature, but not productivity or geometry, predicts elevational diversity gradients in ants across spatial grains. *Glob. Ecol. Biogeogr.* **2007**, *16*, 640–649. [CrossRef]

45. Cammell, M.E.; Way, M.J.; Paiva, M.R. Diversity and structure of ant communities associated with oak, pine, eucalyptus and arable habitats in Portugal. *Insectes Sociaux* **1996**, *43*, 37–46. [CrossRef]

46. Stuble, K.; Kirkman, L.K.; Carroll, C.R. Patterns of abundance of fire ants and native ants in a native ecosystem. *Ecol. Entomol.* **2009**, *34*, 520–526. [CrossRef]

47. LeBrun, E.; Feener, D. When trade-offs interact: Balance of terror enforces dominance discovery trade-off in a local ant assemblage. *J. Anim. Ecol.* **2007**, *76*, 58–64. [CrossRef]

48. Ellison, A.M.; Record, S.; Arguello, A.; Gotelli, N.J. Rapid inventory of the ant assemblage in a temperate hardwood forest: Species composition and assessment of sampling methods. *Environ. Entomol.* **2007**, *36*, 766–775. [CrossRef]

49. Longino, J.T.; Colwell, R.K. Biodiversity assessment using structured inventory: Capturing the ant fauna of a tropical rain forest. *Ecol. Appl.* **1997**, *7*, 1263–1277. [CrossRef]

50. King, J.R. Size–abundance relationships in Florida ant communities reveal how ants break the energetic equivalence rule. *Ecol. Entomol.* **2010**, *35*, 287–298. [CrossRef]

51. Gotelli, N.J.; Ellison, A.M. Biogeography at a regional scale: Determinants of ant species density in new england bogs and forests. *Ecology* **2002**, *83*, 1604–1609. [CrossRef]

52. Lubertazzi, D. The Biology and Natural History of *Aphaenogaster rudis*. *Psyche* **2012**, *2012*, 752815. [CrossRef]

53. Nuhn, T.P.; Wright, C.G.; Farrier, M.H. Notes on the biology of the ant Paratrechina faisonensis (Hymenoptera: Formicidae). *J. Elisha Mitchell Sci. Soc.* **1992**, *108*, 11–18.

54. Kaspari, M. Testing resource-based models of patchiness in four Neotropical litter ant assemblages. *Oikos* **1996**, *76*, 443–454. [CrossRef]

55. McGlynn, T.P.; Hoover, J.R.; Jasper, G.S.; Kelly, M.S.; Polis, A.M.; Spangler, C.M.; Watson, B.J. Resource heterogeneity affects demography of the Costa Rican ant Aphaenogaster araneoides. *J. Trop. Ecol.* **2002**, *18*, 231–244. [CrossRef]

56. Beugnon, G.; Fourcassie, V. How Do Red Wood Ants Orient during Diurnal and Nocturnal Foraging in a 3 Dimensional System. 2. Field Experiments. *Insectes Sociaux* **1988**, *35*, 106–124. [CrossRef]

57. Hillery, A.E.; Fell, R.D. Chemistry and Behavioral significance of rectal and accessory gland contents in Camponotus pennsylvanicus (Hymenoptera: Formicidae). *Ann. Entomol. Soc. Am.* **2000**, *93*, 1294–1299. [CrossRef]

58. Torres, J.A.; Snelling, R.R.; Jones, T.H. Distribution, ecology and behavior of Anochetus kempfi (Hymenoptera: Formicidae) and description of the sexual forms. *Sociobiology* **2000**, *36*, 505–516.

59. Retana, J.; Cerdá, X. Patterns of diversity and composition of Mediterranean ground ant communities tracking spatial and temporal variability in the thermal environment. *Oecologia* **2000**, *123*, 436–444. [CrossRef]

60. Stuble, K.L.; Jurić, I.; Cerdá, X.; Sanders, N.J. Dominance hierarchies are a dominant paradigm in ant ecology (Hymenoptera: Formicidae), but should they be? And what is a dominance hierarchy anyways? *Myrmecol. News* **2017**, *24*, 71–81. [CrossRef]

61. Lessard, J.P.; Dunn, R.R.; Sanders, N.J. Temperature-mediated coexistence in forest ant communities. *Insectes Sociaux* **2009**, *56*, 149–156. [CrossRef]

62. Tschinkel, W.R.; Hess, C.A. Arboreal ant community of a pine forest in northern Florida. *Ann. Entomol. Soc. Am.* **1999**, *92*, 63–70. [CrossRef]

63. Vepsalainen, K.; Pisarski, B. Assembly of Island Ant Communities. *Ann. Zool. Fenn.* **1982**, *19*, 327–335.

64. Colley, W.N. Colley's Bias Free College Football Ranking Method: The Colley Matrix Explained. Available online: http://www.colleyrankings.com/matrate.pdf (accessed on 1 May 2009).

65. Longino, J.T.; Coddington, J.; Colwell, R.K. The ant fauna of a tropical rain forest: Estimating species richness three different ways. *Ecology* **2002**, *83*, 689–702. [CrossRef]

66. Kaspari, M.; Weiser, M.D. The size–grain hypothesis and interspecific scaling in ants. *Funct. Ecol.* **1999**, *13*, 530–538. [CrossRef]

67. Geraghty, M.J.; Dunn, R.R.; Sanders, N.J. Body size, colony size, and range size in ants (Hymenoptera: Formicidae): Are patterns along elevational and latitudinal gradients consistent with Bergmann's Rule? *Myrmecol. News* **2007**, *10*, 51–58.

68. Gotelli, N.J.; Ellison, A.M. *A Primer of Ecological Statistics*; Sinauer Associates, INC.: Sunderlan, MA, USA, 2004.

69. Chesson, P. Mechanisms of Maintenance of Species Diversity. *Annu. Rev. Ecol. Syst.* **2000**, *31*, 343–366. [CrossRef]

70. Chesson, P. Updates on mechanisms of maintenance of species diversity. *J. Ecol.* **2018**, *106*, 1773–1794. [CrossRef]

71. Hölldobler, B.; Wilson, E. *The Ants*; Belknap: Cambridge, MA, USA, 1990.

72. Davidson, D.W. Some Consequences of Diffuse Competition in a Desert Ant Community. *Am. Nat.* **1980**, *116*, 92–105. [CrossRef]

73. Davidson, D.W. An Experimental Study of Diffuse Competition in Harvester Ants. *Am. Nat.* **1985**, *125*, 500–506. [CrossRef]

74. Jeanne, R.L. A Latitudinal Gradient in Rates of Ant Predation. *Ecology* **1979**, *60*, 1211–1224. [CrossRef]
75. Fowler, D.; Lessard, J.-P.; Sanders, N.J. Niche filtering rather than partitioning shapes the structure of temperate forest ant communities. *J. Anim. Ecol.* **2014**, *83*, 943–952. [CrossRef]
76. Lessard, J.-P.; Borregaard, M.K.; Fordyce, J.A.; Rahbek, C.; Weiser, M.D.; Dunn, R.R.; Sanders, N.J. Strong influence of regional species pools on continent-wide structuring of local communities. *Proc. R. Soc. B Biol. Sci.* **2012**, *279*, 266–274. [CrossRef]
77. Stuble, K.L.; Rodriguez-Cabal, M.A.; McCormick, G.L.; Jurić, I.; Dunn, R.R.; Sanders, N.J. Tradeoffs, competition, and coexistence in eastern deciduous forest ant communities. *Oecologia* **2013**, *171*, 981–992. [CrossRef] [PubMed]
78. Albrecht, M.; Gotelli, N.J. Spatial and temporal niche partitioning in grassland ants. *Oecologia* **2001**, *126*, 134–141. [CrossRef]
79. Dunn, R.R.; Parker, C.R.; Sanders, N.J. Temporal patterns of diversity: Assessing the biotic and abiotic controls on ant assemblages. *Biol. J. Linn. Soc.* **2007**, *91*, 191–201. [CrossRef]
80. Cerdá, X.; Arnan, X.; Retana, J. Is competition a significant hallmark of ant (Hymenoptera: Formicidae) ecology? *Myrmecol. News* **2013**, *18*, 131–147.
81. LeBrun, E.G.; Tillberg, C.V.; Suarez, A.V.; Folgarait, P.J.; Smith, C.R.; Holway, D.A. An experimental study of competition between Fire Ants and Argentine Ants in their native range. *Ecology* **2007**, *88*, 63–75. [CrossRef]
82. Stuble, K.L.; Kirkman, L.K.; Carroll, C.R.; Sanders, N.J. Relative effects of disturbance on Red Imported Fire Ants and native ant apecies in a longleaf pine ecosystem. *Conserv. Biol.* **2011**, *25*, 618–622. [CrossRef]
83. Menke, S.B.; Fisher, R.N.; Jetz, W.; Holway, D.A. Biotic and abiotic controls of Argentine Ant invasion success at local and landscape scales. *Ecology* **2007**, *88*, 3164–3173. [CrossRef]

Publisher's Note: MDPI stays neutral with regard to jurisdictional claims in published maps and institutional affiliations.

Review

Eciton Army Ants—Umbrella Species for Conservation in Neotropical Forests

Sílvia Pérez-Espona [1,2]

1   Royal (Dick) School of Veterinary Studies, The University of Edinburgh, Roslin, Scotland EH25 9RG, UK; silvia.perez-espona@ed.ac.uk
2   The Roslin Institute, The University of Edinburgh, Roslin, Scotland EH25 9RG, UK

Abstract: Identification of priority areas for conservation is crucial for the maintenance and protection of biodiversity, particularly in tropical forests where biodiversity continues to be lost at alarming rates. Surveys and research on umbrella species can provide efficient and effective approaches to identify potential areas for conservation at small geographical scales. Army ants of the genus *Eciton* are keystone species in neotropical forests due to their major role as top predators and due to the numerous vertebrate- and invertebrate associated species that depend upon their colonies for survival. These associates range from the iconic army ant-following birds to a wide range of arthropod groups, some of which have evolved intricate morphological, behavioural and/or chemical strategies to conceal their presence and integrate into the colony life. Furthermore, *Eciton* colonies require large forested areas that support a diverse leaf litter prey community and several field-based and genetic studies have demonstrated the negative consequences of forest fragmentation for the long-term maintenance of these colonies. Therefore, *Eciton* species will not only act as umbrella for their associates but also for many other species in neotropical forests, in particular for those that require a large extent of forest. This review summarises past and recent accounts of the main taxonomic groups found associated with *Eciton* colonies, as well research assessing the impact of forest fragmentation on this army ant, to encourage the adoption of *Eciton* army ants as umbrella species for the identification of priority areas for conservation and assessments of the effect of disturbance in neotropical forests.

Keywords: antbird; army ant; biodiversity; biological indicator; conservation; deforestation; habitat fragmentation; myrmecophiles; mimicry; species interactions; tropics

Citation: Pérez-Espona, S. *Eciton* Army Ants—Umbrella Species for Conservation in Neotropical Forests. *Diversity* **2021**, *13*, 136. https://doi.org/10.3390/d13030136

Academic Editors: Alan N. Andersen and Luc Legal

Received: 14 February 2021
Accepted: 13 March 2021
Published: 22 March 2021

Publisher's Note: MDPI stays neutral with regard to jurisdictional claims in published maps and institutional affiliations.

1. Introduction

Disturbance as a consequence of anthropogenic activities continues to be a major driver of biodiversity loss [1–4] and it negatively impacts ecosystems functioning and ecosystem services [5–7]. In tropical forests, anthropogenically-caused disturbance as a result of logging, land clearance for agriculture and hunting [8,9] are ongoing threats to biodiversity and ecosystem function [10,11]. Therefore, to devise effective conservation actions and sustainable management of tropical forests resources, it is important to determine and predict the effects of anthropogenic disturbance in these biodiversity-rich areas [12,13]. However, the assessment of anthropogenic disturbance on biodiversity in tropical forests is challenging due to the lack of knowledge on a large number of taxa in these areas, including those not yet discovered and described [14], and the limited budgets and timeframes for conducting biodiversity assessments [15]. These challenges have partly been overcome by conducting surveys of a reduced number of well-known and identifiable taxonomic groups (or species) that can act as biological indicators to determine the level or magnitude of anthropogenic disturbance [16–19], although not without caveats [20–23].

Insects represent a large part of the biodiversity found in tropical forests and play a central role in ecosystem function and services [24,25]. Owing to their short life cycles and

325

habitat disturbance sensitivity, insects are considered suitable biological indicators for monitoring the effects of fragmentation and other environmental changes on biodiversity [26,27] as well as for monitoring the sustainable use of tropical forests [28]. Furthermore, insects are also used as biodiversity indicators, with their diversity used as surrogate information for species richness estimates in a given area or to identify priority areas for conservation [22]. Among insects, ants comprise the largest fraction of animal biomass in the tropics and are frequently used as environmental, ecological and biodiversity indicators [29–33]. In the Neotropics, army ants are dominant components in forests and as top predators of arthropod communities play a major role in leaf litter community structure [34–36]. The most conspicuous army ants in neotropical forests are those within the genus *Eciton*, due to their large hunting raids and the numerous invertebrate and vertebrate species associated with them that exploit the different environments and conditions provided by these ants' colony life [29,36–38]. Furthermore, *Eciton* colonies are particularly sensitive to landscape fragmentation and deforestation, with large areas of continuous forests required for their long-term survival [34,39].

Monitoring of biodiversity and ecosystems function is essential for the maintenance and sustainable use of neotropical forests [28,40]. Biodiversity conservation through area protection or other effective area-based conservation approaches requires the identification of areas of high biodiversity, high irreplaceability, large intact habitats with a high degree ecological integrity and/or high vulnerability due to the impact of anthropogenic activities [12,13]. Surveys assessing the presence and abundance of *Eciton* colonies in a particular area could provide additional information on the selection of conservation areas in neotropical forests. Although *Eciton* army ants have been used as biodiversity or habitat integrity indicators [41], they have not yet been implicitly considered as umbrella species for biodiversity conservation in neotropical forests or considered as biological indicators to assess the impacts of forest disturbance.

This review aims to encourage the adoption of *Eciton* army ants as potential umbrella species for neotropical forest biodiversity conservation. To this end, I build on previous detailed accounts of *Eciton* colony associates [37,38,42] to provide further insights derived from research conducted in the last decade, including the significant contributions of genetic and genomic studies to our understanding on the relationship of some of these associates with their *Eciton* hosts. Furthermore, this review also summarises research on the effect of deforestation or other land-use changes on *Eciton* colonies, to illustrate the potential use of these army ants in applied research assessing the impact of disturbance or success of restoration projects in neotropical forests.

2. *Eciton* Army Ants

Several ant genera within the subfamily Dorylinae are referred to as army ants, with the genera *Cheliomyrmex*, *Eciton*, *Labidus*, *Neivamyrmex* and *Nomamyrmex* forming a monophyletic group and found in tropical and subtropical regions in the American continent [43,44]. Army ants are characterised by presenting the "army ant syndrome" which includes the evolution of behavioural and reproductive traits that have allowed them to be key predators in neotropical forests. This suite of traits includes obligate collective foraging, nomadic lifestyle, and highly specialised queens which are characterised by being wingless and having large expanding abdomens capable of lying thousands of eggs [38,45–47]. Colony founding of *Eciton* army ants is by fission. When colonies attain a large size, a sexual brood is produced, containing a small number of new queens and hundreds of males. The parental colony will then divide into two daughter colonies, one of them headed by a new queen and the other one headed either by the old queen or another of the new queens [36,48]. Initial phylogenetic analyses based on data derived from three nuclear genes, one mitochondrial gene, and morphological analyses, suggested that the evolution of the "army ant syndrome" in army ants species in the Americas and Afro-Eurasia was as a result of inheritance of these traits from a unique Gondwanan common ancestor [46]. However, a recent comprehensive phylogenomic study based on 2,166 loci has indicated multiple origins for the "army ant syndrome" in different continents [44]; therefore, supporting initial views of convergent evolution of this syndrome [38].

The genus is distributed from Mexico to Argentina [38], with 12 *Eciton* species, and several subspecies, described [49]. Genomic analyses using a large number of loci (> 4,000,000) for 146 specimens, collected from a wide distribution range and representing those nine species for which workers and queens are known, strongly supported interspecific relationships congruent with taxonomic classifications [50]. However, further taxonomic revisions are required to delimit species and subspecies boundaries. For example, a phylogeographic study based on mitochondrial DNA sequences indicated that genetic divergence of *E. burchellii foreli* and *E. b. parvispinum* colonies was higher than between other morphologically described species [51].

Foraging and emigration of *Eciton* army ants takes place above ground [36,52]. The species *E. burchellii* and *E. hamatum* are considered truly epigaeic, as they bivouac above ground. Other *Eciton* species such as *E. mexicanum*, *E. vagans* and *E. dulcium* tend to bivouac in more sheltered and darker places, inside or beneath logs, or in underground erosion chambers [36]. In these more hipogaeic species, raids and emigration columns can take place underground but only for short distances and when near the bivouac [53]. Knowledge on *Eciton* army ants is primarily based on research focused on *E. burchellii* and to less extent on *E. hamatum*, as these two species are the most epigaeic diurnal *Eciton* species. Most *Eciton* species are column raiders (e.g. *E. hamatum*, *E. vagans*), with hunting taking place by ants moving initially in narrow columns that later divide into several diverging branches. The size of the raid varies among species, from sparse raid front such as those in *Eciton dulcium*, often narrower than 2 m, to swarm fronts wider than 5 m in in the swarm raider *E. burchellii* [38,53,54].

A large-scale survey conducted at La Selva Biological Station (Costa Rica) examining the diets of sympatric neotropical army ants, including the species *E. burchellii*, *E. dulcium*, *E. hamatum*, *E. lucanoides*, *E. mexicanum* and *E. vagans* revealed high specialisation and small spatio-temporal niche overlap [55]. Using DNA barcoding to identify ant prey to the species level revealed differences in raiding preferences, with *E. dulcium* and *E. mexicanum* preferentially raiding ground-nesting ants and *E. hamatum* and *E. burchellii* arboreal-nesting ants. Furthermore, *E. hamatum* and *E. burchellii* colonies were found to have more diurnal raids whereas most of the other *Eciton* species had primarily nocturnal raids [55].

All *Eciton* species are predominantly predators of other ants (mainly their brood) although they specialise on different ant genera. *E. hamatum* predominantly preys on the brood of *Acromyrmex* leaf-cutting ants, *E. dulcium* on *Pachycondyla* and *Odontomachus* ants, and *E. burchellii* on *Camponotus* ants [14–20]. Although the diet of *E. burchellii* is more varied and includes a diverse range of non-ant litter arthropods, in particular the brood of other social insects [56–61], this non-ant component of the prey might be comparatively low [55]. The large colony size of *Eciton* species, with estimates ranging from 150,000 to 500,000 workers in *E. hamatum* and 500,000 to 2 million workers *in E. burchellii* [38], and the relatively large body size of of some of the workers imply high colony-energy demands [62]. Colonies of *Eciton* army ants have a nomadic-statary cycle, with periods of in which the bivouac (temporary nest) remains in a fixed location where the queen lays thousands of eggs [63]. During this period, the workers hunt prey locally in a regular manner. The statary phase is followed by a nomadic phase coinciding with the eggs hatching into larvae and, during this phase, the colony emigrates daily after the raids over the period of 2–3 weeks [36,64]. In the most studied species, *E. burchellii*, colonies concentrate their raids on invertebrate-rich forest patches [65] and a colony could capture c. 30,000 items of prey in a single day [66]. The nomadic-statary cycle of *Eciton* colonies is interpreted as an adaptation to avoid prey depletion by promoting population recovery of their prey [56]. The pattern of raids, as part of the nomadic lifestyle of *Eciton* army ants, therefore, exerts top-down regulation of leaf litter arthropod communities and creates a mosaic of habitat patches at different ecological succession stages which increases local species diversity [35,65,67].

3. Associates of *Eciton* Army Ants

Eciton army ants play an important role for the maintenance of biodiversity in neotropical forests not only due to the mosaic habitat that they create as top predators but also due to the hundreds of associate species that depend on these army ants for their survival [36,37,68,69]. The next sections will provide an overview of the main taxonomic groups found with *Eciton* army ant colonies, with a particular focus on arthropods, as this is the most abundant taxonomic group of associates (Figure 1).

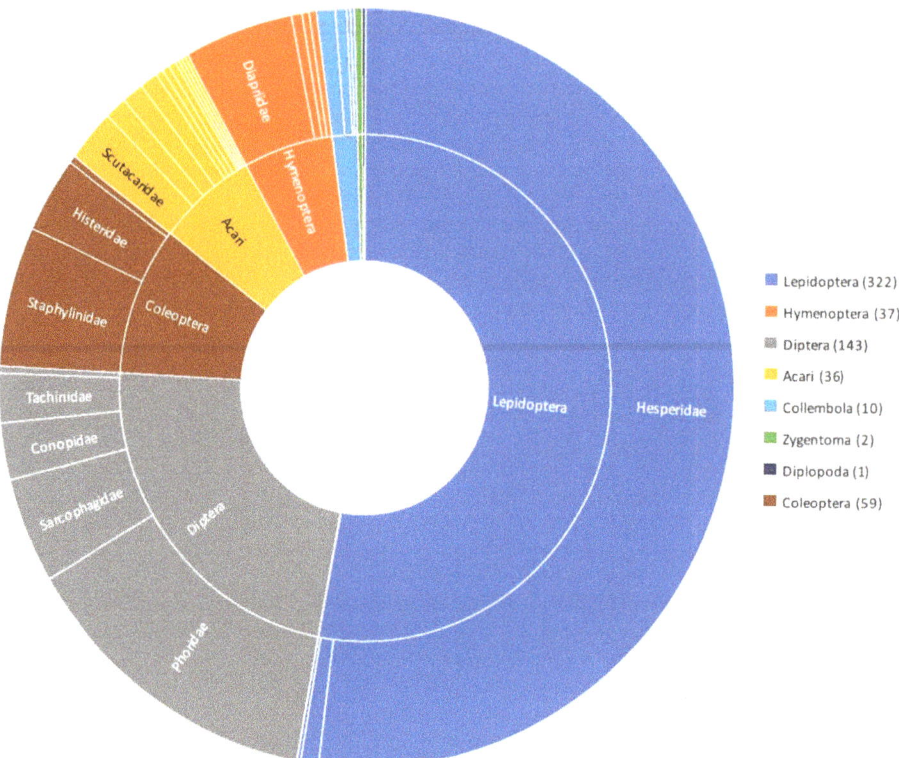

Figure 1. Diagram illustrating the different arthropod taxonomic groups found as associates of *Eciton* army ants. Numbers in brackets indicated number of species reported as *Eciton burchellii* associates; data from Rettenmeyer [69]. See Table S2 for further details of families with lower representation within each of the main arthropod taxonomic groups.

3.1. Birds

The majority of studies on associates of *Eciton* army ants have been conducted on the iconic antbirds (Table S1). Army ants provide important foraging resources for antbirds as these primarily feed on arthropods flushed during the ant raids [36,70]. The most commonly reported army ant following birds are within the families Cuculidae, Formicariidae, Furnariidae, Thamnophilidae and Thraupidae, although species from other bird families can also be found at army ant raids [68]; see Table S1. Phylogenetic studies of the Thamnophilidae have revealed that this army ant-following specialisation has evolved three times in this bird family and that this specialization is ancient, with an origin in the late Miocene [71]. Furthermore, the most likely evolution progression of specialisation has been indicated to be from occasional to regular visits, to an obligate association with the army ants (Figure 2).

Figure 2. The Bicolored Antbird (*Gymnopithys leucaspis*, family Thamnophilidae) is an obligate army ant follower of *Eciton burchellii* colonies. Birds of this species are found at high frequencies at swarms and check bivouac sites. Photo taken at Parque Nacional Darién, Panama (Matt Deres at English Wikipedia, CC BY 3.0).

Antbirds are more commonly found with epigaeic diurnal swarm raid-forming army ants such as *E. burchellii* and *Labidus praedator* [68,69]. Although other *Eciton* species characterised by having column raids, straggling swarm or semi-nocturnal habits might not be important resources for diurnal ant-following birds, reports indicate that antbirds can also be found associated with column-raid army ants when the raids are spread [72]. Antbirds are also occasionally reported with straggling swarms such as those of *E. rapax* [73] and the mainly nocturnal species *E. mexicanum* [72] and *E. vagans* [74]. Screech owls (*Otus* spp.) have also been reported to follow nocturnal raiding army ants in some instances [68].

In addition to the flushed arthropods during army ant raids, antbirds have also been reported to benefit by capturing larger prey such as *Anolis* lizards [75]. The capture of fleeing snakes by Collared forest falcons (*Micrastur semitorquatus*) during *E. burchellii parvispininum* raids has also recently been reported [76]. The association of antbirds with army ants range from occasional visitors to obligate attendants of the raids [68,77–79], see Table S1. Numerous antbirds appear after an army ant raid starts and will then follow the hunting swarm as it advances. Obligate species are likely to track the location of the army ant colonies and some antbirds have been reported to visit statary bivouacs, potentially as a way to monitor colony activity [68,80]. Antbirds are considered parasites of the army ants, as exclusion experiments have indicated a detrimental effect of the antbird foraging on the success rate of prey capturing by ants [81]. Although some field observations [82,83] and gut content analyses [84] have reported the consumption of army ants by antbirds, it is not yet clear if consumption of army ants by antbirds can be intentional in some instances or if this consumption is accidental as most reports indicate [68].

3.2. Other Vertebrates

Observations of mammals associated with *Eciton* army ants are limited to marmosets, with five species having been reported to take prey flushed by *Eciton burchellii* in Brasil [85,86]. *Callithrix humelifer* and *C. kuhli* are the species more frequently reported in *E. burchellii* raids, with *C. humelifer* observed to spend several hours picking prey including that being carried by the ants [85]. Other vertebrate species such as the toad *Bufo marinus*, teid lizards of the genus *Ameiva* and *Kentropyx,* and the iguanid *Anolis frenatus* have also been reported to found associated with *Eciton* army ant raids [68].

3.3. Mites

The most numerous associates of *Eciton* colonies, in terms of number of individuals, are mites. It is estimated that myrmecophilous mites found in colonies of army ants within the tribe Ecitonini can be in a ratio > 100:1 to other myrmecophilous arthropods and c. 1:100 to ants in the colony, although in some colonies the ratio can be as high as 1:4 [37]. At least 126 species belonging to 33 different genera of mites have been described in association with army ants in the tribe Ecitonini, 21 of the genera found to be unique to them [87], and 41 taxa found in *E. burchellii* colonies [69]. However, further research is needed to determine if some of these taxa are associated with the army ants or their prey [87].

Some of the mites associated with *Eciton* species can be found either in refuse deposits or middens taking advantage of this food resource (refuse, discarded prey, ant corpses or fungi). However, those thought to be more intimately associated with the army ants are found in the bivouacs and/or emigrating columns [37,69]. Although the nature of the association of the mite taxa with the army ants is yet not well understood, this can potentially be parasitic, mutualistic, phoretic or considered as predators of other taxa visiting the middens [88]. Most of the mite taxa more closely associated with *Eciton* colonies are phoretic, i.e. using the ants as transport agents at least for one stage of their life cycle dispersal, with the phoretic relationship being either facultative or obligatory [38]. Life cycles for mites associated with army ants remain largely unknown but it is likely that they are synchronised with the emigration intervals in *Eciton* host colonies [38]. Phoretic mites have adaptations to be able to ride along with the emigration columns or attach to different parts of the queen, males, different worker castes, and larvae [37,89]. Phoretic mites have also evolved strategies to hitchhike on the ants, such as cuticular ridges on the lateral body flanges developed on adult *Antennequesoma* mites allowing them to attach to *Eciton* workers' antennae or legs [89], or *Planodiscus* mites that their body is sculptured to resemble structures of their *Eciton* hosts to attach to their legs [88,90,91]. Some mites associated with *Eciton* army ants are ectoparasites of brood or adult ants, and feed on the host's hemolymph or integumental secretions, with some species having evolved striking adaptations to attach on different body parts of workers and integrate into the colony [37,87,88]. An example of ectoparasitic mites is *Macrocheles rettenmeyeri* which is found attached on the membrane of hind legs' pulvilli of *Eciton dulcium* workers [92]. The mite's curve hind leg resembles the ant's tarsal claws and, therefore, assumes the function of this body part when the ants hang to one another forming a bivouac [88]. *Rettenmeyerius carli* is another example of ectoparasitic mite, in this case, found attached to the mouthparts of majors [37].

The effects of parasitic mites on *Eciton* colonies are unknown. Inspection of *E. burchellii* workers collected during raid columns from 20 colonies in Barro Colorado (Panama) indicated that mite species diversity was high, with a greater number of mites observed in ants collected in raids during the statary phase [93]. Mite prevalence was of c. 5% (from a total of 3146 ants analysed), with infection found to be higher in larger workers (majors and submajors), and *R. carli* the most commonly found mite on *E. burchellii* workers [93]. However, it is important to highlight that workers in raid columns represent only one-third of the colony [66] and, therefore, prevalence and infection rates might differ in army ants staying in the bivouac [93]. According to mite groups feeding preference classifications [87,94], most of the mites observed in this study were suggested to be relatively harmless for the ants (e.g. fungivore, bacterivore), with only 2 out of the 31 taxa classified with confidence as parasitic (*R. carli,* and an unknown taxon within the family Circocyllibanidae).

3.4. Flies

Flies are the second most numerous associates in terms of individuals reported on army ant *Eciton* colonies, with flies within the family Phoridae being the most abundant.

Phorid (or scuttle) flies are found in swarm raids, ant columns, bivouacs and refuse deposits [37,69]. In addition to phorid flies, blowflies (Calliphoridae), thick-headed flies (Conopidae), flesh flies (Sarcophagidae) and tachinid flies (Tachinidae) are also found during *Eciton* raids, in particular in the front or in advance of the swarm fronts. Flies found in *Eciton* raids either hover constantly over the raid or rest on objects with the aim to parasitise insects flashed during the raids. These flies avoid being near the army ant columns or swarms and, similar to other associates, can be either facultative or obligatory [95]. Frequently reported parasitic flies are species within the genera *Androeurops*, *Apocephalus* or *Stylogaster* or *Calodexia* [69,95,96]. Different species of *Androeurops* have been reported attacking crickets and cockroaches flushed during *E. hamatum* and *E. burchellii* raids [37,96]. These flies might be attracted to *Eciton* raids when they detect distress pheromones released from the ants' victims such as found for the phorid *Apocephalus paraponerae* that parasitises the bullet ant *Paraponera clavata* [97]. Phorid flies of the genus *Cremersia* and *Dacnophora* have been shown to parasitise non-*Eciton* neotropical army ants (*Neivamyrmex*, *Nomamyrmex*, and *Labidus*) but there are no records of parasitic flies of *Eciton* army ants [95], although other phorid flies found near the raids might parasitise some *Eciton* associates [95]. In addition to those flies found over or near *Eciton* raids, species within the families Phoridae, Muscidae and Sarcophagidae are commonly found in ants refuse deposits and might be flying between bivouacs to locate these food resources [37].

3.5. Beetles

A diversity range of beetles, in particular of the those belonging to the families Staphylinidae and Histeridae, are found associated with *Eciton* colonies either in refuse deposits or in emigration columns [37]. These beetles range from generalists and are found with more than one genus of army ant, e.g. species within the genus *Alloiodites* [37], to host-specific and only found with colonies of a single *Eciton* species, e.g. *Ecitophya*, *Ecitomorpha* [51]. Some beetles are found in refuse deposits feeding either on dead workers or booty refuse, with reports of flying between bivouacs (e.g. *Ecitophora*, *Omalodes*) and/or running in emigration columns (e.g. *Mesynodites*, *Phelister*). Staphylinidae beetles are common in *Eciton* emigration columns, most frequently located at the edges of the column. Beetles found in emigration columns do not attack the ants, except for some predators such as *Tetradonia terminalis* which predates on adult workers of *E. burchellii*, *E. hamatum* and *E. lucanoides* [37,98]. Integrative taxonomic studies, including DNA barcoding approaches, on *Eciton* colonies in La Selva Biological Station (Costa Rica), identified five *Tetradonia* species, including two novel species. Host specificity varied among species, with the generalist *T. laticeps* parasitising six species of *Eciton* to the specialised *T. lizonae* mainly associated with *E. hamatum* colonies [99].

During the procession of associates following *Eciton* emigration columns, the number of beetles is particularly abundant when the brood is carried away from the bivouac. At this phase of the emigration is when the diversity of beetles more intimately associated with *Eciton* colonies can be observed, such as species of the genera *Euxinister*, *Vatesus*, *Cephaloplectus*, *Limulodes*, *Ecitophya* and *Ecitomorpha* [37,100,101]. Among this array of *Eciton*-following beetles, a number of histerid and staphylinid beetles have evolved morphological, chemical and/or behavioural strategies for further integration in the colony. Some species have evolved adaptations to enable them to ride on the different castes. For example, *Pulvinister nevermannii* has been observed riding under the major heads of *E. hamatum* [42,102], *Limulodes* species riding on queens, and some histerids observed riding on males when present in the colony. Some beetle species are found riding on different worker castes, such as *Euxenister wheeleri* found with *E. hamatum* emigration columns, as well as pupae and larvae carried by workers such as *Euxenister caroli* found in emigration columns of *E. burchellii* [37,42,102,103].

Two genera of staphylinids, Ecitomorpha and Ecitophya (Figure 3) are often found in raids, as well as emigration columns of Eciton army ants. These beetles are highly specialised to the nomadic life of Eciton colonies and mimic their media workers [104].

These two genera of staphilinids have evolved similar body modifications to moderately resemble their host, including their colouration [104–106]. The presence of Ecitomorpha and Ecitophya is rare on Eciton colonies, with an estimate presence of less than one beetle per 1000 workers. These beetles are found in raiding columns feeding on dropped prey or booty catches, and, hence, these beetles are considered hunting guests of the army ants [106]. Ecitomorpha and Ecitophya are adapted to the nomadic life of Eciton colonies and are also found in emigration columns, walking in the centre of the column or riding on ant pupae or prey captured by the ants [105,106]. Early taxonomic work on the genera Ecitomorpha and Ecitophya indicated single host-specificity due to their parallel colouration to that of their host [107,108]. However, later taxonomic treatments lumped several species into a single one, due to the difficulties in finding consistent morphological characters, including colouration, to separate species [104,106]. Using an evolutionary and population genetics framework, analyses of mitochondrial DNA sequences of Eciton hosts and their associated Ecitomorpha and Ecitophya, collected in different areas in east and west Panama, revealed that the speciation pattern of these two myrmecophiles was congruent with specialization to a single species (or subspecies) of Eciton host [51]. Molecular clock analyses conducted in this study suggested that Eciton host diversification pre-dated that of their hunting guests, and that Ecitophya have been associated with Eciton for a longer period of evolutionary time than Ecitomorpha [51]. Furthermore, patterns of diversification of the two hunting guests were also consistent across broad geographical areas but not at small geographical scales, with evidence of gene flow between colonies even across large water features [51]. The specialisation of Ecitophya and Ecitomorpha with their Eciton hosts has also been further confirmed by analyses of cuticular hydrocarbons (CHCs) and behavioural analyses of individuals collected at La Selva Biological Station in Costa Rica [109]. In contrast, similar analyses in the staphylinid ant predator Tetradonia indicated that as a generalist, their chemical profile was more dissimilar to their Eciton hosts, compared to that of the specialists Ecitophya and Ecitomorpha [109].

Figure 3. The rove beetle *Ecitophya simulans* (family Staphylinidae) moderately mimics of *Eciton burchellii* media workers and is considered a hunting guest. Photo of *Ecitophya simulans* next to an *Eciton burchellii foreli* media worker (Taku Shimada).

Beetles of the genus *Vatesus* (Staphylinidae) are also considered to be intimately associated with *Eciton* army ants. These large and conspicuous limuloid (horseshoe crab-shaped) beetles, and in particular the species *V. clypeatus*, have been observed in *E. burchellii*, *E. hamatum*, *E. vagans*, *E. dulcium* and possibly *E. mexicanum* colonies [42,100,110]. The life cycle of *Vatesus* beetles was suspected to be tightly linked to their *Eciton* hosts, with reproduction and larval development observed to be synchronised to the reproduction and nomadic cycles of their *Eciton* hosts during field studies [37,101,110]. The full life cycle of *Vatesus*, however, was only reconstructed in a relatively recent study using DNA barcoding approaches in which high *Eciton* host-specificity was confirmed and cryptic diversity within the species *V. clypeatus* identified [111].

3.6. Other Arthropods

Springtails (Collembola) are abundant in refuse deposits and are represented by a large number of species that might primarily feed on fungi or mould [37,42]. Some species are considered myrmecophilous and also found within the bivouac, in particular those belonging to the subfamilies Cyphoderinae and Paraollinae. The most common myrmecophilous springtails found in *Eciton* colonies belong to the genus *Cyphoda* [37], and two species of *Cyphoderus* have also been reported from *E. burchellii* and *E. hamatum* refuse deposits and suggested as myrmecophiles. However, a large number of springtail taxa found in bivouacs and refuse deposits of *E. burchellii* remain to be identified [37,69].

Millipedes (Diplopoda) of the genera *Calymmodesma* and *Rettenmeyeria* have been reported in *Labidus* and *Nomamyrmex* army ant columns but not in *Eciton* [42,112,113]. Only one species, *Prionodesmus fulgens*, has been reported near bivouac detritus of *E. burchellii* colonies [69].

Silverfish (Zygentoma; previously Thysanura) have been observed in bivouacs, emigration and raid columns of *Eciton* species, in particular species within the genus *Trichatelura* [37]. The species *T. manni* and *T. borgmeiri* are frequently observed running in the centre of the emigration columns when brood is carried from the bivouac, or during nocturnal raid columns where *T. manni* has been also observed to ride on large larvae or booty carried by the ants [37,42]. *Trichatelura manni* has been found associated with different *Eciton* species and observed cleaning and riding on the queens of *E. vagans* and *E. dulcium crassinode* [42]. Observations of *T. manni* in the laboratory conditions confirmed this species as strigilator, feeding on secretions or particles from surfaces of army ant adults, larvae and booty [37,52].

Small wasps of the family Diapriidae are found as associates in *Eciton* colonies [37,114,115], with 17 Diapriidae genera found associated with *Eciton burchellii* colonies [69]. Similar to the mimicry observed in myrmecophilous staphylinids, some of these parasitic wasps have also evolved strategies to integrate into *Eciton* colonies by modifying its form and colour to resemble their *Eciton* host media workers. For example, *Mimopria ecitophila* mimics media workers of *E. hamatum* and it is found in raid and emigration columns of this army ant [42]. These army ant-mimicking wasps are frequently found running in the ant columns and are probably parasites of the army ants or some or their guests [37,42,103,114]. Other Hymenoptera families have also been observed over or near *E. burchellii* colonies or refuse deposits, including other parasitic wasp families such as Scelionidae, Proctotrupidae, and Pompilidae; the latter called spider wasps and suggested to follow the army ant swarms to find spider victims to parasitise [69].

The latest survey of *E. burchellii* associates conducted by Rettemeyer et al. [69] indicated that over 300 butterfly (Lepidoptera) species within the families Hesperiidae, Nymphalidae and Papilionidae were sampled on bird droppings and/or using lures at swarm raids of *E. burchellii*. These butterflies might extract uric acid or partly digested proteins from the bird droppings to be used for egg production [42]. Early field observations of these butterflies, in particular skippers, suggested that *Eciton* colony odours during the raids could resemble male mating pheromones and, therefore, explain the presence of only female skippers [116,117]. However, the recording of male skippers in *E. burchellii* swarm raids casted doubt on the previous hypothesis [118]. A more plausible explanation is that skippers, such as those within the genus *Mechanitis*, are attracted to *E. burchellii* swarm raid odours as a method of colony detection where they can search for the antbird droppings [119–121]. After observations of the species *Mechanitis polymnia ishtmia*, *M. lysimmia doryssus*, and *Melinea lilis imitata* feeding on droppings from antbirds, in particular from the family Formicariidae, during *E. burchellii* raids, led to coining the term antbutterflies to highlight the close association of these butterflies with this army ant colonies [120]. In addition to the presence of butterflies in *Eciton* raids, microlepidoptera caterpillars have also been reported from refuse deposits [37].

3.7. Microbes

The application of DNA metabarcoding approaches has also facilitated recent studies assessing the gut microbiota of army ants. The first initial survey of army ants gut microbiota showed that some ant-specific bacteria clades (e.g. Entomoplasmatales) were shared between three species of neotropical army ants, *E. burchellii*, *Nomamyrmex hartigii*, and *Cheliomyrmex* sp. [122]. This study also found that these bacteria were likely to be associated with the predatory lifestyle of the army ants; although no essential nutritional benefits could be attributed to the presence of these bacteria [122]. A more recent study survey of gut microbiota in army ants including samples from three army ant genera (*Eciton*, *Labidus*, and *Nomamyrmex*) further confirmed a degree of specialisation of certain gut microbial communities in army ants, with the most abundant microbes assigned to Unclassified Firmicutes and Unclassified Entomoplasmatales [123]. Further analyses of these data indicated that gut microbiota variation was found at different levels, between and within species of *E. burchellii* and *L. praedator*, as well as between sympatric colonies *E. burchellii* in Monteverde (Costa Rica), Henri Pittier National Park (Venezuela), and Cocahoatan (Mexico). Variation of gut microbiota composition found in army ants would agree with studies indicating that landscape and local environment can shape gut microbiota in wild insect populations, therefore, explaining geographical variation on microbial profiles [124,125]. One intrinsic challenge faced in DNA metabarcoding of gut contents is our inability to distinguish signatures (in this case DNA sequences) derived from the target species and those of their prey [126]. Therefore, future analyses would benefit from a comparison of gut microbiota of *Eciton* host to those of their potential prey.

3.8. Further Research on Eciton Associates

Although a multitude of vertebrate and invertebrate species associated with *Eciton* army ants have been described, substantial research is still required to identify many of the arthropod taxa collected or observed in previous studies [69]. Interdisciplinary integrative taxonomy-focused projects would be crucial to advance the identification and classification of these taxa, the discovery of new species, and elucidation of host specificity [99,127]. Furthermore, reports and studies of associates of *Eciton* army ants have primarily focused on *E. burchellii*. Therefore, further research on associates of other *Eciton* species with different lifestyles, foraging patterns and behaviours (e.g. hypogaeic lifestyle, nocturnal raids) is likely to identify an additional suite of vertebrate and invertebrate associate taxa to those already reported with *E. burchellii*. Beyond research on different *Eciton* species, there is also a need for studies to cover wider geographical areas, as the genus *Eciton* has a wide distribution, from Mexico to Argentina, but studies on associates of these army ants have mainly been restricted to Costa Rica and Panama to date. The creation of a network of researchers ensuring the participation and coordination of research projects by local researchers would facilitate the collection of samples, funding and collaboration needed to increase our understanding of *Eciton* army ants and their associates throughout their distribution range.

4. Deforestation and *Eciton* Army Ants

Deforestation and habitat fragmentation are predicted to have negative effects in particular on species at higher trophic levels, higher mobility, greater ecological specialization and representative early diverging lineages [128]. The effects of habitat fragmentation, therefore, are expected to be more accentuated in top predators such as army ants, with studies on *E. burchellii* indicating negative long-term responses to forest fragmentation [129]. *Eciton* army ants require large forested areas to obtain large amounts of leaf litter prey and maintain their large nomadic colonies, as well as to shelter from higher temperatures encountered in open areas [60,63,130–132]; therefore, making them particularly sensitive to landscape fragmentation and deforestation [34,39]. Simulation studies based on *E. burchellii* colonies on Barro Colorado Island (Panama) estimated that areas of at least 30 ha of continuous forested would be needed to support a single colony of this army ant [39]. This critical area might differ between geographical areas as field-based studies of *E. burchellii* colonies in Mato Grosso (South West Amazonia, Brazil) found that colonies of this species emigrate less frequently than in Barro Colorado Island, possibly due to a larger density of prey in this area [133]. However, other studies on *E. burchellii* colonies have indicated that even when large areas of forest remain after fragmentation, the long-term persistence of colonies is not guaranteed if these patches are not connected by suitable corridors [34,39,56,131,134].

The maintenance and expansion of traditional agrosystems such as shaded coffee and cocoa plantations are considered realistic management actions to address current environmental and socio-economic challenges in neotropical areas [135]. The presence of *Eciton* colonies has been reported in shaded coffee and cocoa plantations, as these habitats provide prey as well as bivouac resources [136–139]. However, the use of shaded coffee and cocoa plantations by *Eciton* colonies might be only temporary and dependent on the availability of non-modified forested areas nearby [136,139,140]. A landscape genetics study of *E. burchellii* colonies in Panama gave further support to the importance of maintaining mature forested areas [141]. This study revealed that deforestation is a major gene flow barrier for *E. burchellii foreli* colonies. In contrast, mature forests were identified as gene flow facilitators. Other forest types, including shaded coffee plantations, were found to act as gene flow facilitators but only when these were located nearby mature forests [141]. The negative effect of deforestation in *E. burchellii* gene flow was also demonstrated in a subsequent landscape genetics study in Costa Rica that revealed that queen relatedness was correlated with land cover, further highlighting the need to maintain connectivity between colonies [142]. Gene flow between colonies was assumed to be primarily by the large winged males when they emerge and fly in large numbers from their colonies in search of mates; *Eciton* queens are wingless and dispersal, therefore, is entirely pedestrian [48]. This male-biased dispersal has also been indicated by genetic studies comparing maternally-inherited mtDNA sequences and biparentally-inherited microsatellite genotyping data from *E. burchellii* colonies [141,143,144]. However, a recent study analysing population structure of *E. b. parvispinum* across years using microsatellite markers indicated that due to the nomadic lifestyle of these army ants, queen dispersal significantly contributes to gene flow and in some instances exceeds that of males [145].

The creation or protection of forested corridors between neighbouring forested fragments has been suggested as possible management actions to facilitate connectivity and longer-term persistence of *E. burchellii* colonies, as thermal ecology studies indicated that high temperatures in open pastures between fragments could prevent the ants to leave their forest patches [131]. These corridors would not only be beneficial to guarantee gene flow between colonies by facilitating queen and colony pedestrian dispersal but also the dispersal of winged males. Although due their nocturnal behaviour, *Eciton* males are less likely to be affected by high temperatures in open areas between forest fragments, flying over open areas increases the risk of predation by nocturnal insectivorous vertebrates [146,147]. Therefore, suitable forest corridors between fragmented areas would facilitate dispersal, and subsequent gene flow, via maternal and paternal lines.

Foraging rates of diurnal epigaeic army ants such *E. burchellii* are lower in open areas compared to forested areas, in particular at lower elevations where these ants might be at their upper limit thermal tolerance threshold [131,132]. Although *E. burchellii* colonies are frequently reported to cross open areas, use remnant forest fragments [131] and heavily overgrown abandoned pastureland [148], even brief exposure to high temperatures can lead to mortality of their workers [131]. Therefore, movement of colonies between forest fragments might be conditioned by temperature in the environment and vegetation characteristics of the less hospitable matrix between the fragments [131]. The crossing of open areas is feasible during cloudy days or in areas where scattered trees and shrubs can provide shade during the passage [131]. However, mortality among workers during these incursions has been observed after cloud clearing during hottest hours of the day (Pérez-Espona, pers. comm.). For the more hypogaeic and nocturnal species of *Eciton*, such as *E. vagans* and *E. mexicanum*, habitat type does not seem to affect foraging rates, with these increasing with elevation [132].

Ecological assemblages are crucial for the maintenance of ecosystem function and services, and the resilience of these is likely to be dependent on local diversity [149,150]. Local diversity has been shown to be strongly reduced by the destruction or disturbance of the natural environment as a result of land-use changes [151]. In tropical areas, the effect of deforestation and habitat fragmentation has led to a decrease in bird species richness [152]. The dependence of antbirds on army ant colonies make them particularly susceptible to habitat fragmentation. The reduction or disappearance of antbirds from certain forests can be explained by a decrease or absence of important resources such as the presence of army ants [34,129,153–157]. Higher species richness, as well as greater attendance of antbirds following army ants is found in continuous forests [78]. In contrast, antbirds are more prone to local extinction in small forest fragments [155,157]. Considering the numerous associates already reported with *Eciton* army ants, the disappearance of their colonies will not only have a negative effect on antbird local diversity but on the diversity of large number of species across different taxonomic groups.

5. *Eciton* Army Ants as Umbrella Species in Neo-Tropical Forests

Data on species richness and abundance is critical to prioritise the establishment of conservation areas [158]. As this information is often absent, and the effort to collect complete datasets would require long-term funded research projects, data on surrogate species – species that provide an indication of biodiversity levels in a particular area – offer an alternative solution to facilitate the establishment of conservation areas [22,159]. The use of surrogate species generally involves gathering data on their diversity and distribution to project the distribution of other less well-known taxa [17,160]. The use of bioindicator groups has often proved successful for identifying patterns of species at large geographical scales [160–162] but their use as predictors for species richness at smaller geographical scales, at which conservation actions take place, has been disputed [17,163,164]. An alternative approach for the delineation of conservation areas at smaller geographical scales is the use of umbrella species. Species categorised as an umbrella require large areas of habitat for their survival and, therefore, by protecting their habitat, many other species will be protected [16,165], including those that share similar ecology [166,167] or those species more directly interacting with them [168]. However, the conservation of these other species is strongly dependent on the appropriate selection of the umbrella species [169,170].

Vertebrates, in particular large predators, have frequently been selected as umbrella species [171]; however, an ongoing debate exists about whether or not the protection of these vertebrates can effectively conserve the habitats and ecosystems where they exist [172]. Invertebrate species have also been used as umbrella species in areas where large iconic vertebrates are absent, with butterflies being the most commonly used in terrestrial ecosystems [168,173]. Furthermore, increasing recognition of the central role of invertebrates, in particular insects, on the functioning of ecosystems has led to their protection as well as to implement their use for the establishment of conservation areas and management plans [21,30,33,168].

It is important to highlight that the value of a particular umbrella species is underpinned by the conservation aims and that, in some instances, multiple species might need to be considered when devising conservation actions to ensure the long-term protection of local biodiversity [168,174]. *Eciton* army ants represent excellent potential umbrella species for the conservation of neotropical forests. A multitude of species, invertebrate and vertebrate, have already been described to be associated with *Eciton* colonies, with many more species yet to be reported and described [69]. Furthermore, large forested areas are needed for the long-term maintenance of their large colonies not only to guarantee a diverse leaf litter prey community but also to protect *Eciton* colonies from temperatures that hinder their dispersal between forest fragments [34,39,131]. The negative effects of deforestation, habitat fragmentation and land-use change on *Eciton* colonies have been reported by field-based studies [60,175] as well as genetic studies [141,142], indicating the importance of the protection of mature forests for the long-term survival of these army ants, and subsequently, for the many species depending on them. Although there are no available studies assessing any negative impact of the disappearance of *Eciton* colonies in non-associates, the designation of *Eciton* as umbrella species would also ensure the protection of other species in neotropical forests coexisting in the same areas, in particular those that require large extents of forested area for the maintenance of their populations. Therefore, conservation actions aimed to maintain the long-term persistence of army ants, either by identifying areas for protection or creation of habitat corridors to maintain or increase connectivity between *Eciton* colonies in fragmented forests would have positive impact on a wide range of neotropical forest species.

Despite their importance as keystone species in neotropical forests, none of the *Eciton* species, or in fact any other army ants, have yet been considered as umbrella species for the delimitation of conservation areas or for the monitoring of sustainable use in neotropical forests, although their significant influence in ecosystem structure, composition and functioning has been widely recognised [34,176]. The selection of *Eciton* species as umbrella species to identify priority conservation areas or monitor sustainable use could represent an effective and efficient approach for the conservation of neotropical forests and the ecosystem services that they provide. Therefore, this review calls for the consideration of *Eciton* army ant species in future conservation planning, design of nature reserves, and monitoring of sustainable use of forests in neotropical areas, by recognising the key role of *Eciton* species for the maintenance of biodiversity. To successfully achieve the long-term protection of neotropical forests, conservation actions must be led by local researchers and environmental agencies and non-governmental organisations with the participation of local communities that depend on forested areas for their livelihood [177–179].

Supplementary Materials: The following are available online at https://www.mdpi.com/1424-2818/13/3/136/s1, Table S1: Antbird species reported to attend *Eciton* army ant raids. Most of the reports refer to *Eciton burchellii*, [P] indicates species that have also been reported attending raid of *E. b. parvispinum*, [r] *Eciton rapax* and [m] *E. mexicanum*. The degree of specialization of antbirds is indicated as Ob (Obligate), R (Regular), and Oc (Occasional). For antbird species for which the behaviour of bivouac checking has been observed this is indicated as [b]. Table S2: Number of species in different arthropod families reported associated with *Eciton burchellii* according to Rettenmeyer et al. (2011).

Funding: This research received no external funding.

Acknowledgments: The academic editor and all the anonymous reviewers are thanked for their constructive criticism and comments on previous versions of the manuscript.

Conflicts of Interest: The authors declare no conflict of interest.

References

1. Sala, O.E.; Chapin, F.S.; Armesto, J.J.; Berlow, E.; Bloomfield, J.; Dirzo, R.; Huber-Sanwald, E.; Huenneke, L.F.; Jackson, R.B.; Kinzig, A.; et al. Global biodiversity scenarios for the year 2100. *Science* **2000**, *287*, 1770–1774. [CrossRef]
2. Fahrig, L. Effects of Habitat Fragmentation on Biodiversity. *Annu. Rev. Ecol. Evol. Syst.* **2003**, *34*, 487–515. [CrossRef]
3. Fischer, J.; Lindenmayer, D.B. Landscape modification and habitat fragmentation: A synthesis. *Glob. Ecol. Biogeogr.* **2007**, *16*, 265–280. [CrossRef]
4. de Chazal, J.; Rounsevell, M.D.A. Land-use and climate change within assessments of biodiversity change: A review. *Glob. Environ. Chang.* **2009**, *19*, 306–315. [CrossRef]

5. Schulze, E.-D.; Mooney, H.A. *Ecosystem Function of Biodiversity: A Summary*; Schulze, E.-D., Mooney, H.A., Eds.; Springer: Berlin, Germany; London, UK, 1994.

6. Cardinale, B.J.; Duffy, J.E.; Gonzalez, A.; Hooper, D.U.; Perrings, C.; Venail, P.; Narwani, A.; MacE, G.M.; Tilman, D.; Wardle, D.A.; et al. Biodiversity loss and its impact on humanity. *Nature* **2012**, *486*, 59–67. [CrossRef]

7. Mitchell, M.G.E.; Bennett, E.M.; González, A. Strong and nonlinear effects of fragmentation on ecosystem service provision at multiple scales. *Environ. Res. Lett.* **2015**, *10*. [CrossRef]

8. Keenan, R.J.; Reams, G.A.; Achard, F.; de Freitas, J.V.; Grainger, A.; Lindquist, E. Dynamics of global forest area: Results from the FAO Global Forest Resources Assessment 2015. *For. Ecol. Manage.* **2015**, *352*, 9–20. [CrossRef]

9. Tyukavina, A.; Hansen, M.C.; Potapov, P.V.; Stehman, S.V.; Smith-Rodriguez, K.; Okpa, C.; Aguilar, R. Types and rates of forest disturbance in Brazilian Legal Amazon, 2000–2013. *Sci. Adv.* **2017**, *3*, 1–16. [CrossRef] [PubMed]

10. Dossa, G.G.O.; Paudel, E.; Schaefer, D.; Zhang, J.L.; Cao, K.F.; Xu, J.C.; Harrison, R.D. Quantifying the factors affecting wood decomposition across a tropical forest disturbance gradient. *For. Ecol. Manage.* **2020**, *468*. [CrossRef]

11. Alroy, J. Effects of habitat disturbance on tropical forest biodiversity. *Proc. Natl. Acad. Sci. USA* **2017**, *114*, 6056–6061. [CrossRef]

12. CBD. Secretariat of the Convention on Biological Diversity (2020). In *Global Biodiversity Outlook 5*; CBD: Montreal, QC, USA; ISBN 9292255398.

13. FAO; UNEP. *The State of the World's Forests 2020. Forests, Biodiversity and People*; FAO and UNEP: Rome, Italy, 2020.

14. Stork, N.E.; Srivastava, D.S.; Eggleton, P.; Hodda, M.; Lawson, G.; Leakey, R.R.B.; Watt, A.D. Consistency of effects of tropical-forest disturbance on species composition and richness relative to use of indicator taxa. *Conserv. Biol.* **2017**, *31*, 924–933. [CrossRef] [PubMed]

15. Gardner, T.A.; Barlow, J.; Araujo, I.S.; Ávila-Pires, T.C.; Bonaldo, A.B.; Costa, J.E.; Esposito, M.C.; Ferreira, L.V.; Hawes, J.; Hernandez, M.I.M.; et al. The cost-effectiveness of biodiversity surveys in tropical forests. *Ecol. Lett.* **2008**, *11*, 139–150. [CrossRef] [PubMed]

16. Noss, R.F. Indicators for monitoring biodiversity: A hierarchical approach. *Conserv. Biol.* **1990**, *4*, 355–364. [CrossRef]

17. Lawton, J.H.; Bignell, D.E.; Bolton, B.; Bloemers, G.F.; Eggleton, P.; Hammond, P.M.; Hodda, M.; Holt, R.D.; Larsen, T.B.; Mawdsley, N.A.; et al. Biodiversity inventories, indicator taxa and effects of habitat modification in tropical forest. *Nature* **1998**, *391*, 72–76. [CrossRef]

18. Lindenmayer, D.B.; Westgate, M.J. Are flagship, umbrella and keystone species useful surrogates to understand the consequences of landscape change? *Curr. Landsc. Ecol. Reports* **2020**, *5*, 76–84. [CrossRef]

19. Lindenmayer, D.B.; Margules, C.R.; Botkin, D.B. Indicators of biodiversity for ecologically sustainable forest management. *Conserv. Biol.* **2000**, *14*, 941–950. [CrossRef]

20. Landres, P.B.; Verner, J.; Thomas, J.W. Ecological uses of vertebrate indicator species: A critique. *Conserv. Biol.* **1988**, *2*, 316–328. [CrossRef]

21. McGeogh, M.A. The selection, testing and application of terrestrial insects as bioindicators. *Biol. Rev.* **1998**, *73*, 181–201. [CrossRef]

22. Caro, T.M.; O'Doherty, G. On the use of surrogate species in conservation biology. *Conserv. Biol.* **1999**, *13*, 805–814. [CrossRef]

23. Lewandowski, A.S.; Noss, R.F.; Parsons, D.R. The effectiveness of surrogate taxa for the representation of biodiversity. *Conserv. Biol.* **2010**, *24*, 1367–1377. [CrossRef] [PubMed]

24. Didham, R.K.; Stork, N.E.; Davis, A.J. Insects in fragmented forests: A functional approach. *Trends Ecol. Evol.* **1996**, *11*, 255–260. [CrossRef]

25. Rands, M.R.W.; Adams, W.M.; Bennun, L.; Butchart, S.H.M.; Clements, A.; Coomes, D.; Entwistle, A.; Hodge, I.; Kapos, V.; Scharlemann, J.P.W.; et al. Biodiversity conservation: Challenges beyond 2010. *Science* **2010**, *329*, 1298–1303. [CrossRef]

26. Kremen, C.; Colwell, R.K.; Erwin, T.L.; Murphy, D.D.; Noss, R.F.; Sanjayan, M.A. Terrestrial arthropod assemblages: Their use in conservation planning. *Conserv. Biol.* **1993**, *7*, 796–808. [CrossRef]

27. Hamilton, A.J.; Basset, Y.; Benke, K.K.; Grimbacher, P.S.; Miller, S.E.; Novotný, V.; Samuelson, G.A.; Stork, N.E.; Weiblen, G.D.; Yen, J.D.L. Quantifying uncertainty in estimation of tropical arthropod species richness. *Am. Nat.* **2010**, *176*, 90–95. [CrossRef]

28. Brown, K.S. Diversity, disturbance, and sustainable use of Neotropical forests: Insects as indicators for conservation monitoring. *J. Insect Conserv.* **1997**, *1*, 25–42. [CrossRef]

29. Andersen, A.N. Using ants as bioindicators: Multiscales issues in ant community ecology. *Conserv. Ecol.* **1997**, *1*. [CrossRef]

30. Andersen, A.N.; Hoffmann, B.D.; Müller, W.J.; Griffiths, A.D. Using ants as bioindicators in land management: Simplifying assessment of ant community responses. *J. Appl. Ecol.* **2002**, *39*, 8–17. [CrossRef]

31. Majer, J.D.; Orabi, G.; Bisevac, L. Ants (Hymenoptera: Formicidae) pass the bioindicator scorecard. *Myrmecol. News* **2007**, *10*, 69–76.

32. Bihn, J.H.; Gebauer, G.; Brandl, R. Loss of functional diversity of ant assemblages in secondary tropical forests. *Ecology* **2010**, *91*, 782–792. [CrossRef]

33. Groc, S.; Delabie, J.H.C.; Fernandez, F.; Petitclerc, F.; Corbara, B.; Leponce, M.; Céréghino, R.; Dejean, A. Litter-dwelling ants as bioindicators to gauge the sustainability of small arboreal monocultures embedded in the Amazonian rainforest. *Ecol. Indic.* **2017**, *82*, 43–49. [CrossRef]

34. Boswell, G.P.; Britton, N.F.; Franks, N.R. Habitat fragmentation, percolation theory and the conservation of a keystone species. *Proc. R. Soc. B Biol. Sci.* **1998**, *265*, 1921–1925. [CrossRef]

35. Kaspari, M.; O'Donnell, S. High rates of army ant raids in the Neotropics and implications for ant colony and community structure. *Evol. Ecol. Res.* **2003**, *5*, 933–939.

36. Schneirla, T. *Army Ants: A Study in Social Organization*; Topoff, H.R., Ed.; Freeman, WH: San Francisco, CA, USA, 1971.

37. Rettenmeyer, C.W. Arthropods associated with neotropical army ants with a review of the behaviour of these ants (Arthropoda; Formicidae; Dorylinae). Ph.D. Thesis, The University of Kansas, Lawrence, KS, USA, 1961.

38. Gotwald, W.H., Jr. *Army Ants: The Biology of Social Predation*; Cornell University Press: Ithaca, NY, USA, 1995.

39. Partridge, L.W.; Britton, N.F.; Franks, N.R. Army ant population dynamics: The effects of habitat quality and reserve size on population size and time to extinction. *Proc. R. Soc. B Biol. Sci.* **1996**, *263*, 735–741. [CrossRef]

40. Secretariat of the Convention on Biological Diversity. *Assessment, Conservation and Sustainable Use of Forest Biodiversity*; CBD Technical Series no. 3; SCBD: Montreal, QC, USA, 2001; 130p.

41. Alonso, L.E.; Persaud, J.; Williams, A. *Biodiversity Assessment Survey of the South Rupununi Savannah, Guyana. BAT Survey Report No 1.*; WWF-Guianas, Guyana Office: Georgetown, Guyana, 2016.

42. Kistner, D.H. The social insects' bestiary. In *Social Insects*; Hermann, H.R., Ed.; Academic Press: London, UK, 1982; Volume 3, pp. 1–244. ISBN 0323148964, 9780323148962.

43. Brady, S.G.; Fisher, B.L.; Schultz, T.R.; Ward, P.S. The rise of army ants and their relatives: Diversification of specialized predatory doryline ants. *BMC Evol. Biol.* **2014**, *14*, 1–14. [CrossRef]

44. Borowiec, M.L. Convergent evolution of the army ant syndrome and congruence in big-data phylogenetics. *Syst. Biol.* **2019**, *68*, 642–656. [CrossRef] [PubMed]

45. Gotwald, W.H., Jr. *Army Ants*; Academic Press, Inc.: London, UK, 1982.

46. Brady, S.G. Evolution of the army ant syndrome: The origin and long-term evolutionary stasis of a complex of behavioral and reproductive adaptations. *Proc. Natl. Acad. Sci. USA* **2003**, *100*, 6575–6579. [CrossRef] [PubMed]

47. Kronauer, D.J.C. *Army Ants: Nature's Ultimate Social Hunters*; Harvard University Press: Cambridge, MA, USA, 2020; ISBN 9780674241558.

48. Franks, N.R.; Hölldobler, B. Sexual competition during colony reproduction in army ants. *Biol. J. Linn. Soc.* **1987**, *30*, 229–243. [CrossRef]

49. Watkins, J.F. *The Identification and Distribution of New World Army Ants*; Baylor University Press: Waco, TX, USA, 1976.

50. Winston, M.E. *Bridging Micro- and Macroevolution in Neotropical Army Ants*; University of Chicago: Chicago, IL, USA, 2017.

51. Pérez-Espona, S.; Goodall-Copestake, W.P.; Berghoff, S.M.; Edwards, K.J.; Franks, N.R. Army imposters: Diversification of army ant-mimicking beetles with their *Eciton* hosts. *Insectes Soc.* **2018**, *65*, 59–75. [CrossRef]

52. Rettenmeyer, C.W. Behavioral studies of army ants. *Univ. Kansas Sci. Bull.* **1963**, *44*, 281–465.

53. Powell, S.; Baker, B. The hidden big predators of the Neotropics: The behaviour, diet, and impact of New World army ants (Ecitoninae). *Insetos Sociais Biol. Apl.* **2008**, *18–37*.

54. Schneirla, T.C. Raiding and other outstanding ohenomena in the behavior of army ants. *Proc. Natl. Acad. Sci. USA* **1934**, *20*, 316–321. [CrossRef]

55. Hoenle, P.O.; Blüthgen, N.; Brückner, A.; Kronauer, D.J.C.; Fiala, B.; Donoso, D.A.; Smith, M.A.; Ospina Jara, B.; von Beeren, C. Species-level predation network uncovers high prey specificity in a Neotropical army ant community. *Mol. Ecol.* **2019**, *28*, 2423–2440. [CrossRef]

56. Franks, N.R.; Fletcher, C.R. Spatial patterns in army ant foraging and migration: *Eciton burchelli* on Barro Colorado Island, Panama. *Behav. Ecol. Sociobiol.* **1983**, *12*, 261–270. [CrossRef]

57. O'Donnell, S.; Jeanne, R.L. Notes on an army ant (*Eciton burchelli*) raid on a social wasp colony (*Agelaia yepocapa*) in Costa Rica. *J. Trop. Ecol.* **1990**, *6*, 507–509. [CrossRef]

58. Silva Vieira, S.; Höfer, H. Prey spectrum of two army ant species in central Amazonia, with special attention on their effect on spider populations. *Andrias* **1994**, *13*, 189–198.

59. Powell, S.; Franks, N.R. Ecology and the evolution of worker morphological diversity: A comparative analysis with *Eciton* army ants. *Funct. Ecol.* **2006**, *20*, 1105–1114. [CrossRef]

60. O'Donnell, S.; Lattke, J.; Powell, S.; Kaspari, M. Army ants in four forests: Geographic variation in raid rates and species composition. *J. Anim. Ecol.* **2007**, *76*, 580–589. [CrossRef]

61. Manubay, J.A.; Powell, S. Detection of prey odours underpins dietary specialization in a Neotropical top-predator: How army ants find their ant prey. *J. Anim. Ecol.* **2020**, *89*, 1165–1174. [CrossRef] [PubMed]

62. Hou, C.; Kaspari, M.; Vander Zanden, H.B.; Gillooly, J.F. Energetic basis of colonial living in social insects. *Proc. Natl. Acad. Sci. USA* **2010**, *107*, 3634–3638. [CrossRef]

63. Schneirla, T.; Brown, R.Z. Army-ant life and behavior under dry-season. 4. Further investigation of cyclic processess in behavioral and reproductive functions. *Bull. Am. Museum Nat. Hist.* **1950**, *95*, 263–354.

64. Franks, N.R.; Sendova-Franks, A.B. Army Ants: A Collective Intelligence: A neural network Seems an apt analogy as a colony of army ants navigates the tropical rain forest. *Am. Sci.* **1989**, *77*, 138–145.

65. Kaspari, M.; Powell, S.; Lattke, J.; O'Donnell, S. Predation and patchiness in the tropical litter: Do swarm-raiding army ants skim the cream or drain the bottle? *J. Anim. Ecol.* **2011**, *80*, 818–823. [CrossRef] [PubMed]

66. Franks, N.R. Reproduction, foraging efficiency and worker polymorphism in army ants. In *Experimental Behavioral Ecology and Sociobiology: In memoriam Karl von Frisch, 1886–1982*; Hölldobler, B., Lindauer, M., Franks, N.R., Eds.; Wiley Online Library: Hoboken, NJ, USA, 1985; Volume Fortschrit, pp. 91–107. ISBN 087893460X.
67. Franks, N.R.; Bossert, W.H. The influence of swarm raiding army ants on the patchiness and diversity of a tropical leaf litter ant community. In *Tropical Rain Forest: Ecology and Management. Special Publication No. 2 of the British Ecological Society*; Sutton, S.L., Whitmore, T.C., Chadwick, A.C., Eds.; Blackwell: Oxford, UK, 1983; pp. 151–163.
68. Willis, E.O.; Oniki, Y. Birds and Army Ants. *Annu. Rev. Ecol. Syst.* **1978**, *9*, 243–263. [CrossRef]
69. Rettenmeyer, C.W.; Rettenmeyer, M.E.; Joseph, J.; Berghoff, S.M. The largest animal association centered on one species: The army ant *Eciton burchellii* and its more than 300 associates. *Insectes Soc.* **2011**, *58*, 281–292. [CrossRef]
70. Johnson, R.A. The behavior of birds attending army ant raids on Barro Colorado Island, Panama Canal Zone. *Proc. Linn. Soc. New York* **1954**, *63–65*, 41–70.
71. Brumfield, R.T.; Tello, J.G.; Cheviron, Z.A.; Carling, M.D.; Crochet, N.; Rosenberg, K.V. Phylogenetic conservatism and antiquity of a tropical specialization: Army-ant-following in the typical antbirds (Thamnophilidae). *Mol. Phylogenet. Evol.* **2007**, *45*, 1–13. [CrossRef]
72. Willis, E. the Behavior of Spotted Antbirds. *Ornithol. Monogr.* **1972**, *10*, 1–162. [CrossRef]
73. Willis, E.O. Studies of the behavior of Lunulated and Salvin's Antbirds. *Condor* **1968**, *70*, 128–148. [CrossRef]
74. Faria, C.M.A.; Rodrigues, M. Birds and army ants in a fragment of the Atlantic Forest of Brazil. *J. F. Ornithol.* **2009**, *80*, 328–335. [CrossRef]
75. Howell, T.R. Birds of a second-growth rain forest area of Nicaragua. *Condor* **1957**, *59*, 73–111. [CrossRef]
76. Driver, R.J.; DeLeon, S.; O'Donnell, S. Novel observation of a raptor, collared forest-falcon (*Micrastur semitorquatus*) depredating a fleeing snake at an army ant (*Eciton burchellii parvispinum*) raid front. *Wilson J. Ornithol.* **2018**, *130*, 792–796. [CrossRef]
77. Chaves-Campos, J.; DeWoody, J.A. The spatial distribution of avian relatives: Do obligate army-ant-following birds roost and feed near family members? *Mol. Ecol.* **2008**, *17*, 2963–2974. [CrossRef]
78. Kumar, A.; O'Donnell, S. Fragmentation and elevation effects on bird–army ant interactions in neotropical montane forest of Costa Rica. *J. Trop. Ecol.* **2007**, *23*, 581. [CrossRef]
79. Willson, S.K. Obligate army-ant following birds: A study of ecology, spatial movement patterns, and behavior in Amazonian Peru. *Ornithol. Monogr.* **2004**, *1*, 1–67. [CrossRef]
80. Swartz, M.B. Bivouac checking, a novel behavior distinguishing obligate from opportunistic species of army-ant-following birds. *Condor* **2001**, *103*, 629–633. [CrossRef]
81. Wrege, P.H.; Wikelski, M.; Mandel, J.T.; Rassweiler, T.; Couzin, I.D. Antbirds parasitizes foraging army ants. *Ecology* **2005**, *86*, 555–559. [CrossRef]
82. Von Ihering, H. Biologie und Verbreitung der brasilianischen Arten von *Eciton*. *Entomol. Mitteilungen* **1912**, *1*, 226–335.
83. Bequaert, J. The predaceous enemies of ants. *Auk* **1923**, *40*, 162. [CrossRef]
84. Chesser, R.T. Comparative diets of obligate ant-following birds at a site in Northern Bolivia. *Biotropica* **1995**, *27*, 382–390. [CrossRef]
85. Rylands, A.B.; Da Cruz, M.A.O.M.; Ferrari, S.F. An association between marmosets and army ants in Brazil. *J. Trop. Ecol.* **1989**, *5*, 113–116. [CrossRef]
86. Martins, M.M. Foraging over army ants by *Callithrix aurita* (Primates: Callitrichidae): Seasonal occurrence? *Rev. Biol. Trop.* **2000**, *48*, 261–262.
87. Eickwort, G.C. Associations of mites with social insects. *Annu. Rev. Entomol.* **1990**, *35*, 469–488. [CrossRef]
88. Gotwald, W.H. Mites that live with army ants: A natural history of some myrmecophilous hitch-hikers, browsers, and parasites. *J. Kansas Entomol. Soc.* **1996**, *69*.
89. Elzinga, R.J. Holdfast mechanisms in certain Uropodine mites (Acarina: Uropodina). *Ann. Entomol. Soc. Am.* **1978**, *71*, 896–900. [CrossRef]
90. Elzinga, R.J.; Rettenmeyer, C.W. Seven new species of *Circocylliba* (Acarina: Uropodina) found on army ants. *Acarologia* **1975**, *16*, 595–611.
91. Elzinga, R.J. Two new species of *Planodiscus* (Acari: Uropodine), range extensions and a synonymy within the genus. *Acarologia* **1990**, *31*, 229–233.
92. Rettenmeyer, C.W. Notes on host specificity and behavior of myrmecophilous macrochelid mites. *J. Kansas Entomol. Soc.* **1962**, *35*, 358–360.
93. Berghoff, S.M.; Wurst, E.; Ebermann, E.; Sendova-Franks, A.B.; Rettenmeyer, C.W.; Franks, N.R. Symbionts of societies that fission: Mites as guests or parasites of army ants. *Ecol. Entomol.* **2009**, *34*, 684–695. [CrossRef]
94. Krantz, G.W. *A Manual of Acarology*, 2nd ed.; Oregon State University Book Stores: Corvallis, OR, USA, 1978.
95. Brown, B.V.; Feener, D.H. Parasitic phorid flies (Diptera: Phoridae) associated with army ants (Hymenoptera: Formicidae: Ecitoninae, Dorylinae) and their conservation biology. *Biotropica* **1998**, *30*, 482–487. [CrossRef]
96. Rettenmeyer, C.W. Observations on the biology and taxonomy of flies found over swarm raids of army ants (Diptera: Tachinidae, Conopidae). *Univ. Kans. Sci. Bull.* **1961**, *42*, 993–1066.
97. Feener, D.H.; Jacobs, L.F.; Schmidt, J.O. Specialized parasitoid attracted to a pheromone of ants. *Anim. Behav.* **1996**, *51*, 61–66. [CrossRef]

98. Santiago-Jiménez, Q.J.; Espinosa De Los Monteros, A. Exploring myrmecophily based on the phylogenetic interrelationships of *Falagonia* Sharp, 1883 (Coleoptera: Staphylinidae: Aleocharinae) and allied genera. *Syst. Entomol.* **2016**, *41*, 794–807. [CrossRef]

99. Von Beeren, C.; Maruyama, M.; Kronauer, D.J.C. Community sampling and integrative taxonomy reveal new species and host specificity in the army ant-associated beetle genus *Tetradonia* (Coleoptera, Staphylinidae, Aleocharinae). *PLoS ONE* **2016**, *11*, 1–19. [CrossRef]

100. Seevers, C.H. A revision of the Vatesini, a tribe of neotropical myrmecophiles (Coleoptera, Staphylinidae). *Rev. Bras. Entomol.* **1958**, *8*, 181–202.

101. Akre, R.D.; Rettenmeyer, C.W. Trail-following by guests of army ants (Hymenoptera: Formicidae: Ecitonini). *J. Kansas Entomol. Soc.* **1968**, *41*, 165–174.

102. Akre, R. The behaviour of *Euxenister* and *Pulvinister*, histerid beetles associated with army ants. (Coleoptera: Histeridae; Hymenoptera: Formicidae: Dorylinae.). *Pan Pacific Entomol.* **1968**, *44*, 87–101.

103. Mann, W.M. Guests of *Eciton Hamatum* (Fab.) Collected by Professor W. M. Wheeler. *Psyche (New York)* **1925**, *32*, 166–177. [CrossRef]

104. Seevers, C.H. The systematics, evolution and zoogeography of staphylinid beetles associated with army ants (Coleoptera, Staphylinidae). *Fieldiana Zool.* **1965**, *47*, 137–351.

105. Akre, R.D.; Rettenmeyer, C.W. Behavior of Staphylinidae associated with army ants (Formicidae: Ecitonini). *J. Kansas Entomol. Soc.* **1966**, *39*, 745–782.

106. Kistner, D.H.; Jacobson, H.R. Cladistic analysis and taxonomic revision of the ecitophilous tribe Ecitocharini with studies of their behavior and evolution (Coleoptera, Staphylinidae, Aleocharinae). *Sociobiology* **1990**, *17*, 333–480.

107. Reichensperger, A. Ecitophilen aus Costa Rica (II), Brasilien und Peru (Staph. Hist. Clavig.). *Rev. Entomol.* **1933**, *3*, 179–194.

108. Reichensperger, A. Beitrag zur Kenntnis der Myrmecophilenfauna Brasiliens und Costa Ricas III. (Col. Staphyl. Hist.). *Arb. über Morphol. Taxon. Entomol. Berlin Dahlem* **1935**, *2*, 188–218.

109. von Beeren, C.; Brückner, A.; Maruyama, M.; Burke, G.; Wieschollek, J.; Kronauer, D.J.C. Chemical and behavioral integration of army ant-associated rove beetles—A comparison between specialists and generalists. *Front. Zool.* **2018**, *15*, 1–15. [CrossRef] [PubMed]

110. Akre, R.; Torgerson, R. Behavior of *Vatesus* beetles associated with army ants (Coleoptera: Staphylinidae). *Pan Pacific Entomol.* **1969**, *45*, 269–281.

111. Von Beeren, C.; Maruyama, M.; Kronauer, D.J.C. Cryptic diversity, high host specificity and reproductive synchronization in army ant-associated *Vatesus* beetles. *Mol. Ecol.* **2016**, *25*, 990–1005. [CrossRef] [PubMed]

112. Loomis, H.F. New myrmecophilous millipeds from Barro Colorado Island, Canal Zone, and Mexico. *Source J. Kansas Entomol. Soc. Kansas Entomol. Soc.* **1959**, *32*, 1–7.

113. Rettenmeyer, C.W. The behavior of millipeds found with neotropical army ants. *J. Kansas Entomol. Soc.* **1962**, *35*, 377–384.

114. Ferrière, C. Nouveaux Diapriides du Brésil, hôtes des *Eciton*. *Zool. Anz.* **1929**, *82*, 156–171.

115. Borgmeier, T. Sobre alguns Diapriideos myrmecophilos, principalmente do Brasil (Hym. Diapriidae). *Rev. Entomol.* **1939**, *10*, 530–535.

116. Zikán, J.F. Myrmekophilie bei Hesperiden? *Entomol. Rundschau* **1929**, *46*, 27–28.

117. Drummond, B.A. Buttterflies associated with an army ant swarm raid in Honduras. *J. Lepid. Soc.* **1976**, *30*, 237–238.

118. Haber, W.A. Evolutionary Ecology of Tropical Mimetic Butterflies (Lepidoptera: Ithomiiae). Ph.D. Thesis, University of Minnesota, Minneapolis, MN, USA, 1978.

119. Young, A.M. Butterflies associated with an army ant swarm raid in Honduras: The "feeding hypothesis" as an alternate explanation. *J. Lepid. Soc.* **1977**, *31*, 190.

120. Ray, T.S.; Andrews, C.C. Antbutterflies: Butterflies that follow army ants to feed on antbird droppings. *Science* **1980**, *210*, 1147–1148. [CrossRef]

121. Austin, G.T.; Brock, J.P.; Mielke, O.H.H. Ants, birds, and skippers. *Trop. Lepidop.* **1993**, *4*, 1–11.

122. Funaro, C.F.; Kronauer, D.J.C.; Moreau, C.S.; Goldman-Huertas, B.; Pierce, N.E.; Russell, J.A. Army ants harbor a host-specific clade of Entomoplasmatales bacteria. *Appl. Environ. Microbiol.* **2011**, *77*, 346–350. [CrossRef]

123. Łukasik, P.; Newton, J.A.; Sanders, J.G.; Hu, Y.; Moreau, C.S.; Kronauer, D.J.C.; O'Donnell, S.; Koga, R.; Russell, J.A. The structured diversity of specialized gut symbionts of the New World army ants. *Mol. Ecol.* **2017**, *26*, 3808–3825. [CrossRef] [PubMed]

124. Chandler, J.A.; Lang, J.; Bhatnagar, S.; Eisen, J.A.; Kopp, A. Bacterial communities of diverse *Drosophila* species: Ecological context of a host-microbe model system. *PLoS Genet.* **2011**. [CrossRef] [PubMed]

125. Tiede, J.; Scherber, C.; Mutschler, J.; McMahon, K.D.; Gratton, C. Gut microbiomes of mobile predators vary with landscape context and species identity. *Ecol. Evol.* **2017**, *7*, 8545–8557. [CrossRef]

126. Paula, D.P.; Linard, B.; Crampton-Platt, A.; Srivathsan, A.; Timmermans, M.J.T.N.; Sujii, E.R.; Pires, C.S.S.; Souza, L.M.; Andow, D.A.; Vogler, A.P. Uncovering trophic interactions in arthropod predators through DNA shotgun-sequencing of gut contents. *PLoS ONE* **2016**, *11*, 1–14. [CrossRef] [PubMed]

127. Padial, J.M.; Miralles, A.; De la Riva, I.; Vences, M. The integrative future of taxonomy. *Front. Zool.* **2010**, *7*, 1–14. [CrossRef] [PubMed]

128. Laurance, W.F. Theory meets reality: How habitat fragmentation research has transcended island biogeographic theory. *Biol. Conserv.* **2008**. [CrossRef]

129. Offerman, H.L.; Dale, V.H.; Pearson, S.M.; Bierregaard, R.O.; O'Neill, R.V. Effects of forest fragmentation on neotropical fauna: Current research and data availability. *Environ. Rev.* **1995**, *3*, 191–211. [CrossRef]

130. Levings, S.C. Seasonal, annual, and among-site variation in the ground ant community of a deciduous tropical forest: Some causes of patchy species distributions. *Ecol. Monogr.* **1983**. [CrossRef]

131. Meisel, J.E. Thermal ecology of the neotropical army ant *Eciton burchellii*. *Ecol. Appl.* **2006**, *16*, 913–922. [CrossRef]

132. Kumar, A.; Longino, J.T.; Colwell, R.K.; Donnell, S.O. Elevational patterns of diversity and abundance of eusocial paper wasps (Vespidae) in Costa Rica. *Biotropica* **2009**, *41*, 338–346. [CrossRef]

133. Willson, S.K.; Sharp, R.; Ramler, I.P.; Sen, A. Spatial movement optimization in amazonian *Eciton burchellii* army ants. *Insectes Soc.* **2011**, *58*, 325–334. [CrossRef]

134. Lozano-Zambrano, F.H.; Ulloa-Chacón, P.; Armbrecht, I. Ants: Species-Area relationship in tropical dry forest fragments. *Neotrop. Entomol.* **2009**, *38*, 44–54. [CrossRef] [PubMed]

135. Perfecto, I.; Vandermeer, J. Biodiversity conservation in tropical agroecosystems: A new conservation paradigm. *Ann. N. Y. Acad. Sci.* **2008**, *1134*, 173–200. [CrossRef]

136. Perfecto, I.; Snelling, R. Biodiversity and the transformation of a tropical agroecosystem: Ants in coffee plantations. *Ecol. Appl.* **1995**, *5*, 1084–1097. [CrossRef]

137. Perfecto, I.; Rice, R.A.; Greenberg, R.; van der Voort, M.E. Shade coffee: Update on a disappearing refuge for biodiversity. *Bioscience* **1996**, *46*, 598–608. [CrossRef]

138. Roberts, D.L.; Cooper, R.J.; Petit, L.J. Use of premontane moist forest and shade coffee agroecosystems by army ants in western Panama. *Conserv. Biol.* **2000**, *14*, 192–199. [CrossRef]

139. Delabie, J.H.C.; Jahyny, B.; Do Nascimento, I.C.; Mariano, C.S.F.; Lacau, S.; Campiolo, S.; Philpott, S.M.; Leponce, M. Contribution of cocoa plantations to the conservation of native ants (Insecta: Hymenoptera: Formicidae) with a special emphasis on the Atlantic Forest fauna of southern Bahia, Brazil. *Biodivers. Conserv.* **2007**, *16*, 2359–2384. [CrossRef]

140. Stouffer, P.C. Survival of the ant followers. *Nat. Hist. Hist.* **1998**, *107*, 40–43.

141. Pérez-Espona, S.; McLeod, J.E.; Franks, N.R. Landscape genetics of a top neotropical predator. *Mol. Ecol.* **2012**, *21*. [CrossRef] [PubMed]

142. Soare, T.W.; Kumar, A.; Naish, K.A.; O'Donnell, S. Genetic evidence for landscape effects on dispersal in the army ant *Eciton burchellii*. *Mol. Ecol.* **2014**, *23*, 96–109. [CrossRef]

143. Berghoff, S.M.; Kronauer, D.J.C.; Edwards, K.J.; Franks, N.R. Dispersal and population structure of a New World predator, the army ant *Eciton burchellii*. *J. Evol. Biol.* **2008**, *21*, 1125–1132. [CrossRef]

144. Jaffé, R.; Moritz, R.F.A.; Kraus, F.B. Gene flow is maintained by polyandry and male dispersal in the army ant *Eciton burchellii*. *Popul. Ecol.* **2009**, *51*, 227–236. [CrossRef]

145. Soare, T.W.; Kumar, A.; Naish, K.A.; O'Donnell, S. Multi-year genetic sampling indicates maternal gene flow via colony emigrations in the army ant *Eciton burchellii parvispinum*. *Insectes Soc.* **2020**, *67*, 155–166. [CrossRef]

146. Schneirla, T. Army-ant life and behavior under dry-season conditions with special reference to reproductive functions. II. The appearance and fate of the males. *Zoologica* **1948**, *33*, 89–112.

147. Baldridge, R.S.; Rettenmeyer, C.W.; Watkins, J.F., II. Seasonal, nocturnal and diurnal flight periodicities of Nearctic army ant males (Hymenoptera: Formicidae). *J. Kansas Entomol. Soc.* **1980**.

148. Stouffer, P.C.; Bierregaard, R.O. Use of Amazonian forest fragments by understory insectivorous birds. *Ecology* **1995**, *76*, 2429–2445. [CrossRef]

149. Hooper, D.U.; Chapin, F.S.; Ewel, J.J.; Hector, A.; Inchausti, P.; Lavorel, S.; Lawton, J.H.; Lodge, D.M.; Loreau, M.; Naeem, S.; et al. Effects of biodiversity on ecosystem functioning: A consensus of current knowledge. *Ecol. Monogr.* **2005**, *75*, 3–35. [CrossRef]

150. Hooper, D.U.; Adair, E.C.; Cardinale, B.J.; Byrnes, J.E.K.; Hungate, B.A.; Matulich, K.L.; Gonzalez, A.; Duffy, J.E.; Gamfeldt, L.; Connor, M.I. A global synthesis reveals biodiversity loss as a major driver of ecosystem change. *Nature* **2012**, *486*, 105–108. [CrossRef]

151. Newbold, T.; Hudson, L.N.; Hill, S.L.L.; Contu, S.; Lysenko, I.; Senior, R.A.; Börger, L.; Bennett, D.J.; Choimes, A.; Collen, B.; et al. Global effects of land use on local terrestrial biodiversity. *Nature* **2015**, *520*, 45–50. [CrossRef]

152. Stratford, J.A.; Robinson, W.D. Gulliver travels to the fragmented tropics: Geographic variation in mechanisms of avian extinction. *Front. Ecol. Environ.* **2005**, *3*, 91–98. [CrossRef]

153. Lovejoy, T.E.; Bierregaard, R.O.; Rylands, A.B. Edge and other effects of isolation on Amazon forest fragments. In *Conservation Biology: The Science of Scarcity and Diversity*; Soulé, M.E., Ed.; Sinauer Associates: Sunderland, MA, USA, 1986; pp. 257–285.

154. Harper, L.H. The persistence of ant-following birds in small forest fragments. *Acta Amaz.* **1989**, *19*, 249–283. [CrossRef]

155. Sekercioglu, Ç.H.; Ehrlich, P.R.; Daily, G.C.; Aygen, D.; Goehring, D.; Sandí, R.F. Disappearance of insectivorous birds from tropical forest fragments. *Proc. Natl. Acad. Sci. USA* **2002**, *99*, 263–267. [CrossRef] [PubMed]

156. Koh, L.P.; Dunn, R.R.; Sodhi, N.S.; Colwell, R.K.; Proctor, H.C.; Smith, V.S. Species coextinctions and the biodiversity crisis. *Science* **2004**, *305*, 1632–1634. [CrossRef] [PubMed]

157. Lees, A.C.; Peres, C.A. Habitat and life history determinants of antbird occurrence in variable-sized Amazonian forest fragments. *Biotropica* **2010**, *42*, 614–621. [CrossRef]

158. Margules, C.R.; Pressey, R.L.; Willias, P.H. Representing biodiversity: Data and procedures for identifying priority areas for conservation. *J. Biosci.* **2002**, *27*, 309–326. [CrossRef] [PubMed]

159. Simberloff, D. Flagships, umbrellas, and keystones: Is single-species management passé in the landscape area? *Biol. Conserv.* **1998**, *83*, 247–257. [CrossRef]
160. Ricketts, T.H.; Dinerstein, E.; Olson, D.M.; Loucks, C. Who's where in North? Patterns of species richness and the utility of indicator taxa. *Bioscience* **1999**, *49*, 369–381. [CrossRef]
161. Pearson, D.L.; Cassola, F. World-wide species richness patterns of Tiger beetles (Coleoptera: Cicindelidae): Indicator taxon for biodiversity and conservation studies. *Conserv. Biol.* **1992**, *6*, 376–391. [CrossRef]
162. Myers, N.; Mittermeier, R.A.; Mittermeier, C.G.; da Fonseca, G.A.B.; Kent, J. Biodiversity hotspots for conservation priorities. *Nature* **2000**, *403*, 853–858. [CrossRef]
163. Reid, W.V. Biodiversity hotspots. *Trends Ecol. Evol.* **1998**, *13*, 275–280. [CrossRef]
164. Van Jaarsveld, A.S.; Freitag, S.; Chown, S.L.; Muller, C.; Koch, S.; Hull, H.; Bellamy, C.; Krüger, M.; Endrödy-Younga, S.; Mansell, M.W.; et al. Biodiversity assessment and conservation strategies. *Science* **1998**, *279*, 2106–2108. [CrossRef] [PubMed]
165. Frankel, O.H.; Soulé, M.E. *Conservation and Evolution*; Cambridge University Press: Cambridge, UK, 1981; ISBN 0521232759.
166. Martikainen, P.; Kaila, L.; Haila, Y. Threatened Beetles in White-Backed Woodpecker Habitats. *Conserv. Biol.* **1998**, *12*, 293–301. [CrossRef]
167. Swengel, S.R.; Swengel, A.B. Correlations in abundance of grassland songbirds and prairie butterflies. *Biol. Conserv.* **1999**, *90*, 1–11. [CrossRef]
168. Launer, A.E.; Murphy, D.D. Umbrella species and the conservation of habitat fragments: A case of threatened butterfly and a vanishing grassland ecosystem. *Biol. Conserv.* **1994**, *69*, 145–153. [CrossRef]
169. Berger, J. Population constraints associated with the use of Black rhinos as an umbrella species for desert herbivores. *Conserv. Biol.* **1997**, *11*, 69–78. [CrossRef]
170. Caro, T.M. Umbrella species: Critique and lessons from East Africa. *Anim. Conserv.* **2003**, *6*, 171–181. [CrossRef]
171. Zhang, C.; Zhu, R.; Sui, X.; Chen, K.; Li, B.; Chen, Y. Ecological use of vertebrate surrogate species in ecosystem conservation. *Glob. Ecol. Conserv.* **2020**, *24*, e01344. [CrossRef]
172. Sergio, F.; Caro, T.; Brown, D.; Clucas, B.; Hunter, J.; Ketchum, J.; McHugh, K.; Hiraldo, F. Top predators as conservation tools: Ecological rationale, assumptions, and efficacy. *Annu. Rev. Ecol. Evol. Syst.* **2008**, *39*, 1–19. [CrossRef]
173. Whiteman, N.K.; Sites, R.W. Aquatic insects as umbrella species for ecosystem protection in Death Valley National Park. *J. Insect Conserv.* **2008**, *12*, 499–509. [CrossRef]
174. Lambeck, R.J. Focal species: A multi-species umbrella for nature conservation. *Conserv. Biol.* **1997**, *11*, 849–856. [CrossRef]
175. Kumar, A.; O'Donnell, S. Elevation and forest clearing effects on foraging differ between surface—And subterranean—Foraging army ants (Formicidae: Ecitoninae). *J. Anim. Ecol.* **2009**, *78*, 91–97. [CrossRef] [PubMed]
176. Kaspari, M. A primer on ant ecology. In *Ants: Standard Methods for Measuring and Monitoring Biodiversity*; Agosti, D., Majer, J.D., Alonso, L.E., Schultz, T.R., Eds.; Smithsonian Institution Press: Washington, DC, USA, 2000; pp. 9–24.
177. Barbour, W.; Schlesinger, C. Who's the boss? Post-colonialism, ecological research and conservation management on Australian Indigenous lands. *Ecol. Manag. Restor.* **2012**, *13*, 36–41. [CrossRef]
178. Domínguez, L.; Luoma, C. Decolonising conservation policy: How colonial land and conservation ideologies persist and perpetuate indigenous injustices at the expense of the environment. *Land* **2020**, *9*, 65. [CrossRef]
179. Baker, K.; Eichhorn, M.P.; Griffiths, M. Decolonizing field ecology. *Biotropica* **2019**, *51*, 288–292. [CrossRef]

Article

Red Imported Fire Ants Reduce Invertebrate Abundance, Richness, and Diversity in Gopher Tortoise Burrows

Deborah M. Epperson [1], Craig R. Allen [2,*] and Katharine F. E. Hogan [2,*]

[1] U.S. Geological Survey—Wetland and Aquatic Research Center, Gainesville, FL 32653, USA; depperson@usgs.gov

[2] Center for Resilience in Agricultural Working Landscapes, School of Natural Resources, University of Nebraska, Lincoln, NE 68583-0961, USA

* Correspondence: callen3@unl.edu (C.R.A.); katharine.hogan@huskers.unl.edu (K.F.E.H.); Tel.: +1-(308)-258-2829 (K.F.E.H.)

Abstract: Gopher Tortoise (*Gopherus polyphemus*) burrows support diverse commensal invertebrate communities that may be of special conservation interest. We investigated the impact of red imported fire ants (*Solenopsis invicta*) on the invertebrate burrow community at 10 study sites in southern Mississippi, sampling burrows (1998–2000) before and after bait treatments to reduce fire ant populations. We sampled invertebrates using an ant bait attractant for ants and burrow vacuums for the broader invertebrate community and calculated fire ant abundance, invertebrate abundance, species richness, and species diversity. Fire ant abundance in gopher tortoise burrows was reduced by >98% in treated sites. There was a positive treatment effect on invertebrate abundance, diversity, and species richness from burrow vacuum sampling which was not observed in ant sampling from burrow baits. Management of fire ants around burrows may benefit both threatened gopher tortoises by reducing potential fire ant predation on hatchlings, as well as the diverse burrow invertebrate community. Fire-ant management may also benefit other species utilizing tortoise burrows, such as the endangered Dusky Gopher Frog and Schaus swallowtail butterfly. This has implications for more effective biodiversity conservation via targeted control of the invasive fire ant at gopher tortoise burrows.

Keywords: invasion ecology; invasive species; red imported fire ant; commensalism; gopher tortoise; diversity; conservation; burrow commensal

Citation: Epperson, D.M.; Allen, C.R.; Hogan, K.F.E. Red Imported Fire Ants Reduce Invertebrate Abundance, Richness, and Diversity in Gopher Tortoise Burrows. *Diversity* **2021**, *13*, 7. https://doi.org/10.3390/d13010007

Received: 30 November 2020
Accepted: 26 December 2020
Published: 29 December 2020

Publisher's Note: MDPI stays neutral with regard to jurisdictional claims in published maps and institutional affiliations.

1. Introduction

Invertebrates are an integral component of most food webs either directly as predators, prey, or indirectly through nutrient cycling [1]. The introduction of non-native species can decimate invertebrate communities resulting in a loss of native species diversity and the potential loss of ecosystem processes [2]. While data are limited on most invertebrate species, some studies report that two-thirds of invertebrate species have declined by 45% in mean abundance [3], with regional studies sometimes reporting much higher losses [4]. Effective use of limited conservation resources may lie in conserving biodiverse "hotspots" that account for a small percentage of the earth's surface [5–7]. Sometimes, these hotspots are created by the presence of keystone or ecosystem engineer species, such as the gopher tortoise (*Gopherus polyphemus*) of the southeastern United States [8].

Gopher tortoises excavate burrows in uplands on well-drained soils that provide habitat for more than 360 species [8–10]. Gopher tortoise burrows vary in size, but may extend up to 10 meters in length [8] and can persist for decades [11], enabling an invertebrate commensal community time to develop and stabilize [12]. No other North American reptile digs such a large, extensive, and relatively stable burrow [12], which makes the associated community of particular interest. In addition to their longevity in upland habitats, these burrows provide a stable thermal refugia for tortoises and other species [13].

A review of all known literature on the fauna of gopher tortoise burrows listed 60 vertebrate and 302 invertebrate species that use gopher tortoise burrows [10]. In this review, the criteria for commensalism were that "taxa that had at least 10 records of burrow use, or for which anecdotal reports are especially numerous." [10]. Based on these criteria, the invertebrate communities found in burrows contain dozens of commensal species, plus dozens more listed as "frequent users" that may also be commensal [10].

Populations of gopher tortoises in southern Mississippi have lower genetic diversity than populations from the eastern part of their range [14] and may be more vulnerable to disturbances and of higher conservation interest [15]. Additionally, research in Mississippi [16] supports evidence of gopher tortoise burrows as biodiverse communities with distinct commensal elements. While most early tortoise burrow community sampling was in Florida, Mississippi gopher tortoise habitats are different in vegetation and soil characteristics [16]. Thirty-seven species in 11 families and 5 orders of insects, and one species of tick were collected from burrows in Mississippi, many of which had not been recorded in earlier research [16]. The presence of diverse, but different invertebrate communities utilizing gopher tortoise burrows in different vegetation and soil conditions suggests that tortoises actively create desirable habitat for many species across invertebrate taxa. This supports previous evidence that these invertebrate communities are commensal to some degree [10] and highlights the importance of conserving tortoise populations in this region [15].

Lago (1991) was the first to record red imported fire ants (*Solenopsis invicta* Buren) as present in tortoise burrows, attacking a beetle at a burrow entrance [16]. The fire ant arrived at the Port of Mobile, Alabama from South America in the 1930s and has since spread across the United States [17] as far west as California [18]. Fire ants are aggressive, generalist predators that consume a variety of prey, including many other invertebrate species and gopher tortoise hatchlings [2,19]. Fire ants have reduced native ant diversity [20,21] in the Southeastern United States, as well as affected native invertebrate communities and associated ecosystem processes [2,22,23]. Once established, fire ants are extremely difficult to eradicate [24], and along with other invasive species are one of the greatest threats to native biodiversity and ecosystems [25,26].

In providing a unique microhabitat for many invertebrate species, gopher tortoise burrows also provide foraging opportunities for other species. An abundance of prey, soil disturbance around the burrow (called an "apron"), and the location of gopher tortoise burrows in sunny areas may provide suitable foraging and nesting habitat for fire ants [27]. One study found that 50% of all burrow aprons sampled in Mississippi had active fire ant mounds present [28], while another study found them present at 33% of burrows [29].

Although the impacts of fire ants on native biodiversity are generally well documented [22,30–32], the impacts of fire ants on invertebrates within gopher tortoise burrows are not. In this study, we assessed the impact of fire ants on commensal invertebrate communities within gopher tortoise burrows by manipulating fire ant densities on large replicated plots. Using two methods (burrow baits and vacuuming), we determined the relative abundance of fire ants and other burrow invertebrates as well as species richness and species diversity in the burrow system before and after treatments to reduce fire ant populations. We conclude by briefly discussing implications for conservation and fire ant control.

2. Materials and Methods

The study was conducted at Camp Shelby Joint Forces Training Center (CSJFTC) in southern Mississippi, USA, which is the nation's largest Army National Guard training site (Figure 1). It covers approximately 54,471 hectares, of which 47,561 hectares are U.S. National Forest land.

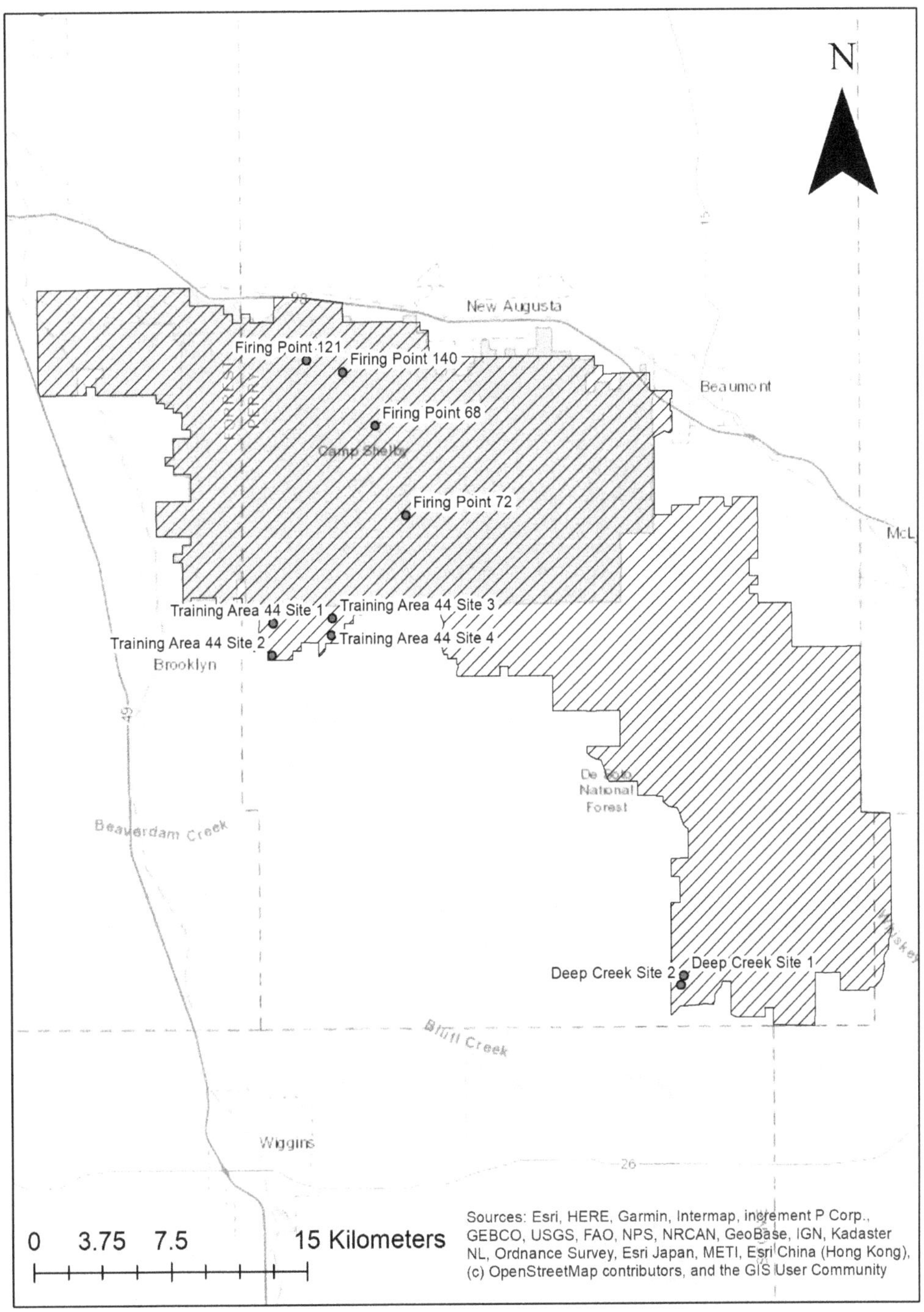

Figure 1. Map of Camp Shelby Joint Forces Training Center (CSJFTC). Study sites are labelled points.

In the spring of 1997, we selected ten sites (20–40 ha each) within CSJFTC for this study (Table A1, Appendix A). The ten sites were paired based on habitat similarities and gopher tortoise densities which allowed us to collect data from 10 burrows within each site, for a total of 100 burrows sampled multiple times. Four of the sites (2 pairs) are National Guard firing points (locations where heavy artillery shoots onto firing ranges) that are mowed and are considered ruderal habitat. Since they were paired by habitat, two of these sites (Firing Points 140 and 68) were located 3.36 km apart, and the remaining two (Firing Points 121 and 72) were 9.88 km apart. Three of the four firing points are surrounded by longleaf pine (*Pinus palustris*), and the fourth firing point is surrounded by slash pine (*Pinus elliottii*). Four sites (2 pairs) were in a gopher tortoise refuge (Training Area 44 or T-44, Sites 1–4), with pairs being contiguous (Sites 1–2, and 3–4). These sites are predominantly longleaf pine with some bluestem (*Schizachyrium* spp.) dominated groundcover. The remaining sites (Deep Creek 1 and 2) were also contiguous, located in a longleaf plantation (planted 1986) and an adjacent, more mature longleaf stand.

The National Guard mows the firing points annually between November and March, and the U.S. Forest Service manages the forested study sites and forested areas surrounding the firing points. Prescribed fire is the preferred management tool. However, fire intervals and season of burn are different from the natural burn regime of the area. The military use of these sites also varies, but firing points are the most heavily impacted by military use, including tank maneuvers and heavy artillery firing. The sites in T-44 have activity restricted to foot traffic, and some limited firing on the firing points contained within T-44. The two remaining sites have no military use.

Prior to data collection, we treated one randomly chosen site from each pair of sites with LOGIC® fire ant bait in spring 1998, spring and fall of 1999, and the spring of 2000. We completed broadcast applications of the bait by both manual hand spreaders and mechanized ground equipment (4-wheeler and tractor) at the rate of approximately 1.67 kg per hectare.

To evaluate use of the burrows by ants, we placed a bait attractant approximately 1 meter into the burrow of ten random adult gopher tortoise burrows at every site. The bait attractant apparatus used to sample burrows was a cotton ball infused with Multi-Species Ant Attractant (MSAA; [33]) within a small (30 mL) perforated plastic condiment container. The MSAA is a mixture of de-ionized water, confectionary sugar, and sodium hydroxide. The container was attached to a meter stick with monofilament line and placed inside the burrow. After one hour, we removed the baits, placed ant samples into Ziploc bags and froze them until they could be transported to the United States Department of Agriculture-Agricultural Research Service (USDA-ARS) laboratory in Gainesville, FL, USA for identification. We sorted all samples and identified ants to family, and to genus and species when possible. We repeated this sampling in the spring and fall of 1998, 1999, and 2000 at the same 10 burrows at each site. Data generated for this study are available within the Supplementary Materials.

To evaluate the invertebrate commensal burrow community, we vacuumed the same ten randomly chosen adult burrows at each site with a D-Vac (John W. Hock Co., Gainesville, FL, USA). The vacuum apparatus was placed as far down the burrow as possible (2–5 m) and then the burrow was suctioned as the vacuum was withdrawn slowly from the burrow. We placed samples in Ziploc bags and froze them until they could be transported to the USDA-ARS-Plant Protection and Quarantine laboratory in Gulfport, MS, USA for identification. We sorted all samples and identified invertebrates to family, and to genus and species when possible. We sampled using the burrow vacuum in the spring and fall of 1998, 1999, and 2000 at the same 10 burrows at each site.

To characterize the ant community in tortoise burrows based on the burrow baiting with MSAA, we calculated fire ant abundance, species richness (excluding fire ants), and species diversity (excluding fire ants) at each site for each of the six sampling periods (1 prior to the first treatment, 5 post-treatment). We averaged data from all ten burrows to derive a value for each site. We determined species richness, and calculated Shannon's

Diversity H′ [34]. We compared the pre-treatment sampling period between treated and untreated sites using a randomized block design analysis of variance (ANOVA). Post-treatment data were compared using a repeated measures randomized block analysis of variance. Due to missing data for one time period for two sites, we compared post-treatment fire ant abundance, species richness, and species diversity data using the PROC MIXED function (SAS Institute Inc., Cary, NC, USA, 1999). We tested all data for normality using the Shapiro-Wilk test. All data were normally distributed. We considered a probability level of 0.05 as significant.

To characterize the broader invertebrate communities in tortoise burrows from the vacuuming, we repeated the same analyses by determining fire ant abundance, invertebrate abundance (excluding fire ants), species richness (excluding fire ants), and species diversity (excluding fire ants) at each site for each of the six sampling periods (1 pre-treatment, 5 post-treatment). We compared the pre-treatment sampling period between treated and untreated sites using a randomized block design analysis of variance (ANOVA) (SAS Institute, Inc., 1999), and compared post-treatment data using a repeated measures randomized block ANOVA. We determined species richness and calculated diversity using Shannon Diversity H′ [34]. We tested all data for normality using the Shapiro-Wilk test. All data were normally distributed. We considered a probability level of 0.05 as significant.

3. Results

The repeated treatments with LOGIC® significantly reduced fire ant abundance in our study sites [22] and this was confirmed by the burrow baiting and burrow vacuum samples which had a 98.8% and 99.9% reduction in fire ant abundance, respectively. The invertebrate community in tortoise burrows included 17 invertebrates positively identified to species from 13 genera and 8 families. In many cases, invertebrates could only be identified to class, genus or family resulting in species collected from four classes and ten orders (Table 1). Insects from eight orders were collected, some found only on treated sites including Lepidoptera, Blattaria, and "Hemiptera" (Genus A).

Table 1. Invertebrates collected from all sites using a burrow vacuum for six sampling periods (1 pre-treatment, 5 post-treatment). If possible, specimens were identified to genus and species. "Treated" refers to species found only on sites treated to reduce fire ant populations, "untreated" refers to species found only on untreated sites, and "both" refers to species found on both treated and untreated sites. Species annotated with an "*" were considered "commensal" [10].

Class	Order	Family	Genus	Species	Treatment
CHILOPODA					Both
DIPOPLODA					Both
ARACHNIDA	Araneae	Agelenidae	*Agelenopsis*	*naevia*	Treated
			Agelenopsis		Both
			Coras		Untreated
			Neoscona	.	Untreated
			Tegeraria	*domestica*	Treated
		Araneidae	*Araneus*	.	Treated
			Genus A		Untreated
			Argiope	*aurantia*	Treated
INSECTA		Gnaphosidae	*Gnaphosa*	.	Both
		Hahniidae	Genus A		Untreated
		Linphiidae	Genus A		Treated
		Lycosidae	*Lycosa*	*helluo*	Both
			Lycosa	*avida*	Treated
			Lycosa	*rabida*	Treated
			Lycosa	.	Untreated
			Schizocosa		Untreated
			Sossipus	.	Both

Table 1. *Cont.*

Class	Order	Family	Genus	Species	Treatment
		Mimetidae	*Mimetus*		Treated
		Philodromidae	*Apollophanes*		Treated
		Salticidae	*Evarcha*		Untreated
			Phidippus		Treated
	Opiliones	Phalangidae	Genus A		Treated
	Blattaria	Blatellidae	Genus A		Treated
	Coleoptera	Scarabaeidae	*Aphodius*		Both
		Staphylinidae	Genus A		Both
		Carabidae	*Agonum*		Treated
			Clivinia		Untreated
		Curclionidae	*Pantomorus*	*cervinus*	Untreated
		Nitidulidae	Genus A		Untreated
		Elateridae	Genus A		Treated
	Diptera	Anthomyidae	*Eutrichota*	*gopheri* *	Both
		Chironomidae	Genus A		Untreated
		Culicidae	Genus A		Both
		Dolichopodidae	*Hercostoma*		Both
			Hercostoma		Both
		Sphaeroceridae	Genus A		Both
INSECTA	Hemiptera	Lygaeidae	Genus A		Untreated
		Miridae	Genus A		Treated
		Pentatomidae	*Thylantea*	*calceata*	Untreated
		Cicadellidae	Genus A		Both
			Genus B		Treated
	Hymenoptera	Braconidae	Genus A		Untreated
		Formicidae	*Aphenogaster*	*carolinense*	Treated
			Aphenogaster	*lamellidans*	Both
			Aphenogaster		Both
			Brachymyrmex	*depilis*	Treated
			Cyphomyrmex	*rimosus*	Both
			Dorymyrmex	*bureni*	Both
			Dorymyrmex	*medeis*	Both
			Paratrechina		Both
			Solenopsis	*invicta*	Both
		Halictidae	Genus A		Treated
		Mutillidae	*Dasymutilla*		Treated
					Both
	Lepidoptera				Treated
	Orthoptera	Gryllacridae	*Ceutophilis*	*divergens**	Both
		Gryllidae	*Gryllus*		Treated
		Tettigoniidae			Both

3.1. Burrow Baits

There were few ants present in the burrows besides fire ants and treating for fire ants did not increase overall ant abundance, richness, or diversity in the burrows. Pretreatment fire ant abundance was not significantly different between sites ($F = 3.29$, df = 4, $P = 0.144$). Repeated LOGIC® applications significantly reduced fire ant abundance at treated sites (Figure 2, $F_{1,8} = 24.78$, $P = 0.001$). Pre-treatment ant species diversity (excluding fire ants) was 0 for both treated and untreated sites. Pre-treatment species richness also was low and was not significantly different between treated and untreated sites (Figure 3, $F = 1.0$, df = 4, $P = 0.374$). After reducing fire ant populations, there were no significant differences in ant species richness (Figure 3, $F_{1,8} = 0.01$, $P = 0.935$) or ant species diversity (Figure 4, $F_{1,8} = 0.14$, $P = 0.722$) between treated and untreated sites. We collected thirteen species of ants (including fire ants) using burrow baits (Table 2). Of these, two species were unique to treated sites, and four species unique to untreated sites.

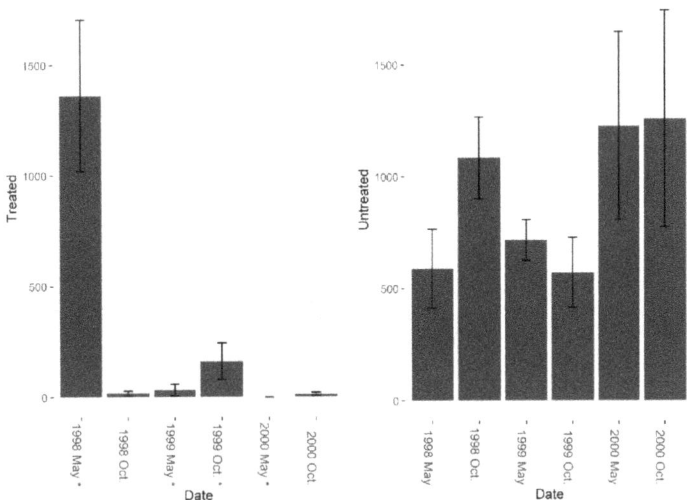

Figure 2. Mean (±1 SE) fire ant abundance from burrow baits at five pairs of treated and untreated study sites at CSJFTC, 1998–2000. Fire ant abundance declined 98.8% in treated sites, and increased 53.2% in untreated sites. Post-treatment data were compared using a repeated measures randomized block ANOVA, and repeated LOGIC® applications significantly reduced fire ant abundance at treated sites ($F_{1,8}$ = 24.78, P = 0.001). Asterisks on the x-axis indicate when treatments occurred.

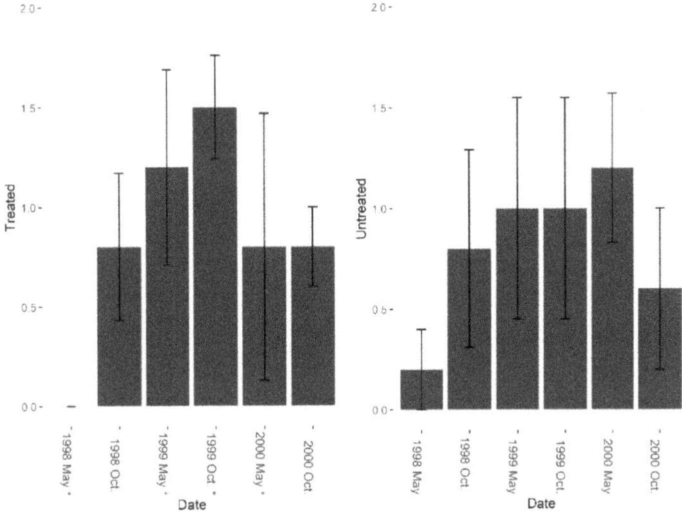

Figure 3. Mean (±1 SE) ant (excluding fire ant) species richness collected from burrow baits at five pairs of treated and untreated study sites at CSJFTC, 1998–2000. There were no significant differences between study sites from a randomized block design ANOVA (F = 1.0, df = 4, P = 0.374). Asterisks on the x-axis indicate when treatments occurred.

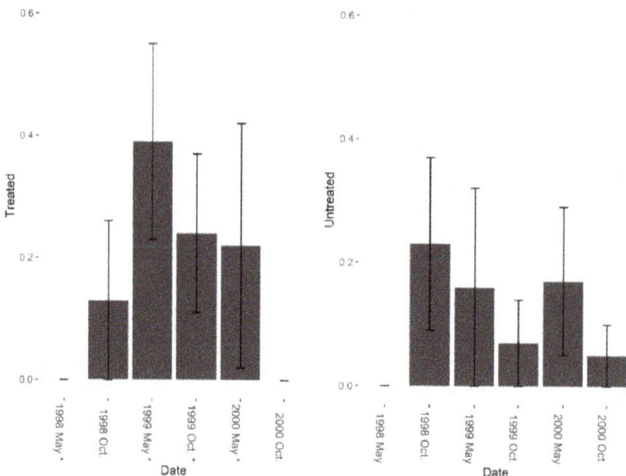

Figure 4. Mean (±1 SE) ant (excluding fire ant) species Shannon diversity indices collected from burrow bait at five pairs of treated and untreated study sites at CSJFTC, 1998–2000. There were no significant differences between study sites from a randomized block design ANOVA ($F_{1,8} = 0.14$, $P = 0.722$). Asterisks on the x-axis indicate when treatments occurred.

Table 2. Ant species (Hymenoptera: Formicidae) found using burrow baits at all study sites on CSJFTC, 1998–2000. "Treated" refers to species found only on sites treated to reduce fire ant populations, "untreated" refers to species found only on untreated sites, and "both" refers to species found on both treated and untreated sites.

Scientific Name	Treatment
Aphenogaster near rudis	both
Brachymyrmex depilis	treated
Dorymyrmex bureni	both
Paratrechina concinna	untreated
Paratrechina longicornis	both
Paratrechina phantasma	untreated
Pheidole dentata	treated
Pheidole dentigula	untreated
Pheidole floridana	both
Pheidole metallescens	both
Pheidole moerens	both
Solenopsis sp.	untreated
Solenopsis invicta	both

3.2. Burrow Vacuums

Data from the burrow vacuums revealed significant increases in overall invertebrate abundance, richness, and diversity post-treatment. Pre-treatment fire ant abundance was not significantly different between sites (Figure 5, $F = 2.01$, $df = 4$, $P = 0.229$). Overall, treatment effects were significant (Figure 5, $F_{1,7} = 7.37$, $P = 0.030$), and fire ant abundance in burrows was significantly reduced by repeated LOGIC® applications. Pre-treatment abundance of burrow invertebrates (excluding fire ants) was significantly different between sites (Figure 6, $F = 14.14$, $df = 4$, $P = 0.019$), and while treatment effects varied over time periods, an overall nearly significant treatment effect was present (Figure 6; $F_{1,7} = 4.01$, $P = 0.085$). Invertebrate abundance was greater on treated sites and did not change from pre-treatment to post-treatment. Species richness (excluding fire ants) was not significantly different between sites pre-treatment (Figure 7, $F = 3.37$, $df = 4$, $P = 0.140$); however, richness was greater on treated sites post-treatment (Figure 7; $F_{1,7}$ = 24.73, P = 0.002). Pre-treatment species diversity (excluding fire ants) was not significantly different between sites (Figure 8, F = 0.44, df = 4, P = 0.543). Post-treatment, a significant treatment effect was observed (Figure 8: $F_{1,7}$ = 9.31, P = 0.019) in species diversity between treated and untreated sites.

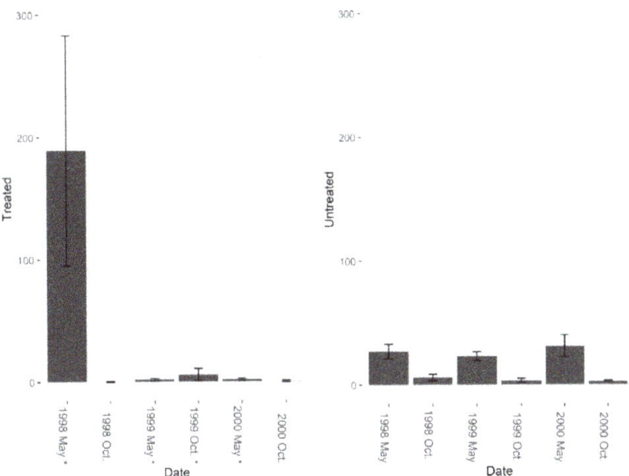

Figure 5. Mean (±1 SE) fire ant abundance collected from burrow vacuuming at five pairs of treated and untreated study sites at CSJFTC, 1998–2000. Fire ant abundance declined 99.9% in treated sites, and decreased 55.6% in untreated sites Pre-treatment fire ant abundance was not significantly different between study sites (F = 2.01, df = 4, P = 0.229). Post-treatment data were compared using a repeated measures randomized block ANOVA, and repeated LOGIC® applications significantly reduced fire ant abundance ($F_{1,7}$ = 7.37, P = 0.030). Asterisks on the x-axis indicate when treatments occurred.

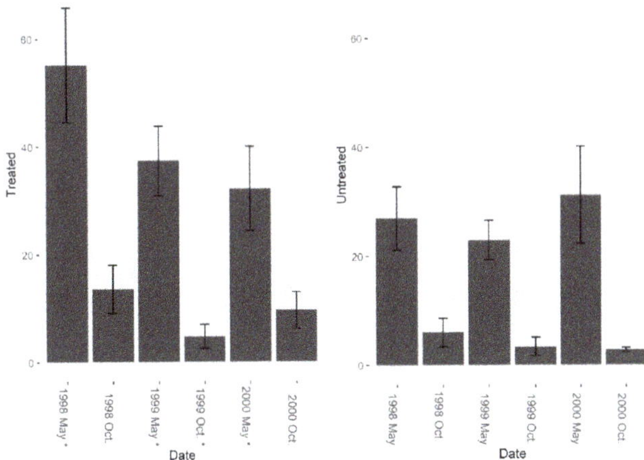

Figure 6. Mean (±1 SE) invertebrate abundance (excluding fire ants) collected from burrow vacuuming at five pairs of treated and untreated study sites at CSJFTC, 1998–2000. Pre-treatment abundance of burrow invertebrates (excluding fire ants) was significantly different between sites (F = 14.14, df = 4, P = 0.019), and a nearly significant treatment effect was present ($F_{1,7}$ = 4.01, P = 0.085). Asterisks on the x-axis indicate when treatments occurred.

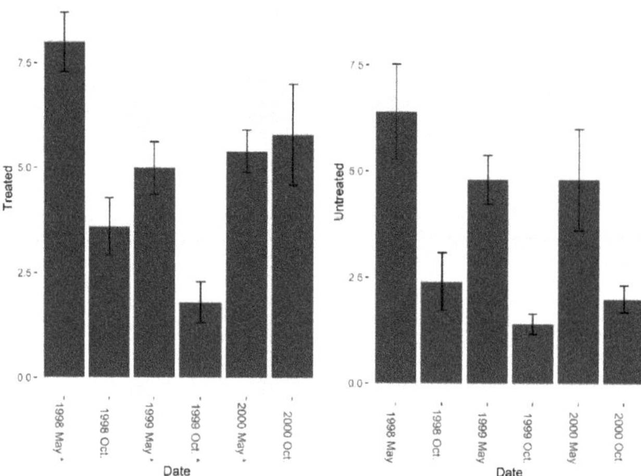

Figure 7. Mean ($\pm$1 SE) species richness (excluding fire ants) collected from burrow vacuuming at five pairs of treated and untreated study sites at CSJFTC, 1998–2000. Species richness (excluding fire ants) was not significantly different between treated and untreated sites pre-treatment (F = 3.37, df = 4, P = 0.140), but species richness was greater on treated sites post-treatment ($F_{1,7}$ = 24.73, P = 0.002). Asterisks on the x-axis indicate when treatments occurred.

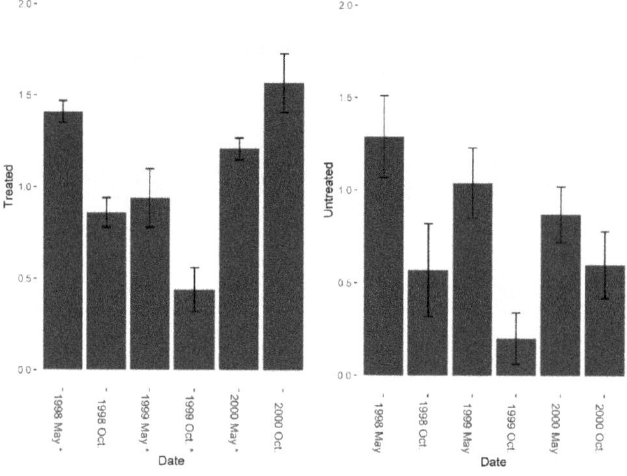

Figure 8. Mean ($\pm$1 SE) Shannon diversity indices (excluding fire ants) collected from burrow vacuuming at five pairs of treated and untreated study sites at CSJFTC, 1998–2000. Pre-treatment species diversity (excluding fire ants) was not significantly different between sites (F = 0.44, df = 4, P = 0.543), but a post-treatment effect was observed ($F_{1,7}$ = 9.31, P = 0.019) between treated and untreated sites. Asterisks on the x-axis indicate when treatments occurred.

4. Discussion

Repeated treatments with LOGIC® reduced fire ant abundance in gopher tortoise burrows. Reductions in fire ant abundance resulted in increased species richness and diversity of burrow arthropods sampled with burrow vacuums on treated sites. We did not, however, observe similar results using MSAA in the burrows for ant species. Ant species diversity and richness did not differ between treated and untreated sites. Sampling with baits may not provide a representative sample of the ant community as fire ants rapidly recruit to the bait and may exclude other species. After four treatments, differences in invertebrate abundance in vacuum samples became more pronounced and an overall treatment effect was observed. A small number of species influenced analysis of abundance data, particularly *Hercostomas* sp., which were particularly abundant during the pre-treatment sampling period. When they were removed from the analysis the treatment effect became more pronounced.

The two most abundant invertebrates collected were *Eutrichota gopheri* and *Ceutophilus divergens*. *Ceutophilis divergens* was the most common invertebrate encountered (excluding fire ants). *Ceutophilis divergens* is a wingless cave or camel cricket and was the second most abundant species encountered in previous surveys [16]; however, the author considered this species an opportunistic inhabitant of the burrow. *Eutrichota gopheri* is a small, copraphagous fly that feeds on tortoise dung found within the burrow [35]. It was previously known as *Pegomyia gopheri*, the gopher fly, and was considered an obligate species in gopher tortoise burrows [36]. *Eutrichota gopheri* was previously the most abundant commensal encountered in surveys of burrow invertebrates in Mississippi [16]. Although it was not the most abundant invertebrate commensal encountered in this study, we recorded it more than 50% of the time and it was most numerous during spring sampling periods.

After treatments, burrow invertebrate species diversity and richness positively responded on treated sites. There was an absence of the Lepidoptera and Blattaria orders on untreated sites. This may be partly because fire ants are known predators during multiple lepidopteran life stages, including on monarch butterfly larvae (*Danaus plexippus*) [37], and were suggested as a driving factor preceding a 50% decline in lepidopteran abundance in Texas post-invasion [38]. Research suggests that fire ants may also be a factor in the decline of the federally endangered Schaus swallowtail butterfly (*Papilio aristodemus ponceanus*) in the Florida Keys [39], and in laboratory experiments with a surrogate swallowtail species researchers documented predation on all immature life stages.

The introduction and spread of fire ants into the upland habitats of the gopher tortoise has the potential to negatively impact a highly diverse commensal invertebrate community. Fire ants may be able to change the burrow invertebrate community through interspecific competition, either exploitative (when fire ants using resources deprives other species of resources) or interference (when a species is harmed by direct fire ant interactions, including predation) [40].

Regardless of the mechanism, the presence of fire ants may result in changes to the larger burrow ecosystem. This may result in changes to the larger burrow ecosystem. Although not investigated in this study, many species that inhabit the burrow system are copraphagous and act as decomposers of tortoise dung found at the terminus of the burrow. If fire ants reduce the diversity and abundance of copraphagous insects, dung may accumulate in burrow systems. In addition, species that feed on copraphagous insects may be negatively impacted, resulting in a cascade of impacts to other burrow invertebrates and ultimately the entire burrow system. The invertebrate burrow community provides increased prey for insectivorous species including other insects and birds [41,42].

However, this evidence suggests that the same characteristics that make gopher tortoise burrows attractive to fire ants also create opportunities for more effective control of fire ants, and thus conservation of the greater burrow system. Repeated LOGIC® applications significantly reduced the abundance of fire ants in both the burrow bait and vacuum sampling, which was followed by a significant increase in overall insect abundance and diversity in the burrow vacuum samples. This suggests that targeted use of similar pesticide applications may be an effective use of limited conservation and management resources. Since fire ants are generalist predators, fire ant control targeted around gopher tortoise burrows could benefit both the diverse insect species within the burrow, as well as the gopher tortoises themselves, as fire ants have depredated nests and killed hatchlings [19]. This might make eradication of fire ants through bait treatments more effective if treatments are timed after imported fire ants are attracted to the nest, but before hatchlings emerge. Given that gopher tortoise burrows in our study area tend to occur in high density colonies [43], and are considered ecosystem engineers that create cascades of processes leading to high local biodiversity [8], this study aligns with others in suggesting the potential of focusing limited conservation resources on biodiversity hotspots [6,7,44].

Supplementary Materials: Supplementary materials are available online at https://www.mdpi.com/1424-2818/13/1/7/s1.

Author Contributions: Conceptualization, D.M.E. and C.R.A.; methodology, D.M.E. and C.R.A.; software, D.M.E. and K.F.E.H.; formal analysis, D.M.E.; investigation, D.M.E.; resources, D.M.E.; writing—original draft preparation, D.M.E.; writing—review and editing, D.M.E. and K.F.E.H.; supervision, C.R.A.; project administration, D.M.E.; funding acquisition, D.M.E. All authors have read and agreed to the published version of the manuscript.

Funding: This research was funded by the U.S. Department of Defense, Mississippi Army National Guard in the form of salary support for D.M.E. Salary support for K.F.E. Hogan was provided by the National Science Foundation NRT-INFEWS: Training in Theory and Application of Cross-scale Resilience in Agriculturally Dominated Social Ecological Systems, Grant No. DGE-1735362. Any opinions, findings, and conclusions or recommendations expressed in this material are those of the author(s) and do not necessarily reflect the views of the National Science Foundation. Any use of trade, firm, or product names is for descriptive purposes only and does not imply endorsement by the U.S. Government.

Institutional Review Board Statement: Not applicable.

Informed Consent Statement: Not applicable.

Data Availability Statement: The data presented in this study are available in the Supplementary Materials.

Acknowledgments: This manuscript was improved by comments from A-M Callcott and L. Yager. We also thank two anonymous reviewers for their feedback.

Conflicts of Interest: The authors declare no conflict of interest.

Appendix A

Table A1. UTM coordinates of all study sites at Camp Shelby Joint Forces Training Center, Mississippi, USA.

Site	Northing	Easting
Deep Creek Site 1	3423586	315624
Deep Creek Site 2	3423153	315479
Training Area 44 Site 1	3440128	296754
Training Area 44 Site 2	3438659	296666
Training Area 44 Site 3	3440288	299548
Training Area 44 Site 4	3439513	299489
Firing Point 68	3449013	301776
Firing Point 72	3444891	303116
Firing Point 121	3452081	298644
Firing Point 140	3451485	300308

References

1. Wilson, E.O. (Ed.) *Biodiversity*; The National Academies Press: Washington, DC, USA, 1988.
2. Porter, S.D.; Savignano, D.A. Invasion of Polygyne Fire Ants Decimates Native Ants and Disrupts Arthropod Community. *Ecology* **1990**, *71*, 2095–2106. [CrossRef]
3. Dirzo, R.; Young, H.S.; Galetti, M.; Ceballos, G.; Isaac, N.J.B.; Collen, B. Defaunation in the Anthropocene. *Science* **2014**, *345*, 401–406. [CrossRef] [PubMed]
4. Hallmann, C.A.; Sorg, M.; Jongejans, E.; Siepel, H.; Hofland, N.; Schwan, H.; Stenmans, W.; Müller, A.; Sumser, H.; Hörren, T.; et al. More than 75 percent decline over 27 years in total flying insect biomass in protected areas. *PLoS ONE* **2017**, *12*, e0185809. [CrossRef] [PubMed]

5. Banks-Leite, C.; Pardini, R.; Tambosi, L.R.; Pearse, W.D.; Bueno, A.A.; Bruscagin, R.T.; Condez, T.H.; Dixo, M.; Igari, A.T.; Martensen, A.C.; et al. Using ecological thresholds to evaluate the costs and benefits of set-asides in a biodiversity hotspot. *Science* **2014**, *345*, 1041–1045. [CrossRef] [PubMed]
6. Poynton, J.C.; Loader, S.P.; Sherratt, E.; Clarke, B.T. Biodiversity hotspots for conservation priorities. *Biodivers. Conserv.* **2007**, *16*, 853–858. [CrossRef]
7. Roberts, C.M.; McClean, C.J.; Veron, J.E.N.; Hawkins, J.P.; Allen, G.R.; McAllister, D.E.; Mittermeier, C.G.; Schueler, F.W.; Spalding, M.; Wells, F.; et al. Marine biodiversity hotspots and conservation priorities for tropical reefs. *Science* **2002**, *295*, 1280–1284. [CrossRef]
8. Kinlaw, A.; Grasmueck, M. Evidence for and geomorphologic consequences of a reptilian ecosystem engineer: The burrowing cascade initiated by the Gopher Tortoise. *Geomorphology* **2012**, *157–158*, 108–121. [CrossRef]
9. Alexy, K.J.; Brunjes, K.J.; Gassett, J.W.; Miller, K.V.; Alex, K.J.; Brunjes, K.J.; Gassett, J.W.; Miller, K.V. Continuous Remote Monitoring of Gopher Tortoise Burrow Use. *Wildl. Soc. Bull.* **2003**, *31*, 1240–1243.
10. Jackson, D.R.; Milstrey, E.G. The fauna of gopher tortoise burrows. In Proceedings of the Gopher Tortoise Relocation Symposium, Tallahassee, FL, USA; 1989; pp. 86–98.
11. Guyer, C.; Hermann, S.M. Patterns of size and longevity of gopher tortoise (*Gopherus polyphemus*) burrows: Implications for the longleaf pine ecosystem. *Chelonian Conserv. Biol.* **1997**, *2*, 507–513.
12. Milstrey, E.G. Ticks and invertebrate commensals in Gopher Tortoise burrows: Implications and importance. In Proceedings of the 5th Annual. Meeting Gopher Tortoise Council, Gainesville, FL, USA; 1986; pp. 4–15.
13. Pike, D.A.; Mitchell, J.C. Burrow-dwelling ecosystem engineers provide thermal refugia throughout the landscape. *Anim. Conserv.* **2013**, *16*, 694–703. [CrossRef]
14. Ennen, J.R.; Kreiser, B.R.; Qualls, C.P. Low Genetic Diversity in Several Gopher Tortoise (Gopherus polyphemus) Populations in the Desoto National Forest, Mississippi. *Herpetologica* **2010**, *66*, 31–38. [CrossRef]
15. Clostio, R.W.; Martinez, A.M.; Leblanc, K.E.; Anthony, N.M. Population genetic structure of a threatened tortoise across the south-eastern United States: Implications for conservation management. *Anim. Conserv.* **2012**, *15*, 613–625. [CrossRef]
16. Lago, P.K. A survey of arthropods associated with Gopher Tortoise burrows in Mississippi. *Entomol. News* **1991**, *102*, 1–13.
17. Callcott, A.-M.A.; Collins, H.L. Invasion and Range Expansion of Imported Fire Ants (Hymenoptera: Formicidae) in North America from 1918–1995. *Florida Entomol.* **1996**, *79*, 240–251. [CrossRef]
18. The University of Georgia EDDMapS: Red Imported Fire Ant Solenopsis Invicta (Buren). Available online: https://www.eddmaps. org/distribution/viewmap.cfm?sub=4675 (accessed on 23 November 2020).
19. Epperson, D.M.; Heise, C.D. Nesting and hatchling ecology of Gopher Tortoises (*Gopherus polyphemus*) in southern Mississippi. *J. Herpetol.* **2003**, *37*, 315–324. [CrossRef]
20. Jusino-Atresino, R.; Phillips, S.A. *Impact of Red Imported Fire Ants on the Ant Fauna of Texas*; Westview Press: Boulder, CO, USA, 1994.
21. Morris, J.R.; Steigman, K.L. Southwestern Association of Naturalists Effects of Polygyne Fire Ant Invasion on Native Ants of a Blackland Prairie in Texas. *Southwest. Nat.* **1993**, *38*, 136–140. [CrossRef]
22. Epperson, D.M.; Allen, C.R. Red Imported Fire Ant Impacts on Upland Arthropods in Southern Mississippi. *Am. Midl. Nat.* **2010**, *163*, 54–63. [CrossRef]
23. Kenis, M.; Auger-Rozenberg, M.A.; Roques, A.; Timms, L.; Péré, C.; Cock, M.J.W.; Settele, J.; Augustin, S.; Lopez-Vaamonde, C. Ecological effects of invasive alien insects. *Biol. Invasions* **2009**, *11*, 21–45. [CrossRef]
24. Holway, D.A.; Lach, L.; Suarez, A.V.; Tsutsui, N.D.; Case, T.J. The causes and consequences of ant invasions. *Annu. Rev. Ecol. Syst.* **2002**, *33*, 181–233. [CrossRef]
25. Boivin, N.L.; Zeder, M.A.; Fuller, D.Q.; Crowther, A.; Larson, G.; Erlandson, J.M.; Denhami, T.; Petraglia, M.D. Ecological consequences of human niche construction: Examining long-term anthropogenic shaping of global species distributions. *Proc. Natl. Acad. Sci. USA* **2016**, *113*, 6388–6396. [CrossRef]
26. Mack, R.N.; Simberloff, D.; Lonsdale, W.M.; Evans, H.; Clout, M.; Bazzaz, F.A. Biotic Invasions: Causes, Epidemiology, Global Consequences, and Control. *Ecol. Appl.* **2000**, *10*, 689–710. [CrossRef]
27. Dziadzio, M.C.; Long, A.K.; Smith, L.L.; Chandler, R.B.; Castleberry, S.B. Presence of the red imported fire ant at gopher tortoise nests. *Wildl. Soc. Bull.* **2016**, *40*, 202–206. [CrossRef]
28. Jennings, R.; Fritts, T. *The Status of the Gopher Tortoise, Gopherus polyphemus Daudin*; Unpublished Final Report; United States Fish and Wildlife Service: Jackson, MS, USA, 1983.
29. Wetterer, J.K.; Moore, J.A. Red Imported Fire Ants (Hymenoptera: Formicidae) At Gopher Tortoise (Testudines: Testudinidae) Burrows. *Fla. Entomol.* **2005**, *88*, 349–354. [CrossRef]
30. Allen, C.R.; Epperson, D.M.; Garmestani, A.S. Red Imported Fire Ant Impacts on Wildlife: A Decade of Research. *Am. Midl. Nat.* **2004**, *152*, 88–103. [CrossRef]
31. Stoker, R.L.; Grant, W.E.; Vinson, S.B. Solenopsis invicta (Hymenoptera: Formicidae) effect on invertebrate decomposers of carrion in central Texas. *Environ. Entomol.* **1995**, *24*, 817–822. [CrossRef]
32. Vinson, S.B. Impact of the invasion of the imported fire ant. *Insect Sci.* **2013**, *20*, 439–455. [CrossRef]
33. Wojcik, D.P.; Allen, C.R.; Brenner, R.J.; Forys, E.A.; Jouvenaz, D.P.; Lutz, R.S. Red Imported Fire Ants: Impact on Biodiversity. *Am. Entomol.* **2001**, *47*, 16–23. [CrossRef]
34. Magurran, A.E. *Ecological Diversity and Its Measurement*; Princeton University Press: Princeton, NJ, USA, 1988.

35. Hubbard, H.G. Additional Notes on the insect guests of the Florida land tortoise. *Proc. Entomol. Soc. Wash.* **1896**, *3*, 299–302.
36. Cox, J.; Inkley, D.; Kautz, R. *Ecology and Protection needs of Gopher Tortoise (Gopherus polyphemus) Populations Found on Lands Slated for Large-Scale Development in Florida*; Nongame Wildlife Program Technical Report No. 4; Florida Game and Fresh Water Fish Commission: Tallahassee, FL, USA, 1987.
37. Calvert, W.H. Fire ant predation on monarch larvae (Nymphalidae: Danainae) in a central Texas prairie. *J. Lepid. Soc.* **1996**, *50*, 149–151.
38. Calvert, W.H. Patterns in the spatial and temporal use of the Texas milkweeds (Asclepiadaceae) by the monarch butterfly (*Danus plexippus* L.) during fall, 1996. *J. Lepid. Soc.* **1999**, *53*, 37–44.
39. Forys, E.A.; Quistorff, A.; Allen, C.R. Potential Fire Ant (Hymenoptera: Formicidae) Impact on the Endangered Schaus Swallowtail. *Fla. Entomol.* **2001**, *84*, 254–258. [CrossRef]
40. Morrison, L.W. Long-term impacts of an arthropod-community invasion by the imported fire ant, Solenopsis invicta. *Ecology* **2002**, *83*, 2337–2345. [CrossRef]
41. Dziadzio, M.C.; Smith, L.L. Vertebrate Use of Gopher Tortoise Burrows and Aprons. *Southeast. Nat.* **2016**, *15*, 586–594. [CrossRef]
42. White, K.N.; Tuberville, T.D. Birds and Burrows: Avifauna Use and Visitation of Burrows of Gopher Tortoises at Two Military Sites in the Florida Panhandle. *Wilson J. Ornithol.* **2017**, *129*, 792–803. [CrossRef]
43. Richter, S.C.; Jackson, J.A.; Hinderliter, M.; Epperson, D.; Theodorakis, C.W.; Adams, S.M. Conservation genetics of the largest cluster of federally threatened gopher tortoise (gopherus polyphemus) colonies with implications for species management. *Herpetologica* **2011**, *67*, 406–419. [CrossRef]
44. Collins, C.D.; Banks-Leite, C.; Brudvig, L.A.; Foster, B.L.; Cook, W.M.; Damschen, E.I.; Andrade, A.; Austin, M.; Camargo, J.L.; Driscoll, D.A.; et al. Fragmentation affects plant community composition over time. *Ecography* **2017**, *40*, 119–130. [CrossRef]

MDPI

St. Alban-Anlage 66

4052 Basel

Switzerland

Tel. +41 61 683 77 34

Fax +41 61 302 89 18

www.mdpi.com

Diversity Editorial Office

E-mail: diversity@mdpi.com

www.mdpi.com/journal/diversity